Studies in Big Data

Volume 18

Series editor

Janusz Kacprzyk, Polish Academy of Sciences, Warsaw, Poland
e-mail: kacprzyk@ibspan.waw.pl

About this Series

The series "Studies in Big Data" (SBD) publishes new developments and advances in the various areas of Big Data- quickly and with a high quality. The intent is to cover the theory, research, development, and applications of Big Data, as embedded in the fields of engineering, computer science, physics, economics and life sciences. The books of the series refer to the analysis and understanding of large, complex, and/or distributed data sets generated from recent digital sources coming from sensors or other physical instruments as well as simulations, crowd sourcing, social networks or other internet transactions, such as emails or video click streams and other. The series contains monographs, lecture notes and edited volumes in Big Data spanning the areas of computational intelligence incl. neural networks, evolutionary computation, soft computing, fuzzy systems, as well as artificial intelligence, data mining, modern statistics and Operations research, as well as self-organizing systems. Of particular value to both the contributors and the readership are the short publication timeframe and the world-wide distribution, which enable both wide and rapid dissemination of research output.

More information about this series at http://www.springer.com/series/11970

Ali Emrouznejad
Editor

Big Data Optimization: Recent Developments and Challenges

Editor
Ali Emrouznejad
Aston Business School
Aston University
Birmingham
UK

ISSN 2197-6503 ISSN 2197-6511 (electronic)
Studies in Big Data
ISBN 978-3-319-30263-8 ISBN 978-3-319-30265-2 (eBook)
DOI 10.1007/978-3-319-30265-2

Library of Congress Control Number: 2016933480

Printed on acid-free paper

This Springer imprint is published by Springer Nature
The registered company is Springer International Publishing AG Switzerland

Preface

The increased capacity of contemporary computers allows the gathering, storage and analysis of large amounts of data which only a few years ago would have been impossible. These new data are providing large quantities of information, and enabling its interconnection using new computing methods and databases. There are many issues arising from the emergence of big data, from computational capacity to data manipulation techniques, all of which present challenging opportunities. Researchers and industries working in various different fields are dedicating efforts to resolve these issues. At the same time, scholars are excited by the scientific possibilities offered by big data, and especially the opportunities to investigate major societal problems related to health, privacy, economics, business dynamics and many more. These large amounts of data present various challenges, one of the most intriguing of which deals with knowledge discovery and large-scale data mining. Although these vast amounts of digital data are extremely informative, and their enormous possibilities have been highlighted on several occasions, issues related to optimization remain to be addressed. For example, formulation of optimization problems of unprecedented sizes (millions or billions of variables) is inevitable.

The main objective of this book is to provide the necessary background to work with big data by introducing some novel optimization algorithms and codes capable of working in the big data setting as well as introducing some applications in big data optimization for interested academics and practitioners, and to benefit society, industry, academia and government.

To facilitate this goal, chapter "Big Data: Who, What and Where? Social, Cognitive and Journals Map of Big Data Publications with Focus on Optimization" provides a literature review and summary of the current research in big data and large-scale optimization. In this chapter, **Emrouznejad and Marra** investigate research areas that are the most influenced by big data availability, and on which aspects of large data handling different scientific communities are working. They employ scientometric mapping techniques to identify who works on what in the area of big data and large-scale optimization problems. This chapter highlights a major effort involved in handling big data optimization and large-scale data mining which has led to several algorithms that have proven to be more efficient, faster and more accurate than earlier solutions.

This is followed by a comprehensive discussion on setting up a big data project in chapter "Setting Up a Big Data Project: Challenges, Opportunities, Technologies and Optimization" as discussed by **Zicari, Rosselli, Ivanov, Korfiatis, Tolle, Niemann and Reichenbach**. The chapter explains the general value of big data analytics for the enterprise and how value can be derived by analysing big data. Then it introduces the characteristics of big data projects and how such projects can be set up, optimized and managed. To be able to choose the optimal big data tools for given requirements, the relevant technologies for handling big data, such as NoSQL and NewSQL systems, in-memory databases, analytical platforms and Hadoop-based solutions, are also outlined in this chapter.

In chapter "Optimizing Intelligent Reduction Techniques for Big Data", **Pop, Negru, Ciolofan, Mocanu, and Cristea** analyse existing techniques for data reduction, at scale to facilitate big data processing optimization. The chapter covers various areas in big data including: data manipulation, analytics and big data reduction techniques considering descriptive analytics, predictive analytics and prescriptive analytics. Cyber-Water cast study is also presented by referring to: optimization process, monitoring, analysis and control of natural resources, especially water resources to preserve the water quality.

Li, Guo and Chen in the chapter "Performance Tools for Big Data Optimization" focus on performance tools for big data optimization. The chapter explains that many big data optimizations have critical performance requirements (e.g., real-time big data analytics), as indicated by the velocity dimension of 4Vs of big data. To accelerate the big data optimization, users typically rely on detailed performance analysis to identify potential performance bottlenecks. To alleviate the challenges of performance analysis, various performance tools have been proposed to understand the runtime behaviours of big data optimization for performance tuning.

Further to this, **Valkonen**, in chapter "Optimising Big Images", presents a very good application of big data optimization that is used for analysing big images. Real-life photographs and other images, such as those from medical imaging modalities, consist of tens of million data points. Mathematically based models for their improvement—due to noise, camera shake, physical and technical limitations, etc.—are moreover often highly non-smooth and increasingly often non-convex. This creates significant optimization challenges for application of the models in quasi-real-time software packages, as opposed to more ad hoc approaches whose reliability is not as easily proven as that of mathematically based variational models. After introducing a general framework for mathematical image processing, this chapter presents the current state-of-the-art in optimization methods for solving such problems, and discuss future possibilities and challenges.

As another novel application **Rajabi and Beheshti**, in chapter "Interlinking Big Data to Web of Data", explain interlinking big data to web of data. The big data problem can be seen as a massive number of data islands, ranging from personal, shared, social to business data. The data in these islands are becoming large-scale, never ending and ever changing, arriving in batches at irregular time intervals. In this context, it is important to investigate how the linked data approach can enable big data optimization. In particular, the linked data approach has recently facilitated accessibility, sharing and enrichment of data on the web. This chapter discusses the advantages of applying the linked data approach, toward optimization of big data in the linked open data (LOD) cloud by: (i) describing the impact of linking big data to LOD cloud; (ii) representing various interlinking tools for linking big data; and (iii) providing a practical case study: linking a big data repository to DBpedia.

Topology of big data is the subject of chapter "Topology, Big Data and Optimization" as discussed by **Vejdemo-Johansson and Skraba**. The idea of using geometry in learning and inference has a long history going back to canonical ideas such as Fisher information, discriminant analysis and principal component analysis. The related area of topological data analysis (TDA) has been developing in the past decade, which aims to extract robust topological features from data and use these summaries for modelling the data. A topological summary generates a coordinate-free, deformation invariant and a highly compressed description of the geometry of an arbitrary data set. This chapter explains how the topological techniques are well suited to extend our understanding of big data.

In chapter "Applications of Big Data Analytics Tools for Data Management", **Jamshidi, Tannahill, Ezell, Yetis and Kaplan** present some applications of big data analytics tools for data management. Our interconnected world of today and the advent of cyber-physical or system of systems (SoS) are a key source of data accumulation—be it numerical, image, text or texture, etc. SoS is basically defined as an integration of independently operating, non-homogeneous systems for a certain duration to achieve a higher goal than the sum of the parts. Recent efforts have developed a promising approach, called "data analytics", which uses statistical and computational intelligence (CI) tools such as principal component analysis (PCA), clustering, fuzzy logic, neuro-computing, evolutionary computation, Bayesian networks, data mining, pattern recognition, etc., to reduce the size of "big data" to a manageable size. This chapter illustrates several case studies and attempts to construct a bridge between SoS and data analytics to develop reliable models for such systems.

Optimizing access policies for big data repositories is the subject discussed by **Contreras** in chapter "Optimizing Access Policies for Big Data Repositories: Latency Variables and the Genome Commons". The design of access policies for large aggregations of scientific data has become increasingly important in today's data-rich research environment. Planners routinely consider and weigh different policy variables when deciding how and when to release data to the public. This chapter proposes a methodology in which the timing of data release can be used to balance policy variables and thereby optimize data release policies. The global aggregation of publicly-available genomic data, or the "genome commons" is used as an illustration of this methodology.

Achieving the full transformative potential of big data in this increasingly digital and interconnected world requires both new data analysis algorithms and a new class of systems to handle the dramatic data growth, the demand to integrate structured and unstructured data analytics, and the increasing computing needs of massive-scale analytics. **Li**, in chapter "Big Data Optimization via Next Generation Data Center Architecture", elaborates big data optimization via next-generation data centre architecture. This chapter discusses the hardware and software features of High Throughput Computing Data Centre architecture (HTC-DC) for big data optimization with a case study at Huawei.

In the same area, big data optimization techniques can enable designers and engineers to realize large-scale monitoring systems in real life, by allowing these systems to comply with real-world constrains in the area of performance, reliability and reliability. In chapter "Big Data Optimization Within Real World Monitoring Constraints", **Helmholt and der Waaij** give details of big data optimization using several examples of real-world monitoring systems.

Handling big data poses a huge challenge in the computer science community. Some of the most appealing research domains such as machine learning, computational biology and social networks are now overwhelmed with large-scale databases that need computationally demanding manipulation. Smart sampling and optimal dimensionality reduction of big data using compressed sensing is the main subject in chapter "Smart Sampling and Optimal Dimensionality Reduction of Big Data Using Compressed Sensing" as elaborated by **Maronidis, Chatzilari, Nikolopoulos and Kompatsiaris**. This chapter proposes several techniques for optimizing big data processing including computational efficient implementations like parallel and distributed architectures. Although Compressed Sensing (CS) is renowned for its capability of providing succinct representations of the data, this chapter investigates its potential as a dimensionality reduction technique in the domain of image annotation.

Another novel application of big data optimization in brain disorder rehabilitation is presented by **Brezany, Štěpánková, Janatoá, Uller and Lenart** in chapter "Optimized Management of BIG Data Produced in Brain Disorder Rehabilitation". This chapter introduces the concept of scientific dataspace that involves and stores numerous and often complex types of data, e.g. primary data captured from the application, data derived by curation and analytics processes, background data including ontology and workflow specifications, semantic relationships between dataspace items based on ontologies, and available published data. The main contribution in this chapter is applying big data and cloud technologies to ensure efficient exploitation of this dataspace, namely novel software architectures, algorithms and methodology for its optimized management and utilization.

This is followed by another application of big data optimization in maritime logistics presented by **Berit Dangaard Brouer, Christian Vad Karsten and David Pisinge** in chapter "Big data Optimization in Maritime Logistics". Large-scale maritime problems are found particularly within liner shipping due to the vast size of the network that global carriers operate. This chapter introduces a selection of large-scale planning problems within the liner shipping industry. It is also shown how large-scale optimization methods can utilize special problem structures such as separable/independent sub-problems and give examples of advanced heuristics using divide-and-conquer paradigms, decomposition and mathematical programming within a large-scale search framework.

On more complex use of big data optimization, chapter "Big Network Analytics Based on Nonconvex Optimization" focuses on the use of network analytics which can contribute to networked big data processing. Many network issues can be modelled as non-convex optimization problems and consequently they can be addressed by optimization techniques. **Gong, Cai, Ma and Jiao**, in this chapter, discuss the big network analytics based on non-convex optimization. In the pipeline of non-convex optimization techniques, evolutionary computation gives an outlet to handle these problems efficiently. Since network community discovery is a critical research agenda of network analytics, this chapter focuses on the evolutionary computation-based non-convex optimization for network community discovery. Several experimental studies are shown to demonstrate the effectiveness of optimization-based approach for big network community analytics.

Large-scale and big data optimization based on Hadoop is the subject of chapter "Large-Scale and Big Optimization Based on Hadoop" presented by **Cao and Sun**. As explained in this chapter, integer linear programming (ILP) is among the most popular optimization techniques found in practical applications, however, it often faces computational issues in modelling real-world problems. Computation can easily outgrow the computing power of standalone computers as the size of problem increases. The modern distributed computing releases the computing power constraints by providing scalable computing resources to match application needs, which boosts large-scale optimization. This chapter presents a paradigm that leverages Hadoop, an open-source distributed computing framework, to solve a large-scale ILP problem that is abstracted from real-world air traffic flow management. The ILP involves millions of decision variables, which is intractable even with the existing state-of-the-art optimization software package.

Further theoretical development and computational approaches in large-scale unconstrained optimization is presented by **Babaie-Kafaki** in chapter "Computational Approaches in Large–Scale Unconstrained Optimization". As a topic of great significance in nonlinear analysis and mathematical programming, unconstrained optimization is widely and increasingly used in engineering, economics, management, industry and other areas. In many big data applications, solving an unconstrained optimization problem with thousands or millions of variables is indispensable. In such situations, methods with the important feature of low memory requirement are helpful tools. This chapter explores two families of methods for solving large-scale unconstrained optimization problems: conjugate gradient methods and limited-memory quasi-Newton methods, both of them are structured based on the line search.

This is followed by explaining numerical methods for large-scale non-smooth optimization (NSO) as discussed by **Karmitsa** in chapter "Numerical Methods for Large-Scale Nonsmooth Optimization". NSO refers to the general problem of minimizing (or maximizing) functions that are typically not differentiable at their minimizers (maximizers). NSO problems are in general difficult to solve even when the size of the problem is small and the problem is convex. This chapter recalls two numerical methods, the limited memory bundle algorithm (LMBM) and the diagonal bundle method (D-BUNDLE), for solving large-scale non-convex NSO problems.

Chapter "Metaheuristics for Continuous Optimization of High-Dimensional Problems: State of the Art and Perspectives" presents a state-of-the-art discussion of metaheuristics for continuous optimization of high-dimensional problems. In this chapter, **Trunfio** shows that the age of big data brings new opportunities in many relevant fields, as well as new research challenges. Among the latter, there is the need for more effective and efficient optimization techniques, able to address problems with hundreds, thousands and even millions of continuous variables. In order to provide a picture of the state of the art in the field of high-dimensional continuous optimization, this chapter describes the most successful algorithms presented in the recent literature, also outlining relevant trends and identifying possible future research directions.

Finally, **Sagratella** discusses convergent parallel algorithms for big data optimization problems in chapter "Convergent Parallel Algorithms for Big Data Optimization Problems". When dealing with big data problems it is crucial to design methods able to decompose the original problem into smaller and more manageable pieces. Parallel methods lead to a solution by concurrently working on different pieces that are distributed among available agents, so as to exploit the computational power of multi-core processors and therefore efficiently solve the problem. Beyond gradient-type methods, which can of course be easily parallelized but suffer from practical drawbacks, recently a convergent decomposition framework for the parallel optimization of (possibly non-convex) big data problems was proposed. Such framework is very flexible and includes both fully parallel and fully sequential schemes, as well as virtually all possibilities in between. This chapter illustrates the versatility of this parallel decomposition framework by specializing it to different well-studied big data optimization problems such as LASSO, logistic regression and support vector machines training.

January 2016 Ali Emrouznejad

Acknowledgments

First among these are the contributing authors—without them, it was not possible to put together such a valuable book, and I am deeply grateful to them for bearing with my repeated requests for materials and revisions while providing the high-quality contributions. I am also grateful to the many reviewers for their critical review of the chapters and the insightful comments and suggestions provided. Thanks are also due to *Professor Janusz Kacprzyk*, the Editor of this Series, for supporting and encouraging me to complete this project. The editor would like to thank *Dr. Thomas Ditzinger* (Springer Senior Editor, Interdisciplinary and Applied Sciences & Engineering), *Ms. Daniela Brandt* (Springer Project Coordinator, Production Heidelberg), *Ms. Gajalakshmi Sundaram* (Springer Production Editor, Project Manager), *Mr. Yadhu Vamsi* (in the Production team, Scientific Publishing Services Pvt. Ltd., Chennai, India) for their excellent editorial and production assistance in producing this volume. I hope the readers will share my excitement with this important scientific contribution to the body of knowledge in Big Data.

Ali Emrouznejad

Contents

About the Editor

Ali Emrouznejad is Professor and Chair in Business Analytics at Aston Business School, UK. His areas of research interest include performance measurement and management, efficiency and productivity analysis as well as data mining. Dr. Emrouznejad is Editor of *Annals of Operations Research*, Associate Editor of *Socio-Economic Planning Sciences*, Associate Editor of *IMA journal of Management Mathematics*, Associate Editor of *RAIRO—Operations Research*, Senior Editor of *Data Envelopment Analysis journal*, and member of editorial boards or guest editor in several other scientific journals. He has published over 100 articles in top ranked journals; he is the author of the book "Applied Operational Research with SAS", editor of books on "Performance Measurement with Fuzzy Data Envelopment Analysis" (Springer), "Managing Service Productivity" (Springer), and "Handbook of Research on Strategic Performance Management and Measurement" (IGI Global). He is also co-founder of Performance Improvement Management Software (PIM-DEA), see http://www.Emrouznejad.com.

Big Data: Who, What and Where? Social, Cognitive and Journals Map of Big Data Publications with Focus on Optimization

Ali Emrouznejad and Marianna Marra

Abstract Contemporary research in various disciplines from social science to computer science, mathematics and physics, is characterized by the availability of large amounts of data. These large amounts of data present various challenges, one of the most intriguing of which deals with knowledge discovery and large-scale data-mining. This chapter investigates the research areas that are the most influenced by big data availability, and on which aspects of large data handling different scientific communities are working. We employ scientometric mapping techniques to identify who works on what in the area of big data and large scale optimization problems.

1 Introduction

The increased capacity of contemporary computers allows the gathering, storage and analysis of large amounts of data which only a few years ago would have been impossible. These new data are providing large quantities of information, and enabling its interconnection using new computing methods and databases. There are many issues arising from the emergence of big data, from computational capacity to data manipulation techniques, all of which present challenging opportunities. Researchers and industries working in various different fields are dedicating efforts to resolve these issues. At the same time, scholars are excited by the scientific possibilities offered by big data, and especially the opportunities to investigate major societal problems related to health [19], privacy [21], economics [38] and business dynamics [10]. Although these vast amounts of digital data are

A. Emrouznejad (✉)
Aston Business School, Aston University, Birmingham, UK
e-mail: a.emrouznejad@aston.ac.uk

M. Marra
Essex Business School, Essex University, Southend-on-Sea, UK
e-mail: marramarianna@essex.ac.uk

A. Emrouznejad (ed.), *Big Data Optimization: Recent Developments and Challenges*, Studies in Big Data 18, DOI 10.1007/978-3-319-30265-2_1

extremely informative, and their enormous possibilities have been highlighted on several occasions, issues related to their measurement, validity and reliability remain to be addressed [22].

Schönberger and Cukier [28] point out that 'there is no rigorous definition of big data'. The term *big data* appeared or the first time in a paper by Cox and Ellsworth [11] describing the challenges facing computer systems when data are too large to be easily stored in the main memory, or local or remote disks.

Big data is a label that indicates a large dataset that cannot be stored conveniently in a traditional Structured Query Language (SQL) database and requires a NoSQL (not only SQL) database which is able to handle large amounts of data [38]. Open-source tools such as Hadoop, Bigtable, MapReduce have been made available by several companies including Google which is among the most active in this respect. These tools have been shown to be efficient for both storing large amounts of data and querying and manipulating them.

This chapter discusses some fundamental issues related to big data analysis which have emerged in publications produced by the scientific communities active in this area of research. We retrieved 4,308 published works from the ISI Web of Science (WoS) academic database which we analysed using scientometrics mapping techniques. We identified the cognitive, social and journal maps of publications dealing with big data and large scale optimization methods.

The study reported by [1] highlights the following five areas requiring scholarly attention: (1) scalable big/fast data infrastructures; (2) the problem of diversity in the management of data; (3) end-to-end processing and understanding of data; (4) cloud services; and (5) managing the diverse roles of people in the data life cycle. Dealing with large quantities of new data involves optimization problems that often are difficult to resolve in reasonable computing time. In this chapter we describe some of the efforts and solutions proposed by authors working in this field, to improve these processes.

2 Methodology

Studying paper citations networks using a scientometric approach and Social Network Analysis (SNA) has become popular in recent years and provides an understanding of various dynamics such as collaboration among researchers [14, 24], knowledge patterns [9] and emerging knowledge trends within disciplines [15, 20].

In this chapter, we combine insights from scientometric mapping techniques to study the journals, and the cognitive and social maps within studies of big data and large scale optimization. We use a scientometric mapping technique known as *overlay mapping*, which has been presented as a strategic intelligence tool for the governance of emerging technologies [34], to investigate different dimensions of the emergence process. The idea underpinning overlay mapping is to project data representing a focal subject area, over a *basemap*, which represents the totality of contemporary activity in scientific outputs. In the present study, we use basemap to trace the emergence across cognitive space, which provides a cognitive mapping of the data on publications referring to big data and its large scale optimization. The basemap represents the totality of research areas, grouped according to 19 factors ranging from social studies, to mathematical methods and computer science. The 19 factors are: mathematical methods; computer science; physics; mechanical engineering; chemistry, environment science and technology; materials science; geoscience; ecology; agriculture; biomed science; infectious diseases; psychological science; health and social issues; clinical psychology; clinical medicine; social studies; business and management; economics politics; and geography. Each factor is represented by a node in the map and is assumed to proxy for a scientific discipline. The areas are detected using the 225 WoS subject categories, which classify journals included in the Science Citation Index (SCI) into disciplinary and sub-disciplinary structures. This allows a visualization of how the publications in a particular field (in our case 'Big Data') relate to different scientific disciplines. The term *cognitive map* refers to the projection of data on published works, onto an *overlay* showing the cognitive space that is the universe of contemporary research areas. On the resulting cognitive map, the node size is proportional to the number of publications related to the given topic, published in the given discipline represented by the node [25, 34]. Different colours are used to represent the 19 factors and to enable an immediate visual understanding. The cognitive map provides a classification of publications into research areas [40]. Although the classification of articles and journals into disciplinary categories using the ISI classification has been questioned [31], Rafols et al. [32] show that the resulting map is relatively robust to classification errors, and several studies have applied this approach to show its usefulness [25, 32, 34]. As Rotolo et al. [34] highlight, mapping emergence in the cognitive space can reveal a number of features. These include the directions of diffusion of a given topic across the key knowledge areas involved in its emergence, how these areas interact, and in which domain actors' knowledge production processes are positioned.

The cognitive map is integrated with the mapping of the journals publishing research on big data, and with the social map, that is, the co-authorship network resulting from our sample. In these co-authorship networks, the connections among authors are the channels through which they gain access to knowledge and generate innovative research outcomes. Co-authorship networks depict direct and intentional forms of knowledge exchange in which authors engage in real collaborative activities [24, 30], thus, we use co-author data to depict the emergence of scientific communities [12, 16].

Table 1 Document type

Document type	Number of documents
Proceedings	1,826
Published articles	1,793
Editorial materials	335
Reviews	125
Book reviews	17
Other documents	213

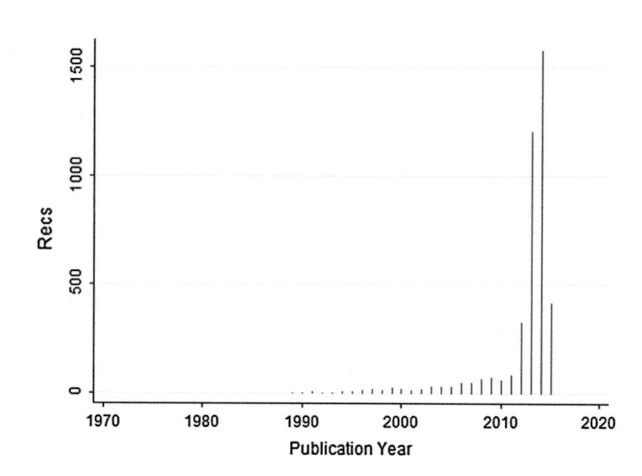

Fig. 1 Distribution of publications per year

3 Data and Basic Statistics

The ISI WoS is the data source for this study. Using the keywords 'big data' and 'large scale optimization' we obtained a sample of 4,308 published works comprising 1,826 proceedings, 1,793 published articles, 335 editorial pieces, 125 reviews, 16 book reviews and 213 documents (Table 1). Figure 1 depicts the yearly

Table 2 Top 10 institutions ranked by number of documents published

	Institution	Number of documents
1	Chinese Academy of Science	85
2	Massachusetts Institute of Technology (MIT)	52
3	Harvard University	50
4	Stanford University	46
5	University of California Los Angeles (UCLA)	45
6	Carnegie Mellon University	41
7	Northwestern University	40
8	University of South California	36
9	Tsinghua University	35
10	University of Illinois	34

Table 3 Top 10 institutions ranked by TLCS

	Institution	Total Local Citations Score (TLCS)
1	Northwestern University	198
2	Stanford University	109
3	Harvard University	99
4	Massachusetts Institute of Technology (MIT)	73
5	University of Colorado	73
6	Rice University	56
7	Catholic University of Nijmegen	47
8	University of Maryland	24
9	University of California Berkeley	42
10	Georgia State University	41

distribution of publications, showing that the majority were published between 2012 and May 2014. These pieces 1990s publications focus mainly on optimization problems and solutions. Tables 2 and 3 present the ranking for the top 10 institutions for number of documents owned (Table 2) and number of citations received within the sample (Table 3), based on the Total Local Citations Score (TLCS). This refers to how many times the papers in the sample have been cited by other papers in the same sample. In both cases, the ranking is based on the institution of the first author. The Chinese Academy of Science owns the highest number of documents (85), followed by the Massachusetts Institute of Technology (52 documents), Harvard University (50 documents), Stanford University (46), and University of California Los Angeles (45 documents). The most cited documents are owned by Northwestern University with a TLCS equal to 198, followed by Stanford University (TLCS = 109), Harvard University (TLCS = 99), Massachusetts Institute of Technology and University of Colorado (TLCS = 73).

4 Results

4.1 Mapping the Cognitive Space

Different dynamics can be traced across the cognitive space by creating an overlay of publications on basemaps of science. The projection (overlay) across the (base) map of science defined by the 225 WoS categories results in the cognitive map displayed in Fig. 2 where each node is a WoS category and proxies for a scientific discipline (labels). The node size is proportional to the number of publication in the given scientific discipline. Links among disciplines are represented by lines whose thickness is proportional to the extent to which the two disciplines cite each other. The cognitive map (Fig. 2) shows the diffusion of big data studies and large scale optimizations across many disciplines. The biggest shape, of computer science and

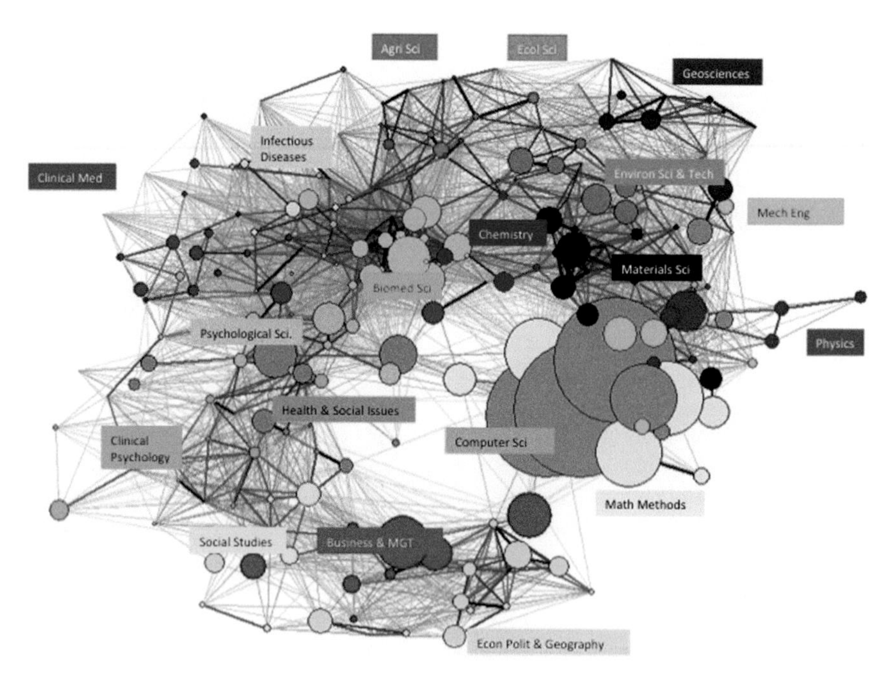

Fig. 2 Cognitive map

mathematical methods, indicates that these are the most active research areas. Most interesting is the contamination of a great number of disciplines, although with a lower number of publications, such as health, physics, biomedical science and material science. We can interpret this as the interest from researchers in biomedical science in opportunity provided by analysis of big data (i.e. mapping the human genome) and the challenges associated with the treatment and optimization of big data.

We can also make inferences based on the mapping of academic journals in Figs. 3 and 4 which provide two different visualizations of most active journals. Figure 3 shows journals aggregated by density. The colour of each item (journal) in the map is related to the density of the items at the point. The red area indicates local high density of journals, and blue area indicates low density. The most dense area is at the top right side of the Fig. 3. It refers to the area occupied by journals dealing with operational research and mathematical methods. The most frequent are *European Journal of Operational Research* (60), *Future Generation Computer Systems—The International journal of Grid Computing and Escience* (29) and *SIAM Journal of Optimization* (28). From Fig. 3 we can identify the second most dense area of research, which is coloured yellow and includes *Nature* (26), *Plos One* (21) and *Science* (17).

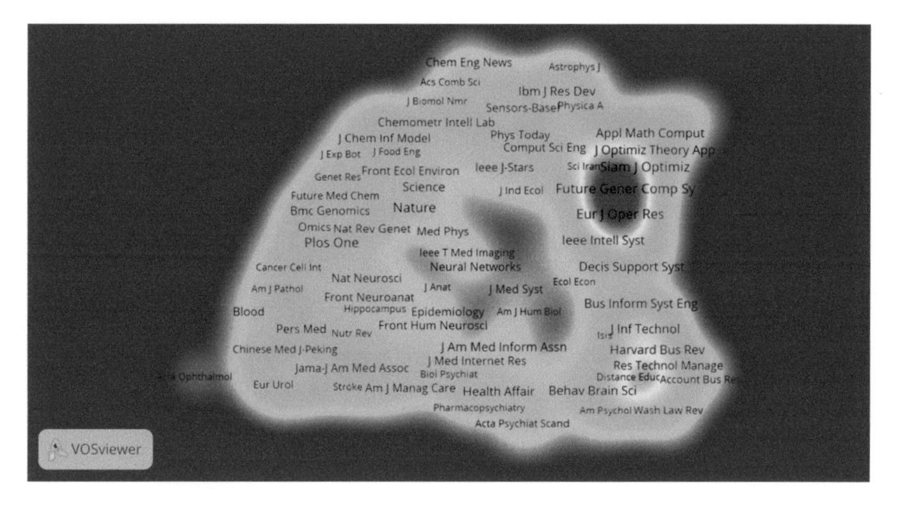

Fig. 3 Journal map (density)

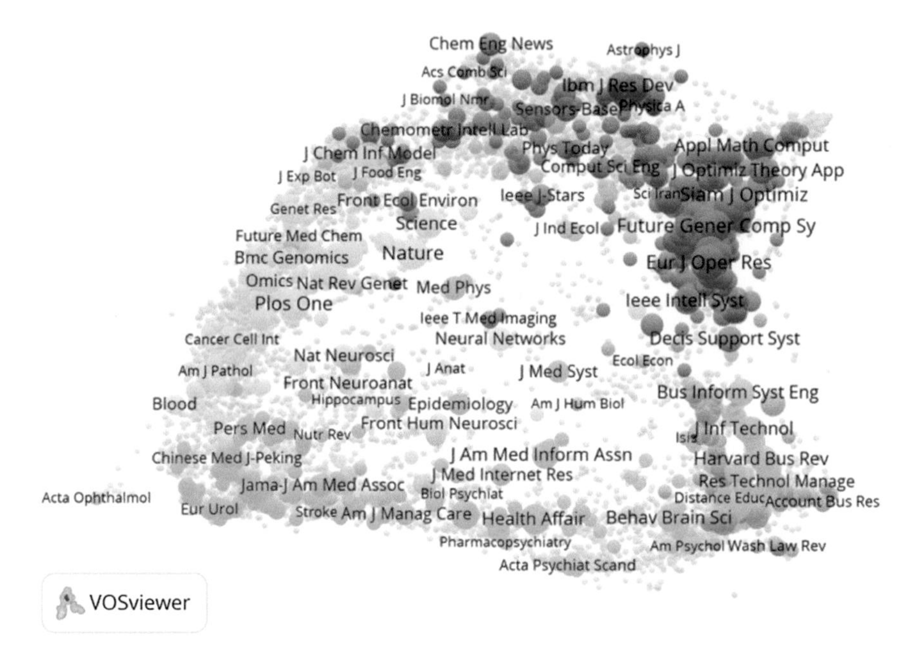

Fig. 4 Journal map (titles)

4.2 Mapping the Social Space

Analysis of the co-author networks shows the strong (number of co-authored works) and successful (highly cited works) connections among collaborating

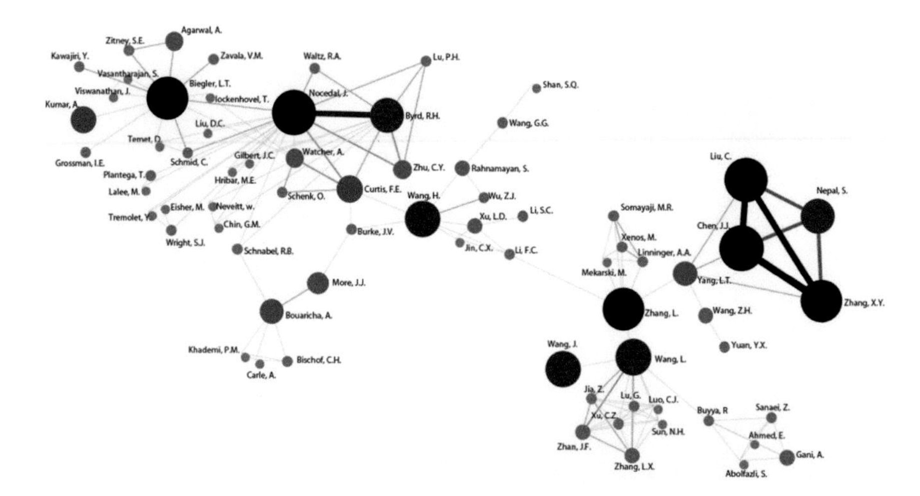

Fig. 5 Scientific community (co-author) working on optimization methods and big data

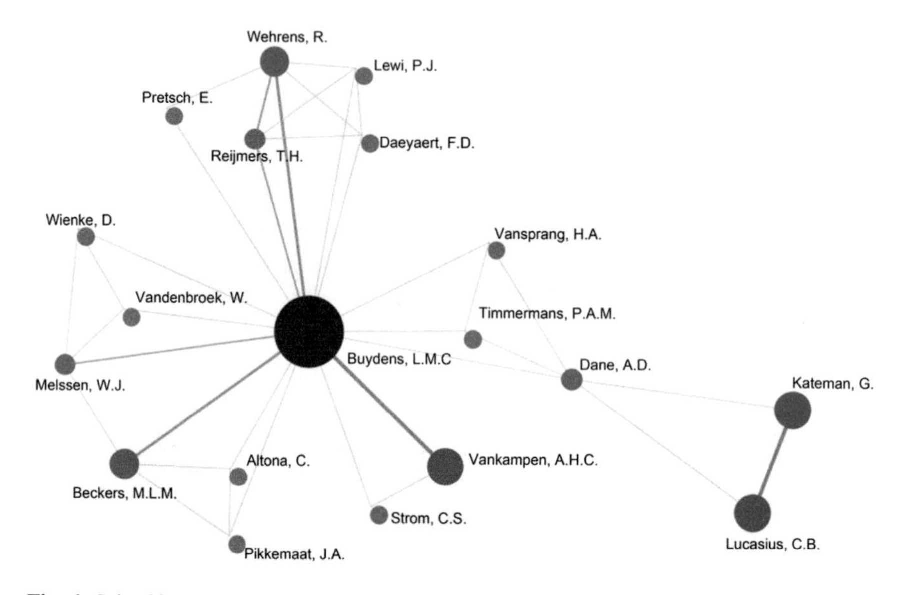

Fig. 6 Scientific community (co-authors) working on analytic chemistry and chemometrics

researchers. The links across the networks in Figs. 5, 6, 7, 8, 9 and 10 are channels of knowledge, and the networks highlight the scientific communities engaged in research on big data and large scale optimization. We describe the emergence of collaborations among researchers from different fields and discuss their work on big data and large scale optimization.

The largest network (Fig. 5) shows the scientific communities of scholars working on different aspects and applications of optimization methods and big data

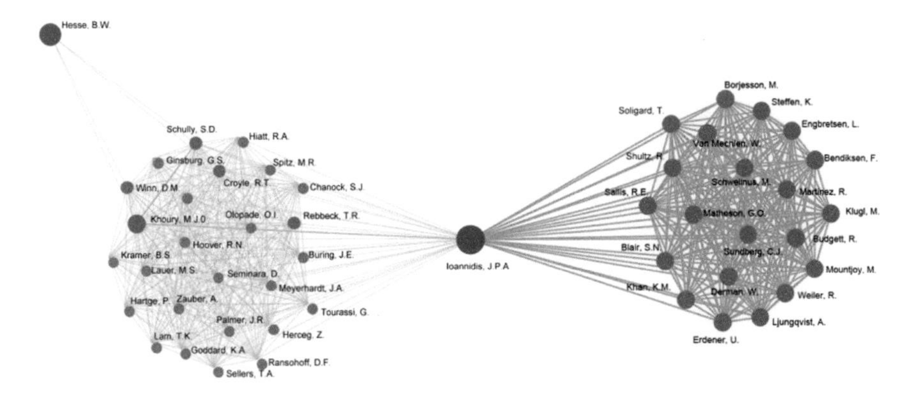

Fig. 7 Scientific community (co-authors) working on health, prevention and epidemiology

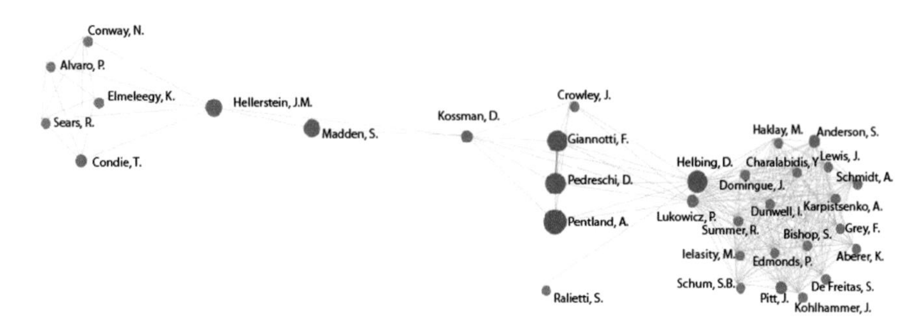

Fig. 8 Scientific community (co-authors) working on

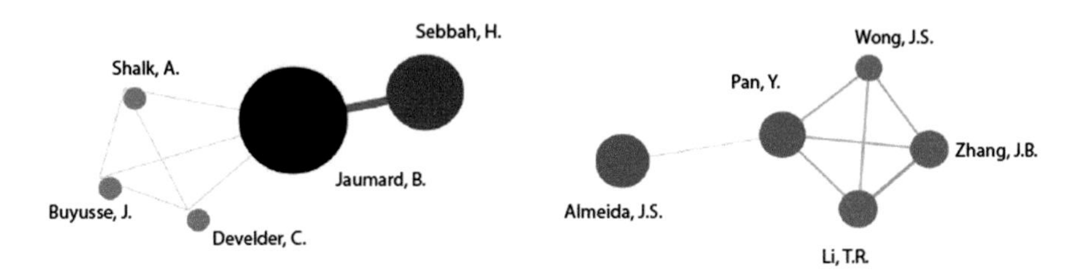

Fig. 9 Scientific community (co-authors) working on optimization solutions

analysis. Most of the authors on the left-hand side of the figure are from Northwestern University, which owns the papers receiving the highest number of citations within our sample. The left side of the figure includes leading authors such as Biegler, Nocedal and Byrd, who in the 1990s were working on large nonlinear optimization problems [6–8] and proposed solutions such as the reduced Hessian algorithm for large scale equality constrained optimization [2, 6]. More recent works were co-authored with Curtis on the topic of an algorithm for large scale

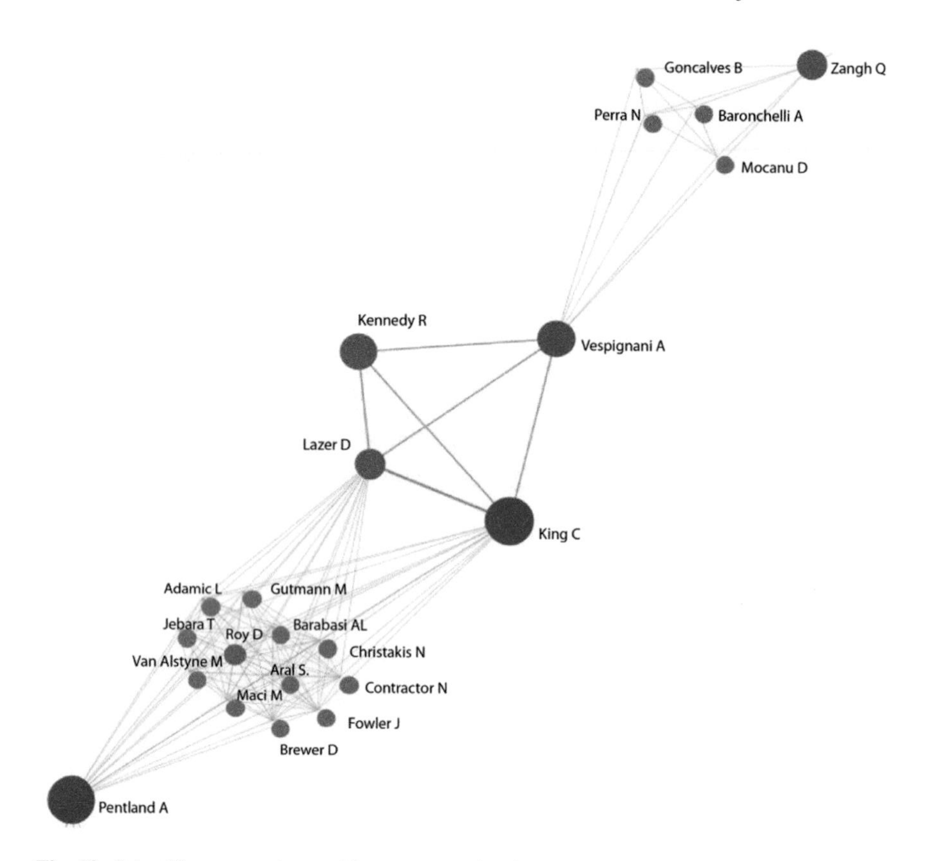

Fig. 10 Scientific community working on network science

equality constrained optimization [4] and its improvement for non-convex problems. They use a method that allows for the presence of negative curvature without requiring information on the inertia of the primal-dual iteration matrix [5]. They also work on problems with (near) rank-deficient Jacobian matrices where inexact step computations are allowed and inertia information is not required to solve non-convex problems [13].

Curtis's work links to Wang, working at the North Carolina A&T State University. They collaborated to solve an issue that arises when dealing with large-scale optimization problems related to solving exact penalty sub-problems on product sets [3]. Curtis and Wang proposed a two matrix-free method, one an Iterative Reweighting Algorithm (IRWA), and the other based on Alternating Direction Augmented Lagrangian (ADAL) technology applied to the given setting. They proved the efficiency of both algorithm, although in some cases the IRWA is more efficient than ADAL.

Wang et al. [41] propose a Generalized Opposition-Based Learning (GOBL) algorithm to improve the performance of Differential Evolution (DE) to solve high dimensional optimizations problems efficiently. The new algorithm, GODE, has been proven to outperform traditional DE on 19 high-dimensional problems with D = 50, 100, 200, 500, 1,000.

Another improved algorithm proposed by the community working with Wang is a genetic algorithm known as the Multi-Stage Composite Genetic Algorithm (MSC-GA) [26]. Its advantages are: (i) the gradual reduction of the optimization-search range, (ii) better convergence and less chance of being trapped into premature states.

The left side of Fig. 5 to which Wang is linked through Li, presents scholars working on big data. For example, the paper by Zhong et al. [47] proposes a practical application of big data handling in the context of water environment monitoring. They propose a new big data processing algorithm, which is based on a fast fuzzy C-means clustering algorithm to analyze water monitoring data. The advantage of fast clustering is that it allows rapid assessment of water quality from the samples.

Nepal, Zhang, Liu, Chen are also part of this network. Their bigger shapes and thicker lines indicate that they co-authored a large number of the papers in our sample and received a large number of citations relative to their publication dates. Two main features characterize the work of this scientific community: data handling in the cloud, and privacy. They address the problem of data handling in cloud computing, focusing on the detection of errors in processing sensor data in the cloud [42]. Error detection is based on a scale-free network topology; most operations are conducted in limited temporal or spatial data blocks rather than on the whole big data set, to enable faster error detection.

These authors also address the problem of privacy, one of the most discussed aspect of big data analysis. Zhang et al. [46] propose a scalable and cost-effective framework related to preserving privacy in relation to big data stored in the cloud. Zhang et al. [45] investigate the scalability of sub-tree anonymization of big data in the cloud. They try to resolve a major shortcoming in previous approaches to sub-tree anonymization—that of parallelization capability, which leads to lack of scalability when handling big data in the cloud. They combine two existing approaches, Top–Down Specialization (TDS) and Bottom–Up Generalization (BUG), and design MapReduce algorithms for each to achieve greater scalability.

The cloud has proven an ideal platform for big data processing, Yang et al. [43] propose a new multi-filter strategy for querying streaming XML big data on the cloud. This multi-filter strategy effectively shares and reduces the filtering space and the amount of time needed by exploiting the scalability of the cloud.

Figure 6 refers to the scientific community working in the area of analytic chemistry and chemometrics and focusing on Genetic Algorithms (GAs). GAs belong to the class of global optimization algorithms [17, 18]. Buydens is one of the main authors who collaborates with other researchers. She is currently Professor at Radboud University working in the Faculty of Analytic Chemistry. Chemometrics research has been affected by the explosion of big data, leading to the need to extract as much information as possible from the newly available biological data. The scientific community depicted in Fig. 6 is characterized by the application of

GAs to solve complex optimization problems such as the construction of phylogenetic trees to find the optimal tree [33].

Figure 7 refers to the scientific community working in health care and medicine. Authors in this network work on epidemiology and disease prevention. The opportunity to analyse big data on population health characteristics and diseases is attractive to many scholars; it provides an understanding of the hidden dynamics of how diseases spread, and helps in the design of prevention measures. Ioannidis, who works in the Department of Medicine at the Stanford Prevention Research Centre, is in the centre between two networks. The one on the left includes researchers working on cancer epidemiology [19]; the one on the right includes researchers working on the opportunities provided by the recently available Electronic Medical Records (EMR) which generate huge datasets of rich information on patients with non-communicable chronic disease (NCD) [27]. The results of these bodies of work have no practical application at present, but constitute propositions for future work.

The work of the scientific community depicted in Fig. 8 is characterized by the use of complex public data, such as open data, and analytic models from complexity science that positively influence societal wellbeing. Researchers in this community have proposed a new techno-social ecosystem—the Planetary Nervous System (PNS)—to compute and monitor new indices of social well-being.

Figure 9 is comprised of two social networks. B. Jaumard is the most prolific author in the left-hand side network and is a co-author of many of the papers in our sample. She works on large scale optimization at Concordia University. Her work deals with methods of protection in survivable Wavelength Division Multiplexing (WDM) networks [35] and Protected Working Capacity Envelope (PWCE) using Column Generation (CG) techniques to design survivable WDM networks based on p-cycle PWCE [36, 37].

The right-hand side network includes authors working on data mining and knowledge discovery, and proposing parallel large-scale rough set based methods for knowledge acquisition using MapReduce. These methods have been shown to be efficient on different platforms, Hadoop, Phoenix and Twister [44].

The scientific community or 'network science' (Fig. 10) includes authors working on social modelling, and large scale analysis of digital data such as microblogging posts [29]. Researchers in this community are trying to understand human and social dynamics using a quantitative approach to the analysis of large amounts of data [23]. They emphasize the opportunities associated with access to new data on human interactions such as e-mail, Twitter, Facebook content, Global Positioning System traffic and mobile telephony for modelling complex networks [39] and studying the behaviour of large-scale aggregates. They emphasize the terrific potential impact of these analyses, from simulations of pandemics to accurate epidemiological forecasting and insights into the management of catastrophic events such as hurricanes.

5 Conclusion

In this chapter we offer some informed perspectives on the emergence of big data and large scale optimization studies across social and cognitive spaces.

We analysed different types of informed perspectives: (i) the main scientific disciplines involved in the emergence of big data and large scale optimization studies; (ii) topics of interest in the scientific communities; (iii) interdisciplinary collaboration among researchers; and (iv) the distribution of publications across journals.

The literature highlights a major effort involved in handling large-scale data mining. This has led to several algorithms that have proven to be more efficient, faster and more accurate than earlier solutions. Their application is useful in the chemometrics, cloud computing, environment and privacy protection fields.

We showed that many publications point to the enormous terrific potential of big data, but that a lot remains to be done to achieve various health sector objectives including epidemiology issues.

Concerns remain over the use of big data in relation to the high risks for governments associated with potential incorrect predictions, which might lead to unnecessary costs and unethical social controls over citizens' privacy and access to their private data.

References

1. Abadi, D., Agrawal, R., Ailamaki, A., Balazinska, M., Bernstein, P. A.: The Beckman report on database research. Sigmod Rec. **43**, 61–70 (2014). doi:10.1145/2694428.2694441
2. Biegler, L.T., Nocedal, J., Schmid, C., Ternet, D.: Numerical experience with a reduced Hessian method for large scale constrained optimization. Comput. Optim. Appl. **15**, 45–67 (2000). doi:10.1023/A:1008723031056
3. Burke, J.V., Curtis, F.E., Wang, H., Wang, J.: Iterative reweighted linear least squares for exact penalty subproblems on product sets. SIAM J. Optim. (2015)
4. Byrd, R.H., Curtis, F.E., Nocedal, J.: An inexact SQP method for equality constrained optimization. SIAM J. Optim. **19**, 351–369 (2008)
5. Byrd, R. H., Curtis, F., E.Nocedal, J.: An inexact Newton method for nonconvex equality constrained optimization. Math Program **122**(2), 273-299 (2008). doi:10.1007/s10107-008-0248-3
6. Byrd, R.H., Lu, P., Nocedal, J., Zhu, C.: A limited memory algorithm for bound constrained optimization. SIAM J. Sci. Comput. **16**, 1190–1208 (1995). doi:10.1137/0916069
7. Byrd, R.H., Nocedal, J., Zhu, C.: Towards a discrete Newton method with memory for large-scale optimization. Nonlinear Optim. Appl. 1–13 (1996a)
8. Byrd, R.H., Nocedal, J., Zhu. C.: Nonlinear Optimization and Applications. Springer, Boston (1996b)
9. Calero-Medina, C., Noyons, E.C.M.: Combining mapping and citation network analysis for a better understanding of the scientific development: the case of the absorptive capacity field. J. Informetr. **2**, 272–279 (2008). doi:10.1016/j.joi.2008.09.005
10. Chen, H., Chiang, R.H.L., Storey, V.C.: Business intelligence and analytics: from big data to big impact. MIS Q. **36**, 1165–1188 (2012)

11. Cox, M., Ellsworth, D.: Application-controlled demand paging for out-of-core visualization, pp. 235–ff (1997)
12. Crane, D.: Invisible Colleges: Diffusion of Knowledge in Scientific Communities. The University of Chicago Press, Chicago (1972)
13. Curtis, F.E., Nocedal, J., Wächter, A.: A matrix-free algorithm for equality constrained optimization problems with rank-deficient Jacobians. SIAM J. Optim. **20**, 1224–1249 (2010). doi:10.1137/08072471X
14. De Stefano, D., Giordano, G., Vitale, M.P.: Issues in the analysis of co-authorship networks. Qual. Quant. **45**, 1091–1107 (2011). doi:10.1007/s11135-011-9493-2
15. Emrouznejad, A., Marra, M.: Ordered weighted averaging operators 1988–2014: a citation-based literature survey. Int. J. Intell. Syst. **29**, 994–1014 (2014). doi:10.1002/int.21673
16. Glänzel, W., Schubert, A.: Analyzing scientific networks through co-authorship. Handbook of Quantitative Science and Technology Research, pp. 257–276. Kluwer Academic Publishers, Dordrech (2004)
17. Goldberg, D.E.: Genetic Algorithms in Search Optimization and Machine Learning. Addison-Wesley Longman Publishing Co., Inc, Boston, MA (1989)
18. Holland, J.H.: Adaptation in Natural and Artificial Systems. The MIT Press, Cambridge, MA (1975)
19. Khoury, M.J., Lam, T.K., Ioannidis, J.P.A., Hartge, P., Spitz, M.R., Buring, J.E., Chanock, S. J., Croyle, R.T., Goddard, K.A., Ginsburg, G.S., Herceg, Z., Hiatt, R.A., Hoover, R.N., Hunter, D.J., Kramer, B.S., Lauer, M.S., Meyerhardt, J.A., Olopade, O.I., Palmer, J.R., Sellers, T.A., Seminara, D., Ransohoff, D.F., Rebbeck, T.R., Tourassi, G., Winn, D.M., Zauber, A., Schully, S.D.: Transforming epidemiology for 21st century medicine and public health. Cancer Epidemiol. Biomarkers Prev. **22**, 508–516 (2013). doi:10.1158/1055-9965.EPI-13-0146
20. Lampe, H.W., Hilgers, D.: Trajectories of efficiency measurement: a bibliometric analysis of DEA and SFA. Eur. J. Oper. Res. **240**, 1–21 (2014). doi:10.1016/j.ejor.2014.04.041
21. Lane, J., Stodden, V., Bender, S., Nissenbaum, H.: Privacy, Big Data, and the Public Good. Cambridge University Press, New York (2014). doi:http://dx.doi.org/10.1017/CBO9781107590205
22. Lazer, D., Kennedy, R., King, G., Vespignani, A.: The parable of Google flu: traps in big data analysis. Science (80-.). **343**, 1203–1205 (2014). doi:10.1126/science.1248506
23. Lazer, D., Kennedy, R., King, G., Vespignani, A.: Twitter: Big data opportunities response. Science **345**(6193), 148–149 (2014). doi:10.1126/science.345.6193
24. Lee, J.-D., Baek, C., Kim, H.-S., Lee, J.-S.: Development pattern of the DEA research field: a social network analysis approach. J. Product. Anal. **41**, 175–186 (2014). doi:10.1007/s11123-012-0293-z
25. Leydesdorff, L., Carley, S., Rafols, I.: Global maps of science based on the new Web-of-Science categories. Scientometrics **94**, 589–593 (2013). doi:10.1007/s11192-012-0784-8
26. Li, F., Xu, L.Da, Jin, C., Wang, H.: Structure of multi-stage composite genetic algorithm (MSC-GA) and its performance. Expert Syst. Appl. **38**, 8929–8937 (2011). doi:10.1016/j.eswa.2011.01.110
27. Matheson, G.O., Klügl, M., Engebretsen, L., Bendiksen, F., Blair, S.N., Börjesson, M., Budgett, R., Derman, W., Erdener, U., Ioannidis, J.P.A., Khan, K.M., Martinez, R., Mechelen, W. Van, Mountjoy, M., Sallis, R.E., Sundberg, C.J., Weiler, R., Ljungqvist, A.: Prevention and management of non-communicable disease: the IOC consensus statement. Clin. J. Sport Med. 1003–1011 (2013). doi:10.1136/bjsports-2013-093034
28. Mayer-Schönberger, V., Cukier, K.: Big Data: A Revolution That Will Transform How We Live, Work, and Think. Houghton Mifflin Harcourt (2013)
29. Mocanu, D., Baronchelli, A., Perra, N., Gonçalves, B., Zhang, Q., Vespignani, A: The Twitter of Babel: mappingworld languages through microblogging platforms. PloS one **8**, (2013). doi:10.1371/journal.pone.0061981

30. Oh, W., Choi, J.N., Kim, K.: Coauthorship dynamics and knowledge capital: the patterns of cross-disciplinary collaboration in Information Systems research. J. Manag. Inf. Syst. **22**, 266–292 (2006). doi:10.2753/MIS0742-1222220309
31. Pudovkin, A.I., Garfield, E.: Algorithmic procedure for finding semantically related journals. J. Am. Soc. Inf. Sci. Technol. **53**, 1113–1119 (2002). doi:10.1002/asi.10153
32. Rafols, I., Porter, A.L., Leydesdorff, L.: Science overlay maps: a new tool for research policy and library management. J. Am. Soc. Inf. Sci. Technol. **61**, 1871–1887 (2010). doi:10.1002/asi.21368
33. Reijmers, T., Wehrens, R., Daeyaert, F., Lewi, P., Buydens, L.M.: Using genetic algorithms for the construction of phylogenetic trees: application to G-protein coupled receptor sequences. Biosystems **49**, 31–43 (1999). doi:10.1016/S0303-2647(98)00033-1
34. Rotolo, D., Rafols, I., Hopkins, M., Leydesdorff, L.: Scientometric mapping as a strategic intelligence tool for the governance of emerging technologies (Digital Libraries) (2013)
35. Sebbah, S., Jaumard, B.: Differentiated quality-of-recovery in survivable optical mesh networks using p-structures. IEEE/ACM Trans. Netw. **20**, 798–810 (2012). doi:10.1109/TNET.2011.2166560
36. Sebbah, S., Jaumard, B.: An efficient column generation design method of p-cycle-based protected working capacity envelope. Photonic Netw. Commun. **24**, 167–176 (2012). doi:10.1007/s11107-012-0377-8
37. Sebbah, S., Jaumard, B.: PWCE design in survivablem networks using unrestricted shape p-structure patterns. In: 2009 Canadian Conference on Electrical and Computer Engineering, pp. 279–282. IEEE (2009). doi:10.1109/CCECE.2009.5090137
38. Varian, H.R.: Big data: new tricks for econometrics. J. Econ. Perspect. **28**, 3–28 (2014). doi:10.1257/jep.28.2.3
39. Vespignani, A.: Predicting the behaviour of techno-social systems. Science **325**(5939), 425–428 (2009). doi:10.1126/science.1171990
40. Waltman, L., van Eck, N.J.: A new methodology for constructing a publication-level classification system of science. J. Am. Soc. Inf. Sci. Technol. **63**, 2378–2392 (2012). doi:10.1002/asi.22748
41. Wang, H., Wu, Z., Rahnamayan, S.: Enhanced opposition-based differential evolution for solving high-dimensional continuous optimization problems. Soft. Comput. **15**, 2127–2140 (2010). doi:10.1007/s00500-010-0642-7
42. Yang, C., Liu, C., Zhang, X., Nepal, S., Chen, J.: A time efficient approach for detecting errors in big sensor data on cloud. IEEE Trans. Parallel Distrib. Syst. **26**, 329–339 (2015). doi:10.1109/TPDS.2013.2295810
43. Yang, C., Liu, C., Zhang, X., Nepal, S., Chen, J.: Querying streaming XML big data with multiple filters on cloud. In: 2013 IEEE 16th International Conference on Computational Science and Engineering, pp. 1121–1127. IEEE (2013). doi:10.1109/CSE.2013.163
44. Zhang, J., Wong, J.-S., Li, T., Pan, Y.: A comparison of parallel large-scale knowledge acquisition using rough set theory on different MapReduce runtime systems. Int. J. Approx. Reason. **55**, 896–907 (2014). doi:10.1016/j.ijar.2013.08.003
45. Zhang, X., Liu, C., Nepal, S., Yang, C., Dou, W., Chen, J.: A hybrid approach for scalable sub-tree anonymization over big data using MapReduce on cloud. J. Comput. Syst. Sci. **80**, 1008–1020 (2014). doi:10.1016/j.jcss.2014.02.007
46. Zhang, X., Liu, C., Nepal, S., Yang, C., Dou, W., Chen, J.: SaC-FRAPP: a scalable and cost-effective framework for privacy preservation over big data on cloud. Concurr. Comput. Pract. Exp. **25**, 2561–2576 (2013). doi:10.1002/cpe.3083
47. Zhong, Y., Zhang, L., Xing, S., Li, F., Wan, B.: The big data processing algorithm for water environment monitoring of the three Gorges reservoir area. Abstr. Appl. Anal. 1–7 (2014)

Author Biographies

Ali Emrouznejad is a Professor and Chair in Business Analytics at Aston Business School, UK. His areas of research interest include performance measurement and management, efficiency and productivity analysis as well as data mining. Dr Emrouznejad is Editor of *Annals of Operations Research*, Associate Editor of *Socio-Economic Planning Sciences*, Associate Editor of *IMA journal of Management Mathematics*, Associate Editor of *RAIRO —Operations Research*, Senior Editor of *Data Envelopment Analysis journal*, and member of editorial boards or guest editor in several other scientific journals. He has published over 100 articles in top ranked journals; he is author of the book on "Applied Operational Research with SAS", editor of the books on "Performance Measurement with Fuzzy Data Envelopment Analysis" (Springer), "Managing Service Productivity" (Springer), and "Handbook of Research on Strategic Performance Management and Measurement" (IGI Global). He is also co-founder of Performance Improvement Management Software (PIM-DEA), see http://www.Emrouznejad.com.

Marianna Marra is Lecturer in Management Science at Essex Business School, UK. She earned her PhD in Operations Management at Aston Business School. Her research focuses on social network analysis, networks, innovation and knowledge transfer processes. She has published papers on Expert Systems with Applications, Information Sciences and the International Journal of Intelligent Systems.

Setting Up a Big Data Project: Challenges, Opportunities, Technologies and Optimization

Roberto V. Zicari, Marten Rosselli, Todor Ivanov, Nikolaos Korfiatis, Karsten Tolle, Raik Niemann and Christoph Reichenbach

Abstract In the first part of this chapter we illustrate how a big data project can be set up and optimized. We explain the general value of big data analytics for the enterprise and how value can be derived by analyzing big data. We go on to introduce the characteristics of big data projects and how such projects can be set up, optimized and managed. Two exemplary real word use cases of big data projects are described at the end of the first part. To be able to choose the optimal big data tools for given requirements, the relevant technologies for handling big data are outlined in the second part of this chapter. This part includes technologies such as NoSQL and NewSQL systems, in-memory databases, analytical platforms and Hadoop based solutions. Finally, the chapter is concluded with an overview over big data

---

R.V. Zicari · M. Rosselli (✉) · T. Ivanov · N. Korfiatis · K. Tolle · R. Niemann · C. Reichenbach
Frankfurt Big Data Lab, Goethe University Frankfurt, Frankfurt Am Main, Germany
e-mail: rosselli@dbis.cs.uni-frankfurt.de
URL: http://www.bigdata.uni-frankfurt.de; http://www.accenture.com

R.V. Zicari
e-mail: zicari@dbis.cs.uni-frankfurt.de

T. Ivanov
e-mail: todor@dbis.cs.uni-frankfurt.de

N. Korfiatis
e-mail: n.korfiatis@uea.ac.uk

K. Tolle
e-mail: tolle@dbis.cs.uni-frankfurt.de

R. Niemann
e-mail: raik.niemann@iisys.de

C. Reichenbach
e-mail: reichenbach@em.uni-frankfurt.de

M. Rosselli
Accenture, Frankfurt, Germany

N. Korfiatis
Norwich Business School, University of East Anglia, Norwich, UK

A. Emrouznejad (ed.), *Big Data Optimization: Recent Developments and Challenges*, Studies in Big Data 18, DOI 10.1007/978-3-319-30265-2_2

benchmarks that allow for performance optimization and evaluation of big data technologies. Especially with the new big data applications, there are requirements that make the platforms more complex and more heterogeneous. The relevant benchmarks designed for big data technologies are categorized in the last part.

1 How to Set Up a Big Data Project

Data is becoming viewed as a corporate weapon [1].

How to set up and optimize a big data project?

How to set up and optimize a big data project is a challenging task that requires understanding the value of big data, the various processes and steps involved on running the project as well as the big data technologies that can be used to store and process the data. This chapter addresses these questions in three separate parts. The possible value of big data for an organization and the steps of a big data project are described in the first part. In the subsequent part the various big data technologies are classified and described according to their data model and typical use cases allowing for an easy selection of the optimal technologies given the requirements. The chapter concludes introducing big data benchmarking as a way to optimize and fine-tune the performance of the big data technologies.

Why run a big data project in the enterprise?

The most straightforward, perhaps too straightforward answer is that a big data project can offer valuable insights obtained from analyzing data, which in turn may offer a quantifiable competitive advantage to the enterprise or organization. However, beyond this high-level insight, things are less straightforward: big data projects have no single canonic use case or structure. Instead, the applications that could benefit from analyzing big data cut across industries and involve a wide variety of data sources.

As James Kobielus states [2], results from big data projects may be materialize either as revenue gains, cost reductions, or risk mitigation which have an easy and measurable Return on Investment (ROI). Often big data projects are in support of Customer Relationship Management (CRM) initiatives in marketing, customer service, sales, and brand monitoring. But of course, these are only a few examples out of many.

Big data for social good

On another approach, a different, but equally important opportunity for big data is serving the people who are in need globally, helping the society in which we live. When big data is used to improve our society and people's conditions within it, we cannot use the standard "ROI" as an impact measure but perhaps a new term such as: "SROI" or Social Return of Investments for big data?

What is big data?

Understanding and optimizing "big data" projects requires definition of what the term means. One definition that has gained considerable attention considers three distinctive characteristics of big data, namely *Volume*, *Variety* and *Velocity*, first introduced by Laney [3]. Zikopoulos and Eaton [4] summarized the dynamics and

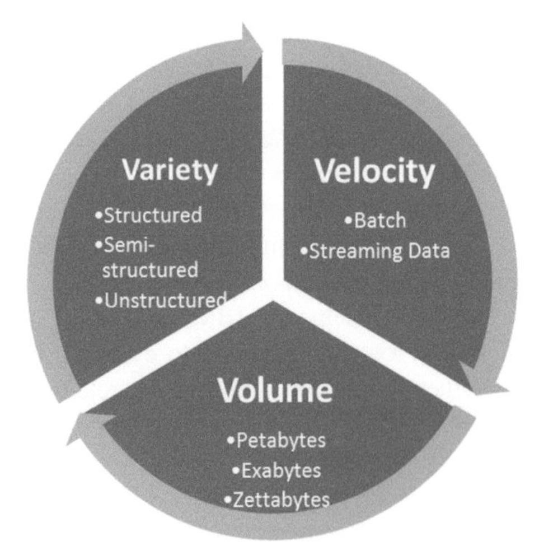

Fig. 1 IBM big data characteristics—3 V. Adopted from [4]

interconnection between these characteristics as a case of interconnected stages, known as the 3 Vs of big data. On the 3V model in Fig. 1, each stage is interconnected and runs as an input to the subsequent one.

The *Volume* represents the ever growing amount of data in petabytes, exabytes or zettabytes that is generated in today's "Internet of things" (IoT), and challenges the current state of storage systems. At the same time the *Variety* of data produced by a multitude of sources like sensors, smart devices and social media in raw, semi-structured, unstructured and rich media formats is further complicating the processing and storage of data. Finally, the *Velocity* aspect describes how quickly the data is retrieved, stored and processed.

From an information processing perspective, these three characteristics describe accurately what big data is and the new challenges that it presents to the current infrastructure. While data *Variability* and *Veracity* are also discussed as additional dimensions on this initial model [3, 4], the core Vs represent the basics for a more complete and systematic big data framework.

The emergence of *new analytical applications* and big data open new challenges [7] which will be addressed later in this chapter.

Why Big Data Analytics?

So what is so special about Big Data Analytics? Werner Vogels, CTO of Amazon.com Inc., once said in an interview [8] that "*one of the core concepts of big data is being able to evolve analytics over time*" and that "*in the new world of data analysis your questions are going to evolve and change over time and as such you need to be able to collect, store and analyze data without being constrained by resources*".

This is the fundamental difference with respect to traditional ways in Business Intelligence (BI), for example. Basically, someone discovers patterns from analyzing data, and then receives the answers to questions that in fact he/she did not ask. This sounds bizarre, but it opens up wider opportunities than simply asking the questions against structured predefined data sets, as was done before.

Jochen L. Leidner, Data Scientist at Thomson Reuters, UK explains that [9]: "*Usually analytics projects happen as an afterthought to leverage existing data created in a legacy process, which means not a lot of change of process is needed at the beginning. This situation changes once there is resulting analytics output, and then the analytics-generating process needs to be integrated with the previous processes*".

How good is the "value" that can be derived by analyzing big data?

First of all, to justify a big data project in an enterprise, in most cases there is a need to identify a *quantitative ROI*.

James Kobielus of IBM suggests [2] the use of Customer Lifetime Value (CLV) as a standard metric to evaluate, through big data analytics, the impact on customer acquisition, onboarding, retention, upsell, cross-sell and other indicators, as well as from corresponding improvements in operational efficiency. This is certainly a possibility.

Cynthia M. Saracco, also of IBM, provides the following scenario [10]: "*For example, if a telecommunications firm is able to reduce customer churn by 10 % as a result of a big data project, what's that worth? If an organization can improve the effectiveness of an email marketing campaign by 20 %, what's that worth? If an organization can respond to business requests twice as quickly, what's that worth? Many clients have these kinds of metrics in mind as they seek to quantify the value they have derived—or hope to derive—from their investment in a big data project.*"

Table 1 CRISP-DM and SEMMA process overview

<table>
<tr><th>CRISP-DM processes</th><th>SEMMA processes</th></tr>
<tr><td>Business Understanding: The process involves an understanding of the business problem and a consultation with domain experts on explaining what the overall goal of the project is and how it is expected to help the business function</td><td>Sample: The process starts with data sampling, e.g., selecting the data set for modeling. The data set should be large enough to contain sufficient information to retrieve, yet small enough to be used efficiently. This phase also deals with data partitioning</td></tr>
<tr><td>Data Understanding: This stage involves collection of initial data and exploratory analysis of the data properties such as separating data into subsets in order to form and evaluate hypotheses</td><td>Explore: This phase covers data understanding through the use of exploratory analytics with the help of data visualization</td></tr>
<tr><td>Data Preparation: This process concerns the preparation of data to be used for the analysis concerning the project goals</td><td>Modify: This stage concerns the extraction, transformation and loading (ETL) of the data to a specialized dataset</td></tr>
<tr><td colspan="2">Modeling: In this stage for both models the focus is on applying various modeling (data mining) techniques on the prepared dataset in order to create models that possibly provide the desired outcome. These can, for example, be predictive or classification models targeted at predicting an output variable (e.g. forecasting sales) or classifying a composite variable (e.g. client profiling) using various features (e.g. client age, region etc.)</td></tr>
<tr><td colspan="2">Evaluation/Assessment: This process involves the evaluation of the reliability, usefulness and the impact of the models created on the business function</td></tr>
<tr><td colspan="2">Deployment: The deployment stage considers the assimilation of the project's output on the production site. This stage is not part of the SEMMA model</td></tr>
</table>

How best to get started with a big data project?

Running a big data project demands an understanding of how the project differs from a typical business intelligence project that might be running concurrently within the same organization. One needs to provide an understanding of the context of the classification of such a project as "big data" using a standard definition. On the other hand, theoretical definitions of what "big data" is and how organizations and enterprises can use it has been a subject of debate [11].

The industry has identified two major methodologies for running a *data oriented project*, namely SEMMA (Sample, Explore, Modify, Model and Assess) and CRISP-DM (Cross Industry Standard Process for Data Mining). SEMMA considers in total five separate stages where in the initial stage no assistance from domain experts is required. CRISP-DM, on the other hand, considers six different stages where assistance from domain experts is expected in the initial understanding of the data. Table 1 lists the steps of the CRISP-DM and SEMMA processes respectively.

However, big data projects pose new challenges that are not covered by existing methodologies. Specifically, if we look at the business decisions that need to be made in order to successfully support a big data project in enterprise, Cynthia M. Saracco [10] and James Kobielus [2] both identify the following critical aspects:

Project's business objectives: "Begin with a clear definition of the project's business objectives and timeline, and be sure that you have appropriate executive sponsorship. The key stakeholders need to agree on a minimal set of compelling results that will impact your business; furthermore, technical leaders need to buy into the overall feasibility of the project and bring design and implementation ideas to the table."

Project scope: **"**Scope the project well to deliver near-term business benefit. Using the nucleus project as the foundation for accelerating future big data projects."

Since big data projects can get pretty complex, it is helpful to segment the work into broad categories and then drill down into each to create a solid plan.

Relevant stakeholders: "The involvement of the relevant stakeholders. It is important that the big data initiative be aligned with key stakeholder requirements. If stakeholders haven't clearly specified their requirements or expectations for your big data initiative, it's not production-ready."

Infrastructure: "In order to successfully support a big data project in the enterprise, you have to make the infrastructure and applications production-ready in your operations." In order to be able to analyze big data (i.e. data is structured and/or not structured at all) you need specific data management technology. One popular big data storage and processing technology is the Hadoop ecosystems of open software tools (see the second part of this chapter).

Skillsets: "The staff needs to have the right skillsets: e.g. database, integration and analytics skills". This point is not trivial and James Kobielus adds [2]: "Data-driven organizations succeed when all personnel—both technical and business—have a common understanding of the core big data best skills, tools and practices. You need all the skills of data management, integration, modeling, and so forth that you already have running your data marts, warehouses, OLAP cubes, and the like."

What are the problems and challenges that need to be faced in many big data projects?

The list of challenges in a big data projects include a combination of the following issues:

- Lack of appropriately scoped objectives,
- lack of required skills,
- the size of big data,
- the non-clearly defined structure of much of the big data,
- the difficulty of enforcing data consistency,
- privacy,
- data management/integration,
- rights management,
- ETL,
- data discovery (how to find high-quality data from the web?),
- data veracity (how can we cope with uncertainty, imprecision, missing details?),
- data verification,
- technical challenges for big data analytics when data is in motion rather than at rest.

Some aspects have been identified as major challenges from industry experts. Paul C. Zikopoulos of IBM comments on data velocity [12]: "*It's not just how fast data is produced or changes, but the speed at which it must be understood, acted upon, turned into something useful.*"

Scott Jarr [1] of VoltDB, a NewSQL data store company, also explains: "There is only so much static data in the world as of today. The vast majority of new data, the data that is said to explode in volume over the next 5 years, is arriving from a high velocity source. It's funny how obvious it is when you think about it. The only way to get big data in the future is to have it arrive in a high velocity rate… We think of big data velocity as data that is coming into the organization at a rate that can't be managed in the traditional database. However, companies want to extract the most value they can from that data as it arrives. We see them doing three specific things: Ingesting relentless feed(s) of data, making decisions on each piece of data as it arrives and using real-time analytics to derive immediate insights into what is happening with this velocity data."

All of these are requirements for choosing a suitable Big Data Analytical Platform.

Choosing the right data platform technology

Which technology can best scale to petabytes? Choosing the right analytics and/or data management platform for a big data project is not a trivial task. In the rest of the chapter we will outline the available data management technologies. A subsequent session on big data benchmarking will be provided as a way to measure the capacity of these technologies. Due to space limitations, we will not cover the analytics aspect in detail.

For a Data Management Platform, the key requirement is the ability to scale. Scalability has three aspects: data volume, hardware size and concurrency.

It is important to note that scale and performance requirements for big data strain conventional relational databases. As a result, new generations of NoSQL database systems, new implementation of relational databases (also called NewSQL), in-memory databases, and graph databases have emerged. In addition to this, there is of course the entire eco-system of open source software related to Hadoop.

Why Hadoop?

John Schroeder, CEO and co-founder of MapR Technologies, explains [13]: "One of the benefits of Hadoop is that you don't need to understand the questions you are going to ask ahead of time, you can combine many different data types and determine required analysis you need after the data is in place."

Daniel Abadi, Co-founder of Hadapt, further explains [14]: "A lot of people are using Hadoop as a sort of data refinery. Data starts off unstructured, and Hadoop jobs are run to clean, transform, and structure the data. Once the data is structured, it is shipped to SQL databases where it can be subsequently analyzed. This leads to the raw data being left in Hadoop and the refined data in the SQL databases."

> But it's basically the same data—one is just a cleaned (and potentially aggregated) version of the other. Having multiple copies of the data can lead to all kinds of problems. For example, let's say you want to update the data in one of the two locations—it does not get automatically propagated to the copy in the other silo. Furthermore, let's say you are doing some analysis in the SQL database and you see something interesting and want to drill down to the raw data—if the raw data is located on a different system, such a drill down becomes highly nontrivial. Furthermore, data provenance is a total nightmare. It's just a really ugly architecture to have these two systems with a connector between them.

Hybrid architectures

This result is, as James Kobielus comments [2]: "In the larger evolutionary perspective, big data is evolving into a hybridized paradigm under which Hadoop, massively parallel processing (MPP) enterprise data warehouses (EDW), in-memory columnar, stream computing, NoSQL, document databases, and other approaches support extreme analytics in the cloud. Hybrid architectures address the heterogeneous reality of big data environments and respond to the need to incorporate both established and new analytic database approaches into a common architecture. The fundamental principle of hybrid architectures is that each constituent big data platform is fit-for-purpose to the role for which it's best suited."

The "big data platform" that is to be used needs to meet the availability, security and robustness requirements expected of the enterprise infrastructure. This affects the entire big data technology stack: databases, middleware, applications, and tools. The environment has to be designed for modular scaling in order to cope with the growth in data volumes, velocities and varieties.

Is Open Source the answer?

A lot of the software for handling big data is open source, but of course not only. Cynthia M. Saracco of IBM mentions that [10]: "Even if your big data solution uses open source software, there are still expenses involved for designing, developing, deploying, and maintaining your solution. So what did your business gain from that investment? The answer to that question is going to be specific to your application and your business".

Elastic computing in the cloud? How does it relate to Big Data Analytics?
Cloud computing and virtualization of resources are directly related to big data. Paul C. Zikopoulos [12] comments: "This is pretty important because I need the utility-like nature of a Hadoop cluster, without the capital investment. Time to analytics is the benefit here.

"After all, if you're a start-up analytics firm seeking venture capital funding, do you really walk into to your investor and ask for millions to set up a cluster; you'll get kicked out the door. No, you go to Rackspace or Amazon, swipe a card, and get going. IBM is there with its Hadoop clusters (private and public) and you're looking at clusters that cost as low as $0.60 US an hour.

> I think at one time I costed out a 100 node Hadoop cluster for an hour and it was like $34US—and the price has likely gone down. What's more, your cluster will be up and running in 30 min.

Use cases from big data projects
To end the first part of this chapter, we will illustrate two use cases as examples of how big data is used to create value for a company:

Use Case 1: NBCUniversal International
Dr. Bassett, Director of Data Science for NBCUniversal International, UK, explains the process by which they "dig into" data [15]:

> I'm the Director of Data Science for NBCUniversal International. I lead a small but highly effective predictive analytics team. I'm also a "data evangelist"; I spend quite a bit of my time helping other business units realize they can find business value from sharing and analyzing their data sources.
> We predict key metrics for the different businesses—everything from television ratings, to how an audience will respond to marketing campaigns, to the value of a particular opening weekend for the box office. To do this, we use machine learning regression and classification algorithms, semantic analysis, monte-carlo methods, and simulations.
> We start with the insight in mind: What blind-spots do our businesses have, what questions are they trying to answer and how should that answer be presented? Our process begins with the key business leaders and figuring out what problems they have—often when they don't yet know there's a problem. Then we start our feature selection, and identify which sources of data will help achieve our end goal—sometimes a different business unit has it sitting in a silo and we need to convince them to share, sometimes we have to build a system to crawl the web to find and collect it.
> Once we have some idea of what we want, we start brainstorming about the right methods and algorithms we should use to reveal useful information: Should we cluster across a multi-variate time series of market response per demographic and use that as an input for a regression model? Can we reliably get a quantitative measure of a demographics engagement from sentiment analysis on comments? This is an iterative process, and we spend quite a bit of time in the "capturing data/transforming the data" step.
> We predict key metrics for the different businesses—everything from television ratings, to how an audience will respond to marketing campaigns, to the value of a particular opening weekend for the box office. To do this, we use machine learning regression and classification algorithms, semantic analysis, monte-carlo methods, and simulations. For instance, our cinema distribution company operates in dozens of countries. For each day in each one, we need to know how much money was spent and by whom -and feed this information into our machine-learning simulations for future predictions. Each country might have dozens more cinema operators, all sending data in different formats and at different qualities. One

territory may neglect demographics, another might mis-report gross revenue. In order for us to use it, we have to find missing or incorrect data and set the appropriate flags in our models and reports for later. Automating this process is the bulk of our big data operation.
Big data helps everything from marketing, to distribution, to planning. In marketing, we know we're wasting half our money. The problem is that we don't know which half. Big data is helping us solve that age-old marketing problem. We're able to track how the market is responding to our advertising campaigns over time, and compare it to past campaigns and products, and use that information to more precisely reach our audience (a bit how the Obama campaign was able to use big data to optimize its strategy). In cinema alone, the opening weekend of a film can affect gross revenue by seven figures (or more), so any insight we can provide into the most optimal time can directly generate thousands or millions of dollars in revenue.
Being able to distill big data from historical information, audiences responses in social media, data from commercial operators, et cetera, into a useable and interactive simulation completely changes how we plan our strategy for the next 6–15 months.

Use Case 2: Thomson Reuters

Dr. Jochen L. Leidner, Lead Scientist of the London R&D at Thomson Reuters, UK explains [9]:

For the most part, I carry out applied research in information access, and that's what I have been doing for quite a while. I am currently a Lead Scientist with Thomson Reuters, where I am building up a newly-established London site part of our Corporate Research & Development group.
Let me say a few words about Thomson Reuters before I go more into my own activities, just for background. Thomson Reuters has around 50,000 employees in over 100 countries and sells information to professionals in many verticals, including finance & risk, legal, intellectual property & scientific, tax & accounting. Our headquarters are located at 3 Time Square in the city of New York, NY, USA. Most people know our REUTERS brand from reading their newspapers (thanks to our highly regarded 3,000 + journalists at news desks in about 100 countries, often putting their lives at risk to give us reliable reports of the world's events) or receiving share price information on the radio or TV, but as a company, we are also involved in as diverse areas as weather prediction (as the weather influences commodity prices) and determining citation impact of academic journals (which helps publishers sell their academic journals to librarians), or predicting Nobel prize winners.
My research colleagues and I study information access and especially means to improve it, using including natural language processing, information extraction, machine learning, search engine ranking, recommendation system and similar areas of investigations. We carry out a lot of contract research for internal business units (especially if external vendors do not offer what we need, or if we believe we can build something internally that is lower cost and/or better suited to our needs), feasibility studies to de-risk potential future products that are considered, and also more strategic, blue-sky research that anticipates future needs. As you would expect, we protect our findings and publish them in the usual scientific venues.
Thomson Reuters is selling information services, often under a subscription model, and for that it is important to have metrics available that indicate usage, in order to inform our strategy. So another example for data analytics is that we study how document usage can inform personalization and ranking, of from where documents are accessed, and we use this to plan network bandwidth and to determine caching server locations.
For most definitions of "big", yes we do have big data. Consider that we operate a news organization, which daily generates in the tens of thousands of news reports (if we count all languages together). Then we have photo journalists who create large numbers of high-quality, professional photographs to document current events visually, and videos

comprising audio-visual storytelling and interviews. We further collect all major laws, statutes, regulations and legal cases around in major jurisdictions around the world, enrich the data with our own meta-data using both manual expertise and automatic classification and tagging tools to enhance findability. We hold collections of scientific articles and patents in full text and abstracts. We gather, consolidate and distribute price information for financial instruments from hundreds of exchanges around the world. We sell real-time live feeds as well as access to decades of these time series for the purpose of back-testing trading strategies.
Big data analytics lead to cost savings as well as generate new revenues in a very real, monetary sense of the word "value". Because our solutions provide what we call "knowledge to act" to our customers, i.e., information that lets them make better decisions, we provide them with value as well: we literally help our customers save the cost of making a wrong decision.
Even with new projects, product managers still don't think of analytics as the first thing to build into a product for a first bare-bones version, and we need to change that; instrumentation is key for data gathering so that analytics functionality can build on it later on.
In general, analytics projects follow a process of (1) capturing data, (2) aligning data from different sources (e.g., resolving when two objects are the same), (3) pre-processing or transforming the data into a form suitable for analysis, (4) building some model and (5) understanding the output (e.g. visualizing and sharing the results). This five-step process is followed by an integration phase into the production process to make the analytics repeatable.

Where are we now?

The IBM Institute for Business Value did a joint study [16] with Said Business School (University of Oxford) on the adoption of big data technologies in enterprise. They found that 28 % were in the pilot phase, 24 % haven't started anything, and 47 % are planning.

The rest of this chapter is organized as follows. Section (2) provides an overview of big data storage technologies. Section (3) provides an overview of big data benchmarking and its use in assessing different platforms. Section (4) provides conclusions and key points with respect to the contributions of this chapter.

2 Big Data Management Technologies

Given the high volume, velocity and variety of big data, the traditional Data Warehouse (DWH) and Business Intelligence (BI) architectures already existing in companies need to be enhanced in order to meet the new requirements of storing and processing big data. To optimize the performance of the Big Data Analytics pipeline, it is important to select the appropriate big data technology for given requirements. This section contains an overview of the various big data technologies and gives recommendations when to use them.

A survey by Forrester Research [17] indicated that most companies are relying on a mix of different technologies to enable the storage and processing of big data. Besides traditional relational data warehouse and business intelligence technologies that already exist in most companies, a big data architecture often includes non-relational technologies, new-relational technologies, in-memory databases,

numbers of empty entries (null values), which would occur for all the missing timestamp entries, or any missing entries in general. HBase uses a strong consistency model avoiding the problems relating to an eventual consistency model. The data is distributed automatically over different servers with linear scalability and including automatic failover. Being based on HDFS and being part of the Hadoop ecosystem from the beginning, Hadoop/HDFS and MapReduce are supported natively. HBase can be accessed programmatically with Java or via a REST API [31].

Wide column stores are ideal for distributed data storage with or without versioned data, as well as large scale and batch-oriented data processing such as sorting, parsing and transformation of data. They are also suitable for purposes of exploratory and predictive analytics [25].

Graph databases: Even for a NoSQL database, graph databases have a very unique data model and are special in terms of scalability and their use cases. A graph consists of nodes connected by edges forming a network. Graph databases are useful for very complex queries traversing the network for which you would otherwise need to join multiple tables in a traditional relational database. Examples of graph databases are AllegroGraph, Neo4 J, DEX, Sones and HyperGraphDB [32].

To be precise, it is necessary to state that the data model behind the different graph databases might differ from system to system, depending on the supported graph structure. This ranges from allowing attributions to nodes or edges, to the support of hypergraphs (where an edge not only can connect one node with another but can interconnect groups of nodes with each other) [33].

For various applications a graph is the native underlying data structure. In the context of big data, one automatically thinks of social networks where persons are represented by nodes connected to their friends, or the structure of the Web where the links between them are the edges (used for example to calculate *page rank*) [34]. The more someone starts thinking about graphs and their expressive power, the more he/she realizes that there are many more applications where graphs can be used. In fact, the object-oriented world with objects as nodes and their associations building edges suggests that graphs could be used for nearly any modern application. This provides there is a very high potential, nevertheless merging to a graph database involves some effort and it depends on the queries you plan to run on your system whether this pays off or not.

On the level of querying graph databases, there are a number of existing query languages available, including SPARQL [35], Cypher, Gremlin [36].

2.2 NewSQL Systems

NewSQL systems form another category of modern database systems. Databases in this category support the ACID features known from traditional relational databases, but unlike them they are designed with the goal of running in distributed environments with a good horizontal scalability.

VoltDB is a typical example of a NewSQL system. In VoltDB the data and the processing associated with it are distributed over the nodes in a cluster. Each node holds a subset of the data and the corresponding stored procedures to process the data. This makes VoltDB a good choice for OLTP style workloads, which require good scalability, high availability and high throughput. However, VoltDB is not optimized for OLAP workloads and analytical queries where a lot of data has to be called up from the database. This system is thus not a good choice as a database for business intelligence or similar applications [37]. Note that VoltDB also contains in-memory technologies and can thus be labeled as an in-memory database (see the next section) as well.

Other prominent examples of NewSQL databases are FoundationDB and NuoDB.

2.3 In-Memory Databases

The extreme performance potentials of in-memory database management system technologies are very attractive to organizations when it comes to real-time or near real-time processing of large amounts of data. In a report by Gartner [38], in-memory infrastructures are defined as follows: "*In-memory-enabling application infrastructure technologies consist of in-memory database management systems, in-memory data grids, high-performance messaging infrastructures, complex-event processing platforms, in-memory analytics and in-memory application servers. These technologies are being used to address a wide variety of application scenarios requiring a combination of speed of execution, scalability and insightful analytics.*"

A good example of an in-memory database is SAP HANA. All data in HANA resides in the memory, which allows for faster query execution compared to traditional disk-based technologies. In HANA disks are only used for backup and recovery purposes.

Other famous examples of in-memory databases are VoltDB and Aerospike.

2.4 Analytical Platforms

Analytical platforms are solutions specifically designed to meet the requirements of advanced analytics and OLAP workloads for huge amounts of data. Gartner defines advanced analytics as "the analysis of all kinds of data using sophisticated quantitative methods (for example, statistics, descriptive and predictive data mining, simulation and optimization) to produce insights that traditional approaches to Business Intelligence (BI) - such as query and reporting - are unlikely to discover" [39]. Gartner further states that"predictive analytics and other categories of advanced analytics are becoming a major factor in the analytics market" [39].

Commercial analytical platforms, which are also called appliances, usually consist of a predefined combination of hardware and software components that can be used in an out of the box deployment. Equivalent open source implementations mostly consist only of software components that can run on any hardware but need further configuration. Examples of analytical platforms are Teradata Aster, IBM Netezza or HP Vertica.

2.5 Hadoop Based Solutions

The core components of Apache Hadoop are the Hadoop Distributed File System (HDFS) [40], inspired by the Google File System (GFS) [41], and the MapReduce programming framework, based on Googles MapReduce algorithm [42]. Additional Hadoop components such as HBase [30] (a wide column store on top of HDFS), Hive [43] (support for SQL queries) and Pig [44] (support for writing MapReduce programs) have been developed on top of Hadoop and make up the so-called Hadoop ecosystem.

A study by Yahoo [45] describes the architecture of Hadoop as follows: "Every HDFS cluster consists of a single master node (NameNode) and multiple, up to thousands, DataNodes who store the data."

The data is stored in files divided in large blocks (typically 128 MB) which are replicated over multiple DataNodes. The replication factor (usually three) is adjustable and can be specified by the user. User interaction with the file system is done using a HDFS code library. Via this library a user can read, write or delete files and directories within the Hadoop file system without being concerned with the different data locations within the cluster [45].

The MapReduce framework allows for parallel processing of the data in HDFS. The processing of the data is broken down into the map and the reduce phases, which in turn allows parallelization. In the map phase the input data is distributed over the map processes, which are also called tasks. A single map task can process its part of the data independently of the other. The purpose of the reduce tasks is to combine the results from all the map tasks and calculate the overall result [46].

In Hadoop it is possible to store and process petabytes of unstructured data in a batch mode. Since the data is stored in the HDFS blocks, no schema definition is required and new nodes can be added at any time for linear increase of the storage space. This makes Hadoop a good choice for a distributed data hub, for example. When the data arrives it can be stored in the HDFS as a first landing zone, then it can be refined by MapReduce jobs in order to eventually transfer parts of the refined data to the traditional Data Warehousing (DWH) systems for analysis with existing BI tools. Recent developments like Apache Spark [47] enhance Hadoop to allow not only pure batch processing, but also interactive data exploration [48].

2.6 Big Data Streaming Systems

Many big data tasks require access not only to static, rarely-changing data, but also to continuous streams of data that represent live events, such as RSS feeds, microblogging feeds, sensor data, or other bus or network events. When off-line batch processing is not an option, systems with explicit support for stream processing can provide numerous benefits:

- reduced computational overhead [49],
- automatic resource re-use through shared processing [50],
- decreased latency, permitting near-real-time notification delivery to human observers (e.g. via dashboards) or automated actors, and
- query language support for streaming.

The principal differentiating factor and challenge for such streaming big data systems is time: new input tuples are continuously coming in through a family of streams, placing different requirements on semantics and evaluation strategies to those for off-line databases. On the other hand, output may also never finish, so that the output of a streaming big data system is another family of streams. The query processor provides plumbing to connect these streams with operators and functions, and may provide facilities to add and remove computations in flight [51], i.e. without disturbing ongoing processing.

As streams may continue to be processed without interruption for arbitrary lengths of time, stream processing cannot retain a full history of all past input tuples. This limits the available operations; while stream-based systems can easily support, for example, UNION operators or other stream merges, arbitrary joins over the entire history are not necessarily feasible. However, many important queries over input streams do not require full historical knowledge, only knowledge of data from the last hour or day. Contemporary querying approaches exploit this insight through *sliding windows* [50, 52, 53] into data streams, which capture all data from a given stream within a specified time frame relative to the current point in time. As time progresses, the window slides forward and the computation is updated to reflect the new contents of the window. Analogously, aggregation can be interpreted as a rolling aggregate, reflecting the status of aggregation at a given point in time.

Such stream processing requires continuous computations. If these begin to overwhelm an individual node, the management process may launch additional processing nodes to increase processing bandwidth. As with off-line queries, such parallel processing can be realized effectively for associative operators. Furthermore, the system may overlap the computation of multiple sliding windows into the same data stream, but at different timestamps, scheduling the windows on different nodes [54]. Alternatively, some application domains may be amenable to dropping input tuples that are not strictly required for the system to function correctly, deliberately degrading output quality according to Quality-of-Service rules [55].

Several streaming big data systems are available today, among them Open Source solutions such as Apache Spark [56] and Storm [57]. Commercial service-based solutions include Amazon Kinesis [58], which is designed to run as part of the Amazon Web Services framework and Google BigQuery. Forrester Research [59] lists several additional commercial platforms, including offerings from IBM, SAP, Software AG, and Tibco [60].

3 Big Data Benchmarking

In this section we focus on the question of how big data technologies can be optimized and evaluated in terms of the performance requirements of big data applications.

3.1 Why Do We Need Big Data Benchmarking?

Choosing the right big data platform and configuring it properly to provide the best performance for a hosted application is not a trivial task.

Especially with the new big data applications, there are requirements that make the platforms more complex, more heterogeneous, and hard to monitor and maintain. The role of benchmarking becomes even more relevant as a method for evaluating and understanding better the internals of a particular platform. Furthermore, benchmarks are used to compare different systems using both technical and economic metrics that can guide the user in the process of finding the right platform that fits their needs.

Nevertheless, the user has to first identify his needs and then choose the ideal big data benchmark. Big Data Benchmarks are a good way to optimize and fine-tune the performance in terms of processing speed, execution time or throughput of the big data system. A benchmark can also be used to evaluate the availability and fault-tolerance of a big data system. Especially for distributed big data systems a high availability is an important requirement.

While some benchmarks are developed to test particular software platforms, others are technology independent and can be implemented for multiple platforms. Usually the technology specific benchmarks are used to simulate specific types of applications, which will be hosted on the platform and should run in an optimal way.

At the Frankfurt Big Data Lab we use benchmarks to not only evaluate the performance of big data platforms [61, 62], but also to evaluate the availability and fault-tolerance [63].

3.2 Big Data Benchmarking Challenges

For many years the benchmarks specified by the Transaction Processing Performance Council (TPC) [64] have been successfully used as a standard for comparing

OLTP and OLAP systems. Just recently the TPC have formed a new group for the standardization of a big data benchmark [65] along with other similar initiatives like the Big Data Top 100 [66] and the Big Data Benchmarking Community [67]. However, the existing and emerging big data applications and platforms have very different characteristics ("3Vs") compared to the traditional transactional and analytical systems. These new platforms can store various types of data (structured, unstructured or semi-structured) with the schema (schema-on-read) defined just before accessing the data. They also support different types of data processing: batch, real-time or near real-time. The large data volumes force its distribution among multiple nodes and the use of additional fault tolerance techniques to guarantee data reliability. At the same time, many new applications that deal with the data are employed, leading to increased workload diversity on the big data systems.

All these big data challenges can make the systems very complex and difficult to standardize, which is a major objective when defining a benchmark. Furthermore it is still not clear if the benchmark should target single system components by using micro-benchmarks, or, on the contrary, it should include an end-to-end benchmark suite which stress tests the entire platform with all types of workloads. Moreover, the metrics provided by big data benchmarks should be extensive and comparable among the multiple systems under test. Chen et al. [68] outline four unique challenges of systems that hinder the development of big data benchmarks: (i) *system complexity*; (ii) *use case diversity*; (iii) *data scale*, which makes reproducing behavior challenging; and (iv) *rapid system evolution*, which requires the benchmark to keep pace with changes in the underlying system. Similarly, Xiong et al. [69] identify three key considerations that a big data benchmark should meet: (i) a benchmark suite should have workloads that are *representative of a wide range of application domains*; (ii) workloads in a benchmark suite should have *diversity of data characteristics*; and (iii) a benchmark suite *should not have redundant workloads in itself*.

3.3 Big Data Benchmarking Comparison

There are numerous projects that identify and compare the main components of a big data benchmark. In their paper presenting the CloudRank-D benchmark, Luo et al. [70] consider a number of workload characteristics which a benchmark suite should meet. These characteristics are listed in Table 2 and compared with similar benchmark suites.

However, as the CloudRank-D covers various representative applications it cannot really be compared with function-specific benchmarks like WL suite (also called SWIM), which analyzes the workload characteristics based on the number of jobs, arrival pattern and computation using synthesized representative data from real MapReduce traces.

Other important arguments discussed in the paper are the accuracy of the reported benchmark metrics and the target platforms that it can evaluate. In Table 3,

Table 2 Comparison of Different Benchmarks Suites; Adopted from [70]

	Diverse characteristics	MineBench	GridMix	HiBench	WL suite	CloudSuite	CloudRank-D
Representative applications	Basic operations	No	Yes	Yes	Yes	No	Yes
	Classification	Yes	No	Yes	No	Yes	Yes
	Clustering	Yes	No	Yes	No	No	Yes
	Recommendation	No	No	No	No	No	Yes
	Sequence learning	Yes	No	No	No	No	Yes
	Association rule mining	Yes	No	No	No	No	Yes
	Data warehouse operations	No	No	No	No	No	Yes
Workloads description	Submission pattern	No	No	No	Yes	No	Yes
	Scheduling policies	No	No	No	No	No	Yes
	System configurations	No	No	No	No	No	Yes
	Data models	No	No	No	No	No	Yes
	Data semantics	No	No	No	No	No	Yes
	Multiple data inputs	Yes	Yes	No	Yes	No	Yes
	Category of data-centric computation	No	No	No	Yes	No	Yes

Table 3 Different targets and metrics among benchmark suites; adopted from [70]

	Targets	Metrics
MineBench	Data mining algorithm on single-node computers	No
GridMix	Hadoop framework	Number of jobs and running time
HiBench	Hadoop framework	Job running time, the number of tasks completed per minute
WL suite	Hadoop framework	No
CloudSuite	Architecture research	No
CloudRank-D	Evaluating cloud systems at the whole system level	Data processed per second and data processed per joule

the authors compare CloudRank-D with the listed benchmarks. This comparison is also not accurate, because the other benchmarks target only a specific technology such as Hadoop, or have no general metric as in the case of WL suites, MineBench and CloudSuite.

Chen et al. [71, 72] present their Statistical Workload Injector for MapReduce (SWIM), while investigating a number of important characteristics of MapReduce workloads, as part of which they compare the most popular Hadoop benchmarks. Based on this investigation, they identify two design goals used in SWIM:

(i) the workload synthesis and execution framework should be agnostic to hardware/software/configuration choices, cluster size, specific MapReduce implementation, and the underlying system; and
(ii) the framework should synthesize representative workloads with short duration of execution.

However, the proposed benchmark is specifically focused on analyzing the internal dynamics of pure MapReduce applications and how to generate representative synthetic data from real workloads. Apart from that, it does not perform other data and computational tasks typical for big data applications which benchmarks like HiBench and PigMix address. Also it does not report any metrics which can be used to compare the different systems under test.

A recent survey on benchmarks for big data [73] extensively reviews the current big data benchmarks and discusses some of the characteristics and challenges they should address. Table 4 summarizes this set of benchmarks. The benchmarks are compared only by reference to targeted platforms and main application characteristics. The list includes benchmarks such as TPC-C/H/W/DS and SSB, which target only a specific set of workload characteristics. The table looks more like a listing of benchmarks than a real comparison based on specific criteria. An important point that the authors make is the need of a more complete, end-to-end benchmarking suite, including both component-based and application-oriented benchmarks along with critical metrics such as energy consumption.

The paper presenting the BigDataBench suite by Wang et al. [74] also discusses extensively the challenges of developing a real big data benchmark and compares

Table 4 Comparison of existing works on big data benchmarks; Adopted from [73]

Work	Target	Characteristics	Comment
TPC-C	RDBMS	Transaction processing, simple query and update	OLTP
TPC-H	RDBMS, Hadoop Hive	Reporting, decision	OLAP
TPC-W	RDBMS, NoSQL	Web applications	Web OLTP
SSB	RDBMS, Hadoop Hive	Reporting, decision	OLAP
TPC-DS	RDBMS, Hadoop Hive	Reporting query, ad hoc query, iterative query, data mining query	OLAP
TeraSort	RDBMS, Hadoop	Data sorting	Sorting only
YCSB	NoSQL database	Cloud-based data serving	Web OLTP
REF 11	Unstructured data management system	Unstructured data only edge detection, proximity search, data scanning, data fusion	Not representative enough
GRAPH 500	Graph NoSQL database	Graph data processing only	Not representative enough
LinkBench	RDBMS, graph NoSQL database	Modeling Facebook real life application graph data processing only	Not representative enough
DFSIO	Hadoop	File system level benchmark	Not representative enough
Hive performance benchmark	Hadoop Hive	GREP, selection, aggregation, join and UDF aggregation only	Not representative enough
GridMix	Hadoop	Mix of Hadoop jobs	Not representative enough
PUMA	MapReduce	Term-vector, inverted-index, self-join, adjacency-list, k-means, classification, histogram-movies, histogram-ratings, sequence-count, ranked inverted index, Tera-sort, GREP, word-count	Comprehensive workload
MRBench	MapReduce	TPC-H queries	OLAP
HiBench	MapReduce	Micro-benchmarks (sort, word count and TeraSort); Web search (Nutch Indexing and page rank) machine learning (Bayesian classification and k-means clustering); HDFS benchmark (file system level benchmark)	Comprehensive workload
CloudRank-D	RDBMS, Hadoop	Basic operations for data analysis, classification, clustering, recommendation, sequence learning, association rule mining, and data warehouse queries	Comprehensive workload
BigBench	RDBMS, Hadoop	Covers data models of structured, semi-structured and unstructured data; addresses variety, velocity and volume aspects of big data systems	Comprehensive workload

Table 5 Comparison of big data benchmarking efforts; adopted from [74]

Benchmark Efforts	Real-world data sets (Data Set Number)	Data scalability (Volume, Veracity)	Workloads variety	Software stacks	Objects to test	Status
HiBench	Unstructured text data (1)	Partial	Offline Analytics; Real-time Analytics	Hadoop and Hive	Hadoop and Hive	Open source
BigBench	None	N/A	Offline Analytics	DBMS and Hadoop	DBMS and Hadoop	Proposal
AMP Benchmarks	None	N/A	Real-time Analytics	Real-time analytic systems	Real-time analytic systems	Open source
YCSB	None	N/A	Online Services	NoSQL systems	NoSQL systems	Open source
LinkBench	Unstructured graph data (1)	Partial	Online Services	Graph database	Graph database	Open source
CouldSuite	Unstructured text data (1)	Partial	Online Services; Offline Analytics	NoSQL systems, Hadoop, GraphLab	Architectures	Open source
BigDataBench	Unstructured text data (1) Semi-structured text data (1) Unstructured graph data (2) Structured table data (1) Semi-structured table data (1)	Total	Online Services; Offline Analytics; Real-time Analytics	NoSQL systems, DBMS, Real-time Analytics Offline Analytics systems	Systems and architecture; NoSQL systems; Different analytics systems	Open source

the BigDataBench with other existing suites. The resulting list, depicted in Table 5, compares them according to data sets included, data scalability, workload variety, software stack and status. Clearly, BigDataBench leads in all of the characteristics as its goal is to be an end-to-end benchmark, but the other benchmarks included, like AMP Benchmarks [75], YCSB [76] and LinkBench [77], have different functionalities and are platform specific.

In short, none of the above reviewed benchmark comparisons evaluates and categorizes the existing benchmarks suites in an objective way. In order for this to be done, an independent classification, targeting the entire spectrum of big data benchmark types and based on clearly set criteria, should be constructed.

4 Conclusions

In this chapter, we aimed to outline some of the main issues that are relevant when setting up and optimizing a big data project. We concentrated our attention first on the managerial task of setting up a big data project using insights from industry leaders, then we looked in some detail at the available data management and processing technologies for big data, and concluded by looking at the task of defining effective benchmarks for big data. Effective benchmarks for big data help the customers pick the optimal technology, help the vendors improve their products, and finally help researchers understand the differences of big data technologies on their path to optimize organizational and technical processes.

References

1. On Big Data Velocity. Interview with Scott Jarr, ODBMS Industry Watch, 28 Jan 2013. http://www.odbms.org/blog/2013/01/on-big-data-velocity-interview-with-scott-jarr/ (2015). Accessed 15 July 2015
2. How to run a Big Data project. Interview with James Kobielus. ODBMS Industry Watch, 15 May 2014. http://www.odbms.org/blog/2014/05/james-kobielus/ (2015). Accessed 15 July 2015
3. Laney, D.: 3D data management: controlling data volume, velocity and variety. Appl. Deliv. Strateg. File, **949** (2001)
4. Zikopoulos, P., Eaton, C.: Understanding Big Data: Analytics for Enterprise Class Hadoop and Streaming Data, 1st ed. McGraw-Hill Osborne Media (IBM) (2011)
5. Foster, I.: Big Process for Big Data, Presented at the HPC 2012 Conference. Cetraro, Italy (2012)
6. Gattiker, A., Gebara, F.H., Hofstee, H.P., Hayes, J.D., Hylick, A.: Big Data text-oriented benchmark creation for Hadoop. IBM J. Res. Dev., **57**(3/4), 10: 1–10: 6 (2013)
7. Zicari, R.: Big Data: Challenges and Opportunities. In: Akerkar, R. (ed.) Big Data Computing, p. 564. Chapman and Hall/CRC (2013)
8. On Big Data: Interview with Dr. Werner Vogels, CTO and VP of Amazon.com. ODBMS Industry Watch, 02 Nov 2011. http://www.odbms.org/blog/2011/11/on-big-data-interview-with-dr-werner-vogels-cto-and-vp-of-amazon-com/ (2015). Accessed 15 July 2015

9. Big Data Analytics at Thomson Reuters. Interview with Jochen L. Leidner. ODBMS Industry Watch, 15 Nov 2013. http://www.odbms.org/blog/2013/11/big-data-analytics-at-thomson-reuters-interview-with-jochen-l-leidner/ (2015). Accessed 15 July 2015
10. Setting up a Big Data project. Interview with Cynthia M. Saracco. ODBMS Industry Watch, 27 Jan 2014. http://www.odbms.org/blog/2014/01/setting-up-a-big-data-project-interview-with-cynthia-m-saracco/ (2015). Accessed 15 July 2015
11. Jacobs, A.: The pathologies of big data. Commun. ACM **52**(8), 36–44 (2009)
12. On Big Data and Hadoop. Interview with Paul C. Zikopoulos. ODBMS Industry Watch, 10 June 2013. http://www.odbms.org/blog/2013/06/on-big-data-and-hadoop-interview-with-paul-c-zikopoulos/ (2015). Accessed 15 July 2015
13. Next generation Hadoop. Interview with John Schroeder. ODBMS Industry Watch, 07 Sep 2012. http://www.odbms.org/blog/2012/09/next-generation-hadoop-interview-with-john-schroeder/ (2015). Accessed 15 July 2015
14. On Big Data, Analytics and Hadoop. Interview with Daniel Abadi. ODBMS Industry Watch, 05 Dec 2012. http://www.odbms.org/blog/2012/12/on-big-data-analytics-and-hadoop-interview-with-daniel-abadi/ (2015). Accessed 15 July 2015
15. Data Analytics at NBCUniversal. Interview with Matthew Eric Bassett. ODBMS Industry Watch, 23 Sep 2013. http://www.odbms.org/blog/2013/09/data-analytics-at-nbcuniversal-interview-with-matthew-eric-bassett/ (2015). Accessed 15 July 2015
16. Analytics: The real-world use of big data. How innovative enterprises extract value from uncertain data (IBM Institute for Business Value and Saïd Business School at the University of Oxford), Oct 2012
17. Hopkins, B.: The Patterns of Big Data. Forrester Research, 11 June 2013
18. Lim, H., Han, Y., Babu, S.: How to Fit when No One Size Fits. In: CIDR (2013)
19. Cattell, R.: Scalable SQL and NoSql Data Stores. SIGMOD Rec., **39**(4), 27 Dec 2010
20. Gilbert, S., Lynch, N.: Brewer's conjecture and the feasibility of consistent, available, partition-tolerant web services. SIGACT News **33**(2), 51–59 (2002)
21. Haerder, T., Reuter, A.: Principles of transaction-oriented database recovery. ACM Comput. Surv. **15**(4), 287–317 (1983)
22. Bailis, P., Ghodsi, A.: Eventual Consistency Today: Limitations, Extensions, and Beyond. Queue **11**(3), pp. 20:20–20:32, Mar 2013
23. Pritchett, D.: BASE: an acid alternative. Queue **6**(3), 48–55 (2008)
24. Vogels, W.: Eventually consistent. Commun. ACM **52**(1), 40–44 (2009)
25. Moniruzzaman, A.B.M., Hossain, S.A.: NoSQL Database: New Era of Databases for Big data Analytics—Classification, Characteristics and Comparison. CoRR (2013). arXiv:1307.0191
26. Datastax, Datastax Apache Cassandra 2.0 Documentation. http://www.datastax.com/documentation/cassandra/2.0/index.html (2015). Accessed 15 Apr 2015
27. Apache Cassandra White Paper. http://www.datastax.com/wp-content/uploads/2011/02/DataStax-cBackgrounder.pdf
28. MongoDB Inc., MongoDB Documentation. http://docs.mongodb.org/manual/MongoDB-manual.pdf (2015). Accessed 15 Apr 2015
29. Chang, F., Dean, S., Ghemawat, W.C., Hsieh, D.A. Wallach, Burrows, M., Chandra, T., Fikes, A.,Gruber, R.E.:Bigtable: a distributed storage system for structured data. In: Proceedings of the 7th USENIX Symposium on Operating Systems Design and Implementation, vol 7, pp. 15–15. Berkeley, CA, USA (2006)
30. George, L.: HBase: The Definitive Guide, 1st ed. O'Reilly Media (2011)
31. Apache Software Foundation, The Apache HBase Reference Guide. https://hbase.apache.org/book.html
32. Buerli, M.: The Current State of Graph Databases, Dec-2012, http://www.cs.utexas.edu/~cannata/dbms/Class%20Notes/08%20Graph_Databases_Survey.pdf (2015). Accessed 15 Apr 2015
33. Angles, R.: A comparison of current graph database models. In: ICDE Workshops, pp. 171–177 (2012)

34. McColl, R.C., Ediger, D., Poovey, J., Campbell, D., Bader, D.A.: A performance evaluation of open source graph databases. In: Proceedings of the First Workshop on Parallel Programming for Analytics Applications, pp. 11–18. New York, NY, USA (2014)
35. Harris, S., Seaborne, A.: SPARQL 1.1 Query Language. SPARQL 1.1 Query Language, 21-Mar-2013. http://www.w3.org/TR/sparql11-query/ (2013)
36. Holzschuher, F., Peinl, R.: Performance of graph query languages: comparison of cypher, gremlin and native access in Neo4 J. In: Proceedings of the Joint EDBT/ICDT 2013 Workshops, pp. 195–204. New York, NY, USA (2013)
37. VoltDB Inc., Using VoltDB. http://voltdb.com/download/documentation/
38. Pezzini, M., Edjlali, R.: Gartner top technology trends, 2013. In: Memory Computing Aims at Mainstream Adoption, 31 Jan 2013
39. Herschel, G., Linden, A., Kart, L.: Gartner Magic Quadrant for Advanced Analytics Platforms, 19 Feb 2014
40. Borthakur, D.: The hadoop distributed file system: Architecture and design. Hadoop Proj. Website **11**, 21 (2007)
41. Ghemawat, S., Gobioff, H., Leung, S.-T.: The google file system. In: Proceedings of the Nineteenth ACM Symposium on Operating Systems Principles, pp. 29–43. New York, NY, USA (2003)
42. Dean, J., Ghemawat, S.: MapReduce: Simplified Data Processing on Large Clusters. Commun. ACM **51**(1), 107–113 (2008)
43. Thusoo, A., Sarma, J.S., Jain, N., Shao, Z., Chakka, P., Anthony, S., Liu, H., Wyckoff, P., Murthy, R.: Hive: A Warehousing Solution over a Map-reduce Framework. Proc. VLDB Endow. **2**(2), 1626–1629 (2009)
44. Olston, C., Reed, B., Srivastava, U., Kumar, R., Tomkins, A.: Pig latin: a not-so-foreign language for data processing. In: Proceedings of the 2008 ACM SIGMOD international conference on Management of data, pp. 1099–1110 (2008)
45. Shvachko, K., Kuang, H., Radia, S., Chansler, R.: The hadoop distributed file system. In: Proceedings of the 2010 IEEE 26th Symposium on Mass Storage Systems and Technologies (MSST), pp. 1–10. Washington, DC, USA (2010)
46. White, T.: Hadoop: The Definitive Guide, 1st ed. O'Reilly Media, Inc., (2009)
47. Apache Spark Project. http://spark.apache.org/
48. Zaharia, M., Chowdhury, M., Franklin, M.J., Shenker, S., Stoica, I.: Spark: cluster computing with working sets. In: Proceedings of the 2Nd USENIX Conference on Hot Topics in Cloud Computing, pp. 10–10. Berkeley, CA, USA (2010)
49. Cranor, C., Johnson, T., Spataschek, O., Shkapenyuk, V.: Gigascope: a stream database for network applications. In: Proceedings of the 2003 ACM SIGMOD International Conference on Management of Data, pp. 647–651. New York, NY, USA (2003)
50. Arasu, A., Babu, S., Widom, J.: The CQL Continuous Query Language: Semantic Foundations and Query Execution. VLDB J. **15**(2), 121–142 (2006)
51. Chen, J., DeWitt, D.J., Tian, F., Wang, Y.: NiagaraCQ: a scalable continuous query system for internet databases. In: Proceedings of the 2000 ACM SIGMOD International Conference on Management of Data, pp. 379–390. New York, NY, USA (2000)
52. Agrawal, J., Diao, Y., Gyllstrom, D, Immerman, N.: Efficient pattern matching over Event streams. In: Proceedings of the 2008 ACM SIGMOD International Conference on Management of Data, pp. 147–160. New York, NY, USA (2008)
53. Jain, N., Mishra, S., Srinivasan, A., Gehrke, J., Widom, J., Balakrishnan, H., Çetintemel, U., Cherniack, M., Tibbetts, R., Zdonik, S.: Towards a Streaming SQL Standard. Proc VLDB Endow **1**(2), 1379–1390 (2008)
54. Balkesen, C., Tatbul, N.: Scalable data partitioning techniques for parallel sliding window processing over data streams. In: VLDB International Workshop on Data Management for Sensor Networks (DMSN'11). Seattle, WA, USA (2011)
55. Ahmad, Y., Berg, B., Çetintemel, U., Humphrey, M., Hwang, J.-H., Jhingran, A., Maskey, A., Papaemmanouil, O., Rasin, A., Tatbul, N., Xing, W., Xing, Y., Zdonik, S.B.: Distributed operation in the Borealis stream processing engine. In: SIGMOD Conference, pp. 882–884 (2006)

56. Apache Spark. http://spark.apache.org/
57. Apache Storm. http://storm.incubator.apache.org/
58. Amazon Kinesis. http://aws.amazon.com/kinesis/
59. Gualtieri, M., Curran, R.: The Forrester Wave: Big Data Streaming Analytics Platforms, Q3 2014, 17 July 2014
60. Tibco Streambase. http://www.streambase.com
61. Ivanov, T., Niemann, R., Izberovic, S., Rosselli, M., Tolle, K., Zicari, R.V.: Performance evaluationi of enterprise big data platforms with HiBench. presented at the In: 9th IEEE International Conference on Big Data Science and Engineering (IEEE BigDataSE 2015), Helsinki, Finland, 20–22 Aug 2015
62. Ivanov, T., Beer, M.: Performance evaluation of spark SQL using BigBench. Presented at the In: 6th Workshop on Big Data Benchmarking (6th WBDB). Canada, Toronto, 16–17 June 2015
63. Rosselli, M., Niemann, R., Ivanov, T., Tolle, K., Zicari, R.V.: "Benchmarking the Availability and Fault Tolerance of Cassandra", presented at the In 6th Workshop on Big Data Benchmarking (6th WBDB), June 16–17, 2015. Canada, Toronto (2015)
64. TPC, TPC-H - Homepage. http://www.tpc.org/tpch/ (2015). Accessed 15 July 2015
65. TPC Big Data Working Group, TPC-BD - Homepage TPC Big Data Working Group. http://www.tpc.org/tpcbd/default.asp (2015). Accessed 15 July 2015
66. BigData Top100, 2013. http://bigdatatop100.org/ (2015). Accessed 15 July 2015
67. Big Data Benchmarking Community, Big Data Benchmarking | Center for Large-scale Data Systems Research, Big Data Benchmarking Community. http://clds.ucsd.edu/bdbc/ (2015). Accessed 15 July 2015
68. Chen, Y.: We don't know enough to make a big data benchmark suite-an academia-industry view. Proc. WBDB (2012)
69. Xiong, W., Yu, Z., Bei, Z., Zhao, J., Zhang, F., Zou, Y., Bai, X., Li, Y., Xu, C.: A characterization of big data Benchmarks. In: Big Data. IEEE International Conference on **2013**, 118–125 (2013)
70. Luo, C., Zhan, J., Jia, Z., Wang, L., Lu, G., Zhang, L., Xu, C.-Z., Sun, N.: CloudRank-D: benchmarking and ranking cloud computing systems for data processing applications. Front. Comput. Sci. **6**(4), 347–362 (2012)
71. Chen, Y., Ganapathi, A., Griffith, R., Katz, R.: The case for evaluating MapReduce performance using workload suites. In: 2011 IEEE 19th International Symposium on Modeling, Analysis & Simulation of Computer and Telecommunication Systems (MASCOTS), pp. 390–399 (2011)
72. Chen, Y., Alspaugh, S., Katz, R.: Interactive analytical processing in big data systems: a cross-industry study of MapReduce workloads. Proc. VLDB Endow. **5**(12), 1802–1813 (2012)
73. Qin, X., Zhou, X.: A survey on Benchmarks for big data and some more considerations. In: Intelligent Data Engineering and Automated Learning–IDEAL. Springer **2013**, 619–627 (2013)
74. Wang, L., Zhan, J., Luo, C., Zhu, Y, Yang, Q., He, Y., Gao, W., Jia, Z., Shi, Y., Zhang, S.: Bigdatabench: a big data benchmark suite from internet services. arXiv:14011406 (2014)
75. AMP Lab Big Data Benchmark. https://amplab.cs.berkeley.edu/benchmark/ (2015). Accessed 15 July 2015
76. Patil, S., Polte, M., Ren, K., Tantisiriroj, W., Xiao, L., López, J., Gibson, G., Fuchs, A., Rinaldi, B.: Ycsb ++: benchmarking and performance debugging advanced features in scalable table stores. In: Proceedings of the 2nd ACM Symposium on Cloud Computing, p. 9 (2011)
77. Armstrong, T.G., Ponnekanti, V., Borthakur, D., Callaghan, M.: Linkbench: a database benchmark based on the facebook social graph. In: Proceedings of the 2013 international conference on Management of data, pp. 1185–1196 (2013)

Author Biographies

Roberto V. Zicari is Full Professor of Database and Information Systems at Frankfurt University. He is an internationally recognized expert in the field of databases. His interests also expand to Innovation and Entrepreneurship. He is the founder of the Big Data Lab at the Goethe University Frankfurt, and the editor of ODBMS.org web portal and ODBMS Industry Watch Blog. He is also a visiting professor with the Center for Entrepreneurship and Technology within the Department of Industrial Engineering and Operations Research at UC Berkeley. Previously, Roberto served as associate professor at Politecnico di Milano, Italy; visiting scientist at IBM Almaden Research Center, USA, the University of California at Berkeley, USA; visiting professor at EPFL in Lausanne, Switzerland, the National University of Mexico City, Mexico and the Copenhagen Business School, Denmark.

Marten Rosselli is a Senior Consultant at Accenture and an expert for Big Data Management and Analytics. He has worked as a Technology Architect in many industrial projects in different industries such as Financial Services, Media and Telecommunication, Automotive, Public Services and Chemical. His expertise includes Project Management, Big Data Management and Analytics, Data Architectures, Software Engineering and Agile Software Development, Data Warehouse Technologies and Business Intelligence. He received his Master degree in Computer Science from the Goethe University Frankfurt am Main and he is a member and Ph.D. student at the Frankfurt Big Data Lab at the University of Frankfurt, Germany.

Todor Ivanov is a Ph.D. student at the Frankfurt Big Data Lab, Goethe University Frankfurt advised by Professor Roberto V. Zicari. He received his BSc. in Computational Engineering and MSc in Distributed Software Systems from Technical University of Darmstadt. His main interests are in the design and benchmarking of complex distributed software systems for big data and data-intensive applications.

Prior to that, he has worked as a senior software engineer developing Flight Information Display Systems (FIDS) for different international airports, research assistant in the FlashyDB project at the department of Databases and Distributed Systems, TU Darmstadt and IT consultant in business intelligence and database related projects in the bank sector.

Nikolaos Korfiatis is Assistant Professor of Business Analytics and Regulation at Norwich Business School, University of East Anglia (UEA) and faculty member at the Centre for Competition Policy (CCP).He is a an affiliate faculty member at the Big Data Laboratory Frankfurt where he previously acted as co-director responsible for Data Science and Analytics. He is active in the industry as a Senior Data Scientist in Residence for Adastra Germany, developing practical use cases for analytics and big data for the automotive and banking sector. He received his Ph.D. in Information Management at Copenhagen Business School (CBS) Denmark and his Diploma and MSc in Engineering (Specialization Information Retrieval) from Royal Institute of Technology (KTH), Stockholm and Uppsala University.

Karsten Tolle currently holds the position of an "Akademischer Rat" (Assistant Professor) at the Database and Information Systems (DBIS) group of the Goethe University Frankfurt. His current roles and activities include: Managing director of the Frankfurt Big Data Lab; Member of the Nomisma.org steering committee; Chair of the OASIS UBL German Localization Subcommittee. His main interests are in Linked Open Data (LOD) and Graph Databases. He studied mathematics with computer science as a special subject at the University of Hannover. He wrote his Master's Thesis during a SOCRATES-ERASMUS exchange program on Crete (Greece) at the ICS-Foundation of Research and Technology - Hellas (FORTH) and received his Ph.D. from the University of Frankfurt (Germany) in the area of Semantic Web.

Raik Niemann studied computer science at the University of Rostock and at the University of Applied Science Stralsund, Germany. He received a german Dipl.-Ing. (FH) and Master of Science degree. Currently Mr. Niemann is a Ph.D. student at the DBIS (Goethe University Frankfurt/Main) and focuses on the energy efficiency of data management systems with very large datasets. This includes all kinds data organization schemes (relational, object-relational, NoSQL and NewSQL), their performance in terms of response time as well as technical architectures and settings.

Christoph Reichenbach is junior professor for Software Engineering and Programming Languages at Goethe University Frankfurt. His principal expertise is in domain-specific languages, software tools, language run-time systems, and compilers. He received his Ph.D. from the University of Colorado at Boulder, and has previously held the positions of software engineer for search quality at Google and postdoctoral fellow at the University of Massachusetts, Amherst.

Optimizing Intelligent Reduction Techniques for Big Data

Florin Pop, Catalin Negru, Sorin N. Ciolofan, Mariana Mocanu and Valentin Cristea

Abstract Working with big volume of data collected through many applications in multiple storage locations is both challenging and rewarding. Extracting valuable information from data means to combine qualitative and quantitative analysis techniques. One of the main promises of analytics is data reduction with the primary function to support decision-making. The motivation of this chapter comes from the new age of applications (social media, smart cities, cyber-infrastructures, environment monitoring and control, healthcare, etc.), which produce big data and many new mechanisms for data creation rather than a new mechanism for data storage. The goal of this chapter is to analyze existing techniques for data reduction, at scale to facilitate Big Data processing optimization and understanding. The chapter will cover the following subjects: data manipulation, analytics and Big Data reduction techniques considering descriptive analytics, predictive analytics and prescriptive analytics. The CyberWater case study will be presented by referring to: optimization process, monitoring, analysis and control of natural resources, especially water resources to preserve the water quality.

Keywords Big data · Descriptive analytics · Predictive analytics · Prospective analytics · Cyber-Infrastructures

F. Pop (✉) · C. Negru · S.N. Ciolofan · M. Mocanu · V. Cristea
Faculty of Automatic Control and Computers, Computer Science Department,
University Politehnica of Bucharest, Bucharest, Romania
e-mail: florin.pop@cs.pub.ro

C. Negru
e-mail: catalin.negru@cs.pub.ro

S.N. Ciolofan
e-mail: sorin.ciolofan@cs.pub.ro

M. Mocanu
e-mail: mariana.mocanu@cs.pub.ro

V. Cristea
e-mail: valentin.cristea@cs.pub.ro

A. Emrouznejad (ed.), *Big Data Optimization: Recent Developments and Challenges*, Studies in Big Data 18, DOI 10.1007/978-3-319-30265-2_3

1 Introduction

There are a lot of applications that generate Big Data, like: social networking profiles, social influence, SaaS (Software as a Service) and Cloud Apps, public web information, MapReduce scientific experiments and simulations, data warehouse, monitoring technologies, e-government services, etc. Data grow rapidly, since applications produce continuously increasing volumes of both unstructured and structured data. Decision-making is critical in real-time systems and also in mobile systems [1] and has an important role in business [2]. Decision-making uses tailored data as input, obtained after a reduction process applied to the whole data. So, a representative and relevant data set must to be extracted from data. This is the subject of data reduction. On the other hand, recognize crowd-data signifi-cance is another challenges with respect to making sense of Big Data: it means to determine “wrong” information from “disagreeing” information and find metrics to determine certainty [3].

Thomas H. Davenport, Jill Dych in their report “Big Data in Big Companies” [2] name “Analytics 3.0” the new approach that well established big companies had to do in order to integrate Big Data infrastructure into their existing IT infrastructure (for example Hadoop clusters that have to coexist with IBM mainframes). The “variety” aspect of Big Data is the main concern for companies (ability to analyze new types of data, eventually unstructured, and not necessary focus on large volumes).

The most evident benefits of switching to Big Data are the cost reductions. Big data currently represents a research frontier, having impact in many areas, such as business, the scientific research, public administration, and so on. Large datasets are produced by multiple and diverse sources like: www, sensor networks, scientific experiments, high throughput instruments, increase at exponential rate [4, 5] as shown in Fig. 1. UPS stores over 16 petabytes of data, tracking 16 millions packages per day for 9 millions customers. Due to an innovative optimization of navigation on roads they saved in 2011, 8.4 millions of gallons of fuel. A bank saved an order of magnitude by buying a Hadoop cluster with 50 servers and 800 processors compared with a traditional warehouse. The second motivation for companies to use Big Data is time reduction. Macy’s was able to optimize pricing for 73 million items on sale

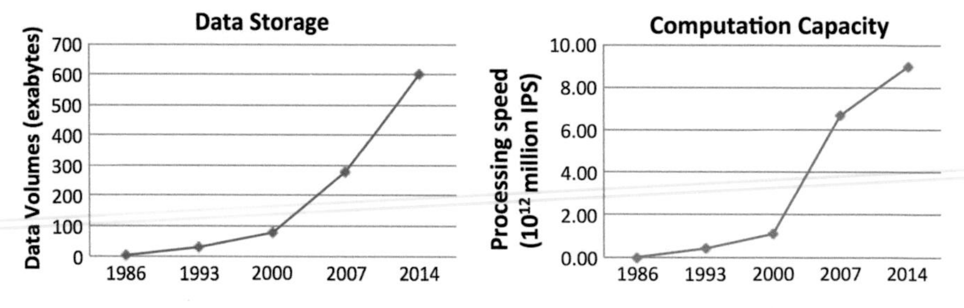

Fig. 1 Data deluge: the increase of data size has surpassed the capabilities of computation in Big Data platforms [4, 5]

from 27 to 1 h. Also, companies can run a lot more (100.000) analytics models than before (10, 20 or 100). Time reduction allows also to real time react to customer habits. The third objective is to create new Big Data specific offerings. LinkedIn launched a set of new services and features such as Groups You May Like, Jobs You May Be Interested In, and Who's Viewed My Profile etc. Google developed Google Plus, Google Apps. Verizon, Sprint and T-Mobile deliver services based on location data provided by mobiles. The fourth advantage offered by Big Data to business is the support in taking internal decision, considering that lot of data coming from customer's interaction is unstructured or semi-structured (web site clicks, voice recording from call center calls, notes, video, emails, web logs etc.). With the aid of natural language processing tools voice can be translated into text and calls can be analyzed.

The Big Data stack is composed of storage, infrastructure (e.g. Hadoop), data (e.g. human genome), applications, views (e.g. Hive) and visualization. The majority of big companies already have in place a warehouse and analytics solution, which needs to be integrated now with the Big Data solution. The challenge is to integrate the legacy ERP, CRM, 3rd party apps, the data warehouse with Hadoop and new types of data (social media, images, videos, web logs, pdfs, etc.) in a way that allows efficient modeling and reporting.

Since we face a large variety of solutions for specific applications and platforms, a thorough and systematic analysis of existing solutions for data reduction models, methods and algorithms used in Big Data is needed [6, 7]. This chapter presents the art of existing solutions and creates an overview of current and near-future trends; it uses a case study as proof of concept for presented techniques.

The chapter is organized as follow. Section 2 presents the data manipulation challenges being focused on spatial and temporal databases, key-value stores and no-SQL, data handling and data cleaning, Big Data processing stack and processing techniques. Section 3 describes the reduction techniques: descriptive analytics, predictive analytics and prescriptive analytics. In Sect. 4 we present a case study focused on CyberWater, which is a research project aiming to create a prototype platform using advanced computational and communications technology for implementation of new frameworks for managing water and land resources in a sustainable and integrative manner. The chapter ends with Sect. 5 presenting conclusions.

2 Data Manipulation Challenges

This section describes the data manipulation challenges: spatial and temporal databases [8], parallel queries processing, key-value stores and no-SQL, Data Cleaning, MapReduce, Hadoop and HDFS [9]. The processing techniques used in data manipulation together with Big Data stack are presented in this section [10–12]. This part also fixes the most important aspects of data handling used in analytics.

2.1 *Spatial and Temporal Databases*

Spatial and temporal database systems are in close collaborative relation with other research area of information technology. These systems integrate with other disciplines such as medicine, CAD/CAM, GIS, environmental science, molecular biology or genomics/bioinformatics. Also, they use large, real databases to store and handle large amount of data [13]. Spatial databases are designed to store and process spatial information systems efficiently. Temporal databases represent attribute of objects that are changing with respect to time. There are different models for spatial/temporal and spatio-temporal systems [14]:

- Snapshot Model—temporal aspects of data time-stamped layers;
- Space-Time composite—every line in space and time is projected onto a Spatial plane and intersected with each other;
- Simple Time Stamping—each object consists of a pair of time stamps representing creation and deletion time of the object;
- Event-Oriented model—a log of events and changes made to the objects are logged into a transaction log;
- History Graph Model—each object version is identified by two timestamp describing the interval of time;
- Object-Relationship (O-R) Model—a conceptual level representation of spatio-temporal DB;
- Spatio-temporal object-oriented data model—is based on object oriented technology;
- Moving Object Data Model—objects are viewed as 3D elements.

The systems designed considering these models manage both space and time information for several classes of applications, like: tracking of moving people or objects, management of wireless communication networks, GIS applications, traffic jam preventions, whether prediction, electronic services (e.g. e-commerce), etc.

Current research focuses on the dimension reduction, which targets spatio-temporal data and processes and is achieved in terms of parameters, grouping or state.[1] Guhaniyogi et al. [15] address the dimension reduction in both parameters and data space. Johnson et al. [16] create clusters of sites based upon their temporal variability. Leininger et al. [17] propose methods to model extensive classification data (land-use, census, and topography). Wu et al. [18] considering the problem of predicting migratory bird settling, propose a threshold vector-autoregressive model for the Conway-Maxwell Poisson (CMP) intensity parameter that allows for regime switching based on climate conditions. Dunstan et al. [19] goal is to study how communities of species respond to environmental changes. In this respect they o classify species into one of a few archetypal forms of environmental response using

[1]These solution were grouped in the Special Issue on "Modern Dimension Reduction Methods for Big Data Problems in Ecology" edited by Wikle, Holan and Hooten, in Journal of Agricultural, Biological, and Environmental Statistics.

regression models. Hooten et al. [20] are concerned with ecological diffusion partial differential equations (PDE's) and propose an optimal approximation of the PDE solver that dramatically improves efficiency. Yang et al. [21] are concerned with prediction in ecological studies based on high-frequency time signals (from sensing devices) and in this respect they develop nonlinear multivariate time-frequency functional models.

2.2 *Key-Value Stores and NoSQL*

NoSQL term may refer two different things, like the data management system is not an SQL-compliant one or accepted one means "Not only SQL", and refers to environments that combine SQL query language with other types of querying and access. Due to the fact that NoSQL databases does not have a fix schema model, they come with a big advantage for developers and data analysts in the process of development and analysis, as is not need to cast every query as a relational table database. Moreover there are different NoSQL frameworks specific to different types of analysis such as key-value stores, document stores, tabular stores, object data stores, and graph databases [22].

Key-value stores represents schema less NoSQL data stores, where values are associated with keys represented by character strings. There are four basic operations when dealing with this type of data stores:

1. *Put(key; value)*—associate a value with the corresponding key;
2. *Delete(key)*—removes all the associated values with supplied key;
3. *Get(key)*—returns the values for the provided key;
4. *MultiGet(key1; key2;. . .; keyn)*—returns the list of values for the provided keys.

In this type of data stores can be stored a large amount of data values that are indexed and can be appended at same key due to the simplicity of representation. Moreover tables can be distributed across storage nodes [22].

Although key-value stores can be used efficiently to store data resulted from algorithms such as phrase counts, these have some drawbacks. First it is hard to maintain unique values as keys, being more and more difficult to generate unique characters string for keys. Second, this model does not provide capacities such as consistency for multiple transactions executed simultaneously, those must be provided by the application level.

Document stores are data stores similar to key-values stores, main difference being that values are represented by "documents" that have some structure and encoding models (e.g. XML, JSON, BSON, etc.) of managed data. Also document stores provide usually and API for retrieving the data.

Tabular stores represent stores for management of structured data based on tables, descending from BigTable design from Google. HBase form Hadoop represents such type on NoSQL data management system. In this type of data store data is stored in a three-dimensional table that is indexed by a row key (that is used in a fashion that is similar to the key-value and document stores), a column key that indicates the specific attribute for which a data value is stored, and a timestamp that may refer to the time at which the row's column value was stored.

Madden defines in [23] Big Data as being data which is "too big (e.g. petabyte order), too fast (must be quickly analyzed, e.g. fraud detection), too hard (requires analysis software to be understood)" to be processed by tools in the context of relational databases. Relational databases, especially commercial ones (Teradata, Netezza, Vertica, etc.) can handle the "too big" problem but fail to address the other two. "Too fast" means handling streaming data that cannot be done efficiently by relational engines. In-database statistics and analytics that are implemented for relational databases does not parallelize efficiently for large amount of data (the "too hard" problem). MapReduce or Hadoop based databases (Hive, HBase) are not solving the Big Data problems but seem to recreate DBMS's. They do not provide data management and show poor results for the "too fast" problem since they work with big blocks of replicated data over a distributed storage thus making difficult to achieve low-latency.

Currently we face with a gap between analytics tools (R, SAS, Matlab) and Hadoop or RDBMS's that can scale. The challenge is to build a bridge either by extending relational model (Oracle Data Mining [24] and Greenplum MadSkills [25] efforts to include data mining, machine learning, statistical algorithms) or extend the MapReduce model (ongoing efforts of Apache Mahout to implement machine learning on top of MapReduce) or creating something new and different from these two (GraphLab from Carnegie Mellon [26] which is a new model tailored for machine learning or SciDB which aims to integrate R and Python with large data sets on disks but these are not enough mature products). All these have problems from the usability perspective [23].

2.3 Data Handling and Data Cleaning

Traditional technologies and techniques for data storage and analysis are not efficient anymore as the data is produced in high-volumes, come with high-velocity, has high-variety and there is an imperative need for discovering valuable knowledge in order to help in decision making process.

There are many challenges when dealing with big data handling, starting from data capture to data visualization. Regarding the process of data capture, sources of data are heterogeneous, geographically distributed, and unreliable, being susceptible to errors. Current real-world storage solutions such as databases are populated in consequence with inconsistent, incomplete and noisy data. Therefore, several data preprocessing techniques, such as data reduction, data cleaning, data integration, and data transformation, must be applied to remove noise and correct inconsistencies and to help in decision-making process [27].

The design of NoSQL databases systems highlights a series of advantages for data handling compared with relational database systems [28]. First data storage and management are independent one from another. The data storage part, called also key-value storage, focus on the scalability and high-performance. For the management part, NoSQL provides low-level access mechanism, which gives the possibility that tasks related to data management can be implemented in the application layer, contrary to relational databases which spread management logic in SQL or stored procedures [29].

So, NoSQL systems provide flexible mechanisms for data handling, and application developments and deployments can be easily updated [28]. Another important design advantage of NoSQL databases is represented by the facts that are schema-free, which permits to modify structure of data in applications; moreover the management layer has policies to enforce data integration and validation.

Data quality is very important, especially in enterprise systems. Data mining techniques are directly affected by data quality. Poor data quality means no relevant results. Data cleaning in a DBMS means record matching, data deduplication, and column segmentation. Three operators are defined for data cleaning tasks: fuzzy lookup that it is used to perform record matching, fuzzy grouping that is used for deduplication and column segmentation that uses regular expressions to segment input strings [30–32].

Big Data can offer benefits in many fields from businesses through scientific field, but with condition to overcome the challenges, which arise in data, capture, data storage, data cleaning, data analysis and visualization.

2.4 Big Data Processing Stack

Finding the best method for a particular processing request behind a particular use remains a significant challenge. We can see the Big Data processing as a big batch process that runs on an HPC cluster by splitting a job into smaller tasks and distributing the work to the cluster nodes. The new types of applications, like social networking, graph analytics, complex business workflows, require data movement and data storage. In [33] is proposed a general view of a four-layer big-data processing stack (see Fig. 2). Storage Engine provides storage solutions (hardware/software) for big data applications: HDFS, S3, Lustre, NFS, etc. Execution Engine provides the reliable and efficient use of computational resources to execute. This layers aggregate YARN-based processing solutions. Programming Model offers support for application development and deployment. High-Level Language allows modeling of queries and general data-processing tasks in easy and flexible languages (especially for non-experts).

The processing models must be aware about data locality and fairness when decide to move date on the computation node or to create new computation nodes neat to data. The workload optimization strategies are the key for a guaranteed profit for resource providers, by using the resource to maximum capacity. For

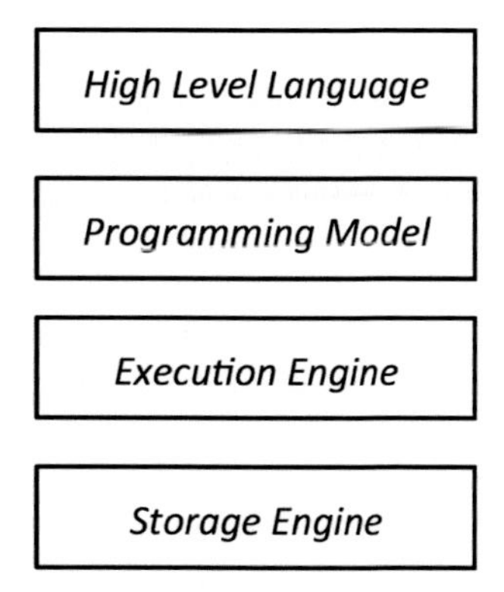

Fig. 2 Big Data processing stack

applications that are both computational and data intensive the processing models combine different techniques like in-memory Big Data or CPU + GPU processing. Figure 3 describes in a general stack used to define a Big Data processing platform.

Moreover, Big Data platforms face with heterogeneous environments where different systems, like Custer, Grid, Cloud, and Peer-to-Peer can offer support for advance processing. At the confluence of Big Data with heterogeneity, scheduling solutions for Big Data platforms consider distributed applications designed for efficient problem solving and parallel data transfers (hide transfer latency) together with techniques for failure management in high heterogeneous computing systems. Handling of heterogeneous data sets becomes a challenge for interoperability through various software systems.

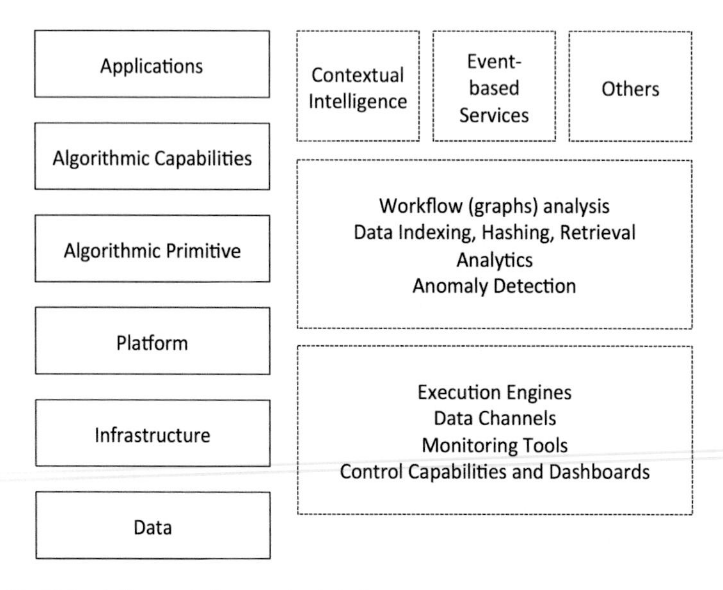

Fig. 3 Big Data platforms stack: an extended view

2.5 Processing in Big Data Platforms

A general Big Data architecture basically consists of two parts: a job manager that coordinates processing nodes and a storage manager that coordinates storage nodes [34]. Apache Hadoop is a set of open source applications that are used together in order to provide a Big Data solution. The two main components mentioned above in Hadoop are, HDFS and YARN. In Fig. 4 is presented the general architecture of a computing platform for Big Data platforms. HDFS Hadoop Distributed File System is organized in clusters where each cluster consists of a name node and several storage nodes. A large file is split into blocks and name node takes care of persisting the parts on data nodes. The name node maintains metadata about the files and commits updates to a file from a temporary cache to the permanent data node. The data node does not have knowledge about the full logical HDFS file; it handles locally each block as a separate file. Fault tolerance is achieved through replication; optimizing the communication by considering the location of the data nodes (the ones located on the same rack are preferred). A high degree of reliability is realized using heartbeat technique (for monitoring), snapshots, metadata replication, checksums (for data integrity), and rebalancing (for performance).

YARN is a name for MapReduce v2.0. This implements a master/slave execution of processes with a JobTracker master node and a pool of Task-Trackers that do the work. The two main responsibilities of the JobTracker, management of resources and job scheduling/monitoring are split. There is a global resource manager (RM) per application, and an Application Master (AM). The slave is a per node entity named Node Manager (NM) which is doing the computations. The AM negotiates with the RM for resources and monitors task progress. Other components

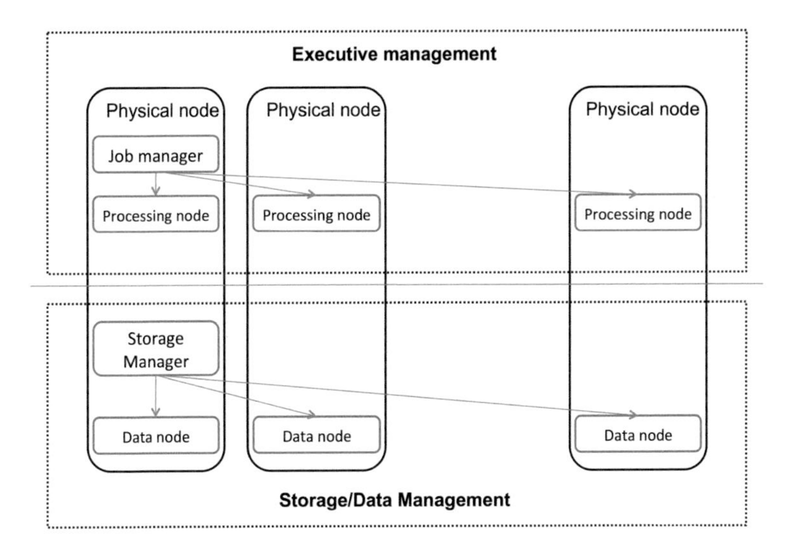

Fig. 4 Typical organization of resources in a big-data platform

are added on top of Hadoop in order to create a Big Data ecosystem capable of configuration management (Zookeeper [35]), columnar organization (HBase [36]), data warehouse querying (Hive [37]), easier development of MapReduce programs (Pig [38]), and machine learning algorithms (Mahout) [39].

3 Big Data Reduction Techniques

This section highlights the analytics for Big Data focusing on reduction techniques. The main reduction techniques are based on: statistical models used in data analysis (e.g. kernel estimation) [40, 41]; machine learning techniques: supervised and un-supervised learning, classification and clustering, k-means [42], multi-dimensional scaling [43]; ranking techniques: PageRank, recursive queries, etc. [44]; latent semantic analysis; filtering techniques: collaborative filtering, multi-objective filtering; self-* techniques (self-tuning, self-configuring, self adaptive, etc.) [45], and data mining techniques [46]. All these techniques are used for descriptive analytics, predictive analytics and prescriptive analytics.

According with [2], we can see three important stages in evolution of analytics methods:

- **Analytics 1.0 (1954–2009)**—data sources small and data was mostly internal, analytics was a batch process that took months;
- **Analytics 2.0 (2005–2012)**—the most important actors are Internet based companies Google, eBay, Yahoo. Data is mostly external, unstructured, huge volume, and it required parallel processing with Hadoop. It is what was named Big Data. The tow of data is much faster than in Analytics 1.0. A new category of skilled employees called data scientists emerged;
- **Analytics 3.0 (2012 present)**—the best of traditional analytics and Big Data techniques are mixed together in order to tackle large volumes of data, both internal and external, both structured and unstructured, obtained in a continuously increasing number of different formats (new sensors are added). Hadoop and Cloud technologies are intensively used not only by online forms but by various companies such as Banks, Retail, HealthCare providers, etc.

3.1 Intelligent Reduction Techniques

In Data Analysis as part of the Qualitative Research for large datasets, in past decades there were proposed Content Analysis (counting the number of occurrences of a word in a text but without considering the context) and Thematic Analysis (themes are patterns that occurs repeatedly in data sets and which are important to the research question). The first form of data reduction is to decide which data from the initial set of data is going to be analyzed (since not all data could be relevant, some of it can be eliminated). In this respect there should be defined some method for categorizing data [47].

Structural coding—code related to questions is then applied to responses to that question in the text. Data can be then sorted using these codes (structural coding acting as labeling).

Frequencies—word counting can be a good method to determine repeated ideas in text. It requires prior knowledge about the text since one should know before the keywords that will be searched. An improvement is to count not words but codes applications (themes).

Co-occurrence—more codes exist inside a segment of text. This allows Boolean queries (and segment with code A AND code B).

Hierarchical Clustering—using co-occurrence matrices (or code similarity matrices) as input. The goal is to derive natural groupings (clusters) in large datasets. A value matrix element $v(i, j) = n$ means that code i and code j co-occurs in n participant files.

Multidimensional Scaling—the input is also similarity matrix and ideas that are considered to be close each to the other are represented as points with a small distance between them. This way is intuitive to visualize graphically the clusters.

Big Data does not raise only engineering concerns (how to manage effectively the large volume of data) but also semantic concerns (how to get meaning information regardless implementation or application specific aspects).

Meaningful data integration process requires following stages, not necessarily in this order [7]:

- Define the problem to be resolved;
- Search the data to find the candidate datasets that meet the problem criteria;
- ETL (Extract, Transform and Load) of the appropriate parts of the candidate data for future processing;
- Entity Resolution checks if data is unique, comprehensive and relevant;
- Answer the problem performs computations to give a solution to the initial problem.

Using the Web of Data, which according to some statistics contains 31 billion RDF triples, is possible to find all data about people and their creations (books, films, musical creations, etc.), translate the data into a single target vocabulary, discover all resources about a specific entity and then integrate this data into a single coherent representation. RDF and Linked Data (such as pre-crawled web data sets BTC 2011 with 2 billion RDF triples or Sindice 2011 with 11 billion RDF triples extracted from 280 million web pages an- notated with RDF) are schema less models that suits Big Data, considering that less than 10 % is genuinely relational data. The challenge is to combine DBMS's with reasoning (the next smart databases) that goes beyond OWL, RIF or SPARQL and for this reason use cases are needed from the community in order to determine exactly what requirements the future DB must satisfy. A web portal should allow people to search keywords in ontologies, data itself and mappings created by users [7].

3.2 Descriptive Analytics

Descriptive analytics is oriented on descriptive statistics (counts, sums, averages, percentages, min, max and simple arithmetic) that summarizes certain groupings or filtered type of the data, which are typically simple counts of some functions, criteria or events. For example, number of post on a forum, number of likes on Facebook or number of sensors in a specific area, etc. The techniques behind descriptive analytics are: standard aggregation in databases, filtering techniques and basic statistics. Descriptive analytics use filters on the date before applying specific statistical functions. We can use geo-filters to get metrics for a geographic region (a country) or temporal filter, to extract date only for a specific period of time (a week). More complex descriptive analytics are dimensional reduction or stochastic variation.

Dimensionality reduction represents an important tool in information analysis. Also scaling down data dimensions is important in process of recognition and classification. Is important to notice that, sparse local operators, which imply less quadratic complexity and faithful multi-scale models make the de sign of dimension reduction procedure a delicate balance between modeling accuracy and efficiency. Moreover the efficiency of dimension reduction tools is measured in terms of memory and computational complexity. The authors provide a theoretical support and demonstrate that working in the natural Eigen-space of the data one could reduce the process complexity while maintaining the model fidelity [48].

A stochastic variation inference is used for Gaussian process models in order to enable the application of Gaussian process (GP) models to data sets containing millions of data points. The key finding in this chapter is that GPs can be decomposed to depend on a set of globally relevant inducing variables, which factorize the model in the necessary manner to perform variation inference. These expressions allow for the transfer of a multitude of Gaussian process techniques to big data [41].

3.3 Predictive Analytics

Predictive analytics, which are probabilistic techniques, refers to: (i) *temporal predictive models* that can be used to summarize existing data, and then to extrapolate to the future where data doesn't exist; (ii) *non-temporal predictive models* (e.g. a model that based on someone's existing social media activity data will predict his/her potential to influence [49]; or sentimental analysis). The most challenging aspect here is to validate the model in the context of Big Data analysis. One example of this model, based on clustering, is presented in the following.

A novel technique for effectively processing big graph data on cloud surpasses the raising challenges when data is processed in heterogeneous environments, such as parallel memory bottlenecks, deadlocks and inefficiency. The data is compressed based on spatial-temporal features, exploring correlations that exist in spatial data. Taking into consideration those correlations graph data is partitioned into clusters where, the workload can be shared by the inference based on time series similarity. The clustering algorithm compares the data streams according to the topology of the streaming data graph topologies from the real world. Furthermore, because the data items in streaming big data sets are heterogeneous and carry very rich order information themselves, an order compression algorithm to further reduce the size of big data sets is developed. The clustering algorithm is developed on the cluster-head. It takes time series set X and similarity threshold as inputs. The output is a clustering result which specifying each cluster-head node and its related leaf nodes [50].

The prediction models used by predictive analytics should have the following properties: simplicity (a simple mathematical model for a time series), flexibility (possibility to configure and extend the model), visualization (the evolution of the predicted values can be seen in parallel with real measured values) and the computation speed (considering full vectorization techniques for array operations). Let's consider a data series V_1; V_2; …; V_n extracted by a descriptive analytics technique. We consider for the prediction problem $P(V_{t+1})$ that denotes the predicted value for the moment t + 1 (next value). This value is:

$$PV_{t+1} = P(V_{t+1}) = f(V_t, V_{t-1}, \ldots, V_{t-window}),$$

where: *window* represents a specific interval with window + 1 values and f can be a linear function such as mean, median, standard deviation or a complex function that uses a bio-inspired techniques (an adaptive one or a method based on neural networks).

The linear prediction can be expressed as follow:

$$PV_{t+1} = f_w = \frac{1}{window+1} \sum_{i=0}^{window} w_i V_{t-i},$$

where $w = [w_i]_{0 \leq i \leq window}$ is a vector with weights. If $\forall i, w_i = 1$ then we have the mean function. It is possible to consider specific distribution of weights as follow: $w_i = t - i$.

The predictive analytics are very useful to make estimation for future behaviors especially when the date is no accessible (it is not possible to obtain or to predict) or is too expensive (e.g. money, time) to measure or to compute the data. The main challenge is to validate the predicted data. One solution is to wait for the real value (in the future) to measure the error, then to propagate it in the system in order to improve the future behavior. Other solution is to measure the impact of the predicted date in the applications that use the data.

3.4 Prescriptive Analytics

Prescriptive analytics predicts multiple futures based on the decision maker's actions. A predictive model of the data is created with two components: actionable (decision making support) and feedback system (tracks the outcome of made decisions). Prescriptive analytics can be used for recommendation systems because it is possible to predict the consequences based on predictive models used. A self-tuning database system is an example that we will present in the following.

Starfish is a self-tuning system for big data analytics, build on Hadoop. This system is designed according to self-tuning database systems [45]. Cohen et al. proposed the acronym MAD (Magnetism, Agility, and Depth) in order to express the features that users expect from a system for big data analytics [51]. Magnetism represents the propriety of a system that attracts all sources of data regardless of different issues (e.g. possible presence of outliers, unknown schema, lack of structure, missing values) keeping many data sources out of conventional data warehouses. Agility represents the propriety of adaptation of systems in sync with rapid data evolution. Depth represents the propriety of a system, which supports analytics needs that go far beyond conventional rollups and drilldowns to complex statistical and machine-learning analysis. Hadoop represents a MAD system that is very popular for big data analytics. This type of systems poses new challenges in the path to self-tuning such as: data opacity until processing, file based processing, and heavy use of programming languages.

Furthermore three more features in addition to MAD are becoming important in analytics systems: data-lifecycle awareness, elasticity, and robustness. Data-lifecycle-awareness means optimization of the movement, storage, and processing of big data during its entire lifecycle by going beyond query execution. Elasticity means adjustment of resource usage and operational costs to the workload and user requirements. Robustness means that this type of system continues to provide service, possibly with graceful degradation, in the face of undesired events like hardware failures, software bugs, and data corruption.

The Starfish system has four levels of tuning: Job-level tuning, Workflow-level tuning, and Workload-level tuning. The novelty in Starfish's approach comes from how it focuses simultaneously on different workload granularities overall workload, workflows, and jobs (procedural and declarative) as well as across various decision points provisioning, optimization, scheduling, and data layout. This approach enables Starfish to handle the significant interactions arising among choices made at different levels [45].

To evaluate a prescriptive analytics model we need a feedback system (to tracks the adjusted outcome based on the action taken) and a model for tacking actions (take actions based on the predicted outcome and based on feedback). We define several metrics for evaluating the performance of prescriptive analytics. *Precision* is the fraction of the data retrieved that are relevant to the user's information need.

Recall is the fraction of the data that are relevant to the query that are successfully retrieved. *Fall-Out* is the proportion of non-relevant data that are retrieved, out of all non-relevant documents available. $F_{measure}$ is the weighted harmonic mean of precision and recall:

$$F_{measure} = \frac{2 * Precision * Recall}{Precision + Recall}.$$

The general formula for this metric is:

$$F_{\beta} = \frac{\left(1 + \beta^2\right) * Precision * Recall}{\beta^2 * (Precision + Recall)}.$$

This metric measures the effectiveness of retrieval with respect to a user who attaches β times as much importance to recall as precision. As a general conclusion we can summarize the actions performed by three types of analytics as follow: descriptive analytics summarize the data (data reduction, sum, count, aggregation, etc.), predictive analytics predict data that we don't have (influence scoring, trends, social analysis, etc.) and prescriptive analytics guide the decision making to a specific outcome.

4 CyberWater Case Study

In this section we present the case study on CyberWater [52], which is a research project aiming to create a prototype platform using advanced computational and communications technology for implementation of new frameworks for managing water and land resources in a sustainable and integrative manner. The main focus of this effort is on acquiring diverse data from various data sources in a common digital platform, which is a lose subject to Big Data, and is subsequently used for routine decision making in normal conditions and for providing assistance in critical situations related to water, such as accidental pollution flooding, which is an analytics subject.

CyberWater system monitors natural resources and water related events, shares data compliant to the INSPIRE Directive, and alerts relevant stakeholders about critical situations. In the area where we conduct the measurements (Someş and Dâmboviţa rivers from Romania) certain types of chemical and physical indicators are of interest: pH, Turbidity, Alkalinity, Conductivity, Total Phenols, dissolved Oxygen, N–NH4, N–NO2, N–NO3, Total N, P–PO4, and Magnesium.

The multi-tier system presented in Fig. 5 is composed of: data layer, processing layer, visualization layer. Water resources management requires the processing of a

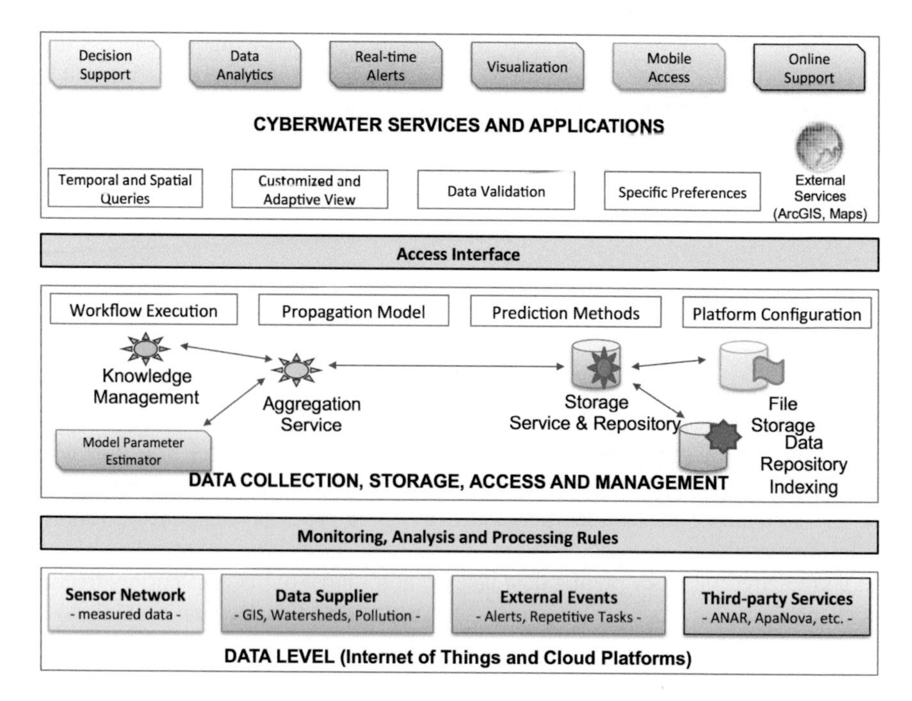

Fig. 5 Multi-tier architecture of CyberWater system

huge amount of information with different levels of accessibility and availability and in various formats. The data collected in CyberWater is derived from various sources: (i) Measured data (from sensors) gathered through the Sensor Observation Service (SOS) standard. The schema is predefined by the OGC standard. Information about sensors is transmitted in the XML based format SensorML and the actual observational data is sent in O&M format; (ii) Predicted data; (iii) Modeled data from the propagation module; (iv) Subscribers data which holds information about subscribers and their associated notification services.

Figure 6 describes the reduction data in the CyberWater system. Data collected from various sources are used to describe the model of propagation of the pollutants (geo-processing tool based on the river bed profile and fluids dynamics equations). This is the Descriptive Analytics phase, were only relevant date are used from the system repository. Phase (2)—Predictive Analytics is represented by the prediction of the next value of a monitored chemical/physical indicator. Then, decision support that triggers alerts to relevant actors of the system is the subject of Perspective Analytics. The main application that generates alerts is a typical Publish/Subscribe application and is the place where the alert services are defined and implemented.

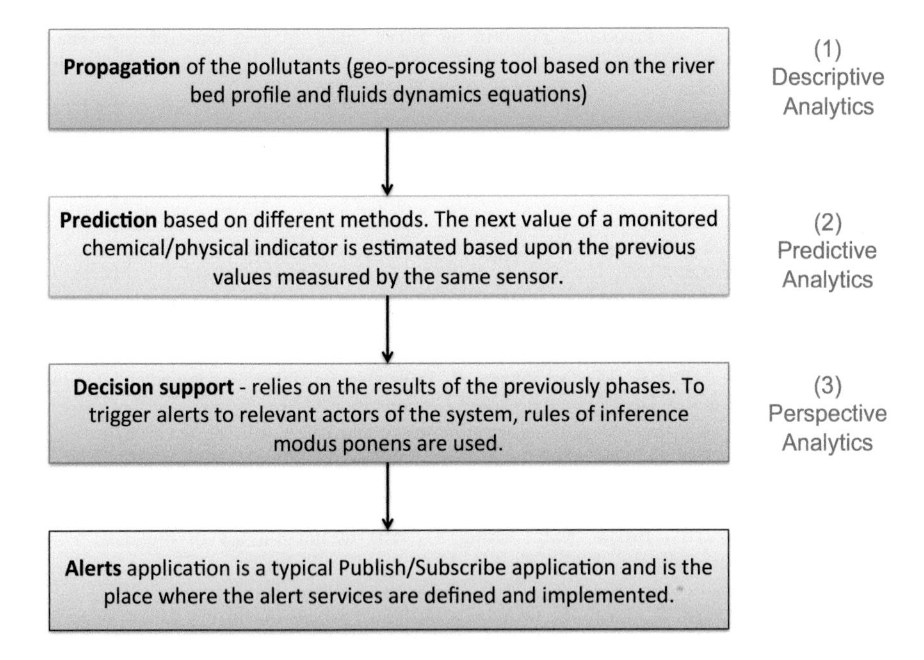

Fig. 6 CyberWater: from descriptive analytics to perspective analytics

5 Conclusion

Qualitative data used to produce relevant information are obtained by reducing the big amount of data collected and aggregated over the time, in different locations. Reduction techniques have the main role to extract as much as possible relevant data that characterize the all analyzed data. In this chapter we made an overview on data manipulation challenges, on reduction techniques: descriptive analytics, predictive analytics and prescriptive analytics. We presented a case study on water resources management that explains the use of reduction techniques for Big Data. As recommendation about the use of reduction techniques for Big Data, we can draw the following line: collect and clean data, extract the relevant information using descriptive analytics, estimate the future using predictive analytics and take decisions to move on using perspective analytics; the loop closes by adding valuable information in the system.

Acknowledgments The research presented in this paper is supported by projects: *CyberWater* grant of the Romanian National Authority for Scientific Research, CNDI-UEFISCDI, project number 47/2012; *CLUeFARM*: Information system based on cloud services accessible through mobile devices, to increase product quality and business development farms—PN-II-PT-PCCA-2013-4-0870;

DataWay: Real-time Data Processing Platform for Smart Cities: Making sense of Big Data - PN-II-RU-TE-2014-4-2731; *MobiWay:* Mobility Beyond Individualism: an Integrated Platform for Intelligent Transportation Systems of Tomorrow—PN-II-PT-PCCA-2013-4-0321.

References

1. Laurila, J.K., Gatica-Perez, D., Aad, I., Bornet, O., Do, T.M.T., Dousse, O., Eberle, J., Miettinen, M.: The mobile data challenge: Big data for mobile computing research. In: Mobile Data Challenge by Nokia Workshop (2012)
2. Davenport, T.H., Dyche, J.: Big data in big companies. Int. Inst. Anal. (2013)
3. Ho, D., Snow, C., Obel, B., Dissing Srensen, P., Kallehave, P.: Unleashing the potential of big data. Technical report, Organizational Design Community (2013)
4. Lynch, C.: Big data: How do your data grow? Nature **455**(7209), pp. 28–29 (2008)
5. Szala, A.: Science in an exponential world. Nature **440**, 2020 (2006)
6. Birney, E.: The making of encode: lessons for big-data projects. Nature **489**(7414), pp. 49–51 (2012)
7. Bizer, C., Boncz, P., Brodie, M.L., Erling, O.: The meaningful use of big data: Four perspectives-four challenges. SIGMOD Rec. **40**(4), pp. 56–60 (2012)
8. Madden, S.: From databases to big data. IEEE Internet Comput. **16**(3), pp. 4–6 (2012)
9. Chen, Y., Alspaugh, S., Katz, R.: Interactive analytical processing in big data systems: a cross-industry study of mapreduce workloads. Proc. VLDB Endow. **5**(12), pp. 1802–1813 (2012)
10. Cuzzocrea, A., Song, I.Y. Davis, K.C.: Analytics over large-scale multidimensional data: the big data revolution! In: Proceedings of the ACM 14th International Workshop on Data Warehousing and OLAP, DOLAP'11, pp. 101–104. ACM, New York, NY, USA (2011)
11. Negru, C., Pop, F., Cristea, V., Bessisy, N., Li, J.: Energy efficient cloud storage service: key issues and challenges. In: Proceedings of the 2013 Fourth International Conference on Emerging Intelligent Data and Web Technologies, EIDWT'13, pp. 763–766. IEEE Computer Society, Washington, DC, USA (2013)
12. Rao, S., Ramakrishnan, R., Silberstein, A., Ovsiannikov, M., Reeves, D.: Sailfish: a framework for large scale data processing. In: Proceedings of the Third ACM Symposium on Cloud Computing, SoCC'12, pp. 4:1–4:14. ACM, New York, NY, USA (2012)
13. Roddick, J.F., Hoel, E., Egenhofer, M.J., Papadias, D., Salzberg, B.: Spatial, temporal and spatio-temporal databases—hot issues and directions for Ph.D. research. SIGMOD Rec. **33**(2), pp. 126–131 (2004)
14. Chen, C.X.: Spatio-temporal databases. In: Shekhar, S., Xiong, H. (eds.) Encyclopedia of GIS, pp. 1121–1121. Springer, USA (2008)
15. Guhaniyogi, R., Finley, A., Banerjee, S., Kobe, R.: Modeling complex spatial dependencies: low-rank spatially varying cross-covariances with application to soil nutrient data. J. Agric. Biol. Environ. Stat. **18**(3), pp. 274–298 (2013)
16. Johnson, D.S., Ream, R.R., Towell, R.G., Williams, M.T., Guerrero, J.D.L.: Bayesian clustering of animal abundance trends for inference and dimension reduction. J. Agric. Biol. Environ. Stat. **18**(3), pp. 299–313 (2013)
17. Leininger, T.J., Gelfand, A.E., Allen, J.M., Silander Jr, J.A.: Spatial regression modeling for compositional data with many zeros. J. Agric. Biol. Environ. Stat. **18**(3), pp. 314–334 (2013)
18. Wu, G., Holan, S.H., Wikle, C.K.: Hierarchical Bayesian spatio-temporal conwaymaxwell poisson models with dynamic dispersion. J. Agric. Biol. Environ. Stat. **18**(3), pp. 335–356 (2013)
19. Dunstan, P.K., Foster, S.D., Hui, F.K., Warton, D.I.: Finite mixture of regression modeling for high-dimensional count and biomass data in ecology. J. Agric. Biol. Environ. Stat. **18**(3), pp. 357–375 (2013)

20. Hooten, M.B., Garlick, M.J., Powell, J.A.: Computationally efficient statistical differential equation modeling using homogenization. J. Agric. Biol. Environ. Stat. **18**(3), pp. 405–428 (2013)
21. Yang, W.-H., Wikle, C.K., Holan, S.H., Wildhaber, M.L.: Ecological prediction with nonlinear multivariate time-frequency functional data models. J. Agric. Biol. Environ. Stat. **18** (3), pp. 450–474 (2013)
22. Loshin, D.: Nosql data management for big data. In: Loshin, D. (ed.) Big Data Analytics, pp. 83–90. Morgan Kaufmann, Boston (2013)
23. Madden, S.: Query processing for streaming sensor data. Comput. Sci. Div. (2002)
24. Hamm, C., Burleson, D.K.: Oracle Data Mining: Mining Gold from Your Warehouse. Oracle In-Focus Series. Rampant TechPress (2006)
25. Hellerstein, J.M., Ré, C., Schoppmann, F., Wang, D.Z., Fratkin, E., Gorajek, A., Ng, K.S., Welton, C., Feng, X., Li, K., Kumar, A.: The MADlib analytics library: or MAD skills, the SQL. Proc. VLDB Endow. **5**(12), pp. 1700–1711 (2012)
26. Low, Y., Bickson, D., Gonzalez, J., Guestrin, C., Kyrola, A., Hellerstein, J.M.: Distributed graphlab: a framework for machine learning and data mining in the cloud. Proc. VLDB Endow. **5**(8), pp. 716–727 (2012)
27. Han, J., Kamber, M.: Data Mining, Southeast Asia Edition: Concepts and Techniques. Morgan kaufmann (2006)
28. Hilbert, M., Lopez, P.: The worlds technological capacity to store, communicate, and compute information. Science **332**(6025), pp. 60–65 (2011)
29. Chang, F., Dean, J., Ghemawat, S., Hsieh, W.C., Wallach, D.A., Burrows, M., Chandra, T., Fikes, A., Gruber, R.E.: Bigtable: a distributed storage system for structured data. ACM Trans. Comput. Syst. (TOCS) **26**(2), 4 (2008)
30. Agrawal, P., Arasu, A., Kaushik, R.: On indexing error-tolerant set containment. In: Proceedings of the 2010 ACM SIGMOD International Conference on Management of Data, SIGMOD'10, pp. 927–938. ACM, New York, NY, USA (2010)
31. Arasu, A., Gotz, M., Kaushik, R.: On active learning of record matching packages. In: Proceedings of the 2010 ACM SIGMOD International Conference on Management of Data, SIGMOD'10, pp. 783–794. ACM, New York, NY, USA (2010)
32. Arasu, A., Re, C., Suciu, D.: Large-scale deduplication with constraints using dedupalog. In: Proceedings of the 2009 IEEE International Conference on Data Engineering, ICDE'09, pp. 952–963. IEEE Computer Society, Washington, DC, USA (2009)
33. Varbanescu, A.L., Iosup, A.: On many-task big data processing: from GPUs to clouds. In: MTAGS Workshop, held in conjunction with ACM/IEEE International Conference for High Performance Computing, Networking, Storage and Analysis (SC), pp. 1–8. ACM (2013)
34. Loshin, D.: Big data tools and techniques. In: Loshin, D. (ed.) Big Data Analytics, pp. 61–72. Morgan Kaufmann, Boston (2013)
35. Hunt, P., Konar, M., Junqueira, F.P., Reed, B.: Zookeeper: Wait-free coordination for internet-scale systems. In: Proceedings of the 2010 USENIX Conference on USENIX Annual Technical Conference, USENIXATC'10, pp. 11–11. USENIX Association, Berkeley, CA, USA (2010)
36. Jiang, Y.: HBase Administration Cookbook. Packt Publishing, Birmingham (2012)
37. Huai, Y., Chauhan, A., Gates, A., Hagleitner, G., Hanson, E.N., O'Malley, O., Pandey, J., Yuan, Y., Lee, R., Zhang, X.: Major technical advancements in apache hive. In: Proceedings of the 2014 ACM SIGMOD International Conference on Management of Data, SIGMOD'14, pp. 1235–1246. ACM, New York, NY, USA (2014)
38. Shang, W., Adams, B., Hassan, A.E.: Using pig as a data preparation language for large-scale mining software repositories studies: an experience report. J. Syst. Softw. **85**(10), pp. 2195–2204 (2012)
39. Owen, S., Anil, R., Dunning, T., Friedman, E.: Mahout in Action. Manning Publications Co., Greenwich, CT, USA (2011)
40. Banerjee, S., Gelfand, A.E., Finley, A.O., Sang, H.: Gaussian predictive process models for large spatial data sets. J. R. Stat. Soc. Series B (Stat. Methodol.) **70**(4), pp. 825–848 (2008)

41. Hensman, J., Fusi, N., Lawrence, N.D.: Gaussian processes for big data (2013). arXiv:1309.6835
42. Feldman, D., Schmidt, M., Sohler, C.: Turning big data into tiny data: constant-size coresets for k-means, pca and projective clustering. In: Proceedings of the Twenty-Fourth Annual ACM-SIAM Symposium on Discrete Algorithms, SODA'13, pp. 1434–1453. SIAM (2013)
43. Aflalo, Y., Kimmel, R.: Spectral multidimensional scaling. Proc. Natl. Acad. Sci. **110**(45), pp. 18052–18057 (2013)
44. Pop, F., Ciobanu, R.-I., Dobre, C.: Adaptive method to support social-based mobile networks using a pagerank approach. In: Concurrency and Computation: Practice and Experience (2013)
45. Herodotou, H., Lim, H., Luo, G., Borisov, N., Dong, L., Cetin, F.B., Babu, S.: Starfish: a self-tuning system for big data analytics. In: CIDR, vol. 11, pp. 261–272 (2011)
46. Kantardzic, M.: Data Mining: Concepts, Models, Methods, and Algorithms, 2nd edn. Wiley-IEEE Press, Hoboken (2011)
47. Namey, E., Guest, G., Thairu, L., Johnson, L.: Data reduction techniques for large qualitative data sets. In: Guest, G., MacQueen, K.M. (eds.) Handbook for Team-Based Qualitative Research, pp. 137–162. AltaMira Press, USA (2007)
48. Aflalo, Y., Kimmel, R., Raviv, D.: Scale invariant geometry for nonrigid shapes. SIAM J. Imaging Sci. **6**(3), pp. 1579–1597 (2013)
49. Cambria, E., Rajagopal, D., Olsher, D., Das, D.: Big social data analysis. In: R. Akerkar (ed.) Big Data Computing, pp. 401–414. Taylor & Francis, New York (2013)
50. Yang, C., Zhang, X., Zhong, C., Liu, C., Pei, J., Ramamohanarao, K., Chen, J.: A spatiotemporal compression based approach for efficient big data processing on cloud. J. Comput. Syst. Sci. **80**(8), pp. 1563–1583 (2014)
51. Cohen, J., Dolan, B., Dunlap, M., Hellerstein, J.M., Welton, C.: Mad skills: new analysis practices for big data. Proc. VLDB Endow. **2**(2), pp. 1481–1492 (2009)
52. Ciolofan, S.N., Mocanu, M., Ionita, A.: Distributed cyberinfrastructure for decision support in risk related environments. In: 2013 IEEE 12th International Symposium on Parallel and Distributed Computing (ISPDC), pp. 109–115 (2013)

Author Biographies

Florin Pop received his Ph.D. in Computer Science at the University POLITEHNICA of Bucharest in 2008. He is Associate Professor within the Computer Science Department and also an active member of Distributed System Laboratory. His research interests are in scheduling and resource management, multi-criteria optimization methods, Grid middleware tools and applications development, prediction methods, self-organizing systems, contextualized services in distributed systems. He is the author or co-author of more than 150 publications. He was awarded with "Magna cum laude" distinction his Ph.D. results, one IBM Faculty Award in 2012, two Prizes for Excellence from IBM and Oracle (2008 and 2009), Best young researcher in software services Award in 2011 and two Best Paper Awards. He worked in several international and national research projects in the distributed systems field as coordinator and member as well. He is a senior member of the IEEE and ACM.

Catalin Negru Computer Science diplomat engineer, finished his master studies in 2011 on Advance Computing Architectures program at University Politehnica of Bucharest, Faculty of Automatic Control and Computers, Computer Science Department. He is a Ph.D. candidate and an active member of Distributed Systems Laboratory. His research interests are in Storage in Distributed Systems Cloud System and Resource Management, especially in cost optimization, SLA assurance, multi-criteria optimization, Cloud middleware tools, complex applications design and implementation. He worked in several international and national research projects in the distributed systems field as young researcher.

Sorin N. Ciolofan is currently Teaching Assistant and Ph.D. candidate at the Computer Science Department, University Politehnica of Bucharest. He finished his MSc at the same university in 2004 and after that participated in over 10 international and national projects for research and private clients (finance, tourism, oil and gas). As a software engineer, consultant and senior software engineer in various companies/institutions (including IBM Dublin Software Laboratory, FORTH Greece—Institute of Computer Science Information Systems Laboratory etc.) he was involved in all stages of software development process working mainly with Java/J2EE, portal and semantic web technologies. His research interests are interdisciplinary, being placed at the intersection of Big Data, Data Mining, Environmental Sciences, GIS, Cloud Computing, Cyber Physical Systems and Semantic Technologies.

Mariana Mocanu is a professor of the Computer Science Department, University Politehnica of Bucharest, and has a long experience in developing information systems for industrial and economic processes, and in project management. She performs teaching for both undergraduate and master's degree in software engineering, systems integration, software services and logic design. At the University of Regensburg, as visiting professor, she thought Process Computers. She worked for ten years in a multidisciplinary research team for vehicles, being co-author of two patents. She participated in numerous research projects, implementing information systems for control/optimization of processes in various areas (transport, environment, medicine, natural resources management). Her results are reflected in articles published in journals, in papers presented at national and international conferences, and books. She is a member of the University Senate, at the faculty she is responsible for quality assurance and is a board member of the department.

Valentin Cristea is a professor of the Computer Science Department of the University Politehnica of Bucharest, and Ph. D. supervisor in the domain of Distributed Systems. His main fields of expertise are large scale distributed systems, cloud computing and e-services. He is co-Founder and Director of the National Center for Information Technology of UPB. He led the UPB team in COOPER (FP6), datagrid@work (INRIA "Associate Teams" project), CoLaborator project for building a Center and collaborative environment for HPC, distributed dependable systems project DEPSYS, etc. He co-supervised the UPB Team in European projects SEE-GRID-SCI (FP7) and EGEE (FP7). The research results have been published in more than 230 papers in international journals or peer-reviewed proceedings, and more than 30 books and chapters. In 2003 and 2011 he received the IBM faculty award for research contributions in e-Service and Smart City domains. He is a member of the Romanian Academy of Technical Sciences.

Performance Tools for Big Data Optimization

Yan Li, Qi Guo and Guancheng Chen

Abstract Many big data optimizations have critical performance requirements (e.g., real-time big data analytics), as indicated by the Velocity dimension of 4Vs of big data. To accelerate the big data optimization, users typically rely on detailed performance analysis to identify potential performance bottlenecks. However, due to the large scale and high abstraction of existing big data optimization frameworks (e.g., Apache Hadoop MapReduce), it remains a major challenge to tune the massively distributed systems in a fine granularity. To alleviate the challenges of performance analysis, various performance tools have been proposed to understand the runtime behaviors of big data optimization for performance tuning. In this chapter, we introduce several performance tools for big data optimization from various aspects, including the requirements of ideal performance tools, the challenges of performance tools, and state-of-the-art performance tool examples.

Keywords MapReduce · Performance analysis · Performance optimization · Auto-tuning

1 What Performance Tool the Users Really Need for Big Data Optimization?

Many big data applications have critical performance requirements, for example, detecting fraud while someone is swiping a credit card, or placing an ad on the website while someone is browsing a specific good. All such scenarios advance that big data analytics to be completed within limited time interval. Therefore, optimizing the performance of big data applications is a major concern for various big data

Y. Li · Q. Guo (✉) · G. Chen
IBM Research, Beijing, China
e-mail: qiguo.work@gmail.com

Y. Li
e-mail: liyannewmail@gmail.com

G. Chen
e-mail: chengc@cn.ibm.com

A. Emrouznejad (ed.), *Big Data Optimization: Recent Developments and Challenges*, Studies in Big Data 18, DOI 10.1007/978-3-319-30265-2_4

practitioners, such as application developers, system administrators, platform developers and system architects, etc. Currently, to improve the performance of big data optimization, users typically have to conduct detailed performance analysis to find potential performance bottlenecks, and then eliminate such bottlenecks through different ways such as expanding the memory capacity if the processing data cannot fit into current memory, or adjusting the parameters of big data platforms to occupy the CPU if the CPU is not fully utilized, etc.

Nevertheless, conducting accurate and efficient performance analysis is a very challenging task for big data optimization. The first challenge is that a big data system may consist of thousands of distributed computing nodes, which indicates that performance problems may exist in a large number of subsystems, such as processors, memory, disks and network, of different nodes. The second challenge is that the entire software/hardware stack is very complicated for big data applications. As shown in Fig. 1, the whole software/hardware stack of an IBM's big data solution contains the hardware server (IBM POWER* Servers [1]), the operating system (IBM PowerLinux* [2]), the JVM (IBM Optimized JVM [3]), the big data platform (IBM BigInsights* [4] or IBM Platform Symphony* [5]), and big data workloads (e.g., predictive/content analytics), from the bottom to the top. Thus, it is relatively hard to accurately identify which layers have severe performance issues compared with traditional applications (e.g., applications that are written in C language) that are running directly on the operating system. The third challenge is that even we can collect enough performance data from different sources, the analysis process is tedious, time-consuming and error-prone without proper performance tools.

Moreover, even we have a performance tool to facilitate the performance data collection, visualization, analysis and optimization process, different kinds of users may have different requirements on such performance tools. The first kind of users should be big data application developers. They pay more attentions on the performance

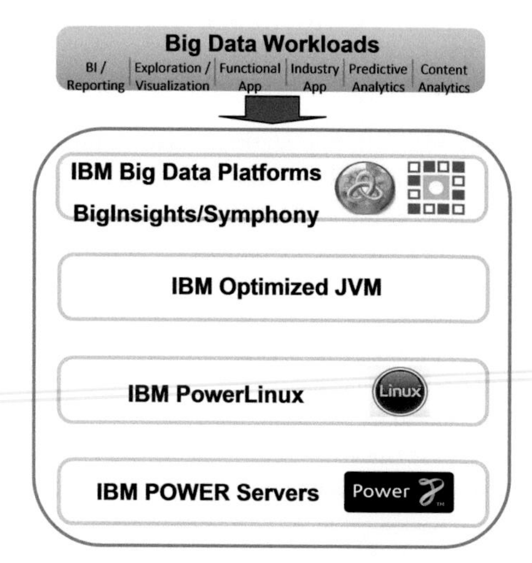

Fig. 1 The whole software/hardware stack of an IBM's big data solution contains the hardware server (IBM POWER* Servers [1]), the operating system (IBM PowerLinux* [2]), the JVM (IBM Optimized JVM [3]), the big data platform (IBM BigInsights* [4] or IBM Platform Symphony* [5]), and big data workloads (e.g., predictive/content analytics)

bottlenecks of developed applications, more specifically, whether there exists inefficient codes in the hand-written big data applications. The second kind of users should be the developers of big data platforms (e.g., contributors of Apache Hadoop source codes). In addition to determining whether or not newly updates or patches of the big data platform is efficient, they also want to propose potential new optimizations for current design. The third kind of users should be the big data system architects. They focus on the optimization of different hardware subsystems, such as CPU, memory, disk and network, to construct a balanced system. Moreover, from another perspective, all the above users can be further grouped into two categories, that is, experienced users and inexperienced users. For the experienced users, they could optimize the performance according to the key insights extracted from the big data execution behaviors. For the inexperienced users, they prefer to achieve the optimal performance gain without too much human investment.

2 Challenges of Ideal Performance Tool

An ideal performance tool should meet all the above requirements of different target users. There exists several challenges to design and implement such an ideal performance tool. We elaborate the difficulties from various aspects, including data collection, data presentation, and data analysis.

2.1 Data Collection

In order to satisfy the requirements of different users, the performance tool should at first collect sufficient information. From our perspective, at least the following information should be collected. For the big data application developers, the execution details of the developed codes (e.g., MapReduce applications) should be collected for further optimization. The basic execution details may include but not limited to the job/task identifier, the total execution time, the begin/end timestamp of interested codes, etc. For the big data platform designers, the execution details of subphases (e.g., map/shuffle/reduce phase of MapReduce paradigm) of big data applications should be tracked and recorded. In addition to the timing information, several engine-specific information should also be collected. For example, the read bytes from HDFS (Hadoop Distributed File System), the input records of the reduce phase, and the compression ratio of the compression phase, etc. Since the interested execution details may vary between platforms and users, the performance tool should also provide API (Application Programming Interface) for the end-users to collect the information of interested code sections. For the big data system architects, the hardware system behaviors, such as CPU/memory utilization, disk and network bandwidth, should be collected as the basic information for conducting system-level analysis and optimization.

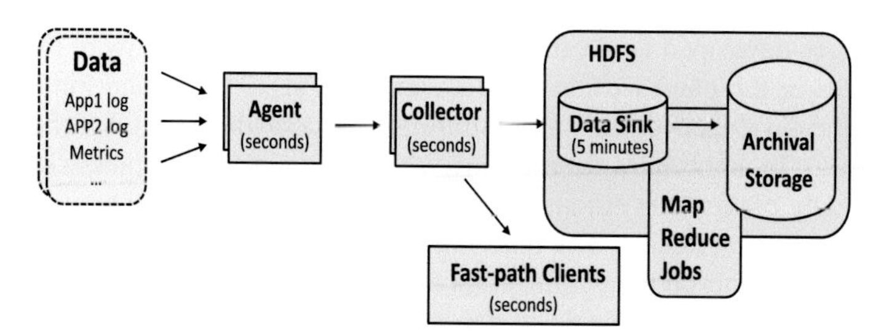

Fig. 2 The overall architecture of a distributed scalable log collector, called *Chukwa* (*Original source* [6])

Since a large amount of information data should be collected for different users, an efficient information collection mechanism is very critical, especially for distributed large-scale clusters. The distributed log collection brings several performance, scalability and reliability challenges. Chukwa is a scalable log collector built upon HDFS and MapReduce, which can be used to reliably collect the execution details for big data performance analysis [6]. Actually, Chukwa has already been used in HiTune [7], which is a dataflow-based performance analysis tool for Hadoop. The basic framework of Chukwa is shown in Fig. 2. In order to cleanly incorporate existing log files as well as communication protocols, the agents on each node are designed to be highly configurable. In more detail, they use *adaptors*, which are dynamically loadable and configurable modules, to read data from the log file and applications. In addition to periodically querying the status of each adaptor, the agent also sends data across the network. To avoid directly writing data to HDFS by agents, which results in a large number of small files, Chukwa uses several collectors to multiplex the data coming from a large number of agents (e.g., 100). The output file of each collector is stored in the *data sink* directory. Chukwa also supports *fast-path* clients for latency-sensitive applications. These fast-path clients send data via TCP (Transmission Control Protocol) sockets to the requesting process. For the *regular* path, collectors outputs data as the standard Hadoop sequence files. To reduce the number of files and to ease later analysis as well, Chukwa includes an *archiving* MapReduce job to group data by the cluster, date and type.

2.2 Data Presentation

Even we can efficiently collect enough information from different nodes and tasks, it is still very challenging to present this information to users to facilitate the performance analysis. For big data application developers, they want to see the comparison between different applications to efficiently identify the most time-consuming ones (e.g., the longest query of several Hive applications). For platform designers, they want to see the comparison between different phases of the runtime engine to

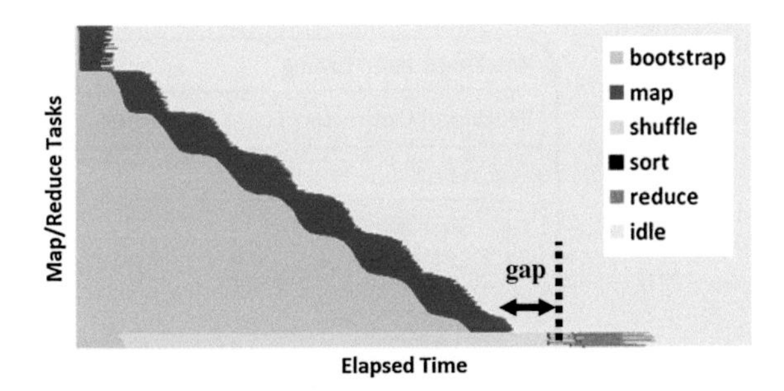

Fig. 3 An example of the visualized view of MapReduce tasks provided by HiTune. The x-axis is the elapse of wall clock time, and each *horizontal line* represents a map or reduce task. It also visualizes the duration of subphases, such as shuffle and sort (*Original source* [7])

propose potential optimization plans. For system architects, in addition to the basic time-varying resource utilization, they also want to see how the resource utilization correlates with different phases, tasks and nodes.

HiTune is a dataflow-based performance analysis tool specifically designed for Hadoop [7]. It can provide visualized view of the Map/Reduce tasks and resource utilization (including CPU, disk, and network utilization). As shown in Fig. 3, which is the dataflow execution of TeraSort application,[1] where the x-axis is the elapse of wall clock time, and each horizontal line represents a map or reduce task. We can clearly see that there exists a gap between the completion of map phase and that of the shuffle phase. Ideally, the shuffle phase can complete as soon as the map tasks complete. Therefore, with such visualization tools we can easily identify the potential performance issues.

2.3 Data Analysis

As shown in Fig. 3, after the visualized execution details of big data applications are presented, only experience experts could efficiently identify the performance issues for optimization. Apparently, the analysis and tuning process is indeed a tedious and iterative process. For example, during the tuning of TeraSort on IBM BigInsights* with IBM POWER7+* system, the performance bottlenecks vary in different tuning cycles, since the change of one parameter may have significant impacts on the others. Actually, such analysis process requires deep understanding of the whole system, and it is not applicable for the inexperienced users. The inexperienced users prefer to get the optimal performance without manual analysis process, which calls for analysis tools that can automatically optimize the performance based on the collected data.

To address this problem, several tools have been proposed. Starfish is one of the most well-known tools, which can automatically tune the vast number of parameters

[1]The application sorts 1TB bytes data consisting of 10 billion 100-byte records.

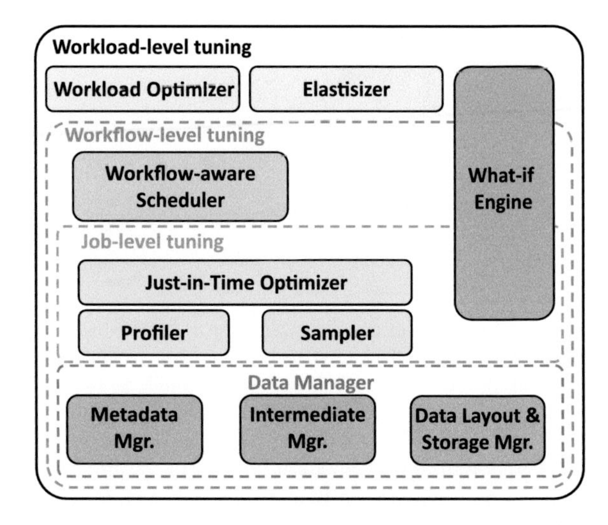

Fig. 4 The overall architecture of Starfish. It consists of three levels tuning, including job-level tuning, workflow-level tuning, and the workload-level tuning (*Original source* [8])

of Hadoop runtime engine to achieve optimal performance [8]. Figure 4 is the overall architecture of Starfish, which contains job-level tuning, workflow-level tuning, and workload-level tuning. In job-level tuning, to find the optimal parameter combinations of the MapReduce job, the *Just-in-Time Optimizer* takes the *Profiler* and *Sampler* as the inputs. The *Profiler* uses instrumentation to construct performance models for the MapReduce job, and the *Sampler* collects the statistics about the input, intermediate, and output key-value spaces of the MapReduce job. Based on the information provided by *Profiler* and *Sampler*, the *What-if engine* predicts the performance of new jobs. Then, the *Just-in-Time Optimizer* efficiently searches for the (near)-optimal Hadoop parameters. In the workflow-level tuning, a *Workflow-aware Scheduler* is provided to schedule tasks in a "data-local" fashion by moving the computation to the underlying distributed filesystems (e.g., HDFS). To achieve this goal, the scheduler closely works with the *What-if Engine* and the *Data Manager*, which manages the metadata, intermediate data, and the data layout and storage. In the workload-level tuning, given a workload containing several workflows, the *Workload Optimizer* produces an equivalent, but optimized workflows. Then, the *Workflow-aware Scheduler* is in charge of scheduling of these optimized workflows.

In summary, due to the above challenges, it is very hard to build a one-size-fit-all performance tool to meet all the requirements of different users. Hence, researchers have proposed several different tools to tackle different challenges. In the following parts, we will introduce two performance tools targeting experienced experts and inexperienced users, respectively.

3 A Performance Tool for Tuning Experts: SONATA

In this section, we introduce a performance tool called as SONATA [9, 10], which has already been integrated into IBM's big data product, i.e., IBM Platform Symphony*. SONATA is mainly designed for facilitating the experienced experts to efficiently tune the performance of big data optimization, which is built upon the MapReduce paradigm.

3.1 Target Users

The main target users of SONATA include the application developers, runtime engine designers, and system architects. The application developers can easily identify whether the user-defined codes (e.g., user-defined map or reduce functions) are efficient. The runtime engine designers mainly focus on the execution efficiency of internal phases (e.g., the shuffle process from map tasks to reduce tasks) of the runtime engine. The system architects primarily pay attentions to the resource utilization of the whole system to determine potential optimization on current system design. A simple example is that once the architects notice that the available memory is very low for running big data applications, expanding the memory capacity has the potential to significantly improve the performance.

3.2 Design Considerations

To meet all the above requirements of different users, SONATA provides four different analysis views to present more insights of the execution of big data applications. The first view is shown in Fig. 5, which is the *overall view* to present the execution timelines of all map and reduce tasks. The overall view reveals execution information of all tasks, for example, how many tasks were running at a given time point and the number of map task waves. The second view is called as *resource view* as shown in Fig. 6, which presents the usage of CPU and memory, the bandwidth of disk and network of each node. From the resource view, abnormal usage of specific nodes can be easily identified. The third view is *breakdown view* to show the execution breakdown of map and reduce tasks, as shown in Fig. 7. The critical phases of one (map/reduce) task can be identified from this view. Moreover, the above three views can be correlated to facilitate identifying the performance bottlenecks. For instance, the user can first select a task in the *overall view*, and then the tasks running on the same node and the resource usage information in the overall view and resource view, respectively, are highlighted. Also, the execution breakdown of the selected task is

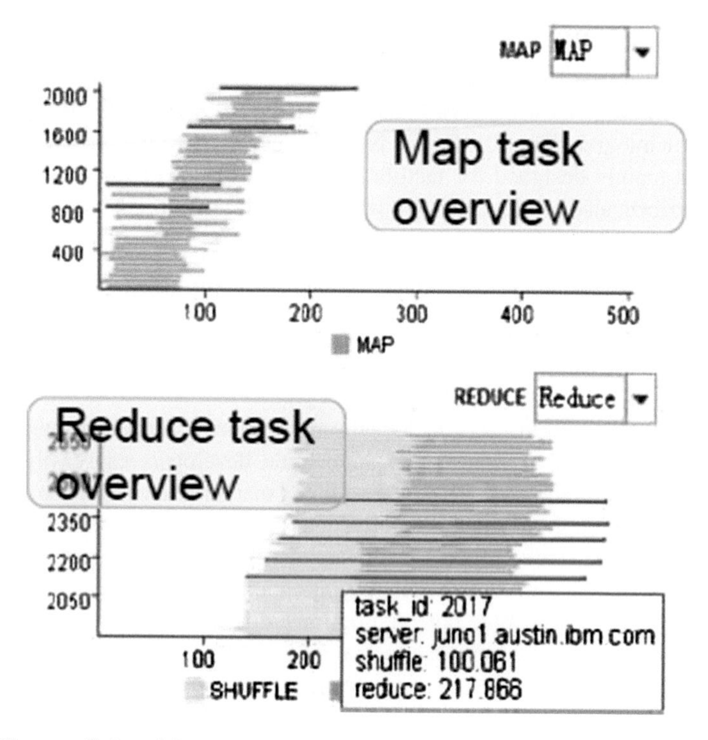

Fig. 5 The *overall view* of SONATA to present the execution timelines of all map and reduce tasks. The overall view reveals execution information of all tasks, for example, how many tasks were running at a given time point and the number of map task waves

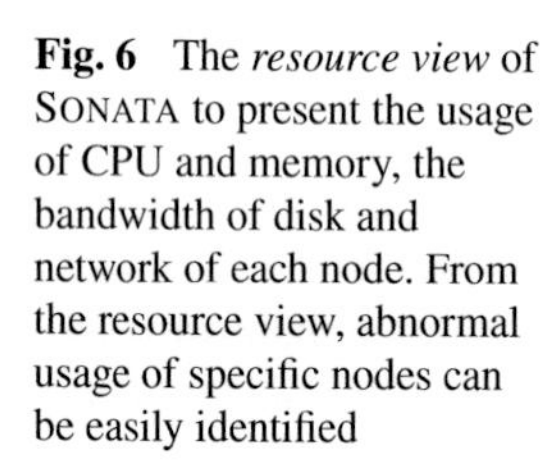

Fig. 6 The *resource view* of SONATA to present the usage of CPU and memory, the bandwidth of disk and network of each node. From the resource view, abnormal usage of specific nodes can be easily identified

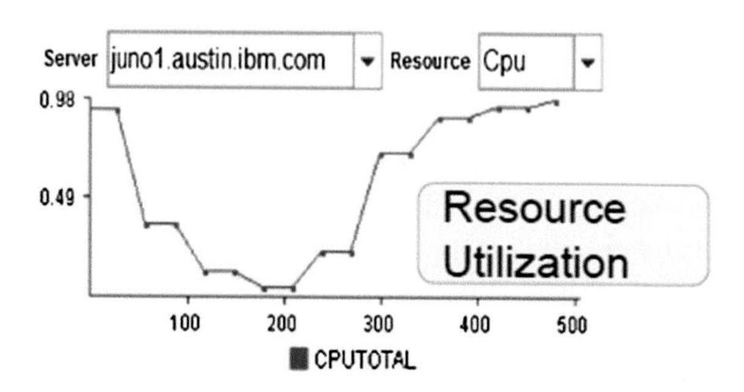

also shown in the breakdown view. The fourth view is the *Statistical Data* as shown in Fig. 8, which lists the detailed statistical information of the entire job, such as, the job status, the average execution time of all reduce tasks, the written bytes of HDFS, etc.

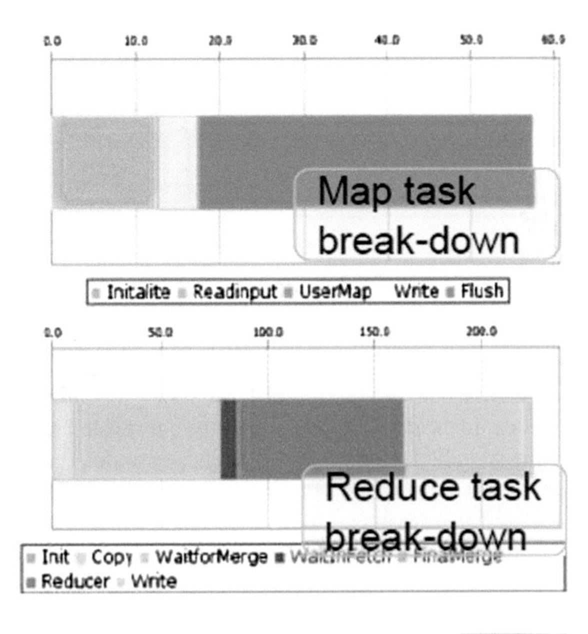

Fig. 7 The *breakdown view* of SONATA to show the execution breakdown of map and reduce tasks, where critical phases of one (map/reduce) task can be easily identified

Category Map

task duration	63650
longest map	128116
shortest map	36557
initialize duration	5
readinput duration	12486
usermap duration	0
write duration	4805
flush duration	40202
spill times	0
sort time	0
spill time	0

Fig. 8 The *Statistical Data* of SONATA that lists the detailed statistical information of the entire job, such as, the job status, the average execution time of all reduce tasks, the written bytes of HDFS, etc.

3.3 Overall Architecture

The overall architecture of SONATA is shown in Fig. 9. SONATA contains four phase, that is, *data collection*, *data loading*, *performance visualization* and *optimization*

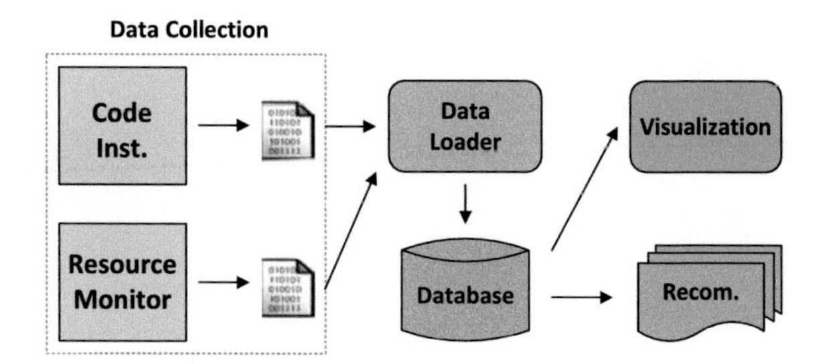

Fig. 9 The overall architecture of SONATA (*Original source* [9])

recommendation. In the *data collection* phase, the execution details and resource usage at each node are collected, and then such information is aggregated at the master node. In the *data loading* phase, the aggregated statistics are periodically retrieved by the so-called *data loader* at the master node. The data loader then write such statistics into the database. In the *performance visualization* phase, first collected runtime statistics are retrieved from the database, and then such information is displayed in a correlated mode through web-based performance visualizer. In the *optimization recommendation* phase, the potential critical outliers can also be recognized. In addition, the corresponding optimization suggestions will also be generated.

3.4 Implementation Details

3.4.1 Data Collection

The *Data Collection* consists of two main components, i.e., *monitor* and *aggregator*, as shown in Fig. 10.

Each node has a monitor for gathering both the execution details of MapReduce applications and the hardware resource usage information. The collection of execution details relies on MapReduce's built-in*counters*, and they are useful mechanism to gather execution statistics and supported by many MapReduce implementations such as Apache Hadoop and IBM Platform Symphony*. More specifically, we first define several new counters for interested execution phases, for instance, the potential critical execution phases, of the MapReduce engine. Then, corresponding codes are instrumented to the source codes of the runtime engine to update the counters at runtime. On the other hand, the hardware resource usage information, e.g., CPU/Memory usage, disk and network bandwidth, is collected by utilizing a lightweight performance monitoring tool called as Ganglia [11]. The original information collected by Ganglia are processed by the monitor to generate organized data for processing by the aggregator.

On the master node, the aggregator collects the data generated from monitors at all slave nodes. The data are organized in the XML format and then stored into a history file sequentially. Figure 11 shows an XML entry of the collected execution

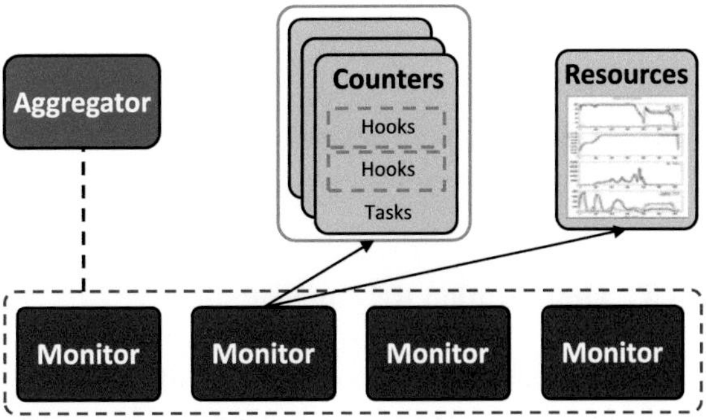

Fig. 10 The components for data collection (*Original source* [9])

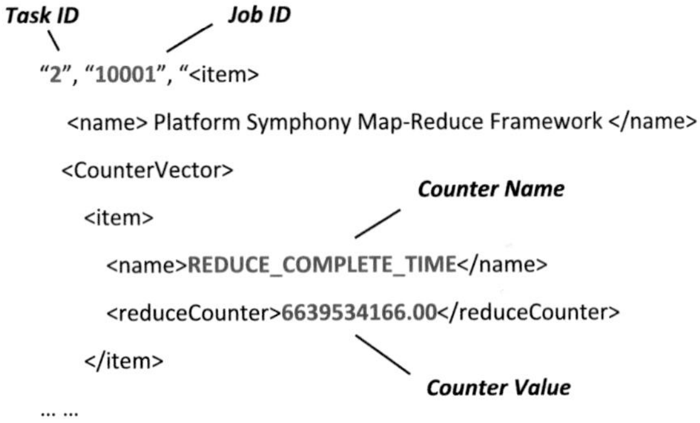

Fig. 11 The REDUCE_COMPLETE_TIME counter is stored in XML format for task 2 in job 10001

```
Task ID          Job ID
"2", "10001", "<item>
  <name> Platform Symphony Map-Reduce Framework </name>
  <CounterVector>                    Counter Name
    <item>
      <name>REDUCE_COMPLETE_TIME</name>
      <reduceCounter>6639534166.00</reduceCounter>
    </item>                          Counter Value
... ...
```

details in the history file. Moreover, as the size of history file increases along with execution time, the aggregator periodically flushed the data in current history files into out-of-date files to reduce the size of the history files.

3.4.2 Data Loading

As the filesystem is not scalable to handle a large amount of data for performance analysis, SONATA further stores the collected data in the history file to the underlying database by using *data loader*. The data loader runs as a daemon on the master node, and it periodically retrieves XML items from the history file and then writes the corresponding records into the database. The data loader starts reading from the last accessed offset of the history file when it is invoked, avoiding writing duplicate records to the database.

In addition, the data loader also collects the data on-the-fly from the *aggregator*, and writes them into the database. In more detail, the collected data is appended with the current timestamp to generate a new record, and the new record is written into the database. As a result, such dynamically updated information can facilitate the on-the-fly visualization and performance monitoring.

3.4.3 Performance Visualization

As stated in Sect. 3.2, in performance visualization, four analysis views can be generated to present more insights of the execution of programs, that is, *overall view*, *resource view*, *breakdown view*, and *statistical data*.

3.4.4 Optimization Recommendation

Although expert users can only rely on the performance visualization to efficiently conduct performance analysis, it still would be very helpful to automatically identify the performance outliers and thus offer optimization recommendations for inexperienced users. To achieve this goal, we embedded several empirical rules into the optimization engine to automatically recommend tuning guidelines. An illustrative example is shown in Fig. 12. By correlated different tasks, we can easily find that only the tasks on the node JUNO2 (e.g., task 2584) are performance outliers, and it implies that there are performance issues on node JUNO2. To determine the root cause of this performance problem, the tasks on JUNO2 can be correlated with its resource usage. As a result, we observe that the CPU usage on JUNO2 is overutilized (i.e., >99 %) while the network access is underutilized (i.e., maximal disk bandwidth is 170 MB/s) compared against other nodes. The observation implies that the CPU of JUNO2 is the bottleneck of the entire execution. Therefore, SONATA offers recommendation that the CPU-related configurations, e.g., the SMT (Simultaneous Multithreading) setting, hardware prefetching, and CPU frequency, etc., on JUNO2 should be checked. By checking such configurations, eventually we determine that this performance problem is caused by setting the number of hardware threads (i.e., SMT) as 2 on JUNO2 while on other nodes the number of hardware threads is set as 4. Figure 13 shows the corresponding empirical rules that are derived from such domain knowledge.

3.5 *User Cases of Performance Analysis*

In this subsection, we introduce several user cases when deploying the performance tool on a production environment. The employed cluster is listed in Table 1. In more detail, the cluster consists of up to 10 POWER7R2 servers (the name of each node begins with JUNO, i.e., from JUNO0 to JUNO9), each contains 2 IBM POWER7* chips, and each chip has 32 hardware threads.

3.5.1 Diagnosing System Problems

In addition to the example in Fig. 12 that shows how SONATA help find misconfigurations of SMT setting, we also present an example that SONATA can also facilitate

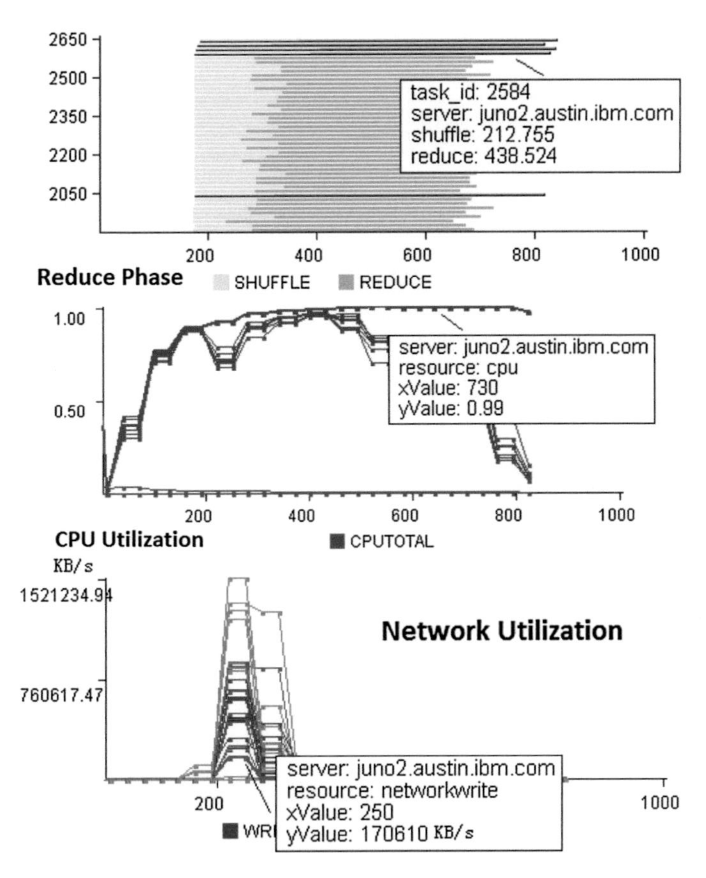

Fig. 12 An illustrative example to diagnosis the hardware configuration problem (*Original source* [9])

```
#define THRESHOLD = 0.2
if (avg_exec_time (NODE_ID) – avg_exec_time(ALL_NODES) / avg_exec_time (ALL_NODES) > THRESHOLD &&
   cpu_util(NODE_ID) > cpu_util(ALL_OTHER_NODES) &&
   mem_util(NODE_ID) < mem_util(ANY_OTHER_NODE) &&
   disk_util(NODE_ID) < disk_util(ANY_OTHER_NODE) &&
   net_util(NODE_ID) < net_util(ANY_OTHER_NODE)) {
      Check the configuration of CPU such as SMT/Prefetching/Frequency setting on Node ID.
}
```

Fig. 13 The example in Fig. 12 can be summarized as an empirical rule for automatic optimization recommendation (*Original source* [9])

Table 1 Evaluated IBM POWER7* cluster

Components	Configuration
Cluster	10 PowerLinux 7R2 Servers
CPU	16 processor cores per server (160 in total)
Memory	128 GB per server (1280 GB in total)
Internal storage	6 * 600 GB internal SAS drives per server (36TB in total)
Expansion storage	24 * 600 GB SAS drives in IBM EXP24S SFF Gen2-bay Drawer per server (144TB in total)
Network	2 10Gbe connections per server
Switch	BNT BLADE RackSwitch G8264

detecting an abnormal disk behavior. As shown in Fig. 14, from the *overall view*, we observe that several reduce tasks on node JUNO1 are outliers that determine the overall execution time. Then, we correlate the tasks with the disk usage of different nodes, and we found that the total disk bandwidth on JUNO1 (i.e., <1.72 GB/s) is much larger than that of other nodes (e.g., <864 MB/s on node JUNO9).

By investigating the execution status of the cluster, we finally identify that some other users had run a disk benchmark on node JUNO1 at the same time, as the entire cluster is shared by multiple teams. Then, we obtain exclusive access to this cluster to obtain accurate performance benchmark results.

3.5.2 Tuning Runtime Configurations

Traditionally, there exists many adjustable configurations in MapReduce platforms such as Apache Hadoop and IBM Platform Symphony*. Here we show an example that SONATA can help optimize the buffer usage of map tasks, as shown in Fig. 15. One of the critical tasks, task 402, can be easily identified by correlating different tasks. During the execution of task 402, there exists 2 sorts an spills, and the resultant overhead is more than 49 % of the overall map time. Since the overhead of sorts and spills is larger than a predefined threshold (e.g., 20 %), SONATA offers recommendation that the buffer-related parameters (e.g., *io.sort.mb*), should be increased to reduce the sorts and spills. Thus, by increasing this parameter, we completely eliminate sorts and spills during the map execution, and the overall performance is also significantly improved.

3.5.3 Improving Runtime Efficiency

SONATA is also used for optimizing the implementation of MapReduce runtime engines. In traditional Apache Hadoop runtime engine, the map and reduce slots are only used for running the map and reduce tasks, respectively. Similar to Apache

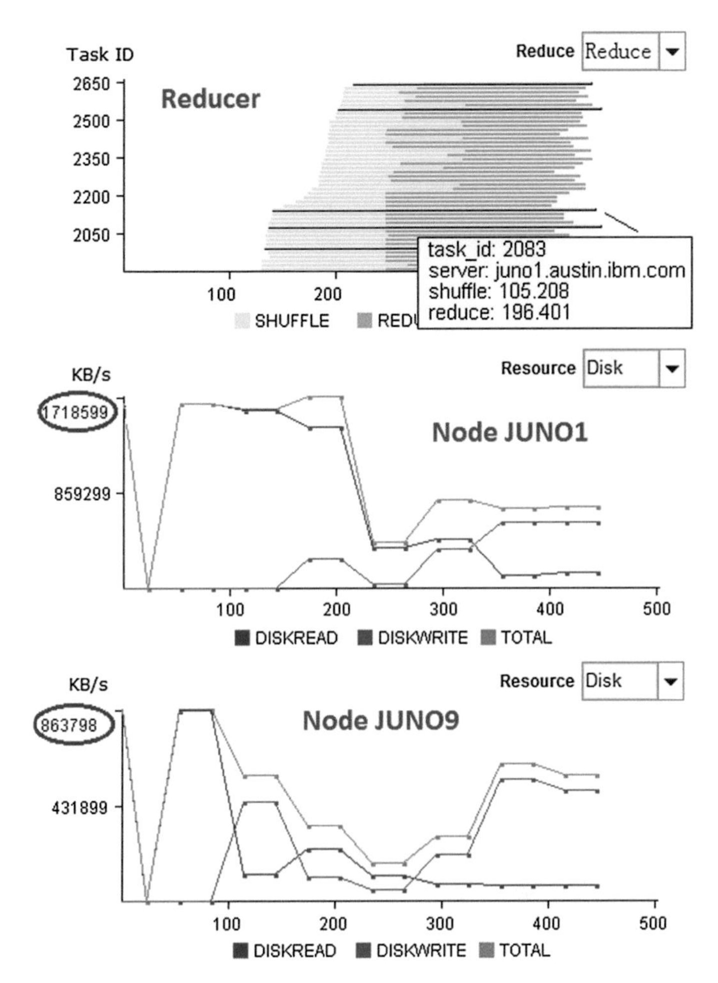

Fig. 14 SONATA can be used to diagnose disk problem. In this example, the disk bandwidth of node JUNO1 is much higher than that of other nodes such as node JUNO9 (*Original source* [9])

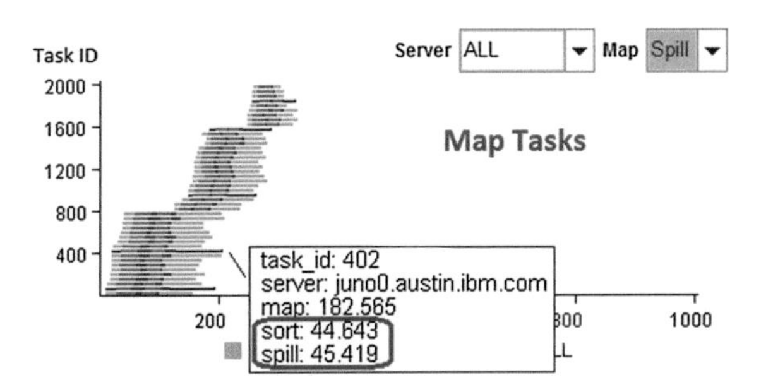

Fig. 15 SONATA can be used for optimizing runtime parameters. In this example, multiple costly sorts and spills occur in the outlier map task (i.e., task 402). Such sorts and spills are eliminated by enlarging the runtime parameters such as *io.sort.mb* (*Original source* [9])

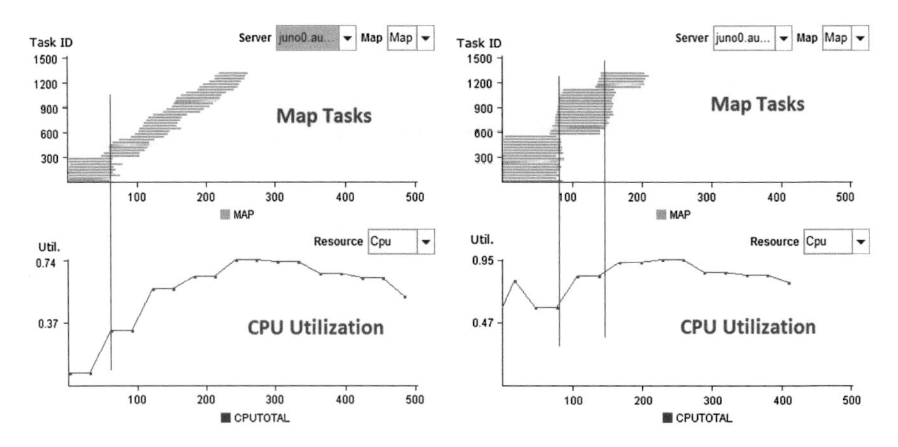

Fig. 16 Terasort running with separate and shared slots, respectively (*Original source* [9])

Hadoop, IBM Platform Symphony* also does not allow slots shared between the map and reduce tasks by default. Under this circumstance, we set the number of map and reduce slots as 32 for each node with 64 hardware threads, and the corresponding execution behaviors of Terasort are shown in the left part of Fig. 16. During the first wave of map tasks, we observe that the CPU utilization is relatively low, for example, the CPU utilization of node JUNO0 is less than 37 %.

To improve the CPU utilization, in addition to default *separate* slot setting, IBM Platform Symphony* also provides the *generic* slot that can be shared between the map and reduce tasks. By setting all the slots as generic slots, the CPU utilization can be improved to more than 50 % for node JUNO0, as shown in the right part of Fig. 16. Overall, the generic slots reduce the execution time of this job by 12.99 % than the separate slots (422s vs. 485s).

3.6 Other Performance Analysis Tools

In this subsection, we introduce several other performance analysis tools and systems.

Mantri can monitor MapReduce tasks and cull outliers by using cause- and resource-aware techniques [12]. Mantri classifies the root causes for MapReduce outliers into three categories, that is, *machine characteristics*, *network characteristics*, and *work imbalance*. After identifying the root causes, Mantri can efficiently handle outliers by evaluating the costs of specific actions such as restart and placement. By deploying Mantri on Microsoft Bing's production clusters, Mantri can reduce job completion time by 32 %.

As stated before, HiTune is a scalable, lightweight and extensible performance analyzer for Hadoop [7]. HiTune is built upon distributed instrumentations and dataflow-driven performance analysis. More specifically, HiTune correlates concurrent performance activities of different tasks and computation nodes, reconstructs the dataflow-based process of the MapReduce application, and relates the resource usage to the dataflow model. HiTune is mainly used for post-execution performance analysis and optimization.

Theia is a visualization tool to present the performance of MapReduce jobs in a compact representation [13]. More specifically, Theia provides *visual signatures* that contain the task durations, task status, and data consumption of jobs, and they can be used to facilitate troubleshooting performance problems such as hardware failure, data skew, and software bugs.

Hadoop Vaidya is a rule-based performance diagnostic tool for MapReduce jobs [14]. Vaidya analyzes the collected execution statistics from the job history and configuration files by using several predefined test rules. Such rules can not only identify root causes of performance problems, but also provide optimization suggestions to the end-users.

PerfXPlain is a system that enables end-users to ask performance-related questions such as the relative performances (in terms of execution time) of pairs of MapReduce jobs [15]. Given a specific performance question, PerfXPlain automatically produces an corresponding explanation. PerfXPlain works by using a new query language to form the performance-related query and an algorithm to generate explanations from logs of past MapReduce executions.

4 An Easy-of-Use Full-System Auto-Tuning Tool: Turbo

In this section, we discuss a performance tool, called Turbo, that can *automatically* conduct performance optimization without human intervention, which mainly targets to the inexperienced users.

4.1 Target Users

In contrast to SONATA targeting to tuning experts, Turbo is designed for the users that do not have a deep understanding of the detailed mechanism of deployed big data platforms. For example, programmers of big data query may not know the implementation details of different big data platforms, even such platforms provide the same application-programming interfaces (API) for compatible purpose. Moreover, even for the experienced tuning experts, the performance tuning process is also very tedious and time-consuming, which also calls for an automatic tool to speedup the tuning process, especially in the early tuning stages. Actually, with the help of Turbo, near-optimal performance can be achieved efficiently without human intervention.

4.2 Design Considerations

There are two design considerations for this tool. First, compared with the performance analysis tool introduced in the previous sections, inexperienced users may expect a tool to automatically tune the performance of big data optimization without detailed analysis on the execution behaviors. Second, the performance gain should be notable enough compared with baseline after the deployment of this tool. From our perspective, the performance of big data optimization is determined by configurations at various levels, including, MapReduce runtime engine, JVM (Java Virtual Machine), operating system, virtual machine and underlying hardware. Therefore, it is expected that a performance tool, which can automatically optimize the adjustable parameters at different levels, would be well received for the users.

4.3 Overall Architecture

The overall architecture of the proposed full-system performance auto-tuning tool is shown in Fig. 17. The tool receives the profiling data from previously executed jobs as inputs. Then, it automatically optimizes the whole system at different levels. The first level is the system-level performance tuning, where the system tuner optimizes the underlying hardware (e.g., SMT/hardware prefetcher/frequency/network adaptor, etc.) and operating system (e.g., scheduling policy, huge page, etc.). The second level is the JVM-level performance tuning, where the JVM tuner automatically appends the optimal JVM configurations (e.g., garbage collection policy, JIT policy, and lock contention, etc.) to the JVM of MapReduce jobs. The third level is the runtime engine level, which consists of a performance model and an optimizer.

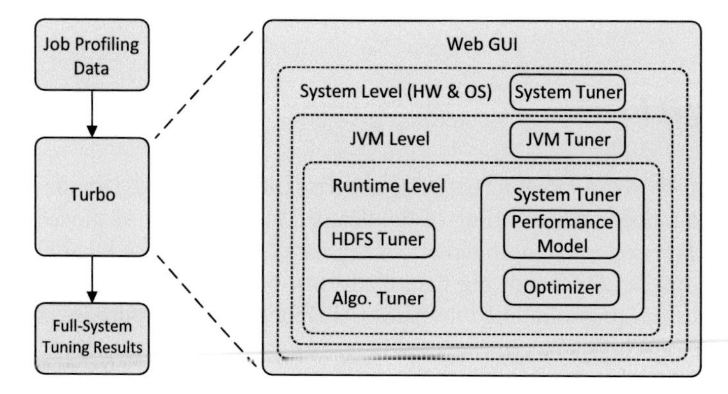

Fig. 17 The overall architecture of a full-system performance auto-tuning tool, which is called as Turbo

The performance model is to mimic the behavior of MapReduce execution in real clusters, and the optimizer is to efficiently find the optimal configurations based on the predicted results of the performance model.

4.4 Implementation Details

We introduce the detailed implementation of the tuning policy for different levels, including the system level, the JVM level and the runtime level. The basic idea is that the profiled data of previous jobs can be utilized to determine the optimal configurations for later execution of jobs.

4.4.1 System Level

At the system level, Turbo focuses on tuning the adjustable parameters of hardware and operating systems. The underlying hardware always provides different configurations for advanced users. For example, existing processors employ hardware prefetching to hide the memory access latency, which would be critical to the performance of big data applications. Hardware prefetchers have different perfecting policies, e.g., IBM POWER7* processor provides a user-programmable register to configure more than 10 different prefetching configurations.[2] According to Victor et al. [16], dynamically changing the prefetching policy of the IBM POWER7* processor can result in as much as 2.7x performance speedup. To exploit the advantage of configurable hardware prefetching, in the profiling run, the execution statistics (e.g., cache miss rate, memory accesses, instruction-per-cycle, etc.) are collected by using system-provided performance counters. Then, some empirical rules are employed to determine the optimal prefetching setting based on execution statistics. After that, in the later executions, the optimal prefetching setting can be applied to improve the execution performance.

There also exists some other hardware settings, such as SMT, which provides multiple independent threads to improve the resource utilization of modern superscalar processors. However, sometimes SMT could be detrimental when performance is limited by application scalability or when there is significant contention for processor resources [17]. In other words, the performance gains from SMT vary greatly for different applications.[3] Therefore, it is very necessary to determine the optimal SMT setting for different applications. Actually, the above methodology used for finding the optimal configuration for hardware prefetching can also be applied to SMT optimization.

[2]Intel processors also provide different prefetching policies such as DPL (Data Prefetch Logic prefetcher) and ACL (Adjacent Cache Line prefetcher).

[3]The performance gains from SMT vary from 0.5x to 2.5x for evaluated applications in [17].

At the operating system level, the page-based virtual memory has significant impacts on the performance of big data applications. Some big data applications spend 51 % of the execution cycles to service the TLB (Translation Lookaside Buffer) misses with a small size of page [18]. To improve the performance, Turbo selects the large size of page for big data applications, and automatically determines the total amount of memory configured with large page mode according to the size of input data.

4.4.2 JVM Level

It is non-trivial to tune the Java runtime system due to a large number of settings of JVM. Actually, the settings related with the the memory management and garbage collection, such as the size of Java Heap, GC algorithm, and GC generation etc., are most important to the performance of JVM. In order to find the optimal JVM settings, Turbo employs iterative optimization techniques [19] to search for the best possible combination of JVM setting for a given workload. Traditionally, iterative optimization always requires a larger number of training runs for finding the best combination. To reduce the training overheads, Turbo distributes tasks with different JVM settings to different computing nodes, collects the runtime statistics at the master, and then evolves to the optimal combination of JVM setting. Therefore, the optimal JVM setting can be found with a small number profiling runs, which makes the Turbo practical in industrial usage.

4.4.3 Runtime Level

At the runtime level, Turbo considers to optimize the MapReduce runtime engine, which typically contains more than 70 performance-critical parameters [20]. In order to efficiently explore the huge parameter space, Turbo is built upon a cost-based optimization to automatically determine the optimal configurations [21]. To achieve this, Turbo consists of three main components, that is, statistic profiler, performance model, and optimizer.

The statistic profiler collects the job profiles, and a job profile consists of two parts, as dataflow and cost statistics. The dataflow statistics include the information regarding the number of bytes and key-value pairs flowing throw different tasks and phases of a MapReduce job execution, e.g., the number of map output bytes. The cost statistics contain the execution timing, e.g., the average time to read a record from file systems. Such statistic information is served as the inputs of the performance model, which is specifically designed for IBM Platform Symphony*. The overall process of the performance model is shown in Fig. 18, which uses the profiling information of job j to estimate the performance of a new job j'. At first, both the profiling information of job j and the information (i.e., input data, resources and configurations information) are treated as inputs. The basic intuition is that the dataflow of the job's phases is proportional to the input data size, which indicates that the dataflow

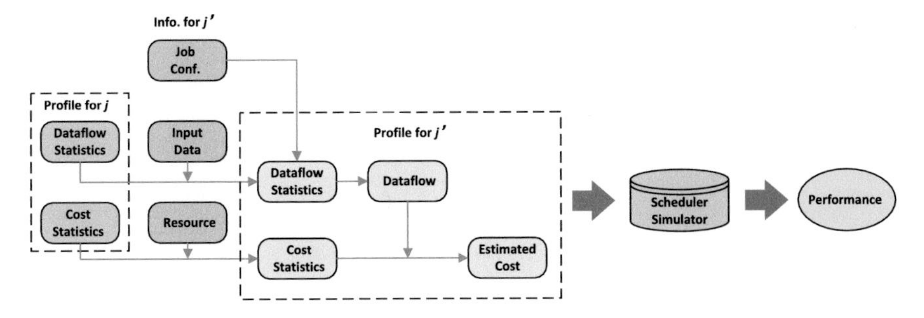

Fig. 18 The overall process to predict the performance of virtual jobs

statistics of job j' can be scaled from that of job j according to their input data. Once the costs of job j' are obtained, the task- and phase-level timing information can also be derived. To obtain the overall execution time, a task scheduler simulator is employed to simulate the scheduling and execution of map and reduce tasks of j'.

Based on the predicted performance of jobs under different configurations, the optimal configuration could be found by searching the entire parameter space. However, since the space is exponentially large, Turbo employs some heuristics searching algorithms, e.g., recursive random search (RSS), to efficiently find the optimal configuration.

4.5 Industrial Use Cases

Here we present a real industrial case to demonstrate the advantage of the autotuning performance tool. In this case, the hardware information is listed in Table 2. The testing application is a real application from the telecommunication companies, which is used for analyzing the web access behaviors of mobile devices.

In this cluster, the proposed tool can automatically achieve 2.4x performance speedup compared with native execution, as shown in Fig. 19. The baseline performance of the job is 845 s. After running the autotuning tool, the performance is

Table 2 Hardware configuration of a real industrial case

Components	Configuration
Cluster	4 PowerLinux 7R1 Servers
CPU	8 processor cores per server (32 in total)
Memory	64 GB per server (256 GB in total)
Storage	24 * 600 GB SAS drives in IBM EXP24S SFF Gen2-bay Drawer per server (144TB in total)
Network	2 10Gbe connections per server

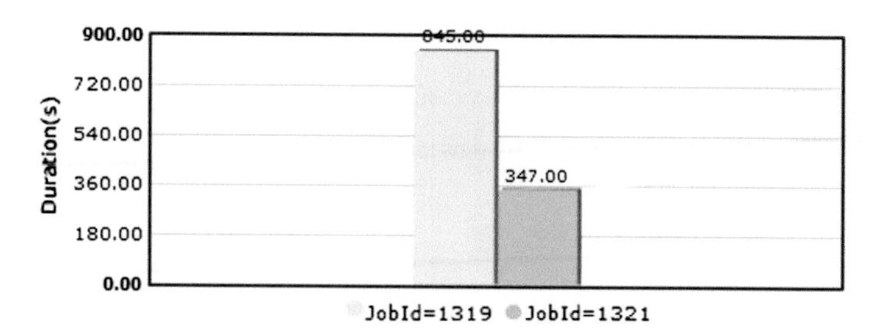

Fig. 19 The tuning results of Turbo. The baseline performance of the job is 845 s, and the automatically tuned performance is 347 s. Turbo achieves 2.3x performance speedup, and this result is about 91 % of the expert's tuning

Table 3 Optimal configurations found at different levels

System level	
SMT setting	SMT4
Prefetching setting	0x1D
MTU	9000
Hugh page	50 GB per Node
JVM level	
-server -Xnoclassgc -Xgcpolicy:gencon -Xjit:optLevel=hot	
-Xjit:disableProfiling -Xgcthreads4 -XlockReservation	
Runtime engine level	
pmr.ondemand.2nd.sort.mb	true
mapred.reduce.slowstart.completed.maps	0.84
io.sort.mb	268
mapred.reduce.tasks	768
compression codec	LZ4

improved to 347 s. Actually, the tuning results with expert's help is 317 s. Therefore, the tunning tool can achieve about 91 % of expert's tuning.

In addition, Table 3 shows the optimal configurations found at different levels. At the system level, the SMT and hardware prefetching should be set as 4 and 0x1D, respectively. The MTU of network adaptor should be set as 9000 and the huge page of OS should be configured with 50 GB to match the size of input data. At the JVM level, most configurations that are closely related to garbage collection should be set. At the runtime engine level, Turbo also automatically find the optimal configurations for different parameters. For example, the size of *io.sort.mb*[4] should be set as 268 MB, the number of reduce tasks should be set as 768, and LZ4 [22] should be used for compressing the intermediate data.

[4]The total amount of buffer memory to use while sorting files, in megabytes.

4.6 Other Auto-Tuning Tools

Autotuning tool for the performance optimization of big data applications have received increasing attentions from both academic and industrial practitioners. Here we briefly introduce some other autotuning tools.

Babu proposed a competition-based approach to automate the setting of tuning parameters for MapReduce applications. This basic idea of this approach is to start multiple instances of the same task with different setting of parameters, and then the configurations of the best instance will be identified [23].

Starfish is a self-tuning system that allows Hadoop users to obtain good performance without deep understanding of many tuning parameters. Starfish features the automatic setting of near-optimal parameters for Hadoop jobs. Actually, the engine-level tuning of Turbo extends Starfish's optimization framework for IBM Platform Symphony*, since Starfish mainly focuses on the tuning of the parameters of Apache Hadoop runtime framework [8]. In addition to the job-level autotuning, as stated before, Starfish also contains workflow-level tuning and workload-level tuning. The workflow-level tuning focuses on how to schedule tasks in a *data-local* fashion by moving the computation to the underlying distributed filesystems (e.g., HDFS). The workload-level tuning produces an equivalent, but optimized workflows, and passes them to the workflow-aware scheduler for execution, given a workload consisting of a collection of workflow.

Intel has proposed a machine-learning based auto-tunning tool for MapReduce applications [24]. This approach leverages support vector regression model (SVR) to learn the relationship between the job parameters and job execution time, and then uses smart search algorithm to explore the parameter space. The proposed autotuning flow involves two phases: building and parameter optimization. In the first phase, training data are used to *learn* the black-box performance model. By evaluating various machine learning techniques (e.g., linear regression, artificial neural networks, model trees, etc.), the authors claim that support vector regression model (SVR), has both good accuracy and computational performance. To reduce the number of training samples to build the model, smart sampling is employed. In the second phase, since the performance model has already been built, the auto-tuner generates a parameter search space and explores it using smart search techniques, and it uses the performance model to find the parameter configuration that has the best predicted performance.

Zhang et al. have proposed an approach, called *AutoTune*, to optimize the workflow performance through adjusting the number of reduce tasks [25]. *AutoTune* contains two key components: *the ensemble of performance models* to estimate the duration of a MapReduce program for processing different datasets, and *optimization strategies* that are used for determining the numbers of reduce tasks of jobs in the MapReduce workflow to achieve specific performance objective. According to the experiments conducted on realistic MapReduce applications as TPC-H queries and programs mining a collection of enterprise web proxy logs, *AutoTune* can significantly improve the performance compared with *rules of thumb* of the setting of reducer numbers.

5 Conclusion

Many performance tools have been proposed to accelerate big data optimization. In this chapter, we first introduce the requirements of ideal performance tools. Then, we present the challenges of design and deployment of performance tools in practice. Finally, we show two examples of state-of-the-art performance tools for big data optimization. The first performance tool is called as SONATA that targets at tuning experts. The second performance tool is called as Turbo that can automatically optimize MapReduce applications without human intervention. Both tools can significantly improve the runtime efficiency of big data optimization applications.

*Trademark, service mark, or registered trademark of International Business Machines Corporation in the United States, other countries, or both.

References

1. IBM Power Server. http://www-03.ibm.com/systems/power/hardware/. Accessed June 2015
2. IBM Power Linux. http://www-03.ibm.com/systems/power/software/linux/. Accessed June 2015
3. IBM Java Development Kit. http://www.ibm.com/developerworks/java/jdk/. Accessed June 2015
4. IBM InfoSphere BigInsight. http://www-01.ibm.com/software/data/infosphere/biginsights/. Accessed June 2015
5. IBM Platform Symphony. http://www-03.ibm.com/systems/platformcomputing/products/symphony/. Accessed June 2015
6. Chukwa. https://chukwa.apache.org/. Accessed June 2015
7. Dai, J., Huang, J., Huang, S., Huang, B., Liu, Y.: HiTune: dataflow-based performance analysis for big data cloud. In: Proceedings of the 2011 USENIX Annual Technical Conference (USENIX ATC'11) (2011)
8. Herodotou, H., Lim, H., Luo, G., Borisov, N., Dong, L., Cetin, F.B., Babu, S.: Starfish: a self-tuning system for big data analytics. In: Proceedings Biennial Conference on Innovative Data Systems Research (CIDR'11), pp. 261–272 (2011)
9. Guo, Q., Li, Y., Liu, T., Wang, K., Chen, G., Bao, X., Tang, W.: Correlation-based performance analysis for full-system MapReduce optimization. In: Proceedings of the IEEE International Conference on Big Data (BigData'13), pp. 753–761 (2013)
10. Li, Y., Wang, K., Guo, Q., Zhang, X., Chen, G., Liu, T., Li, J.: Breaking the boundary for whole-system performance optimization of big data. In: Proceedings of the International Symposium on Low Power Electronics and Design (ISLPED'13), pp. 126–131 (2013)
11. Ganglia. http://ganglia.sourceforge.net/. Accessed June 2015
12. Ananthanarayanan, G., Kandula, S., Greenberg, A., Stoica, I., Lu, Y., Saha, B., Harris, E.: Reining in the outliers in map-reduce clusters using Mantri. In: Proceedings of the 9th USENIX Conference on Operating Systems Design and Implementation (OSDI'10), pp. 1–16 (2010)
13. Garduno, E., Kavulya, S.P., Tan, J., Gandhi, R., Narasimhan, P.: Theia: visual signatures for problem diagnosis in large hadoop clusters. In: Proceedings of the 26th International Conference on Large Installation System Administration: Strategies, Tools, and Techniques (LISA'12), pp. 33–42 (2012)
14. http://hadoop.apache.org/docs/r1.2.1/vaidya.html. Accessed June 2015
15. Khoussainova, N., Balazinska, M., Suciu, D.: PerfXplain: debugging MapReduce job performance. Proc. VLDB Endow. **5**(7), 598–609 (2012)

16. Jimnez, V., Cazorla, F.J., Gioiosa, R., Buyuktosunoglu, A., Bose, P., O'Connell, F.P., Mealey, B.G.: Adaptive prefetching on POWER7: improving performance and power consumption. ACM Trans. Parallel Comput. (TOPC) **1**(1), Article 4 (2014)
17. Funston, J.R., El Maghraoui, K., Jann, J., Pattnaik, P., Fedorova, A.: An SMT-selection metric to improve multithreaded applications' performance. In: Proceedings of the 2012 IEEE 26th International Parallel and Distributed Processing Symposium (IPDPS'12), pp. 1388–1399 (2012)
18. Basu, A., Gandhi, J., Chang, J., Hill, M.D., Swift, M.M.: Efficient virtual memory for big memory servers. In: Proceedings of the 40th Annual International Symposium on Computer Architecture (ISCA'13), pp. 237–248 (2013)
19. Chen, Y., Fang, S., Huang, Y., Eeckhout, L., Fursin, G., Temam, O., Wu, C.: Deconstructing iterative optimization. ACM Trans. Archit. Code Optim. (TACO) **9**(3), Article 21 (2012)
20. Li, M., Zeng, L., Meng, S., Tan, J., Zhang, L., Butt, A.R., Fuller, N.: MRONLINE: MapReduce online performance tuning. In: Proceedings of International Symposium on High-performance Parallel and Distributed Computing (HPDC'14), pp. 165–176 (2014)
21. Herodotou, H., Babu, S.: Profiling, what-if analysis, and cost-based optimization of MapReduce programs. Proc. VLDB Endow. **4**(11), 1111–1122 (2011)
22. LZ4. https://code.google.com/p/lz4/. Accessed June 2015
23. Babu, S.: Towards automatic optimization of MapReduce programs. In: Proceedings of the ACM Symposium on Cloud Computing (SoCC'10), pp. 137–142 (2010)
24. Yigitbasi, N., Willke, T.L., Liao, G., Epema, D.: Towards machine learning-based Auto-tuning of MapReduce. In: Proceedings of the 2013 IEEE 21st International Symposium on Modelling, Analysis and Simulation of Computer and Telecommunication Systems (MASCOTS'13), pp. 11–20 (2013)
25. Zhang, Z., Cherkasova, L., Loo, B.T.: AutoTune: optimizing execution concurrency and resource usgae in MapReduce workflows. In: Proceedings of the International Conference on Autonomic Computing (ICAC'13), pp. 175–181 (2013)

Author Biographies

Yan Li is the senior program manager in Microsoft. She is responsible for Microsoft big data product feature definition and customer relationship management. Before 2014, she has been working for 8 years in IBM China Research Lab as research staff member. Her research interests are focusing on big data/distributed system/parallel computation and programming languages. She especially has strong interest on open source big data projects, like Hadoop/Spark/Rayon. Her paper on big data has been accepted by top conference like IPDPS, IOD and ISLPED and also published on international journals.

Qi Guo is currently a Postdoctoral Research Associate at Department of Electric and Computer Engineering, Carnegie Mellon University. He obtained the Ph.D. degree from Institute of Computing Technology, Chinese Academy of Sciences in 2012. His main research interests are big data system, computer architecture, and high performance computing.

Guancheng Chen is a Research Staff Member in IBM Research—China. His research interest in computer architecture lies in improving efficiency, programmability and reliability of massive-scale distributed systems which are composed of multicore, GPU, FPGA and other hardware accelerators running big data workloads such as Hadoop, Spark etc. These interests branch into performance analysis tools, parallel programming models and other related system software including cloud, VM, containers, operating systems, compiler, run-time, middleware etc.

Optimising Big Images

Tuomo Valkonen

Abstract We take a look at big data challenges in image processing. Real-life photographs and other images, such ones from medical imaging modalities, consist of tens of million data points. Mathematically based models for their improvement—due to noise, camera shake, physical and technical limitations, etc.—are moreover often highly non-smooth and increasingly often non-convex. This creates significant optimisation challenges for the application of the models in quasi-real-time software packages, as opposed to more ad hoc approaches whose reliability is not as easily proven as that of mathematically based variational models. After introducing a general framework for mathematical image processing, we take a look at the current state-of-the-art in optimisation methods for solving such problems, and discuss future possibilities and challenges.

1 Introduction: Big Image Processing Tasks

A photograph taken with current state-of-the-art digital cameras has between 10 and 20 million pixels. Some cameras, such as the semi-prototypical Nokia 808 PureView have up to 41 million sensor pixels. Despite advances in sensor and optical technology, technically perfect photographs are still elusive in demanding conditions—although some of the more artistic inclination might say that current cameras are already too perfect, and opt for the vintage. With this in mind, in low light even the best cameras however produce noisy images. Casual photographers also cannot always hold the camera steady, and the photograph becomes blurry despite advanced shake reduction technologies. We are thus presented with the challenge of improving the photographs in post-processing. This would desirably be an automated process, based on mathematically well understood models that can be relied upon to not introduce undesired artefacts, and to restore desired features as well as possible.

T. Valkonen (✉)
Department of Applied Mathematics and Theoretical Physics,
University of Cambrige, Cambrige, UK
e-mail: tuomo.valkonen@iki.fi

A. Emrouznejad (ed.), *Big Data Optimization: Recent Developments and Challenges*, Studies in Big Data 18, DOI 10.1007/978-3-319-30265-2_5

The difficulty with real photographs of tens of millions of pixels is that the resulting optimisation problems are huge, and computationally very intensive. Moreover, state-of-the-art image processing techniques generally involve non-smooth *regularisers* for the modelling of our prior assumptions of what a good photograph or image looks like. This causes further difficulties in the application of conventional optimisation methods. State-of-the-art image processing techniques based on mathematical principles are only up to processing tiny images in real time. Further, choosing the right parameters for simple Tikhonov regularisation models can be difficult. Parametrisation can be facilitated by computationally more difficult iterative regularisation models [70] with easier parametrisation, or through parameter learning [39, 41, 57, 77]. These processes are computationally very intensive, requiring processing the data for multiple parameters in order to find the optimal one. The question now is, can we design fast optimisation algorithms that would make this and other image processing tasks tractable for real high-resolution photographs?

Besides photography, big image processing problems can be found in various scientific and medical areas, such magnetic resonance imaging (MRI). An example is full three-dimensional diffusion tensor MRI, and the discovery of neural pathways, as illustrated in Fig. 1. I will not go into physical details about MRI here, as the focus of the chapter is in general-purpose image processing algorithms, not in particular applications and modelling. It suffices to say that diffusion tensor imaging [127] combines multiple diffusion weighted MRI images (DWI) into a single tensor image u. At each point, the tensor $u(x)$ is the 3×3 covariance matrix of a Gaussian probability distribution for the diffusion direction of water molecules at x.

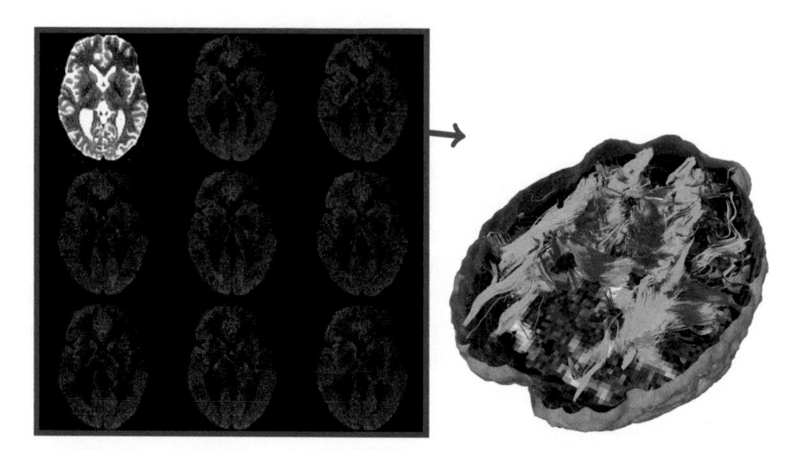

Fig. 1 Illustration of diffusion tensor imaging and tractography process. Multiple diffusion-weighted MRI images with different diffusion-sensitising gradients are first taken (*left*). After processing a tensor field that can be produced from these, neural pathways can be discovered through the tractography process (*right*). (*The author would like to thank Karl Koschutnig for the raw data, and Kristian Bredies for producing with DSI Studio the tractography image with from the diffusion tensor image computed by the author*)

Current MRI technology does not have nearly as high resolution as digital cameras; a $256 \times 256 \times 64$ volume would be considered to have high resolution by today's standards. However, each tensor $u(x)$ has six elements. Combined this gives 25 million variables. Moreover, higher-order regularisers such as TGV^2 [19], which we will discuss in detail in Sect. 3, demand additional variables for their realisation; using the PDHGM (Chambolle-Pock) method, one requires 42 variables per voxel x [135], bringing the total count to 176 million variables. Considering that a double precision floating point number takes eight bytes of computer memory, this means 1.4 GB of variables. If the resolution of MRI technology can be improved, as would be desirable from a compressed sensing point of view [1], the computational challenges will grow even greater.

Due to sparsity, modelled by the geometric regularisers, imaging problems have structure that sets them apart from general big data problems. This is especially the case in a compressed sensing setting. Looking to reconstruct an image from, let's say, partial Fourier samples, there is actually very little *source data*. But the solution that we are looking for is *big*, yet, in a sense, *sparse*. This, and the poor separability of the problems, create a demand for specialised algorithms and approaches. Further big data challenges in imaging are created by convex relaxation approaches that seek to find global solutions to non-convex problems by solving a relaxed convex problem in a bigger space [34, 79, 107, 109, 110, 112]. We discuss such approaches in more detail in Sect. 7.1.

Overall, in this chapter, we review the state of the art of optimisation methods applicable to typical image processing tasks. The latter we will discuss shortly. In the following two sections, Sects. 2 and 3, we then review the typical mathematical *regularisation of inverse problems* approach to solving such imaging problems. After this, we look at optimisation methods amenable to solving the resulting computational models. More specifically, in Sect. 4 we take a look at first-order methods popular in the mathematical imaging community. In Sect. 5 we look at suggested second-order methods, and in Sect. 6 we discuss approaches for the two related topics of problems non-linear forward operators, and iterative regularisation. We finish the chapter in Sect. 7 with a look at early research into handling larger pictures through decomposition and preconditioning techniques, as well as the big data challenges posed by turning small problems into big ones through convex relaxation.

1.1 Types of Image Processing Tasks

What kind of image processing tasks there are? At least mathematically the most basic one is the one we began with, *denoising*, or removing noise from an image. For an example, see Fig. 2. In photography, noise is typically the result of low light conditions, which in modern digital cameras causes the CCD (charge-coupled device) sensor array of the camera to not be excited enough. As a result the sensor images the electric current passing through it. Within the context of photography, another

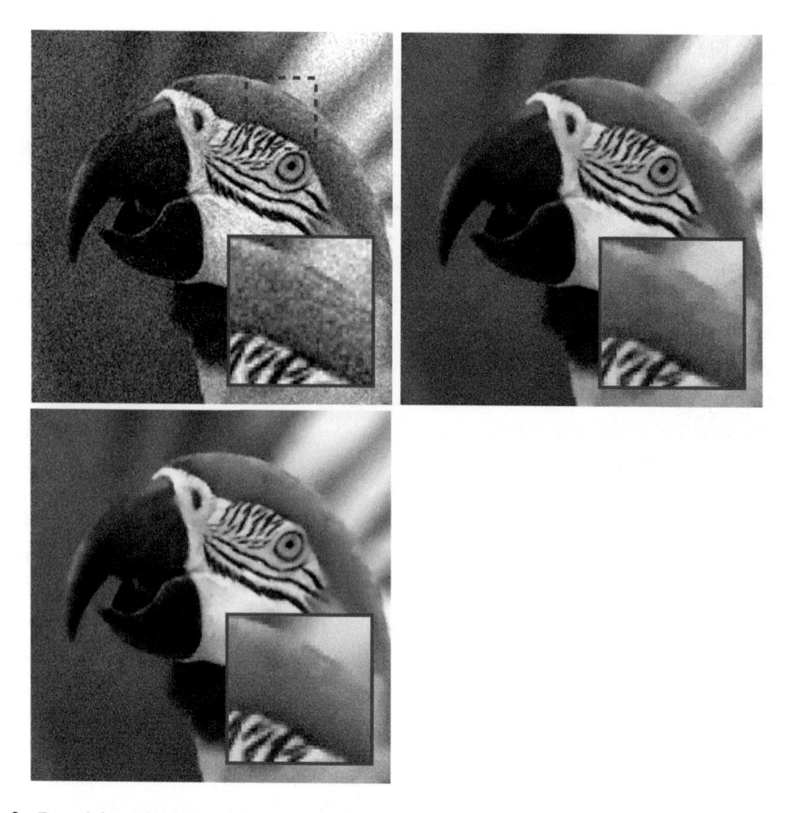

Fig. 2 Denoising of a noisy photograph (*top-left*) by the geometric regularisers TV (*top-right*) and TGV^2 (*bottom-left*). Observe the stair-casing effect exhibited by TV. (*Freely available Kodak stock photo*)

important task is *deblurring* or *deconvolution*, see Fig. 3. Here one seeks to create a sharp image out of an unsharp image, which might be the result of camera shake—something that can also be avoided to some extent in sufficient light conditions by mechanical shake reduction technologies. In *dehazing* one seeks to remove translucent objects—clouds, haze—that obscure parts of an image and make it unsharp; see [49] for an approach fitting our variational image processing framework and further references. Another basic task is *regularised inversion*. This involves the computation of an image from data in a different domain, such as the frequency domain. When only partial data is available, we need to add additional information into the problem in terms of the aforementioned regularisers. Problems of this form can be found in many medical imaging modalities such as magnetic resonance imaging (MRI, [11, 72, 73, 135]), positron emission tomography (PET, [119, 141]), electrical impedance tomography (EIT, [92]), computed tomography (CT, [101]), and diffuse optical tomography (DOT, [5, 74])—the references providing merely a few starting points. Related imaging modalities can be found in the earth and planetary sciences, for example seismic tomography [86] and synthetic aperture radar (SAR,

Fig. 3 Deblurring example. Could we do better, and faster? (*top*) Lousy photograph. (*middle*) TGV^2 deblurring. $\min_u \frac{1}{2}\|f - \rho_\varepsilon * u\| + \mathrm{TGV}^2(u)$ for blur kernel ρ_ε. (*bottom*) Ideal photograph. (*Author's own photograph. All rights withheld. Reprinted with permission*)

[32]). In *image fusion* one seeks to combine the data from many such modalities in order to obtain a better overall picture [15, 45, 71, 123]. In many computer vision tasks, including the automated understanding of medical imaging data, a task of paramount importance is *segmentation* [4, 34, 93, 109, 138]. Here, we seek to differentiate or detect objects in an image in order understand its content by higher-level algorithms. This may also involve *tracking* of the objects [90, 143], and in turn provides a connection to video processing, and tasks such as *optical flow* computation [6, 30, 33, 68, 134].

2 Regularisation of Inverse Problems

We consider image processing tasks as *inverse problems* whose basic setup is as follows. We are presented with data or measurements f, and a *forward operator* A that produced the data f, possibly corrupted by noise ν, from an unknown $\hat{u}$ that we wish to recover. Formally $f = A\hat{u} + \nu$. In imaging problems $\hat{u}$ is the uncorrupted ideal image that we want, and f the corrupted, transformed, or partial image that we have. The operator A would be the identity for denoising, a convolution operator for deblurring, and a (sub-sampled) Fourier, Radon, or other transform operator for regularised inversion. Besides the noise ν, the difficulty in recovering $\hat{u}$ is that the operator A is ill-conditioned, or simply not invertible. The overall problem is ill-posed. We therefore seek to add some prior information to the problem, to make the it well-posed. This comes in terms of a regularisation functional R, which should model our domain-specific prior assumptions of what the solution should look like. Modelling the noise and the operator equation by a fidelity functional G, the *Tikhonov regularisation* approach then seeks to solve

$$\min_u G(u) + \alpha R(u) \tag{P_α}$$

for some *regularisation parameter* $\alpha > 0$ that needs to be determined. Its role is to balance between regularity and good fit to data. If the noise ν is Gaussian, as is often assumed, we take

$$G(u) := \frac{1}{2}\|f - Au\|_2^2. \tag{1}$$

The choice of the regulariser R depends heavily on the problem in question; we will shortly discuss typical and emerging choices of R for imaging problems.

A major difficulty with the Tikhonov approach (P_α) is that the regularisation parameter α is difficult to choose. Moreover, with the L^2-squared fidelity (1), the scheme suffers from loss of contrast, as illustrated in [11]. If the noise level $\bar{\sigma}$ is known, an alternative approach is to solve the constrained problem

$$\min_u R(u) \quad \text{subject to} \quad G(u) \leq \bar{\sigma}. \tag{$\mathrm{P}^{\bar{\sigma}}$}$$

Computationally this problem tends to be much more difficult than (P_α). An approach to *estimate* solutions to this is provided by *iterative regularisation* [46, 70, 120], which we discuss in more detail in Sect. 6.2. The basic idea is to take a suitably chosen sequence $\alpha_k \searrow 0$. Letting $k \to \infty$, one solves (P_α) for $\alpha = \alpha_k$ to obtain u^k, and stops when $F(u^k) \leq \bar{\sigma}$. This stopping criterion is known as Morozov's *discrepancy principle* [91]. Various other a priori and a posteriori heuristics also exist. Besides iterative regularisation and heuristic stopping rules, another option for facilitating the choice of α is computationally intensive parameter learning strategies [39, 41, 57, 77], which can deal with more complicated noise models as well.

3 Non-smooth Geometric Regularisers for Imaging

The regulariser R should try to restore and enhance desired image features without introducing artefacts. Typical images feature smooth parts as well as non-smooth geometric features such as edges. The first "geometric regularisation" models in this context have been proposed in the pioneering works of Rudin-Osher-Fatemi [118] and Perona-Malik [106]. In the former, total variation (TV) has been proposed as a regulariser for image denoising, that is $R(u) = \mathrm{TV}(u)$. Slightly cutting corners around distributional intricacies, this can be defined as the one-norm of the gradient. The interested reader may delve into all the details by grabbing a copy of [3]. In the typical case of *isotropic* TV that does not favour any particular directions, the pointwise or pixelwise base norm is the two-norm, so that

$$\mathrm{TV}(u) := \|\nabla u\|_{2,1} := \int_\Omega \|\nabla u(x)\|_2 dx$$

The Rudin-Osher-Fatemi (ROF) model is then

$$\min_u \frac{1}{2}\|f - u\|_2^2 + \alpha \mathrm{TV}(u), \tag{2}$$

where $u \in L^1(\Omega)$ is our unknown image, represented by a function from the domain $\Omega \subset \mathbb{R}^n$ into intensities in $\mathbb{R}$. Typically Ω is a rectangle in $\mathbb{R}^2$ or a cube in $\mathbb{R}^3$, and its elements represent different points or coordinates $x = (x_1, \ldots, x_n)$ within the n-dimensional image. For simplicity we limit ourselves in this introductory exposition to greyscale images with intensities in $\mathbb{R}$. With $\mathscr{D} := C_c^\infty(\Omega; \mathbb{R}^n)$, the total variation may also be written

$$\mathrm{TV}(u) = \sup \left\{ \int_\Omega \nabla^* \phi(x) u(x) dx \,\middle|\, \phi \in \mathscr{D},\ \sup_{x \in \Omega} \|\phi(x)\|_2 \leq 1 \right\}, \tag{3}$$

which is useful for primal-dual and predual algorithms. Here $\nabla^* = -\mathrm{div}$ is the conjugate of the gradient operator.

The ROF model (2) successfully eliminates Gaussian noise and at the same time preserves characteristic image features like edges and cartoon-like parts. It however has several shortcomings. A major one is the staircasing effect, resulting in blocky images; cf. Fig. 2. It also does not deal with texture very well. Something better is therefore needed. In parts of the image processing community coming more from the engineering side, the BM3D block-matching filter [36] is often seen as the state-of-the-art method for image denoising specifically. From the visual point of view, it indeed performs very well with regard to texture under low noise levels. Not based on a compact mathematical model, such as those considered here, it is however very challenging to analyse, to prove its reliability. It, in fact, appears to completely break down under high noise, introducing very intrusive artefacts [48]. In other parts of the image and signal processing community, particularly in the context of compressed sensing, promoting sparsity in a wavelet basis is popular. This would correspond to a regulariser like $R(u) = \|Wu\|_1$, for W a wavelet transform. The simplest approaches in this category also suffer from serious artefacts, cf. [124, p. 229] and [11].

To overcome some of these issues, second- (and higher-) order geometric regularisers have been proposed in the last few years. The idea is to intelligently balance between features at different scales or orders, correctly restoring all three, smooth features, geometric features such as edges, and finer details. Starting with [87], proposed variants include total generalised variation (TGV, [19]), infimal convolution TV (ICTV, [25]), and many others [22, 28, 37, 42, 104]. Curvature based regularisers such as Euler's elastica [27, 122] and [12] have also recently received attention for the better modelling of curvature in images. Further, non-convex total variation schemes have been studied in the last few years for their better modelling of real image gradient distributions [62, 65, 66, 69, 99], see Fig. 4. In the other direction, in order to model texture in images, "lower-order schemes" have recently been proposed, including Meyer's G-norm [88, 139] and the Kantorovich-Rubinstein discrepancy [78]. (Other ways to model texture include non-local filtering schemes

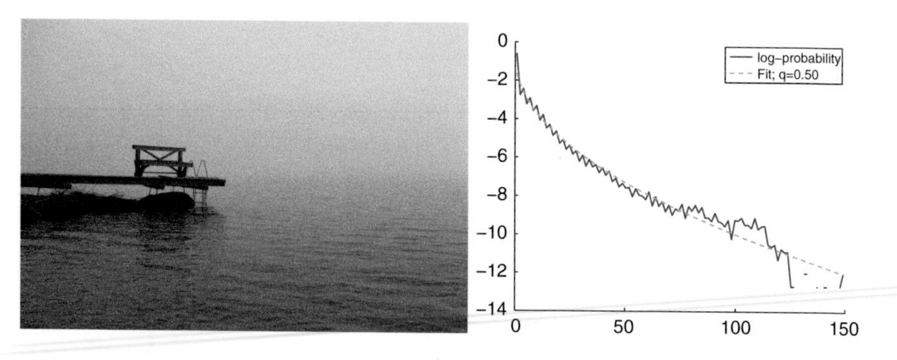

Fig. 4 Illustration of image gradient statistics. (*left*) Original image. (*right*) Log-probability (*vertical axis*) of gradient magnitude (*horizontal axis*) and optimal $t \mapsto \alpha t^q$ model fit. The optimal $q = 0.5$ causes $R(u) = \int_\Omega \|\nabla u(x)\|^q dx$ to become non-convex. (*Author's own photograph. All rights withheld. Reprinted with permission*)

such as BM3D and NL-means [21, 36].) These models have in common that they are generally non-smooth, and increasingly often non-convex, creating various optimisation challenges. The Mumford-Shah functional [93] in particular, useful as a choice of R for segmentation, is computationally extremely difficult. As a result, either various approximations [4, 109, 138] or convex relaxation techniques are usually employed. We will take a brief look at the latter in Sect. 7.1.

Due to its increasing popularity, simplicity, and reasonably good visual performance, we concentrate here on second-order *total generalised variation* (TGV [19], pun intended) as our example higher-order regulariser. In the *differentiation cascade form* [18, 20], it may be written for two parameters $(\beta, \alpha) > 0$ as

$$\mathrm{TGV}^2_{(\beta,\alpha)}(u) := \min_{w \in L^1(\Omega;\mathbb{R}^n)} \alpha\|\nabla u - w\|_{2,1} + \beta\|\mathscr{E}w\|_{F,1}.$$

Here $\mathscr{E}w$ is the symmetrised gradient, defined as

$$\mathscr{E}w(x) := \frac{1}{2}(\nabla u(x) + [\nabla u(x)]^T) \in \mathbb{R}^{n\times n}.$$

The norm in

$$\|\mathscr{E}w\|_{F,1} := \int_\Omega \|\mathscr{E}w(x)\|_F dx,$$

is based on the pointwise Frobenius norm, which makes the regulariser rotationally invariant [135]. Again we slightly cut corners with distributional intricacies.

The idea in TGV^2 is that the extra variable w, over which we minimise, extracts features from u that are rather seen as second-order features. The division between first-order and second-order features is decided by the ratio β/α. If β is very large, TGV^2 essentially becomes TV, i.e., a first-order regulariser, while if β is small, all features of order larger than zero are *gratis*. In other words, only singularities, such as edges, are penalised. The use of the symmetrised gradient demands an explanation. A rationale is that if $w = \nabla v$ is already the gradient of a smooth function v, then $\nabla w = \nabla^2 v$ is symmetric. This connects TGV^2 to ICTV, which can be formulated as

$$\mathrm{ICTV}_{(\beta,\alpha)}(u) := \min_{v \in L^1(\Omega)} \alpha\|\nabla u - \nabla v\|_{2,1} + \beta\|\nabla^2 v\|_{F,1}.$$

Indeed,

$$\mathrm{TGV}^2_{(\beta,\alpha)}(u) \leq \mathrm{ICTV}_{(\beta,\alpha)}(u) \leq \alpha\mathrm{TV}(u),$$

so that TGV^2 penalises higher-order features less than ICTV or TV.

A simple comparison of TGV^2 versus TV, showing how it avoids the stair-casing effect of TV, can be found in Fig. 2. While quite a bit is known analytically about the artefacts introduced and features restored by TV [23, 29, 44, 114], a similar study of TGV^2 and other advanced regularisers is a challenging ongoing effort [20, 40, 102, 103, 132, 133]. A more complete analytical understanding would be desirable towards the reliability of any regularisation method in critical real-life applications.

4 First-Order Optimisation Methods for Imaging

Popular, well-performing, optimisation methods in the imaging community tend to be based on variations of operator splitting and proximal (backward) steps. These include the primal-dual method of Chambolle-Pock(-Bischof-Cremers) [26, 109], the alternating directions method of multipliers (ADMM) and other Augmented Lagrangian schemes [53], as well as FISTA [9, 10, 84]. While asymptotic convergence properties of these methods are, in general, comparable to the gradient descent method, in special cases they reach the $O(1/N^2)$ rate of Nesterov's optimal gradient method [94]. Folklore also tells us that they tend to reach a visually acceptable solution in fewer iterations. The performance of the methods unfortunately decreases as the problems become increasingly ill-conditioned [85].

In all of the methods of this section, it is crucial to be able to calculate a proximal map, which we will shortly introduce. We gradually move from methods potentially involving difficult proximal maps to ones that ease or partially eliminate their computation. Generally the ones with difficult proximal maps are more efficient, if the map can be computed efficiently. We first look at primal methods, especially FISTA and its application to TV denoising in Sect. 4.2. We then study primal-dual methods in Sect. 4.3, concentrating on the PDHGM, and Sect. 4.4, where we concentrate on the GIST. First we begin with a few remarks about notation and discretisation, however.

4.1 Remarks About Notation and Discretisation

Remark 1 (Discretisation) The methods considered in this section are in principle for finite-dimensional problems, and stated in this way. We therefore have to discretise our ideal infinite-dimensional models in Sect. 3. We take a cell width $h > 0$, and set

$$\Omega_h := h\mathbb{Z}^n \cap \Omega.$$

Then, if $u : \Omega_h \to \mathbb{R}$, we define

$$\mathrm{TV}(u) := \sum_{x \in \Omega_h} h^n \|\nabla_h u(x)\|_2,$$

for ∇_h a suitable discrete gradient operator on Ω_h, e.g., a forward-differences operator. Similarly to the dual form (3), we also have

$$\mathrm{TV}(u) = \sup \left\{ \sum_{x \in \Omega_h} h^n \nabla^* \phi(x) u(x) \,\middle|\, \phi \in \mathscr{D}_h,\ \sup_{x \in \Omega_h} \|\phi(x)\|_2 \leq 1 \right\},$$

where $\mathscr{D}_h$ denotes the set of all functions $\phi : \Omega_h \to \mathbb{R}$. Likewise, we replace the operator $A : L^1(\Omega) \to Z$ in (1), for any given space $Z \ni f$, by a discretisation A_h, discretising Z if necessary. Often in regularised inversion, Z is already discrete, however, as we have a finite number of measurements $f = (f_1, \dots, f_m)$.

In the following, we generally drop the subscript h and, working on an abstract level, making no distinction between the finite-dimensional discretisations, and the ideal infinite-dimensional formulations. The algorithms will always be applied to the discretisations.

Remark 2 (Notation) In the literature more on the optimisation than imaging side, often the primal unknown that we denote by u is denoted x, and the dual unknown that we denote by p is denoted by y. We have chosen to use u for the unknown image, common in the imaging literature, with x standing for a coordinate, i.e., the location of a pixel within an image. Likewise, sometimes the role of the operators A and K are interchanged, as is the role of the functionals F and G. With regard to these, we use the convention in [26]. K is then always an operator occurring in the saddle-point problem ($\mathrm{P}_{\mathrm{saddle}}$), and A occurs within the functional G, as in (1). These notations are exactly the opposite in [84].

The spaces X and Y are always suitable finite-dimensional Hilbert spaces (isomorphic to $\mathbb{R}^k$ for some k), usually resulting from discretisations of our ideal infinite-dimensional image space and its predual.

4.2 Primal: FISTA, NESTA, etc.

Perhaps the best-known primal method for imaging problems is FISTA, or the *Fast Iterative Shrinkage-Thresholding Algorithm* [10]. It is based on the *forward-backward splitting* algorithm [82, 105]. A special case of this method has been derived in the literature multiple times through various different means—we refer to [38] for just one such derivation—and called the Iterative Shrinkage-Thresholding Algorithm or ISTA. FISTA adds to this an acceleration scheme similar to Nesterov's optimal gradient method [94]. The method solves a general problem of the form

$$\min_{u \in X} G(u) + F(u), \tag{$\mathrm{P}_{\mathrm{primal}}$}$$

where X is a finite-dimensional Hilbert space, e.g., a discretisation of our image space $L^1(\Omega)$. The functional $F : X \to (-\infty, \infty]$ is convex but possibly non-smooth, and $G : X \to \mathbb{R}$ is continuous with a Lipschitz continuous gradient. It is naturally assumed that the problem ($\mathrm{P}_{\mathrm{primal}}$) has a solution.

We describe FISTA in Algorithm 1. A basic ingredient of the method is the *proximal map* or *resolvent* $P_{F,\tau}$ of F. This may for a parameter $\tau > 0$, be written as

$$P_{F,\tau}(u') := \arg\min_u \left\{ F(u) + \frac{1}{2\tau} \|u - u'\|_2^2 \right\}.$$

Algorithm 1 FISTA [10] for (P_{primal})

Require: L_f Lipschitz constant of ∇f.
1: Initialise $v^1 = u^0 \in X$, $t_1 := 1$, and $\tau := 1/L_f$. Set $k := 1$.
2: **repeat**
3: Compute $u^k := P_{F,\tau}(v^k - \tau \nabla G(v^k))$,
4: $$t_{k+1} := \frac{1 + \sqrt{1 + 4t_k^2}}{2},$$
5: $$v^{k+1} := u^k + \frac{t_k - 1}{t_{k+1}}(u^k - u^{k-1}).$$
6: Update $k := k + 1$.
7: **until** A stopping criterion is fulfilled.

Alternatively

$$P_{F,\tau}(u') = (I + \tau \partial F)^{-1}(u'),$$

for ∂F the subgradient of F in terms of convex analysis; for details we refer to [67, 116]. More information about proximal maps may be found, in particular, in [115]. The update

$$u^{k+1} := P_{F,\tau}(u^k)$$

with step length τ is known as the *backward* or *proximal step*. Roughly, the idea in FISTA is to take a gradient step with respect to G, and a proximal step with respect to F. This is done in Step 3 of Algorithm 1. However, the gradient step does not use the main iterate sequence $\{u^k\}_{k=1}^{\infty}$, but an alternative sequence $\{v^k\}_{k=1}^{\infty}$, which is needed for the fast convergence. Steps 4 and 5 are about acceleration. Step 4 changes the step length parameter for the additional sequence $\{v^k\}$, while Step 5 updates it such that it stays close to the main sequence; indeed v^{k+1} is an over-relaxed or *inertia* version of u^k—a physical interpretation is a heavy ball rolling down a hill not getting stuck in local plateaus thanks to its inertia. The sequence $t_k \to \infty$, so that eventually

$$v^{k+1} \approx 2u^k - u^{k-1}.$$

In this way, by using two different sequences, some level of second order information can be seen to be encoded into the first-order algorithm.

FISTA is very similar to Nesterov's optimal gradient method [94, 95], however somewhat simplified and in principle applicable to a wider class of functions. Step 3 is exactly the same, and the only difference is in the construction of the sequence $\{v^{k+1}\}_{k=1}^{\infty}$. In Nesterov's method a more general scheme that depends on a longer history is used. NESTA [8], based on Nesterov's method, is effective for some compressed sensing problems, and can also be applied to constrained total variation minimisation, that is the problem ($P^{\bar{\sigma}}$) with $R = \mathrm{TV}$ and $G(u) = \|f - Au\|_2^2$.

In principle, we could apply FISTA to the total variation denoising problem (2). We would set $G(u) = \frac{1}{2}\|f - u\|_2^2$, and $F(u) = \|\nabla u\|_1$. However, there is a problem. In order for FISTA to be practical, the proximal map $P_{\tau,F}$ has to be computationally cheap. This is not the case for the total variation seminorm. This direct approach to using FISTA is therefore not practical. The trick here is to solve the *predual problem* of (2). (In the discretised setting, it is just the dual problem.) This may be written

$$\min_{\phi\in\mathscr{D}} \frac{1}{2}\|f-\nabla^*\phi\|_2^2 \quad \text{subject to} \quad \|\phi(x)\|_2 \le \alpha \text{ for all } x \in \Omega. \tag{4}$$

We set

$$G(\phi) := \frac{1}{2}\|f-\nabla^*\phi\|_2^2, \quad \text{and} \quad F(\phi) := \delta_{B_\alpha^\infty}(\phi),$$

for

$$B_\alpha^\infty(\phi) := \{\phi \in \mathscr{D} \mid \sup_{x\in\Omega} \|\phi(x)\|_2 \le \alpha\}.$$

(We recall that for a convex set B, the indicator function $\delta_B(\phi)$ is zero if $\phi \in B$, and $+\infty$ otherwise.) Now, the proximal map $P_{F,\tau}$ is easy to calculate—it is just the pixelwise projection onto the ball $B(0,\alpha)$ in $\mathbb{R}^n$. We may therefore apply FISTA to total variation denoising [9].

One might think of using the same predual approach to solving the more difficult TGV^2 denoising problem

$$\min_u \frac{1}{2}\|f-u\|_2^2 + \text{TGV}^2_{(\beta,\alpha)}(u). \tag{5}$$

The predual of this problem however has a difficult non-pointwise constraint set, and the resulting algorithm is not efficient [19]. Therefore, other approaches are needed.

4.3 Primal-Dual: PDHGM, ADMM, and Other Variants on a Theme

Both (2) and (5), as well as many more problems of the form

$$\min_u \frac{1}{2}\|f-Au\|_2^2 + R(u), \tag{6}$$

for $R = \alpha\text{TV}$ or $R = \text{TGV}^2_{(\beta,\alpha)}$ can in their finite-dimensional discrete forms be written as saddle-point problems

$$\min_{u\in X} \max_{p\in Y} G(u) + \langle Ku, p\rangle - F^*(p). \tag{P$_\text{saddle}$}$$

Here $G : X \to (-\infty,\infty]$ and $F^* : Y \to (-\infty,\infty]$ are convex, proper, and lower semicontinuous, and $K : X \to Y$ is a linear operator. The functional F^* is moreover assumed to be the convex conjugate of some F satisfying the same assumptions. Here the spaces X and Y are again finite-dimensional Hilbert spaces. If $P_{G_0,\tau}$ is easy to calculate for $G_0(u) := \frac{1}{2}\|f - Au\|_2^2$, then for $R = \alpha\mathrm{TV}$, we simply transform (6) into the form ($\mathrm{P}_{\mathrm{saddle}}$) by setting

$$G = G_0, \quad K = \nabla, \quad \text{and} \quad F^*(p) = \delta_{B_\alpha^\infty}(p). \tag{7}$$

For $R = \mathrm{TGV}^2_{(\beta,\alpha)}$, we write $u = (v, w)$, $p = (\phi, \psi)$, and set

$$G(u) = G_0(v), \quad Ku = (\nabla v - w, \mathscr{E}w), \quad \text{and} \quad F(p) = \delta_{B_\alpha^\infty}(\phi) + \delta_{B_\beta^\infty}(\psi). \tag{8}$$

Observe that G in (7) for TV is strongly convex if the nullspace $\mathscr{N}(K) = \{0\}$, but G in (8) is never strongly convex. This has important implications.

Namely, problems of the form ($\mathrm{P}_{\mathrm{saddle}}$) can be solved by the Chambolle-Pock (-Bischof-Cremers) algorithm [26, 109], also called the *modified primal dual hybrid-gradient method* (PDHGM) in [47]. In the presence of strong convexity of either F^* or G, a Nesterov acceleration scheme as in FISTA can be employed to speed up the convergence to $O(1/N^2)$. The unaccelerated variant has rate $O(1/N)$. Therefore the performance of the method for TGV^2 denoising is theoretically significantly worse than for TV. We describe the two variants of the algorithm, accelerated and unaccelerated, in detail in Algorithm 2 and Algorithm 3, respectively.

Algorithm 2 PDHGM [26] for ($\mathrm{P}_{\mathrm{saddle}}$)

Require: L a bound on $\|K\|$, over-relaxation parameter θ ($\theta = 1$ usually, for convergence proofs to hold), primal and dual step lengths $\tau, \sigma > 0$ such that $\tau\sigma L^2 < 1$.

1: Initialise primal and dual iterate $u^1 \in X$ and $p^1 \in Y$. Set $k := 1$.
2: **repeat**
3: Compute $u^{k+1} := P_{G,\tau}(u^k - \tau K^* p^k)$,
4: $\bar{u}^{k+1} := u^{k+1} + \theta(u^{k+1} - u^k)$,
5: $p^{k+1} := P_{F^*,\sigma}(p^k + \sigma K\bar{u}^{k+1})$.
6: Update $k := k + 1$.
7: **until** A stopping criterion is fulfilled.

Algorithm 3 Accelerated PDHGM [26] for ($\mathrm{P}_{\mathrm{saddle}}$)

Require: L a bound on $\|K\|$, $\gamma > 0$ factor of strong convexity of G or F^*, initial primal and dual step lengths $\tau_1, \sigma_1 > 0$ such that $\tau_1\sigma_1 L^2 < 1$.

1: Initialise primal and dual iterate $u^1 \in X$ and $p^1 \in Y$. Set $k := 1$.
2: **repeat**
3: Compute $u^{k+1} := P_{G,\tau_k}(u^k - \tau_k K^* p^k)$,
4: $\theta_k := 1/\sqrt{1 + 2\gamma\tau_k}$, $\tau_{k+1} := \theta_k\tau_k$, and $\sigma_{k+1} := \sigma_k/\theta_k$,
5: $\bar{u}^{k+1} := u^{k+1} + \theta_k(u^{k+1} - u^k)$,
6: $p^{k+1} := P_{F^*,\sigma_{k+1}}(p^k + \sigma_{k+1}K\bar{u}^{k+1})$.
7: Update $k := k + 1$.
8: **until** A stopping criterion is fulfilled.

The method is based on proximal or backward steps for both the primal and dual variables. Essentially one holds u and p alternatingly fixed in (P_{saddle}), and takes a proximal step for the other. However, this scheme, known as PDHG (primal-dual hybrid gradient method, [147]), is generally not convergent. That is why the *over-relaxation* or *inertia* step $\bar{u}^{k+1} := u^{k+1} + \theta(u^{k+1} - u^k)$ for the primal variable is crucial. We also need to take $\theta = 1$ for the convergence results to hold [26]. Naturally inertia step on the primal variables u could be replaced by a corresponding step on the dual variable p.

It can be shown that the PDHGM is actually a preconditioned proximal point method [59], see also [117, 131]. (This reformulation is the reason why the ordering of the steps in Algorithm 2 is different from the original one in [26].) Proximal point methods apply to general monotone inclusions, not just convex optimisation, and the inertial and splitting ideas of Algorithm 2 have been generalised to those [83].

The PDHGM is very closely related to a variety other algorithms popular in image processing. For $K = I$, the unaccelerated version of the method reduces [26] to the earlier *alternating direction method of multipliers* (ADMM, [53]), which itself is a variant of the classical *Douglas-Rachford splitting algorithm* (DRS, [43]), and an approximation of the *Augmented Lagrangian method*. The idea here is to consider the primal problem corresponding to (P_{saddle}), that is

$$\min_u G(u) + F(Ku).$$

Then we write this as

$$\min_{u,p} G(u) + F(p) \quad \text{subject to} \quad Ku = p.$$

The form of the Augmented Lagrangian method in Algorithm 4 may be applied to this. If, in the method, we perform Step 3 first with respect to u and then respect to p, keeping the other fixed, and keep the *penalty parameter* μ^k constant, we obtain the ADMM. For $K = I$, this will be just the PDHGM. For $K \neq I$, the PDHGM can be seen as a preconditioned ADMM [47].

The ADMM is further related to the *split inexact Uzawa method*, and equals on specific problems the *alternating split Bregman method*. This is again based on a proximal point method employing in $P_{G,\tau}$, instead of the standard L^2-squared distance, alternative so-called Bregman distances related to the problem at hand; see [121] for an overview. We refer to [26, 47, 121] for even further connections.

Generalising, it can be said that FISTA performs better than PDHGM when the computation of the proximal mappings it requires can be done fast [26]. The PDHGM is however often one of the best performers, and often very easy to implement thanks to the straightforward linear operations and often easy proximal maps. It can be applied to TGV^2 regularisation problems, and generally outperforms FISTA, which was still used for TGV^2 minimisation in the original TGV paper [19]. The problem is that the proximal map required by FISTA for the predual formulation of TGV^2 denoising is too difficult to compute. The primal formulation would be even more

Algorithm 4 Augmented Lagrangian method for $\min_u F(u)$ subject to $Au = f$

Require: A sequence of *penalty parameters* $\mu^k \searrow 0$, initial iterate $u^0 \in X$, and initial Lagrange multiplier $\lambda^1 \in Y$.
1: Define the *Augmented Lagrangian*

$$\mathscr{L}(u, \lambda, \mu) := F(u) + \langle \lambda, Au - f \rangle + \frac{1}{2\mu} \|Au - f\|_2^2.$$

2: **repeat**
3: Compute $u^k := \arg\min_u \mathscr{L}(u, \lambda^k; \mu^k)$ starting from u^{k-1}, and
4: $\lambda^{k+1} := \lambda^k - (Au^k - b)/\mu^k$.
5: Update $k := k + 1$.
6: **until** A stopping criterion is fulfilled.

difficult, being of same form as the original problem. This limits the applicability of FISTA. But the PDHGM is also not completely without these limitations.

4.4 When the Proximal Mapping is Difficult

In typical imaging applications, with $R = \text{TV}$ or $R = \text{TGV}^2$, the proximal map $P_{F^*,\sigma}$ corresponding to the regulariser is easy to calculate for PDHGM—it consists of simple pointwise projections to unit balls. But there are many situations, when the proximal map $P_{G_0,\tau}$ corresponding to the data term is unfeasible to calculate on every iteration of Algorithm 2. Of course, if the operator $A = I$ is the identity, this is a trivial linear operation. Even when $A = S\mathscr{F}$ is a sub-sampled Fourier transform, the proximal map reduces to a simple linear operation thanks to the *unitarity* $\mathscr{F}^*\mathscr{F} = \mathscr{F}\mathscr{F}^* = I$ of the Fourier transform. But what if the operator is more complicated, or, let's say

$$G_0(v) = \frac{1}{2}\|f - Av\|_2^2 + \delta_C(v),$$

for some difficult constraint set C? In a few important seemingly difficult cases, calculating the proximal map is still very feasible. This includes a pointwise positive semi-definiteness constraint on a diffusion tensor field when $A = I$ [135]. Here also a form of *unitary invariance* of the constraint set is crucial [81]. If $A \neq I$ with the positivity constraint, the proximal mapping can become very difficult. If A is a pointwise (pixelwise) operator, this can still be marginally feasible if special *small data* interior point algorithms are used for its pointwise computation [130, 136]. Nevertheless, even in this case [136], a reformulation tends to be more efficient. Namely, we can rewrite

$$G_0(v) = \sup_\lambda \langle Av - f, \lambda \rangle - \frac{1}{2}\|\lambda\|_2^2 + \delta_C(v).$$

Then, in case of TV regularisation, we set $p = (\phi, \lambda)$, and

$$G(u) := \delta_C(u), \quad Ku := (\nabla u, Au), \quad \text{and} \quad F^*(p) := \delta_{B_\alpha^\infty}(\phi) + \langle f, \lambda \rangle + \frac{1}{2}\|\lambda\|_2^2.$$

The modifications for TGV^2 regularisation are analogous. Now, if the projection into C is easy, and A and A^* can be calculated easily, as is typically the case, application of Algorithm 2 becomes feasible. The accelerated version is usually no longer applicable, as the reformulated G is not strongly convex, and F^* usually isn't.

However, there may be better approaches. One is the GIST or *Generalised Iterative Soft Thresholding* algorithm of [84], whose steps are laid out in detail in Algorithm 5. As the name implies, it is also based on the ISTA algorithm as was FISTA, and is a type of forward-backward splitting approach. It is applicable to saddle-point problems ($\mathrm{P}_{\mathrm{saddle}}$) with

$$G(u) = \frac{1}{2}\|f - Au\|^2. \tag{9}$$

In essence, the algorithm first takes a forward (gradient) step for u in the saddle-point formulation ($\mathrm{P}_{\mathrm{saddle}}$), keeping p fixed. This is only used to calculate the point where to next take a proximal step for p keeping u fixed. Then it takes a forward step for u at the new p to actually update u. In this way, also GIST has a second over-relaxation type sequence for obtaining convergence. If $\|A\| < \sqrt{2}$ and $\|K\| < 1$, then GIST converges with rate $O(1/N)$. We recall that forward-backward splitting generally has rather stronger requirements for convergence, see [128] as well as [61, 121] for an overview and relevance to image processing. Also, in comparison to the PDHGM, the calculation of the proximal map of G is avoided, and the algorithm requires less variables and memory than PDHGM with the aforementioned "dual transportation" reformulation of the problem.

Algorithm 5 GIST [84] for ($\mathrm{P}_{\mathrm{saddle}}$) with (9)

1: Initialise primal and dual iterate $u^1 \in X$ and $p^1 \in Y$. Set $k := 1$.
2: **repeat**
3: Compute $\bar{u}^{k+1} := u^k + A^T(f - Au^k) - K^T p^k$,
4: $p^{k+1} := P_{F^*,1}(p^k + K\bar{u}^{k+1})$,
5: $u^{k+1} := u^k + A^T(f - Au^k) - K^T p^{k+1}$.
6: Update $k := k + 1$.
7: **until** A stopping criterion is fulfilled.

5 Second-Order Optimisation Methods for Imaging

Although second-order methods are more difficult to scale to large images, and the non-smoothness of typical regularisers R causes complications, there has been a good amount of work into second-order methods for total variation regularisation, in particular for the ROF problem (2). Typically some smoothing of the problem is required. The first work in this category is [140]. There, the total variation seminorm $\|\nabla u\|_{2,1}$ is replaced by the smoothed version

$$\widetilde{\mathrm{TV}}_{\varepsilon}(u) := \int_{\Omega} \sqrt{\|\nabla u(x)\|^2 + \varepsilon dx}. \tag{10}$$

Then the Newton method is applied—after discretisation, which is needed for u to live in and $\widetilde{\mathrm{TV}}_{\varepsilon}$ to have gradients in a "nice" space. In the following, we will discuss one further development, primarily to illustrate the issues in the application of second order method to imaging problems, not just from the point of view of big data, but also from the point of view of imaging problems. Moreover, second-order methods generally find high-precision solutions faster than first-order methods when it is feasible to apply one, and are in principle more capable of finding actual local minimisers to non-convex problems. These include non-convex total variation regularisation or inversion with non-linear forward operators.

5.1 Huber-Regularisation

In recent works on second order methods, Huber-regularisation, also sometimes called Nesterov-regularisation, is more common than the smoothing of (10). This has the advantage of only distorting the one-norm of TV locally for small gradients, and has a particularly attractive form in primal-dual or (pre)dual methods. Moreover, Huber-regularisation tends to ameliorate the stair-casing effect of TV. The Huber-regularisation of the two-norm on $\mathbb{R}^n$ may for a parameter $\gamma > 0$ be written as

$$|g|_{\gamma} := \begin{cases} \|g\|_2 - \frac{1}{2\gamma}, & \|g\|_2 \geq 1/\gamma, \\ \frac{\gamma}{2}\|g\|_2^2, & \|g\|_2 < 1/\gamma. \end{cases} \tag{11}$$

Alternatively, in terms of convex conjugates, we have the dual formulation

$$|g|_{\gamma} = \max \left\{ \langle g, \xi \rangle - \frac{1}{2\gamma}\|\xi\|_2^2 \,\middle|\, \xi \in \mathbb{R}^n,\ \|\xi\|_2 \leq 1 \right\}. \tag{12}$$

In other words, the sharp corner of the graph of the two-norm is smoothed around zero—the more the smaller the parameter γ is. (Sometimes in the literature, our γ is replaced by $1/\gamma$, and so smaller value is less regularisation.) In the dual formulation, we just regularise the dual variable. This helps to avoid its oscillation. With (11), we may then define the (isotropic) Huber-regularised total variation as

$$\mathrm{TV}_{\gamma}(u) := \int_{\Omega} |\nabla u(x)|_{\gamma} dx.$$

5.2 A Primal-Dual Semi-smooth Newton Approach

In the infinite-dimensional setting, we add for a small parameter $\varepsilon > 0$ the penalty $\varepsilon\|\nabla u\|_2^2$ to (2), to pose it in a Hilbert space. This will cause the corresponding functional to have "easy" subdifferentials without the measure-theoretic complications of working in the Banach space of functions of bounded variation. With Huber-regularisation, (2) then becomes differentiable, or "semismooth" [31, 111]. A generalised Newton's method can be applied. We follow here the "infeasible active set" approach on the predual problem (4), developed in [64], but see also [76]. In fact, we describe here the extension in [39] for solving the more general problem

$$\min_{u\in H^1(\Omega;\mathbb{R}^N)} \varepsilon\|\nabla u\|_2^2 + \frac{1}{2}\|f - Au\|_2^2 + \sum_{j=1}^{N} \alpha_j \int_\Omega |[K_j u](x)|_\gamma dx, \tag{$\mathrm{P_{SSN}}$}$$

where $A : H^1(\Omega; \mathbb{R}^m) \to L^2(\Omega)$, and $K_j : H^1(\Omega; \mathbb{R}^m) \to L^1(\Omega; \mathbb{R}^{m_j})$, $(j = 1, \dots, N)$, are linear operators with corresponding weights $\alpha_j > 0$. This formulation is applicable to the TGV denoising problem (5) by setting $u = (v, w)$, $Au = v$, $K_1 u = \nabla v - w$, and $K_2 u = \mathcal{E} w$. The first-order optimality conditions for ($\mathrm{P_{SSN}}$) may be derived as

$$-\varepsilon\Delta u + A^*Au + \sum_{i=1}^{N} K_j^* p_j = A^* f, \tag{13a}$$

$$\max\{1/\gamma, |[K_j u](x)|_2\} p_j(x) - \alpha_j [K_j u](x) = 0, \quad (j = 1, \dots, N;\ x \in \Omega). \tag{13b}$$

Here (13b) corresponds pointwise for the optimality of $\xi = p_j(x)/\alpha_j$ for $g = \alpha[K_j u](x)$ and $\gamma' = \gamma/\alpha_j$ in (12). To see why this is right, it is important to observe that $\alpha|g|_\gamma = |\alpha g|_{\gamma/\alpha}$. Even in a finite-dimensional setting, although we are naturally in a Hilbert space, the further regularisation by $\varepsilon\|\nabla u\|_2^2$ is generally required to make the system matrix invertible. If we linearise (13b), solve the resulting linear system and update each variable accordingly, momentarily allowing each dual variable p_j to become infeasible, and then project back into the respective dual ball, we obtain Algorithm 6. For details of the derivation we refer to [39, 64]. Following [125], it can be shown that the method converges locally superlinearly near a point where the subdifferentials of the operator on $(u, p_1, \dots p_N)$ corresponding to (13) are non-singular. Further dampening as in [64] guarantees local superlinear convergence at any point.

Remark 3 If one wants to use a straightforward Matlab implementation of Algorithm 6 with TGV^2 and expect anything besides a computer become a lifeless brick, the system (14) has to be simplified. Indeed B is invertible, so we may solve δu from

$$B\delta u = R_1 - \sum_{j=1}^{N} K_j^* \delta p_j. \tag{15}$$

Algorithm 6 An infeasible semi-smooth Newton method for ($\mathrm{P}_{\mathrm{SSN}}$) [39, 64]

Require: Step length $\tau > 0$.

1: Define the helper functions

$$\mathfrak{m}_j(u)(x) := \max\{1/\gamma, |[K_j u](x)|_2\}, \qquad [\mathfrak{D}(p)q](x) := p(x)q(x),$$
$$\mathfrak{N}(z)(x) := \begin{cases} 0, & |z(x)|_2 < 1/\gamma, \\ \frac{z(x)}{|z(x)|_2}, & |z(x)|_2 > 1/\gamma, \end{cases} \qquad (x \in \Omega).$$

2: Initialise primal iterate u^1 and dual iterates $(p^1, \dots, p^N)$. Set $k := 1$.

3: **repeat**

4: Solve $(\delta u, \delta p_1, \dots, \delta p_N)$ from the system

$$\begin{pmatrix} B, & K_1^* & \dots & K_N^* \\ -\alpha_1 K_1 + \mathfrak{N}(K_1 u^k)^* \mathfrak{D}(p_1) K_1 & \mathfrak{D}(\mathfrak{m}_j(u^k)) & 0 & 0 \\ \vdots & 0 & \ddots & 0 \\ -\alpha_N K_N + \mathfrak{N}(K_N u^k)^* \mathfrak{D}(p_N) K_N & 0 & 0 & \mathfrak{D}(\mathfrak{m}_N(u)) \end{pmatrix} \begin{pmatrix} \delta u \\ \delta p_1 \\ \vdots \\ \delta p_N \end{pmatrix} = R \tag{14}$$

where

$$R := \begin{pmatrix} -Bu^k - \sum_{i=1}^N K_j^* p_j^k + A^* f \\ \alpha_1 K_1 u^k - \mathfrak{D}(\mathfrak{m}_1(u)) p_1^k \\ \vdots \\ \alpha_N K_N u^k - \mathfrak{D}(\mathfrak{m}_N(u^k)) p_N^k \end{pmatrix},$$

and

$$B := -\varepsilon \Delta + A^* A.$$

5: Update

$$(u^{k+1}, \tilde{p}_1^{k+1}, \dots \tilde{p}_N^{k+1}) := (u^k + \tau \delta u, p_1^k + \tau \delta p_1, p_N^k + \tau \delta p_N),$$

6: Project

$$p_j^{k+1} := \mathfrak{P}(\tilde{p}_j^{k+1}; \alpha_j), \quad \text{where} \quad \mathfrak{P}(p; \alpha)(x) := \operatorname{sgn}(p(x)) \min\{\alpha, |p(x)|\},$$

7: Update $k := k + 1$.

8: **until** A stopping criterion is satisfied.

Thus we may simplify δu out of (14), and only solve for $\delta p_1, \dots, \delta p_N$ using a reduced system matrix. Finally we calculate δu from (15).

In [39], the algorithm is compared against the PDHGM (Algorithm 2) both for TV and TGV^2 denoising, (2) and (5), respectively. It is observed that the performance can be comparable to PDHGM for TV with images up to size about 256×256. In case of TGV^2 the method performs significantly worse due to the SSN system (14) being worse-conditioned, and the data size of TGV^2 being far larger through the additional variables w and p_2. For images in the range 512×512 the method is no longer practical on current desktop computers, so definitely not for multi-million megapixel real-life photographs.

5.3 *A Note on Interior Point Methods*

Application of interior point methods to (2)—which can be very easily done with CVX [55, 56]—has similar scalability problems as Algorithm 6. This is mainly due to excessive memory demands. For small problem sizes the performance can be good when high accuracy is desired—especially the commercial MOSEK solver performs very well. However, as is to be expected, the performance deteriorates quickly as problem sizes increase and the interior point formulations become too large to fit in memory [80, 108].

A way forward for second-order methods is to use preconditioning to make the system matrix better conditioned, or to split the problem into smaller pieces using domain decomposition techniques. We will discuss what early progress has been made in this area in Sect. 7.2.

5.4 *Methods for Non-convex Regularisers*

One reason for us introducing Algorithm 6, despite being evidently not up to the processing of big images at this stage, is that the same ideas can be used derive methods for solving non-convex total variation problems [62, 65, 66]

$$\min_u \frac{1}{2}\|f - Au\|_2^2 + \alpha \int_\Omega \psi(\|\nabla u(x)\|)dx. \tag{16}$$

Here $\psi : [0, \infty) \to [0, \infty)$ is a concave energy that attempts to model real gradient distributions in images, recall Fig. 4. Usually $\psi(t) = t^q$ for $q \in (0, 1)$ although this has significant theoretical problems [62]. Alternative, first-order, approaches include the iPiano of [98], which looks a lot like FISTA, allowing F in ($\mathrm{P}_{\mathrm{primal}}$) to be non-convex and modifying the updates a little. The PDHGM has also recently been extended to "semiconvex" F [89]; this includes (16) when the energies $\psi(t) = t^q$ are linearised for small t. No comparisons between the methods are known to the author.

6 Non-linear Operators and Methods for Iterative Regularisation

We now discuss typical and novel approaches for two closely related topics: inverse problems with non-linear forward operators A, and iterative regularisation. Overall, the workhorse algorithms in this category are much less developed than for Tikhonov regularisation with linear forward operators, in which case both data and convex terms are convex.

6.1 Inverse Problems with Non-linear Operators

We now let A be a non-linear operator and set

$$G(u) := \frac{1}{2}\|f - A(u)\|_2^2. \tag{17}$$

Although second-order methods in particular could in principle be applied to smoothed versions of the resulting Tikhonov problem (P_α), in inverse problems research, a classical approach in this case is the Gauss-Newton method, described in Algorithm 7. It is based on linearising A at an iterate u^k and solving the resulting convex problem at each iteration until hopeful eventual convergence. This can be very expensive, and convergence is not generally guaranteed [97], as experiments in [131] numerically confirm. However, for the realisation of the algorithm, it is not necessary that R is (semi-)smooth as with Newton type methods.

Algorithm 7 Gauss-Newton method for (P_α) with (17)

1: Initialise primal iterate $u^1 \in X$. Set $k := 1$.
2: **repeat**
3: Solve for $u^{k+1} := u$ the convex problem

$$\min_u \frac{1}{2}\|f - A(u^k) - \nabla A(u^k)(u - u^k)\|_2^2 + \alpha R(u). \tag{18}$$

4: Update $k := k + 1$.
5: **until** A stopping criterion is fulfilled.

A more recent related development is the *primal-dual hybrid gradient method for non-linear operators* (NL-PDHGM, [131]), which we describe in Algorithm 8. It extends the iterations of the PDHGM (Algorithm 2) to non-linear K in the saddle-point problem ($\mathrm{P}_{\mathrm{saddle}}$). That is, it looks for critical points of the problem

$$\min_{u\in X} \max_{p\in Y} G(u) + \langle K(u), p\rangle - F^*(p), \tag{$\mathrm{P}_{\text{nl-saddle}}$}$$

where now $K \in C^2(X; Y)$, but $G : X \to (-\infty, \infty]$ and $F^* : Y \to (-\infty, \infty]$ are still convex, proper, and lower semicontinuous. Through the reformulations we discussed in Sect. 4.4, it can also be applied when G is as in (17) with nonlinear A. According to the experiments in [131], the NL-PDHGM by far outperforms Gauss-Newton on example problems from magnetic resonance imaging. Moreover, the non-linear models considered improve upon the visual and PSNR performance of earlier linear models in [135, 136] for diffusion tensor imaging, and in [11] for MR velocity imaging, cf. also [146]. The method can be proved to converge locally on rather strict conditions. For one, Huber-regularisation of TV or TGV^2 is required. The second peculiar condition is that the regularisation parameter α and the noise level $\bar{\sigma}$ have to be "small". An approximate linearity condition, as with the is common with the combination of the Gauss-Newton method with iterative regularisation, discussed next, is however not required.

Algorithm 8 NL-PDHGM [131] for ($P_{\text{nl-saddle}}$)

Require: L a local bound on $\|\nabla K(u)\|$ in a neighbourhood of a solution (u^*, p^*), over-relaxation parameter θ (usually $\theta = 1$ for convergence results to hold), primal and dual step lengths $\tau, \sigma > 0$ such that $\tau\sigma L^2 < 1$.
1: Initialise primal and dual iterate $u^1 \in X$ and $p^1 \in Y$. Set $k := 1$.
2: **repeat**
3: Compute $u^{k+1} := P_{G,\tau}(u^k - \tau[\nabla K(u^k)]^* p^k)$,
4: $\bar{u}^{k+1} := u^{k+1} + \theta(u^{k+1} - u^k)$,
5: $p^{k+1} := P_{F^*,\sigma}(p^k + \sigma K(\bar{u}^{k+1}))$.
6: Update $k := k + 1$.
7: **until** A stopping criterion is fulfilled.

6.2 *Iterative Regularisation*

We now briefly consider solution approaches for the constrained problem ($P^{\bar{\sigma}}$), which tends to be much more difficult than the Tikhonov problem (P_α). In some special cases, as we've already mentioned, NESTA [8] can be applied. One can also apply the classical Augmented Lagrangian method [97]. If one minimises R subject to the exact constraint $Au = f$, the method has the form in Algorithm 4 with a suitable rule of decreasing the penalty parameter μ^k. The latter has roughly the same role here as α in the Tikhonov problem (P_α). Thus the Augmented Lagrangian method forms a way of iterative regularisation, if we actually stop the iterations when Morozov's discrepancy principle is violated. (In this case, we do not expect $Au = f$ to have a solution, but require $\|Au - f\| \leq \bar{\sigma}$ to have a solution.) If we fix $\mu^k \equiv 1$ the Augmented Lagrangian method then corresponds [144] to so-called Bregman iterations [54, 100] on (P_α). Another way to view this is that one keeps α in (P_α) fixed, but, iterating its solution, replaces on each iteration the distance $\frac{1}{2}\|Au - f\|_2^2$ by the *Bregman distance*

$$D_G^\lambda(u, u^{k-1}) := G(u) - G(u^{k-1}) - \langle \lambda, u - u^{k-1} \rangle,$$

for $G(u) = \frac{1}{2}\|Au - f\|_2^2$, and $\lambda \in \partial G(u^{k-1})$. This scheme has a marked contrast-enhancing effect compared to the basic Tikhonov approach.

But how about just letting $\alpha \searrow 0$ in (P_α), as we discussed in Sect. 2, and stopping when Morozov's discrepancy principle is violated? This is equally feasible for linear A as the Augmented Lagrangian approach. But what if A is non-linear? The Gauss-Newton approach for solving each of the inner Tikhonov problems results in this case in three nested optimisation loops: one for $\alpha_k \searrow 0$, one for solving the non-convex problem for $\alpha = \alpha_k$, and one for solving (18). Aside from toy problems, this start to be computationally unfeasible. There is some light at the end of the tunnel however: the *Levenberg-Marquardt*, and *iteratively regularised Landweber and Gauss-Newton* (IRGN) methods [14, 70]. Similar approaches can also be devised for Bregman iterations when A is nonlinear [7].

The iteratively regularised Levenberg-Marquardt scheme [46, 58, 70] is the one most straightforward for general regularisers, including the non-smooth ones we are interested in. In the general case, a convergence theory is however lacking to the best of our knowledge, unless the scheme is Bregmanised as in [7]. Bregman distances have indeed been generally found useful for the transfer of various results from Hilbert spaces to Banach spaces [120]. Nevertheless, the Levenberg-Marquardt scheme combines the Gauss-Newton step (18) with the parameter reduction scheme $\alpha_k \searrow 0$ into a single step. It then remains to solve (18) for $\alpha = \alpha_k$ with another method, such as those discussed in Sect. 4. For the simple, smooth, regulariser $R(u) = \|u - u_0\|^2$, not generally relevant to imaging problems, the iteratively regularised Landweber and Gauss-Newton methods can combine even this into a single overall loop. Convergence requires, in general, a degree of *approximate linearity* from A. In the worst case, this involves the existence of $\eta, \rho > 0$ and a solution u^* of $A(u^*) = f$ such that

$$\|A(\widetilde{u}) - A(u) - \nabla A(u)(\widetilde{u} - u)\| \leq \eta \|u - \widetilde{u}\| \|A(u) - A(\widetilde{u})\|, \tag{19}$$

whenever u and $\widetilde{u}$ satisfy $\|u - u^*\|, \|\widetilde{u} - u^*\| \leq \rho$. Although (19) and related conditions can be shown to hold for certain non-linear parameter identification problems, in general it is rarely satisfied [46, 70]. For example, (19) is not satisfied by the operators considered for magnetic resonance imaging (MRI) in [131], where Algorithm 8 was developed.

7 Emerging Topics

We finish this chapter with a brief overlook at a couple of topics that have the potential to improve image optimisation performance, and turn other challenges into big data challenges. The latter, discussed first, is convex relaxation, which transforms difficult non-convex problems into large-scale convex problems. The former is decomposition and preconditioning techniques, which seek to turn large problems into smaller ones.

7.1 Convex Relaxation

The basic idea behind convex relaxation approaches is to lift a non-convex problem into a higher-dimensional space, where it becomes convex. This kind of approaches are becoming popular in image processing, especially in the context of the difficult problem of segmentation, and the Mumford-Shah problem [93]. This may be written

$$\min_u \frac{1}{2}\|f - u\|_2^2 + \alpha \int_\Omega \|\nabla u(x)\|_2^2 dx + \beta \mathcal{H}^{n-1}(J_u). \tag{20}$$

Here u is a *function of bounded variation*, which may have discontinuities J_u, corresponding to boundaries of different objects in the scene f. The final term measures their length, and the middle term forces u to be smooth outside the discontinuities; for details we refer to [3]. The connected components of Ω, as split by J_u, allow us to divide the scene into different segments.

Let us consider trying to solve for $u \in L^1(\Omega)$ and a general non-convex G the problem

$$\min_{u \in L^1(\Omega)} G(u). \tag{21}$$

Any global solution of this problem is a solution of

$$\min_{u \in L^1(\Omega)} \overline{G}(u),$$

where $\overline{G}$ is the convex lower semicontinuous envelope of G, or the greatest lower semicontinuous convex function such that $\overline{G} \leq G$. The minimum values of the functionals agree, and under some conditions, the minimisers of G and $\overline{G}$ agree.

But how to compute $\overline{G}$? It turns out that in some cases [24], it is significantly easier to calculate the convex lower semicontinuous envelope of

$$\mathcal{G}(v) := \begin{cases} G(v), & v = \chi_{\Gamma_u}, \\ \infty, & \text{otherwise}, \end{cases}$$

Here

$$\Gamma_u = \{(x, t) \in \Omega \times \mathbb{R} \mid t < u(x)\}$$

is the lower graph of u, while $v \in L^1(\Omega \times \mathbb{R}; [0, 1])$. Then

$$\overline{G}(u) = \overline{\mathcal{G}}(\chi_{\Gamma_u}),$$

and instead of solving (21), one attempts to solve the convex problem

$$\min_{v \in L^1(\Omega \times \mathbb{R}; [0,1])} \overline{\mathcal{G}}(v).$$

Observe that v lives in a larger space than u. Although the problem has become convex and more feasible to solve globally than the original one, it has become *bigger*.

Often [24], one can write

$$\overline{\mathscr{G}}(v) = \sup_{\phi \in K} \int_{\Omega \times \mathbb{R}} \nabla^* \phi(x,t) v(x,t) d(x,t)$$

for some closed convex set $K \subset C_0(\Omega \times \mathbb{R}; \mathbb{R}^{n+1})$. In some cases, the set K has a numerically realisable analytical expression, although the dimensions of K make the problem even *bigger*.

A particularly important case when K has a simple analytical expression is the for the Mumford-Shah problem (20) [2]. Other problems that can, at least roughly, be handled this way include regularisation by Euler's elastica [16] and multi-label segmentation [107]. Although not exactly fitting this framework, total variation regularisation of discretised manifold-valued data, such as normal fields or direction vectors, can also be performed through convex relaxation in a higher-dimensional space [79]. This approach also covers something as useless, but of utmost mathematical satisfaction, as the smoothing of the path of an ant lost on the Möbius band.

7.2 Decomposition and Preconditioning Techniques

The idea in domain decomposition is to divide a big problem into small subproblems, solve them, and then combine the solutions. This area is still in its infancy within image processing, although well researched in the context of finite element methods for partial differential equations. The current approaches within the field [51, 52, 60, 63] are still proof-of-concept meta-algorithms that have not replaced the more conventional algorithms discussed in Sects. 4 and 5. They pose difficult (but smaller) problems on each sub-domain. These then have to be solved by one of the conventional algorithms multiple times within the meta-algorithm, which within each of its iterations performs *subspace correction* to glue the solutions together. In case of second-order methods, i.e., if high accuracy is desired, even current domain decomposition techniques may however make problems of previously untractable size tractable.

Depending on the operator A, the first-order methods discussed in Sect. 4 are however usually easily parallelised within each iteration on multiple CPU cores or on a graphics processing unit (GPU), cf. e.g., [137]. Any advantages of domain decomposition meta-algorithms are therefore doubtful. Intelligent decomposition techniques could however help to reduce the workload within each iteration. This is, roughly, the idea behind stochastic coordinate descent methods, popular in general big data optimisation. We point in particular to [50] for an approach related to FISTA, to [13, 126] for ones related to the ADMM, and to [35, 75, 96, 113, 129, 142, 145] for just a small selection of other approaches. These methods update on each of their iterations only small subsets of unknown variables, even single variables or pixels, and obtain acceleration from local adaptation of step lengths. All of this is done in a random fashion to guarantee fast *expected convergence* on massive data sets. This type of methods form an interesting possibility for image processing.

Stochastic coordinate descent methods generally, however, demand a degree of separability from the problem, limiting the degree of dependence of each variable from other variables. This is necessary both for parallelisation and to prevent lock-up—to guarantee, statistically, that the randomly chosen variable can be updated without other variables restricting this. This is generally a problem for imaging applications that often lack this level of separability. However, the "coordinate-descent FISTA" of [50], for example, is applicable to the predual formulation (4) of TV denoising. In our preliminary experiments (with Olivier Fercoq and Peter Richtárik), we did not however obtain any acceleration compared to standard FISTA. The problem is that the matrix for the divergence operator in (4) is very uniform. The acceleration features of the current line-up of stochastic gradient descent methods however depend on varying "local curvature" of the problem in terms of local features of the Hessian of the objective. In (4) the Hessian only involves the "uniformly curved" divergence operator, and not the data itself. Therefore, no significant acceleration is obtained, aside from possibly better parallelisation performance, for example on a GPU.

Another alternative to typical domain decomposition techniques is preconditioning—something that has been studied for a long time for general numerical linear algebra, but is still making its inroads into mathematical image processing. Domain decomposition in its per-iteration form can also seen as an approach to preconditioning, of course. Here the idea is to make each iteration cheaper, or, in a sense, to adapt the step sizes spatially. This can be done in the context of the PDHGM, exploiting the proximal point formulation; see [108], where spatial adaptation of the step lengths reportedly significantly improved the performance of the PDHGM. Another recent alternative for which promising performance has been reported, is the use of the conventional Douglas-Rachford splitting method with Gauss-Seidel preconditioning in [17].

8 Conclusions

In this chapter, we have taken a look into the state-of-the-art of optimisation algorithms suitable for solving mathematical image processing models. Our focus has been on relatively simple first-order splitting methods, as these generally provide the best performance on large-scale images. Moving from FISTA and PDHGM to GIST, we have gradually changed the types of proximal mappings that need to be computed, at the cost of expanding the problem size or reducing theoretical convergence rate. We have also taken a brief look at stochastic gradient descent methods, popular for more general big data problems. At the present stage, such methods are, however, unsuitable for imaging problems. There is thus still significant work to be done in this area—can we come up with an optimisation method that would put mathematically-based state-of-the-art image enhancement models on a pocket camera?

Acknowledgments The preparation of this chapter was supported by a Prometeo fellowship of the Senescyt (Ecuadorian Ministry of Education, Science, Technology, and Innovation) while the author was at the Centre for Mathematical Modelling (ModeMat), Escuela Politécnica, Nacional, Quito, Ecuador.

References

1. Adcock, B., Hansen, A.C., Poon, C., Roman, B.: Breaking the coherence barrier: asymptotic incoherence and asymptotic sparsity in compressed sensing. In: Proceedings SampTA 2013 (2013)
2. Alberti, G., Bouchitté, G., Dal Maso, G.: The calibration method for the mumford-shah functional and free-discontinuity problems. Calc. Var. Partial Differ. Equ. **16**(3), 299–333 (2003). doi:10.1007/s005260100152
3. Ambrosio, L., Fusco, N., Pallara, D.: Functions of Bounded Variation and Free Discontinuity Problems. Oxford University Press (2000)
4. Ambrosio, L., Tortorelli, V.M.: Approximation of functional depending on jumps by elliptic functional via t-convergence. Commun. Pure Appl. Math. **43**(8), 999–1036 (1990). doi:10.1002/cpa.3160430805
5. Arridge, S.R., Schotland, J.C.: Optical tomography: forward and inverse problems. Inverse Probl. **25**(12), 123,010 (2009). doi:10.1088/0266-5611/25/12/123010
6. Aubert, G., Kornprobst, P.: Mathematical Problems in Image Processing: Partial Differential Equations and the Calculus of Variations, 2nd edn. Springer, New York (2006)
7. Bachmayr, M., Burger, M.: Iterative total variation schemes for nonlinear inverse problems. Inverse Probl. **25**(10) (2009). doi:10.1088/0266-5611/25/10/105004
8. Becker, S., Bobin, J., Candés, E.: Nesta: a fast and accurate first-order method for sparse recovery. SIAM J. Imaging Sci. **4**(1), 1–39 (2011). doi:10.1137/090756855
9. Beck, A., Teboulle, M.: Fast gradient-based algorithms for constrained total variation image denoising and deblurring problems. IEEE Trans. Image Process. **18**(11), 2419–2434 (2009). doi:10.1109/TIP.2009.2028250
10. Beck, A., Teboulle, M.: A fast iterative shrinkage-thresholding algorithm for linear inverse problems. SIAM J. Imaging Sci. **2**(1), 183–202 (2009). doi:10.1137/080716542
11. Benning, M., Gladden, L., Holland, D., Schönlieb, C.B., Valkonen, T.: Phase reconstruction from velocity-encoded MRI measurements—a survey of sparsity-promoting variational approaches. J. Magn. Reson. **238**, 26–43 (2014). doi:10.1016/j.jmr.2013.10.003
12. Bertozzi, A.L., Greer, J.B.: Low-curvature image simplifiers: global regularity of smooth solutions and Laplacian limiting schemes. Commun. Pure Appl. Math. **57**(6), 764–790 (2004). doi:10.1002/cpa.20019
13. Bianchi, P., Hachem, W., Iutzeler, F.: A stochastic coordinate descent primal-dual algorithm and applications to large-scale composite optimization (2016). Preprint
14. Blaschke, B., Neubauer, A., Scherzer, O.: On convergence rates for the iteratively regularized Gauss-Newton method. IMA J. Numer. Anal. **17**(3), 421–436 (1997)
15. Blum, R., Liu, Z.: Multi-Sensor Image Fusion and Its Applications. Signal Processing and Communications. Taylor & Francis (2005)
16. Bredies, K., Pock, T., Wirth, B.: A convex, lower semi-continuous approximation of euler's elastica energy. SFB-Report 2013–013, University of Graz (2013)
17. Bredies, K., Sun, H.: Preconditioned Douglas-Rachford splitting methods saddle-point problems with applications to image denoising and deblurring. SFB-Report 2014–002, University of Graz (2014)

18. Bredies, K., Valkonen, T.: Inverse problems with second-order total generalized variation constraints. In: Proceedings SampTA 2011 (2011)
19. Bredies, K., Kunisch, K., Pock, T.: Total generalized variation. SIAM J. Imaging Sci. **3**, 492–526 (2011). doi:10.1137/090769521
20. Bredies, K., Kunisch, K., Valkonen, T.: Properties of L^1-TGV^2: the one-dimensional case. J. Math. Anal. Appl. **398**, 438–454 (2013). doi:10.1016/j.jmaa.2012.08.053
21. Buades, A., Coll, B., Morel, J.M.: A non-local algorithm for image denoising. In: IEEE CVPR, vol. 2, pp. 60–65 (2005). doi:10.1109/CVPR.2005.38
22. Burger, M., Franek, M., Schönlieb, C.B.: Regularized regression and density estimation based on optimal transport. AMRX Appl. Math. Res. Express **2012**(2), 209–253 (2012). doi:10.1093/amrx/abs007
23. Caselles, V., Chambolle, A., Novaga, M.: The discontinuity set of solutions of the TV denoising problem and some extensions. Multiscale Model. Simul. **6**(3), 879–894 (2008)
24. Chambolle, A.: Convex representation for lower semicontinuous envelopes of functionals in l^1. J. Convex Anal. **8**(1), 149–170 (2001)
25. Chambolle, A., Lions, P.L.: Image recovery via total variation minimization and related problems. Numer. Math. **76**, 167–188 (1997)
26. Chambolle, A., Pock, T.: A first-order primal-dual algorithm for convex problems with applications to imaging. J. Math. Imaging Vis. **40**, 120–145 (2011). doi:10.1007/s10851-010-0251-1
27. Chan, T.F., Kang, S.H., Shen, J.: Euler's elastica and curvature-based inpainting. SIAM J. Appl. Math. pp. 564–592 (2002)
28. Chan, T., Marquina, A., Mulet, P.: High-order total variation-based image restoration. SIAM J. Sci. Comput. **22**(2), 503–516 (2000). doi:10.1137/S1064827598344169
29. Chan, T.F., Esedoglu, S.: Aspects of total variation regularized L^1 function approximation. SIAM J. Appl. Math. **65**, 1817–1837 (2005)
30. Chen, K., Lorenz, D.A.: Image sequence interpolation based on optical flow, segmentation, and optimal control. IEEE Trans. Image Process. **21**(3) (2012). doi:10.1109/TIP.2011.2179305
31. Chen, X., Nashed, Z., Qi, L.: Smoothing methods and semismooth methods for nondifferentiable operator equations. SIAM J. Numer. Anal. **38**(4), 1200–1216 (2001)
32. Cheney, M., Borden, B.: Problems in synthetic-aperture radar imaging. Inverse Probl. **25**(12), 123,005 (2009). doi:10.1088/0266-5611/25/12/123005
33. Chen, K., Lorenz, D.A.: Image sequence interpolation using optimal control. J. Math. Imaging Vis. **41**, 222–238 (2011). doi:10.1007/s10851-011-0274-2
34. Cremers, D., Pock, T., Kolev, K., Chambolle, A.: Convex relaxation techniques for segmentation, stereo and multiview reconstruction. In: Markov Random Fields for Vision and Image Processing. MIT Press (2011)
35. Csiba, D., Qu, Z., Richtárik, P.: Stochastic dual coordinate ascent with adaptive probabilities (2016). Preprint
36. Dabov, K., Foi, A., Katkovnik, V., Egiazarian, K.: Image denoising by sparse 3-d transform-domain collaborative filtering. IEEE Trans. Image Process. **16**(8), 2080–2095 (2007). doi:10.1109/TIP.2007.901238
37. Dal Maso, G., Fonseca, I., Leoni, G., Morini, M.: A higher order model for image restoration: the one-dimensional case. SIAM J. Math. Anal. **40**(6), 2351–2391 (2009). doi:10.1137/070697823
38. Daubechies, I., Defrise, M., De Mol, C.: An iterative thresholding algorithm for linear inverse problems with a sparsity constraint. Commun. Pure Appl. Math. **57**(11), 1413–1457 (2004). doi:10.1002/cpa.20042
39. de Los Reyes, J.C., Schönlieb, C.B., Valkonen, T.: Optimal parameter learning for higher-order regularisation models (2014). In preparation

40. de Los Reyes, J.C., Schönlieb, C.B., Valkonen, T.: The structure of optimal parameters for image restoration problems (2015). http://iki.fi/tuomov/mathematics/interior.pdf. Submitted
41. de Los Reyes, J.C., Schönlieb, C.B.: Image denoising: Learning noise distribution via PDE-constrained optimization. Inverse Probl. Imaging (2014). To appear
42. Didas, S., Weickert, J., Burgeth, B.: Properties of higher order nonlinear diffusion filtering. J. Math. Imaging Vis. **35**(3), 208–226 (2009). doi:10.1007/s10851-009-0166-x
43. Douglas, J., Rachford, H.H.: On the numerical solution of heat conduction problems in two and three space variables. Trans. Am. Math. Soc. **82**(2), 421–439 (1956)
44. Duval, V., Aujol, J.F., Gousseau, Y.: The TVL1 model: a geometric point of view. Multiscale Model. Simul. **8**, 154–189 (2009)
45. Ehrhardt, M., Arridge, S.: Vector-valued image processing by parallel level sets. IEEE Trans. Image Process. **23**(1), 9–18 (2014). doi:10.1109/TIP.2013.2277775
46. Engl, H., Hanke, M., Neubauer, A.: Regularization of Inverse Problems. Mathematics and Its Applications. Springer, Netherlands (2000)
47. Esser, E., Zhang, X., Chan, T.F.: A general framework for a class of first order primal-dual algorithms for convex optimization in imaging science. SIAM J. Imaging Sci. **3**(4), 1015–1046 (2010). doi:10.1137/09076934X
48. Estrada, F.J., Fleet, D.J., Jepson, A.D.: Stochastic image denoising. In: BMVC, pp. 1–11 (2009). See also http://www.cs.utoronto.ca/strider/Denoise/Benchmark/ for updated benchmarks
49. Fang, F., Li, F., Zeng, T.: Single image dehazing and denoising: a fast variational approach. SIAM J. Imaging Sci. **7**(2), 969–996 (2014). doi:10.1137/130919696
50. Fercoq, O., Richtárik, P.: Accelerated, parallel and proximal coordinate descent (2013). Preprint
51. Fornasier, M., Langer, A., Schönlieb, C.B.: A convergent overlapping domain decomposition method for total variation minimization. Numer. Math. **116**(4), 645–685 (2010). doi:10.1007/s00211-010-0314-7
52. Fornasier, M., Schönlieb, C.: Subspace correction methods for total variation and ℓ_1-minimization. SIAM J. Numer. Anal. **47**(5), 3397–3428 (2009). doi:10.1137/070710779
53. Gabay, D.: Applications of the method of multipliers to variational inequalities. In: Fortin, M., Glowinski, R. (eds.) Augmented Lagrangian Methods: Applications to the Numerical Solution of Boundary-Value Problems, Studies in Mathematics and its Applications, vol. 15, pp. 299–331. North-Holland, Amsterdam (1983)
54. Goldstein, T., Osher, S.: The split bregman method for l1-regularized problems. SIAM J. Imaging Sci. **2**(2), 323–343 (2009). doi:10.1137/080725891
55. Grant, M., Boyd, S.: CVX: Matlab software for disciplined convex programming, version 2.1 (2014). http://cvxr.com/cvx
56. Grant, M., Boyd, S.: Graph implementations for nonsmooth convex programs. In: Blondel, V., Boyd, S., Kimura, H. (eds.) Recent Advances in Learning and Control. Lecture Notes in Control and Information Sciences, pp. 95–110. Springer, Verlag (2008)
57. Haber, E., Horesh, L., Tenorio, L.: Numerical methods for experimental design of large-scale linear ill-posed inverse problems. Inverse Probl. **24**(5), 055,012 (2008). doi:10.1088/0266-5611/24/5/055012
58. Hanke, M.: A regularizing levenberg-marquardt scheme, with applications to inverse groundwater filtration problems. Inverse Probl. **13**(1), 79 (1997). doi:10.1088/0266-5611/13/1/007
59. He, B., Yuan, X.: Convergence analysis of primal-dual algorithms for a saddle-point problem: from contraction perspective. SIAM J. Imaging Sci. **5**(1), 119–149 (2012). doi:10.1137/100814494
60. Hintermüller, M., Langer, A.: Non-overlapping domain decomposition methods for dual total variation based image denoising. SFB-Report 2013–014, University of Graz (2013)

61. Hintermüller, M., Rautenberg, C.N., Hahn, J.: Functional-analytic and numerical issues in splitting methods for total variation-based image reconstruction. Inverse Probl. **30**(5), 055,014 (2014). doi:10.1088/0266-5611/30/5/055014
62. Hintermüller, M., Valkonen, T., Wu, T.: Limiting aspects of non-convex TV^{φ} models (2014). http://iki.fi/tuomov/mathematics/tvq.pdf. Submitted
63. Hintermüller, M., Langer, A.: Subspace correction methods for a class of nonsmooth and nonadditive convex variational problems with mixed l^1/l^2 data-fidelity in image processing. SIAM J. Imaging Sci. **6**(4), 2134–2173 (2013). doi:10.1137/120894130
64. Hintermüller, M., Stadler, G.: An infeasible primal-dual algorithm for total bounded variation-based inf-convolution-type image restoration. SIAM J. Sci. Comput. **28**(1), 1–23 (2006)
65. Hintermüller, M., Wu, T.: Nonconvex TV^q-models in image restoration: analysis and a trust-region regularization-based superlinearly convergent solver. SIAM J. Imaging Sci. **6**, 1385–1415 (2013)
66. Hintermüller, M., Wu, T.: A superlinearly convergent R-regularized Newton scheme for variational models with concave sparsity-promoting priors. Comput. Optim. Appl. **57**, 1–25 (2014)
67. Hiriart-Urruty, J.B., Lemaréchal, C.: Convex analysis and minimization algorithms I-II. Springer (1993)
68. Horn, B.K., Schunck, B.G.: Determining optical flow. Proc. SPIE **0281**, 319–331 (1981). doi:10.1117/12.965761
69. Huang, J., Mumford, D.: Statistics of natural images and models. In: IEEE CVPR, vol. 1 (1999)
70. Kaltenbacher, B., Neubauer, A., Scherzer, O.: Iterative Regularization Methods for Nonlinear Ill-Posed Problems. No. 6 in Radon Series on Computational and Applied Mathematics. De Gruyter (2008)
71. Kluckner, S., Pock, T., Bischof, H.: Exploiting redundancy for aerial image fusion using convex optimization. In: Goesele, M., Roth, S., Kuijper, A., Schiele, B., Schindler, K. (eds.) Pattern Recognition, Lecture Notes in Computer Science, vol. 6376, pp. 303–312. Springer, Berlin Heidelberg (2010). doi:10.1007/978-3-642-15986-2_31
72. Knoll, F., Bredies, K., Pock, T., Stollberger, R.: Second order total generalized variation (TGV) for MRI. Mag. Reson. Med. **65**(2), 480–491 (2011). doi:10.1002/mrm.22595
73. Knoll, F., Clason, C., Bredies, K., Uecker, M., Stollberger, R.: Parallel imaging with nonlinear reconstruction using variational penalties. Mag. Reson. Med. **67**(1), 34–41 (2012)
74. Kolehmainen, V., Tarvainen, T., Arridge, S.R., Kaipio, J.P.: Marginalization of uninteresting distributed parameters in inverse problems-application to diffuse optical tomography. Int. J. Uncertain. Quantif. **1**(1) (2011)
75. Konečný, J., Richtárik, P.: Semi-stochastic gradient descent methods (2013). Preprint
76. Kunisch, K., Hintermüller, M.: Total bounded variation regularization as a bilaterally constrained optimization problem. SIAM J. Imaging Sci. **64**(4), 1311–1333 (2004). doi:10.1137/S0036139903422784
77. Kunisch, K., Pock, T.: A bilevel optimization approach for parameter learning in variational models. SIAM J. Imaging Sci. **6**(2), 938–983 (2013)
78. Lellmann, J., Lorenz, D., Schönlieb, C.B., Valkonen, T.: Imaging with Kantorovich-Rubinstein discrepancy. SIAM J. Imaging Sci. **7**, 2833–2859 (2014). doi:10.1137/140975528, http://iki.fi/tuomov/mathematics/krtv.pdf
79. Lellmann, J., Strekalovskiy, E., Koetter, S., Cremers, D.: Total variation regularization for functions with values in a manifold. In: 2013 IEEE International Conference on Computer Vision (ICCV), pp. 2944–2951 (2013). doi:10.1109/ICCV.2013.366
80. Lellmann, J., Lellmann, B., Widmann, F., Schnörr, C.: Discrete and continuous models for partitioning problems. Int. J. Comput. Vis. **104**(3), 241–269 (2013). doi:10.1007/s11263-013-0621-4

81. Lewis, A.: The convex analysis of unitarily invariant matrix functions. J. Convex Anal. **2**(1), 173–183 (1995)
82. Lions, P.L., Mercier, B.: Splitting algorithms for the sum of two nonlinear operators. SIAM J. Numer. Anal. **16**(6), 964–979 (1979)
83. Lorenz, D.A., Pock, T.: An accelerated forward-backward method for monotone inclusions (2014). Preprint
84. Loris, I., Verhoeven, C.: On a generalization of the iterative soft-thresholding algorithm for the case of non-separable penalty. Inverse Probl. **27**(12), 125,007 (2011). doi:10.1088/0266-5611/27/12/125007
85. Loris, I.: On the performance of algorithms for the minimization of ℓ_1-penalized functionals. Inverse Probl. **25**(3), 035,008 (2009). doi:10.1088/0266-5611/25/3/035008
86. Loris, I., Verhoeven, C.: Iterative algorithms for total variation-like reconstructions in seismic tomography. GEM Int. J. Geomath. **3**(2), 179–208 (2012). doi:10.1007/s13137-012-0036-3
87. Lysaker, M., Lundervold, A., Tai, X.C.: Noise removal using fourth-order partial differential equation with applications to medical magnetic resonance images in space and time. IEEE Trans. Image Process. **12**(12), 1579–1590 (2003). doi:10.1109/TIP.2003.819229
88. Meyer, Y.: Oscillating patterns in image processing and nonlinear evolution equations. AMS (2001)
89. Möllenhoff, T., Strekalovskiy, E., Möller, M., Cremers, D.: The primal-dual hybrid gradient method for semiconvex splittings (2014). arXiv preprint arXiv:1407.1723
90. Möller, M., Burger, M., Dieterich, P., Schwab, A.: A framework for automated cell tracking in phase contrast microscopic videos based on normal velocities. J. Vis. Commun. Image Represent. **25**(2), 396–409 (2014). doi:10.1016/j.jvcir.2013.12.002, http://www.sciencedirect.com/science/article/pii/S1047320313002162
91. Morozov, V.A.: On the solution of functional equations by the method of regularization. Soviet Math. Dokldy **7**, 414–417 (1966)
92. Mueller, J.L., Siltanen, S.: Linear and Nonlinear Inverse Problems with Practical Applications. Society for Industrial and Applied Mathematics, Philadelphia, PA (2012). doi:10.1137/1.9781611972344
93. Mumford, D., Shah, J.: Optimal approximations by piecewise smooth functions and associated variational problems. Commun. Pure Appl. Math. **42**(5), 577–685 (1989). doi:10.1002/cpa.3160420503
94. Nesterov, Y.: A method of solving a convex programming problem with convergence rate $O(1/k^2)$. Soviet Math. Doklady **27**(2), 372–376 (1983)
95. Nesterov, Y.: Smooth minimization of non-smooth functions. Math. Program. **103**(1), 127–152 (2005). doi:10.1007/s10107-004-0552-5
96. Nesterov, Y.: Efficiency of coordinate descent methods on huge-scale optimization problems. SIAM J. Optim. **22**(2), 341–362 (2012). doi:10.1137/100802001
97. Nocedal, J., Wright, S.: Numerical Optimization. Springer Series in Operations Research and Financial Engineering. Springer (2006)
98. Ochs, P., Chen, Y., Brox, T., Pock, T.: iPiano: Inertial proximal algorithm for non-convex optimization (2014). arXiv preprint arXiv:1404.4805
99. Ochs, P., Dosovitskiy, A., Brox, T., Pock, T.: An iterated l1 algorithm for non-smooth non-convex optimization in computer vision. In: IEEE CVPR (2013)
100. Osher, S., Burger, M., Goldfarb, D., Xu, J., Yin, W.: An iterative regularization method for total variation-based image restoration. Multiscale Model. Simul. **4**(2), 460–489 (2005). doi:10.1137/040605412
101. Pan, X., Sidky, E.Y., Vannier, M.: Why do commercial ct scanners still employ traditional, filtered back-projection for image reconstruction? Inverse Probl. **25**(12), 123,009 (2009). doi:10.1088/0266-5611/25/12/123009
102. Papafitsoros, K., Bredies, K.: A study of the one dimensional total generalised variation regularisation problem (2013). Preprint
103. Papafitsoros, K., Valkonen, T.: Asymptotic behaviour of total generalised variation. In: Fifth International Conference on Scale Space and Variational Methods in Computer Vision (SSVM) (2015). http://iki.fi/tuomov/mathematics/ssvm2015-40.pdf, Accepted

104. Papafitsoros, K., Schönlieb, C.B.: A combined first and second order variational approach for image reconstruction. J. Math. Imaging Vis. **48**(2), 308–338 (2014). doi:10.1007/s10851-013-0445-4
105. Passty, G.B.: Ergodic convergence to a zero of the sum of monotone operators in hilbert space. J. Math. Anal Appl. **72**(2), 383–390 (1979). doi:10.1016/0022-247X(79)90234-8
106. Perona, P., Malik, J.: Scale-space and edge detection using anisotropic diffusion. IEEE TPAMI **12**(7), 629–639 (1990). doi:10.1109/34.56205
107. Pock, T., Chambolle, A., Cremers, D., Bischof, H.: A convex relaxation approach for computing minimal partitions. In: IEEE CVPR, pp. 810–817 (2009). doi:10.1109/CVPR.2009.5206604
108. Pock, T., Chambolle, A.: Diagonal preconditioning for first order primal-dual algorithms in convex optimization. In: 2011 IEEE International Conference on Computer Vision (ICCV), pp. 1762–1769 (2011). doi:10.1109/ICCV.2011.6126441
109. Pock, T., Cremers, D., Bischof, H., Chambolle, A.: An algorithm for minimizing the mumford-shah functional. In: 2009 IEEE 12th International Conference on Computer Vision, pp. 1133–1140 (2009). doi:10.1109/ICCV.2009.5459348
110. Pock, T., Cremers, D., Bischof, H., Chambolle, A.: Global solutions of variational models with convex regularization. SIAM J. Imaging Sci. **3**(4), 1122–1145 (2010). doi:10.1137/090757617
111. Qi, L., Sun, J.: A nonsmooth version of newton's method. Math. Program. **58**(1–3), 353–367 (1993). doi:10.1007/BF01581275
112. Ranftl, R., Pock, T., Bischof, H.: Minimizing tgv-based variational models with non-convex data terms. In: Kuijper, A., Bredies, K., Pock, T., Bischof, H. (eds.) Scale Space and Variational Methods in Computer Vision, Lecture Notes in Computer Science, vol. 7893, pp. 282–293. Springer, Berlin Heidelberg (2013). doi:10.1007/978-3-642-38267-3_24
113. Richtárik, P., Takáč, M.: Parallel coordinate descent methods for big data optimization. Math. Program. pp. 1–52 (2015). doi:10.1007/s10107-015-0901-6
114. Ring, W.: Structural properties of solutions to total variation regularization problems. ESAIM Math. Model. Numer. Anal. **34**, 799–810 (2000). doi:10.1051/m2an:2000104
115. Rockafellar, R.T., Wets, R.J.B.: Variational Analysis. Springer (1998)
116. Rockafellar, R.T.: Convex Analysis. Princeton University Press (1972)
117. Rockafellar, R.T.: Monotone operators and the proximal point algorithm. SIAM J. Optim. **14**(5), 877–898 (1976). doi:10.1137/0314056
118. Rudin, L., Osher, S., Fatemi, E.: Nonlinear total variation based noise removal algorithms. Phys. D **60**, 259–268 (1992)
119. Sawatzky, A., Brune, C., Möller, J., Burger, M.: Total variation processing of images with poisson statistics. In: Jiang, X., Petkov, N. (eds.) Computer Analysis of Images and Patterns, Lecture Notes in Computer Science, vol. 5702, pp. 533–540. Springer, Berlin Heidelberg (2009). doi:10.1007/978-3-642-03767-2_65
120. Schuster, T., Kaltenbacher, B., Hofmann, B., Kazimierski, K.: Regularization Methods in Banach Spaces. Radon Series on Computational and Applied Mathematics. De Gruyter (2012)
121. Setzer, S.: Operator splittings, bregman methods and frame shrinkage in image processing. Int. J. Comput. Vis. **92**(3), 265–280 (2011). doi:10.1007/s11263-010-0357-3
122. Shen, J., Kang, S., Chan, T.: Euler's elastica and curvature-based inpainting. SIAM J. Appl. Math. **63**(2), 564–592 (2003). doi:10.1137/S0036139901390088
123. Stathaki, T.: Image Fusion: Algorithms and Applications. Elsevier Science (2011)
124. Strang, G., Nguyen, T.: Wavelets and filter banks. Wellesley Cambridge Press (1996)
125. Sun, D., Han, J.: Newton and Quasi-Newton methods for a class of nonsmooth equations and related problems. SIAM J. Optim. **7**(2), 463–480 (1997). doi:10.1137/S1052623494274970
126. Suzuki, T.: Stochastic dual coordinate ascent with alternating direction multiplier method (2013). Preprint

127. Tournier, J.D., Mori, S., Leemans, A.: Diffusion tensor imaging and beyond. Mag. Reson. Med. **65**(6), 1532–1556 (2011). doi:10.1002/mrm.22924
128. Tseng, P.: Applications of a splitting algorithm to decomposition in convex programming and variational inequalities. SIAM J. Control Optim. **29**(1), 119–138 (1991). doi:10.1137/0329006
129. Tseng, P., Yun, S.: A coordinate gradient descent method for nonsmooth separable minimization. Math. Program. **117**(1–2), 387–423 (2009). doi:10.1007/s10107-007-0170-0
130. Valkonen, T.: A method for weighted projections to the positive definite cone. Optim. (2014). doi:10.1080/02331934.2014.929680, Published online 24 Jun
131. Valkonen, T.: A primal-dual hybrid gradient method for non-linear operators with applications to MRI. Inverse Probl. **30**(5), 055,012 (2014). doi:10.1088/0266-5611/30/5/055012
132. Valkonen, T.: The jump set under geometric regularisation. Part 1: Basic technique and first-order denoising. SIAM J. Math. Anal. **47**(4), 2587–2629 (2015). doi:10.1137/140976248, http://iki.fi/tuomov/mathematics/jumpset.pdf
133. Valkonen, T.: The jump set under geometric regularisation. Part 2: Higher-order approaches (2014). Submitted
134. Valkonen, T.: Transport equation and image interpolation with SBD velocity fields. J. Math. Pures Appl. **95**, 459–494 (2011). doi:10.1016/j.matpur.2010.10.010
135. Valkonen, T., Bredies, K., Knoll, F.: Total generalised variation in diffusion tensor imaging. SIAM J. Imaging Sci. **6**(1), 487–525 (2013). doi:10.1137/120867172
136. Valkonen, T., Knoll, F., Bredies, K.: TGV for diffusion tensors: A comparison of fidelity functions. J. Inverse Ill-Posed Probl. **21**(355–377), 2012 (2013). doi:10.1515/jip-2013-0005, Special issue for IP:M&S, Antalya, Turkey
137. Valkonen, T., Liebmann, M.: GPU-accelerated regularisation of large diffusion tensor volumes. Computing **95**(771–784), 2012 (2013). doi:10.1007/s00607-012-0277-x, Special issue for ESCO, Pilsen, Czech Republic
138. Vese, L.A., Chan, T.F.: A multiphase level set framework for image segmentation using the Mumford and Shah model. Int. J. Comput. Vis. **50**(3), 271–293 (2002). doi:10.1023/A:1020874308076
139. Vese, L.A., Osher, S.J.: Modeling textures with total variation minimization and oscillating patterns in image processing. J. Sci. Comput. **19**(1–3), 553–572 (2003)
140. Vogel, C., Oman, M.: Iterative methods for total variation denoising. SIAM J. Sci. Comput. **17**(1), 227–238 (1996). doi:10.1137/0917016
141. Wernick, M., Aarsvold, J.: Emission Tomography: The Fundamentals of PET and SPECT. Elsevier Science (2004)
142. Wright, S.: Coordinate descent algorithms. Math. Program. **151**(1), 3–34 (2015). doi:10.1007/s10107-015-0892-3
143. Yilmaz, A., Javed, O., Shah, M.: Object tracking: a survey. ACM Comput. Surv. (CSUR) **38**(4), 13 (2006)
144. Yin, W., Osher, S., Goldfarb, D., Darbon, J.: Bregman iterative algorithms for ℓ_1-minimization with applications to compressed sensing. SIAM J. Imaging Sci. **1**(1), 143–168 (2008). doi:10.1137/070703983
145. Zhao, P., Zhang, T.: Stochastic optimization with importance sampling (2014). Preprint
146. Zhao, F., Noll, D., Nielsen, J.F., Fessler, J.: Separate magnitude and phase regularization via compressed sensing. IEEE Trans. Med. Imaging **31**(9), 1713–1723 (2012). doi:10.1109/TMI.2012.2196707
147. Zhu, M., Chan, T.: An efficient primal-dual hybrid gradient algorithm for total variation image restoration. UCLA CAM Report (2008)

Author Biography

Tuomo Valkonen received his Ph.D in scientific computing from the University of Jyväskylä (Finland) in 2008. He has since then worked in well-known research groups in Graz, Cambridge and Quito. Currently in Cambridge, his research concentrates on the mathematical analysis of image processing models, towards studying their reliability, and the development of fast optimisation algorithms for the solution of these models.

Interlinking Big Data to Web of Data

Enayat Rajabi and Seyed-Mehdi-Reza Beheshti

Abstract The big data problem can be seen as a massive number of data islands, ranging from personal, shared, social to business data. The data in these islands is getting large scale, never ending, and ever changing, arriving in batches at irregular time intervals. Examples of these are social and business data. Linking and analyzing of this potentially connected data is of high and valuable interest. In this context, it will be important to investigate how the Linked Data approach can enable the Big Data optimization. In particular, the Linked Data approach has recently facilitated the accessibility, sharing, and enrichment of data on the Web. Scientists believe that Linked Data reduces Big Data variability by some of the scientifically less interesting dimensions. In particular, by applying the Linked Data techniques for exposing structured data and eventually interlinking them to useful knowledge on the Web, many syntactic issues vanish. Generally speaking, this approach improves data optimization by providing some solutions for intelligent and automatic linking among datasets. In this chapter, we aim to discuss the advantages of applying the Linked Data approach, towards the optimization of Big Data in the Linked Open Data (LOD) cloud by: (i) describing the impact of linking Big Data to LOD cloud; (ii) representing various interlinking tools for linking Big Data; and (iii) providing a practical case study: linking a very large dataset to DBpedia.

Keywords Big data · Linked open data · Interlinking optimization

E. Rajabi (✉)
Computer Science Department, Dalhousie University, Halifax, NS, Canada
e-mail: rajabi@dal.ca

Seyed-Mehdi-RezaBeheshti
University of New South Wales, Sydney, Australia
e-mail: sbeheshti@cse.unsw.edu.au

A. Emrouznejad (ed.), *Big Data Optimization: Recent Developments and Challenges*, Studies in Big Data 18, DOI 10.1007/978-3-319-30265-2_6

1 Introduction

The big data problem can be seen as a massive number of data islands, ranging from personal, shared, social to business data. The data in these islands are increasingly becoming large-scale, never-ending, and ever changing; they may also arrive in batches at irregular time intervals. Examples of these are social (the streams of 3,000–6,000 tweets per second in Twitter) and business data. The adoption of social media, the digitalisation of business artefacts (e.g. files, documents, reports, and receipts), using sensors (to measure and track everything), and more importantly generating huge metadata (e.g. versioning, provenance, security, and privacy), for imbuing the business data with additional semantics, generate part of this big data. Wide physical distribution, diversity of formats, non-standard data models, independently-managed and heterogeneous semantics are characteristics of this big data. Linking and analysing of this potentially connected data is of high and valuable interest. In this context, it will be important to investigate how the Linked Data approach can enable the Big Data optimization.

In recent years, the Linked Data approach [1] has facilitated the availability of different kinds of information on the Web and in some senses; it has been part of the Big Data [2]. The view that data objects are linked and shared is very much in line with the goals of Big Data and it is fair to mention that Linked Data could be an ideal pilot place in Big Data research. Linked Data reduces Big Data variability by some of the scientifically less interesting dimensions. Connecting and exploring data using RDF [3], a general way to describe structured information in Linked Data, may lead to creation of new information, which in turn may enable data publishers to formulate better solutions and identify new opportunities. Moreover, the Linked Data approach applies vocabularies which are created using a few formally well-defined languages (e.g., OWL [4]). From searching and accessibility perspective, a lot of compatible free and open source tools and systems have been developed on the Linked Data context to facilitate the loading, querying and interlinking of open data islands. These techniques can be largely applied in the context of Big Data.

In this context, optimization approaches to interlinking Big Data to the Web of Data can play a critical role in scaling and understanding the potentially connected resources scattered over the Web. For example, Open Government establishes a modern cooperation among politicians, public administration, industry and private citizens by enabling more transparency, democracy, participation and collaboration. Using and optimizing the links between Open Government Data (OGD) and useful knowledge on the Web, OGD stakeholders can contribute to provide collections of enriched data. For instance, US government data[1] including around 111,154 datasets, at the time of writing this book, that was launched on May 2009 having

[1] http://data.gov/.

only 76 datasets from 11 government agencies. This dataset, as a US government Web portal provides the public with access to federal government-created datasets and increases efficiency among government agencies. Most US government agencies already work on the codified information dissemination requirements, and 'data.gov' being conceived as a tool to aid their mission delivery. Another notable example, in the context of e-learning, provides linking of educational resources from different repositories to other datasets on the Web.

Optimizing approaches to interconnecting e-Learning resources may enable sharing, navigation and reusing of learning objects. As a motivating scenario, consider a researcher who might explore the contents of a big data repository in order to find a specific resource. In one of the resources, a video on the subject of his interests may catch the researcher's attention and thus follows the provided description, which has been provided in another language. Assuming that the resources in the repository have been previously interlinked with knowledge bases such as DBpedia,[2] the user will be enabled to find more information on the topic including different translations.

Obviously, the core of data accessibility throughout the Web can provide the links between items, as this idea is prominent in literature on Linked Data principles [1]. Indeed, establishing links between objects in a big dataset is based on the assumption that the Web is migrating from a model of isolated data repositories to a Web of interlinked data. One advantage of data connectivity in a big dataset [5] is the possibility of connecting a resource to valuable collections on the Web. In this chapter, we discuss how optimization approaches to interlinking Web of data to Big Data can enrich a Big Dataset. After a brief discussing on different interlinking tools in the Linked Data context, we explain how an interlinking process can be applied for linking a dataset to Web of Data. Later, we experiments an interlinking approach over a sample of Big Dataset in eLearning literature and conclude the chapter by reporting on the results.

2 Interlinking Tools

There exist several approaches for interlinking data in the context of LD. Simperl et al. [6] provided a comparison of interlinking tools based upon some criteria such as use cases, annotation, input and output. Likewise, we explain some of the related tools, by focusing on their need to human contribution (to what extent users have to contribute in interlinking), their automation (to what extent the tool needs human input), and the area (in which environment the tool can be applied).

From a human contribution perspective, User Contributed Interlinking (UCI) [7] creates different types of semantic links such as *owl:sameas* and *rdf:seeAlso*

[2]http://dbpedia.org.

between two datasets relying on user contributions. In this Wiki-style approach, users can add, view or delete links between data items in a dataset by making use of a UCI interface. Games With A Purpose (GWAP) [8] is another software which provides incentives for users to interlink datasets using game and pictures by distinguishing different pictures with the same name. Linkage Query Writer (LinQuer) [9] is also another tool for semantic link discovery [10] between different datasets which allows users to write their queries in an interface using some APIs.

Automatic Interlinking (AI) is another linking approach for interconnecting of data sources applied for identifying semantic links between data sources. Semi-automatic interlinking [11], as an example, is a kind of analyzing technique to assign multimedia data to users using multimedia metadata. Interlinking multimedia (iM) [11] is also a pragmatic way in this context for applying the LD to fragments of multimedia items and presents methods for enabling a widespread use of interlinking multimedia. RDF-IA [12] is another linking tool that carries out matching and fusion of RDF datasets according to the user configuration, and generates several outputs including *owl:sameAs* statements between the data items.

Another semi-automatic approach for interlinking is the Silk Link Discovery Framework [13], which finds the similarities within different LD sources by specifying the types of RDF links via SPARQL endpoints or data dumps. LIMES [14] is also a link discovery software in the LOD that presents a tool in command-line and GUI for finding similarities between two datasets and suggests the results to users based on the metrics automatically. LODRefine [15] is another tool for cleaning, transforming, and interlinking any kinds of data with a web user interface. It has the benefit of reconciling data to the LOD datasets (e.g., Freebase or DBpedia) [15]. The following table briefly summarizes the described tools and mentions the area of application for each one (Table 1).

Table 1 Existing interlinking tools description

Tool	Area
UCI	General data source
GWAP	Web pages, e-commerce offerings, Flickr images, and YouTube
LinQuer	LOD datasets
IM	Multimedia
RDF-IA	LOD datasets
Silk	LOD datasets
LIMES	LOD datasets
LODRefine	General data, LOD datasets

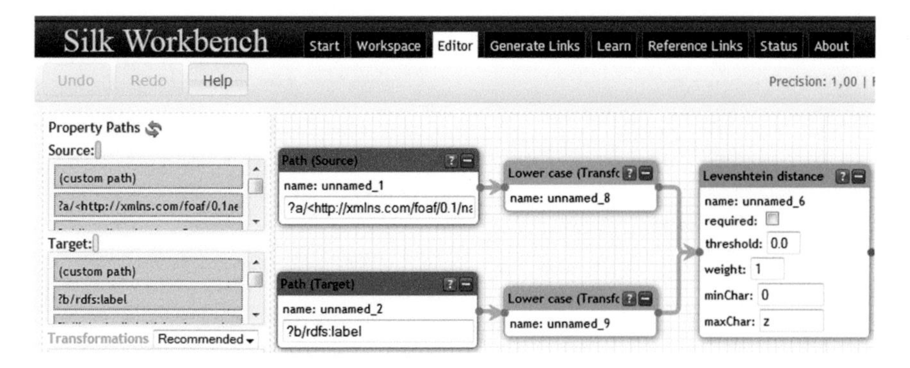

Fig. 1 Silk work-bench interface

To discuss the most used tools in Linked Data context we have selected three software and explain their characteristics and the way that they interlink datasets.

2.1 *Silk*

Silk [13] is an interlinking software that matches two datasets using string matching techniques. It applies some similarity metrics to discover similarities between two concepts. By specifying two datasets as input (SPARQL endpoints or RDF dumps), Silk provides as an output e.g., "sameAs" triples between the matched entities. Silk Workbench, is the web application variant of Silk which allows users to interlink datasets through the process of interlinking different data sources by offering a graphical editor to create link specifications (consider Fig. 1). After performing the interlinking process, the user can evaluate the generated links. A number of projects including DataLift [16] have employed the Silk engine to carry out their interlinking purposes.

2.2 *LIMES*

Link Discovery Framework for Metric Spaces (LIMES) is another interlinking tool which presents a linking approach for discovering relationships between entities contained in Linked Data sources [14]. LIMES leverages several mathematical characteristics of metric spaces to compute pessimistic approximations of the similarity of instances. It processes the strings by making use of suffix-, prefix- and position filtering in a string mapper by specifying a source dataset, a target dataset,

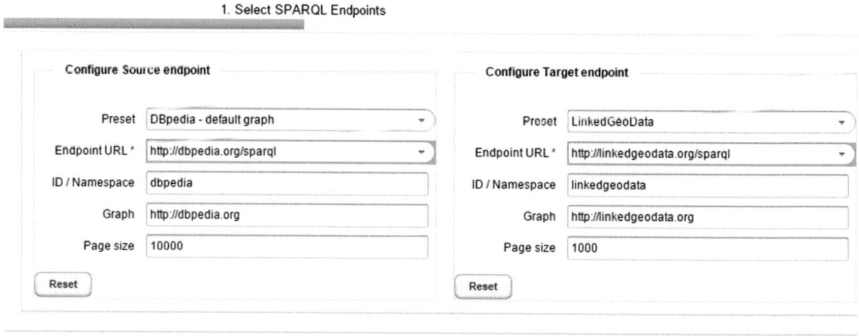

Fig. 2 LIMES web interface

and a link specification. LIMES applies either a SPARQL Endpoint or a RDF dump from both targets. A user can also set a threshold for various matching metrics by which two instances are considered as matched, when the similarity between the terms exceeds the defined value. A recent study [14] evaluated LIMES as a time-efficient approach, particularly when it is applied to link large data collections. Figure 2 depicts the web interface of LIMES (called SAIM[3]) was recently provided by AKSW group.[4]

2.3 LODRefine

LODRefine [15] is another tool in this area that allows data to be loaded, refined, and reconciled. It also provides additional functionalities for dealing with the Linked Open Data cloud. This software discovers similarities between datasets by linking the data items to the target datasets. LODRefine matches similar concepts automatically and suggests the results to users for review. Users also can expand their contents with concepts from the LOD datasets (e.g., DBpedia) once the data has been reconciled. They can also specify the condition for the interlinking. Eventually, LODRefine reports the interlinking results and provides several functionalities for filtering the results. LODRefine also allows users to refine and manage data before starting the interlinking process, which is very useful when the user dataset includes several messy content (e.g., null, unrelated contents) and facilitates the process by reducing the number of source concepts. Figure 3 depicts a snapshot of this tool.

[3]http://saim.aksw.org/.

[4]http://www.aksw.org.

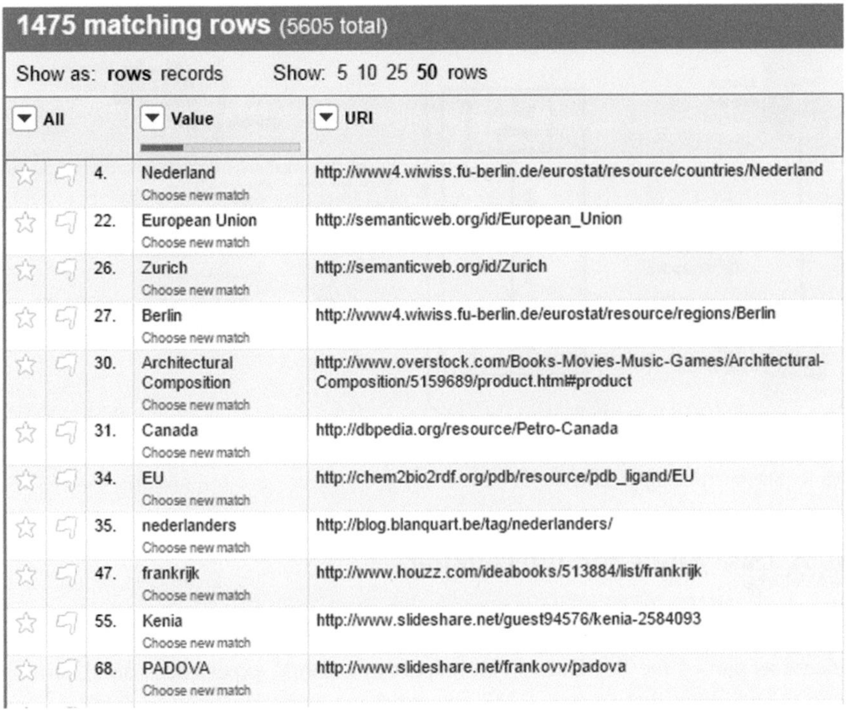

1475 matching rows (5605 total)

Show as: **rows** records Show: 5 10 25 **50** rows

All	Value	URI
4.	Nederland Choose new match	http://www4.wiwiss.fu-berlin.de/eurostat/resource/countries/Nederland
22.	European Union Choose new match	http://semanticweb.org/id/European_Union
26.	Zurich Choose new match	http://semanticweb.org/id/Zurich
27.	Berlin Choose new match	http://www4.wiwiss.fu-berlin.de/eurostat/resource/regions/Berlin
30.	Architectural Composition Choose new match	http://www.overstock.com/Books-Movies-Music-Games/Architectural-Composition/5159689/product.html#product
31.	Canada Choose new match	http://dbpedia.org/resource/Petro-Canada
34.	EU Choose new match	http://chem2bio2rdf.org/pdb/resource/pdb_ligand/EU
35.	nederlanders Choose new match	http://blog.blanquart.be/tag/nederlanders/
47.	frankrijk Choose new match	http://www.houzz.com/ideabooks/513884/list/frankrijk
55.	Kenia Choose new match	http://www.slideshare.net/guest94576/kenia-2584093
68.	PADOVA Choose new match	http://www.slideshare.net/frankovv/padova

Fig. 3 LODRefine interface

3 Interlinking Process

In an ideal scenario, a data island can be linked to a diverse collection of sources on the Web of Data. However, connecting each entity, available in the data island, to an appropriate source is very time-consuming. Particularly when we face a big number of data items, the domain expert needs to explore the target dataset in order to be able to apply queries. As mentioned earlier and to minimize the human contribution, interlinking tools have facilitated the interlinking process by implementing a number of matching techniques. While using an interlinking tool, several issues such as defining the configuration for the linking process, specifying the criteria, and post-processing the output need to be addressed. In particular, the user sets a configuration file in order to specify the criteria under which items are linked in the datasets. Eventually, the tool generates links between concepts under the specified criteria and provides output in order to be reviewed and verified by users. Once the linking process has finished, the user can evaluate the accuracy of the generated links that are close to the similarity threshold. Specifically, the user can verify or reject each link recommended by the tool as the two matching concepts (see Fig. 4).

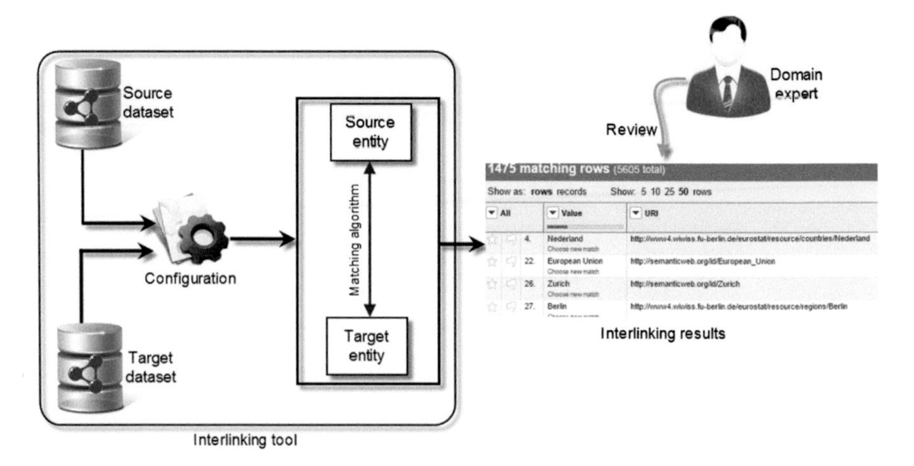

Fig. 4 The interlinking process

4 A Case Study for Interlinking

There exist a wide variety of data sources on the Web of Data that can be considered as part of the Big Data. With respect to authors' experiences on eLearning context and given that around 1,362 datasets have been registered in datahub[5] and tagged as "eLearning datasets", we selected the GLOBE repository,[6] a large dataset with almost 1.2 million learning resources and more than 10 million concepts [5]. The GLOBE is a federated repository that consists of several other repositories, such as OER Commons [17] which includes manually created metadata as well as aggregated metadata from different sources, we selected GLOBE for our case study to assess the possibility of interlinking. The metadata of learning resources in GLOBE are based upon the IEEE LOM schema [18] which is a de facto standard for describing learning objects on the Web. Title, keywords, taxonomies, language, and description of a learning resource are some of the metadata elements in an IEEE LOM schema which includes more than 50 elements. Current research on the use of GLOBE learning resource metadata [19] shows that 20 metadata elements are used consistently in the repository.

To analyze the GLOBE resource metadata, we collected more than 800,000 metadata files via OAI-PMH[7] protocol from the GLOBE repository. Some GLOBE metadata could not be harvested due to validation errors (e.g., LOM extension errors). Particularly, several repositories in GLOBE extended the IEEE LOM by adding new elements without using namespaces, which caused a number of errors

[5]http://datahub.io/.

[6]http://globe-info.org/.

[7]Open Archives Initiative Protocol for Metadata Harvesting." [Online]. Available: http://www.openarchives.org/pmh. [Accessed: 22-February-2014].

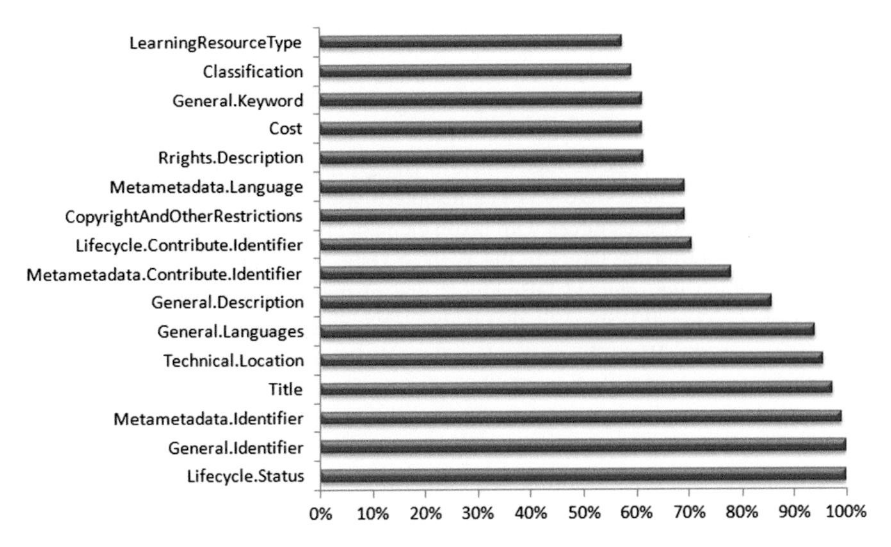

Fig. 5 The usage of metadata elements by GLOBE resources

detected by the ARIADNE validation service.[8] Later, we converted the harvested XML files into a relational database using a JAVA program in order to examine those elements that are more useful for the interlinking purpose. Figure 5 illustrates the metadata elements those used by more than 50 % of learning resources in GLOBE of which title of learning resource, as an example, has been applied by more than 97 % of the GLOBE resources. More than half (around 55 %) of the resources were in English and 99 % of the learning objects were open and free to use. English is the most prominent language in GLOBE [5] and thus the linking elements used as a source in our data scope were limited to English terms of the selected elements, which were represented in more than one language.

Several metadata elements such as *General.Identifier* or *Technical.Location* are mostly included local values provided by each repository (e.g., "ed091288" or "http://www.maa.org/") and thus could not be considered for interlinking. Additionally, constant values (e.g., dates and times) or controlled vocabularies (e.g., "*Contribute.Role*" and "*Lifecycle.Status*") were not suitable for interlinking, as the user could not obtain useful information by linking these elements. Finally, the following metadata elements were selected for the case study, as they were identified as the most appropriate elements for interlinking [20]:

- Title a learning resource (*General.Title*)
- The taxonomy given to a learning resource *(Classification.Taxon)*
- A Keyword or phrase describing the topic of learning objects *(General. Keyword).*

[8] http://ariadne.cs.kuleuven.be/validationService/.

As the GLOBE resources were not available as RDF, we exposed the GLOBE metadata via a SPARQL endpoint.[9] We exposed the harvested metadata, which were converted into a relational database, as RDF using a mapping service (e.g., D2RQ [10]) and set up a SPARQL Endpoint in order to complete the interlinking process. As a result, the GLOBE data was accessible through a local SPARQL endpoint in order to be interlinked to a target dataset. There were 434,112 resources with title, 306,949 resources with Keyword, and 176,439 resources with taxon element all in English language.

To find an appropriate target in the Web of Data, we studies a set of datasets in datahub of which we selected DBpedia,[11] one of the most used datasets [21] and Linked Data version of Wikipedia that makes it possible to link data items to general information on the Web. In particular, the advantage of linking of contents to DBpedia is to make public information usable for other datasets and to enrich datasets by linking to valuable resources on the Web of Data.The full DBpedia dataset features labels and abstracts for 10.3 million unique topics in 111 different languages[12] about persons, places, and organizations. All DBpedia contents have been classified into 900,000 English concepts, and are provided according to SKOS [13], as a common data model for linking knowledge organization systems on the Web of Data. Hence, this dataset was selected for linking keywords and taxonomies of metadata.

When running an interlinking tool like LIMES, the user sets a configuration file in order to specify the criteria under which items are linked in the two datasets. The tool generates links between items under the specified criteria and provides output which defines whether there was a match or a similar term in order to be verified by users. Once the linking process has finished, the user can evaluate the accuracy of the generated links that are close to the similarity threshold. Specifically, the user can verify or reject each record recommended by the tool as two matching concepts. Eventually, we ran LIMES over three elements of GLOBE (title, Keyword, and Taxon) and DBpedia subjects. Table 2 illustrates the interlinking results in which more than 217,000 GLOBE resources linked to 10,676 DBpedia subjects through keywords. In respect to Taxonomy interlinking, around 132,000 resources in GLOBE were connected to 1,203 resources of the DBpedia dataset, while only 443 GLOBE resources matched to 118 DBpedia resources. The low number of matched links for the title element refers to this fact that interlinking long strings does not lead many matched resources, as most of the GLOBE metadata contained titles with more than two or three words.

The following table (Table 3) illustrates some sample results show those GLOBE resources connected to the DBpedia subjects (two results per element). Having the results and reviewing the matched links by data providers, GLOBE can be enriched with new information so that each resource is connected to DBpedia using e.g., *owl:sameAs* relationship.

---

[9]http://www.w3.org/wiki/SparqlEndpoints.

[10]http://d2rq.org/

[11]http://dbpedia.org.

[12]http://blog.dbpedia.org.

[13]https://www.w3.org/2004/02/skos/

Table 2 Interlinking results between GLOBE and DBpedia

Element	Globe resources#	DBpedia resources#	Total links
Title	443	118	443
Keyword	217,026	10,676	623,390
Taxon	132,693	1,203	268,302

Table 3 Sample interlinking results

Phrase	Element	DBpedia resources URI
Bibliography	Title	http://dbpedia.org/resource/Category:Bibliography
Analysis	Title	http://dbpedia.org/resource/Category:Analysis
Plutoniu	Keyword	http://dbpedia.org/resource/Category:Plutonium
Biology	Keyword	http://dbpedia.org/resource/Category:Biology
Transportation	Taxon	http://dbpedia.org/resource/Category:Transportation
Trigonometry	Taxon	http://dbpedia.org/resource/Category:Trigonometry

5 Conclusions and Future Directions

In this chapter we explained the interlinking approach as a way of optimizing and enriching different kinds of data. We have described the impact of linking Big Data to the LOD cloud. Afterward, we explained various interlinking tools used in Linked Data for interconnecting datasets, along with a discussion about the interlinking process and how a dataset can be interlinked to Web of Data. Finally, we have represented a case study where a interlinking tools (LIMES) used for linking the GLOBE repository to DBpedia. Running the tool and examining the results, many GLOBE resources could connect to DBpedia and after an optimization and enrichment step the new information can be added to the source datasets. This process makes the dataset more valuable and the dataset' users can get more knowledge about the learning resources. The enrichment process over one of large datasets in eLearning context have been presented and it was shown that this process can be extend to other types of data: the process does not depend to a specific context. The quality of a dataset is also optimized when it is connected to other related information on the Web. The previous study on our selected interlinking tool (LIMES) [14] is also showed that it is a promising software when it is applied to a large amount of data.

In conclusion, we believe that enabling the optimization of Big Data and the open data is an important research area, which will attract a lot of attention in the research community. It is important as the explosion of unstructured data has created an information challenge for many organizations. Significant research directions in this area includes: (i) Enhancing linked data approaches with semantic information gathered from a wide variety of sources. Prominent examples include the Google Knowledge Graph [22] and the IBM Watson question answering system [23]; (ii) Integration of existing machine learning and natural language processing algorithms into Big Data platforms [24]; and (iii) High-level declarative approaches to assist users in interlinking Big data to open data. A good example of this can be something similar to OpenRefine [25] which can be specialized for the optimization and enrichment of interlinking big data to different types of open source data; e.g. social data such as Twitter. Summarization approaches such as [26] can be also used to interlinking big data to different sources.

References

1. Bizer, C., Heath, T., Berners-Lee, T.: Linked Data—The Story So Far. Int. J. Semantic Web Inf. Syst. (IJSWIS) . **5**(3) 1–22, 33 (2009)
2. Mayer-Schönberger, V., Cukier, K.: Big Data: A Revolution That Will Transform How We Live, Work, and Think, Reprint edn. Eamon Dolan/Houghton Mifflin Harcourt (2013)
3. Klyne, G., Carroll, J.J.: Resource description framework (RDF): concepts and abstract syntax. W3C Recommendation (2004)
4. OWL 2 Web Ontology Language Document Overview, 2nd edn. http://www.w3.org/TR/owl2-overview/. Accessed 19 May 2013
5. Rajabi, E., Sicilia, M.-A., Sanchez-Alonso, S.: Interlinking educational data: an experiment with GLOBE resources. In: Presented at the First International Conference on Technological Ecosystem for Enhancing Multiculturality, Salamanca, Spain (2013)
6. Simperl, E., Wölger, S., Thaler, S., Norton, B., Bürger, T.: Combining human and computation intelligence: the case of data interlinking tools. Int. J. Metadata Semant. Ontol. **7**(2), 77–92 (2012)
7. Hausenblas, M., Halb, W., Raimond, Y.: Scripting user contributed interlinking. In: Proceedings of the 4th workshop on Scripting for the Semantic Web (SFSW2008), co-located with ESWC2008 (2008)
8. Siorpaes, K., Hepp, M.: Games with a purpose for the semantic web. EEE Intell. Syst **23**(3), 50–60 (2008)
9. Hassanzadeh, O., Xin, R., Miller, J., Kementsietsidis, A., Lim, L., Wang, M.: Linkage query writer. In: Proceedings of the VLDB Endowment (2009)
10. Beheshti, S.M.R., Moshkenani, M.S.: Development of grid resource discovery service based on semantic information. In: Proceedings of the 2007 Spring Simulation Multiconference, San Diego, CA, USA, vol. 1, pp. 141–148 (2007)
11. Bürger, T., Hausenblas, M.: Interlinking Multimedia—Principles and Requirements
12. Scharffe, F., Liu, Y., Zhou, C.: RDF-AI: an architecture for RDF datasets matching, fusion and interlink. In: Proceedings of IJCAI 2009 IR-KR Workshop (2009)
13. Volz, J., Bizer, C., Berlin, F.U., Gaedke, M., Kobilarov, G.: Silk—A Link Discovery Framework for the Web of Data. In Proceedings of the 2nd Linked Data on the Web Workshop (LDOW2009), Madrid, Spain, 2009.
14. Ngonga, A., Sören, A.: LIMES—a time-efficient approach for large-scale link discovery on the web of data. In: Presented at the IJCAI (2011)
15. Verlic, M.: LODGrefine—LOD-enabled Google refine in action. In: Presented at the I-SEMANTICS (Posters & Demos) (2012)
16. The Datalift project: A catalyser for the Web of data. http://datalift.org. Accessed 24 Nov 2013
17. OER Commons. http://www.oercommons.org/. Accessed 17 Jun 2013
18. IEEE P1484.12.4^{TM}/D1, Draft recommended practice for expressing IEEE learning object metadata instances using the dublin core abstract model. http://dublincore.org/educationwiki/DCMIIEEELTSCTaskforce?action=AttachFile&do=get&target=LOM-DCAM-newdraft.pdf. Accessed 12 May 2013

19. Ochoa, X., Klerkx, J., Vandeputte, B., Duval, E.: On the use of learning object metadata: the GLOBE experience. In: Proceedings of the 6th European Conference on Technology Enhanced Learning: towards Ubiquitous Learning, pp. 271–284. Berlin, Heidelberg (2011)
20. Rajabi, E., Sicilia, M.-A., Sanchez-Alonso, S.: Interlinking educational resources to web of data through IEEE LOM. Comput. Sci. Inf. Syst. J. **12**(1) (2014)
21. Rajabi, E., Sanchez-Alonso, S., Sicilia, M.-A.: Analyzing broken links on the web of data: an experiment with DBpedia. J. Am. Soc. Inf. Scechnol. JASIST 2014;65(8):1721
22. Official Google Blog: Introducing the knowledge graph: things, not strings https://googleblog.blogspot.ca/2012/05/introducing-knowledge-graph-things-not.html
23. Ferrucci, D.A.: Introduction to this is Watson. IBM J. Res. Dev. **56**(3.4), 1:1–1:15 (2012)
24. Beheshti, S.-M.-R., Venugopal, S., Ryu, S.H., Benatallah, B., Wang, W.: Big data and cross-document coreference resolution: current state and future opportunities (2013). arXiv: 13113987
25. Open-refine—Google Refine, a powerful tool for working with messy data. http://openrefine.org/. Accessed 14 Jul 2013
26. Beheshti, S.-M.-R., Benatallah, B., Motahari-Nezhad, H.: Scalable graph-based OLAP analytics over process execution data. Distributed and Parallel Databases, 1–45 (2015). doi:10.1007/s10619-014-7171-9

Author Biographies

Dr. Enayat Rajabi is a Post-Doctoral Fellow at the NICHE Research Group at Dalhousie University, Halifax, NS, Canada. He got his PhD in Knowledge Engineering at the Computer Science department of University of Alcalá (Spain) under the supervision of Dr. Salvador Sanchez-Alonso and Prof. Miguel-Angel Sicilia. The subject of his PhD was "Interlinking educational data to Web of Data", in which he applied the Semantic Web technologies to interlink several eLearning datasets to the LOD cloud.

Dr. Beheshti is a Lecturer and Senior Research Associate in the Service Oriented Computing Group, School of Computer Science and Engineering (CSE), University of New South Wales (UNSW), Australia. He did his PhD and Postdoc under supervision of Prof. Boualem Benatallah in UNSW Australia.

Topology, Big Data and Optimization

Mikael Vejdemo-Johansson and Primoz Skraba

Abstract The idea of using geometry in learning and inference has a long history going back to canonical ideas such as Fisher information, Discriminant analysis, and Principal component analysis. The related area of Topological Data Analysis (TDA) has been developing in the last decade. The idea is to extract robust topological features from data and use these summaries for modeling the data. A topological summary generates a coordinate-free, deformation invariant and highly compressed description of the geometry of an arbitrary data set. Topological techniques are well-suited to extend our understanding of Big Data. These tools do not supplant existing techniques, but rather provide a complementary viewpoint to existing techniques. The qualitative nature of topological features do not give particular importance to individual samples, and the coordinate-free nature of topology generates algorithms and viewpoints well suited to highly complex datasets. With the introduction of persistence and other geometric-topological ideas we can find and quantify local-to-global properties as well as quantifying qualitative changes in data.

Keywords Applied topology · Persistent homology · Mapper · Euler characteristic curve · Topological Data Analysis

1 Introduction

All data is geometric.

Every data set is characterized by the way individual observations compare to each other. Statistics of data sets tend to describe location (mean, median, mode) or shape of the data. The shape is intrinsically encoded in the mutual distances between

M. Vejdemo-Johansson (✉)
Computer Vision and Active Perception Laboratory, KTH Royal Institute of Technology, 100 44 Stockholm, Sweden
e-mail: mvj@kth.se

P. Skraba
AI Laboratory, Jozef Stefan Institute, Jamova 39, Ljubljana, Slovenia
e-mail: primoz.skraba@ijs.si

A. Emrouznejad (ed.), *Big Data Optimization: Recent Developments and Challenges*, Studies in Big Data 18, DOI 10.1007/978-3-319-30265-2_7

data points, and analyses of data sets extract geometric invariants: statistical descriptors that are stable with respect to similarities in data. A statistic that describes a data set needs to stay similar if applied to another data set describing the same entity.

In most mainstream big data, computationally lean statistics are computed for data sets that exceed the capacity of more traditional methods in volume, variety, velocity or complexity. Methods that update approximations of location measures, or of fits of simple geometric models—probability distributions or linear regressions—to tall data, with high volume or velocity, are commonplace in the field.

We will focus instead on a family of methods that pick up the geometric aspects of big data, and produce invariants that describe far more of the *complexity* in wide data: invariants that extract a far more detailed description of the data set that goes beyond the location and simplified shapes of linear or classical probabilistic models. For the more detailed geometric description, computational complexity increases. In particular worst case complexities tend to be far too high to scale to large data sets; but even for this, a linear complexity is often observed in practice.

Whereas geometric methods have emerged for big data, such as information geometry [7] and geometric data analysis [67], our focus is on topological methods. Topology focuses on an underlying concept of *closeness*, replacing *distance*. With this switch of focus, the influence of noise is dampened, and invariants emerge that are coordinate-free, invariant under deformation and produce compressed representations of the data. The *coordinate-free* nature of topological methods means, inter alia, that the ambient space for data—the width of the data set—is less relevant for computational complexities and analysis techniques than the intrinsic complexity of the data set itself. *Deformation invariance* is the aspect that produces stability and robustness for the invariants, and dampens out the effects of noise. Finally, *compressed representations* of data enables far quicker further analyses and easily visible features in visualizations.

One first fundamental example of a topological class of algorithms is *clustering*. We will develop homology, a higher-dimensional extension of clustering, with persistent homology taking over the role of hierarchical clustering for more complex shape features. From these topological tools then flow coordinatization techniques for dimensionality reduction, feature generation and localization, all with underlying stability results guaranteeing and quantifying the fidelity of invariants to the original data.

Once you are done with this book chapter, we recommend two further articles to boost your understanding of the emerging field of topological data analysis: Topology and Data by Carlsson [24] and Barcodes: the persistent topology of data by Ghrist [57].

We will start out laying down the fundamentals of topology in Sect. 2. After the classical field of topology, we introduce the adaptation from pure mathematics to data analysis tools in Sect. 3. An important technique that has taken off significantly in recent years is Mapper, producing an intrinsic network description of a data set. We describe and discuss Mapper in Sect. 4. In Sect. 5 we explore connections between topology and optimization: both how optimization tools play a large importance in our topological techniques, and how topological invariants and partitions of features help setup and constrain classes of optimization problems. Next in Sect. 6, we go through several classes of applications of the techniques seen earlier in the chapter. We investigate how topology provides the tools to glue local information into global descriptors, various approaches to nonlinear dimensionality reduction with topological tools, and emerging uses in visualization.

2 Topology

Topology can be viewed as geometry where *closeness* takes over the role of *size* from classical geometry. The fundamental notion is closeness expressed through connectedness. This focus on connections rather than sizes means that invariants focus on qualitative features rather than quantitative: features that do not change with the change of units of measurement, and that stay stable in the face of small perturbations or deformations. For an introductory primer, we recommend the highly accessible textbook by Hatcher [61]. Most of what follows are standard definitions and arguments, slightly adapted to our particular needs in this chapter.

In topological data analysis the focus is on compressed combinatorial representations of shapes. The fundamental building block is the *cell complex*, most often the special case of a *simplicial complex*—though for specific applications *cubical complex*es or more general constructions are relevant.

Definition 1 A *convex polytope* (or convex polyhedron) is the convex hull of some collection of points in $\mathbb{R}^d$. The dimension of the polytope P is the largest n such that the intersection of P with some n-dimensional linear subspace of $\mathbb{R}^d$ contains an n-dimensional open ball.

For an n-dimensional polytope, its boundary decomposes into a union of $n-1$-dimensional polytopes. These are called the *facets* of the polytope. Decomposing facets into their facets produces lower dimensional building blocks—this process continues all the way down to vertices. The set of facets of facets etc. are called the *faces* of the polytope. We write P_n for the set of n-dimensional faces of P.

A *cell complex* is a collection of convex polytopes where the intersection of any two polytopes is a face of each of the polytopes.

We illustrate these geometric conditions in Fig. 1.

From a cell complex, we can produce a *chain complex*. This is a collection of vector spaces with linear maps connecting them. C_nP is the vector space spanned by the n-dimensional faces of P: C_nP has one basis vector for each n-dimensional face. The connecting linear maps are called *boundary maps*: the boundary map $\partial_n : C_nP \to C_{n-1}P$ maps the basis vector v_σ corresponding to a face σ to a linear combination of the vectors that correspond to facets of σ. The coefficients of this linear combination depends on the precise way that the polytopes are connected—if

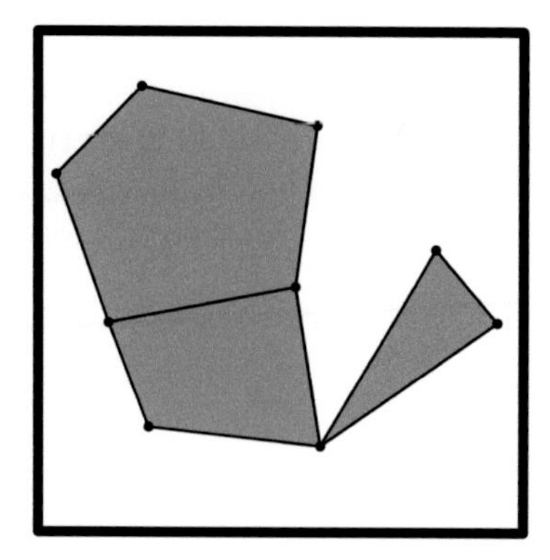
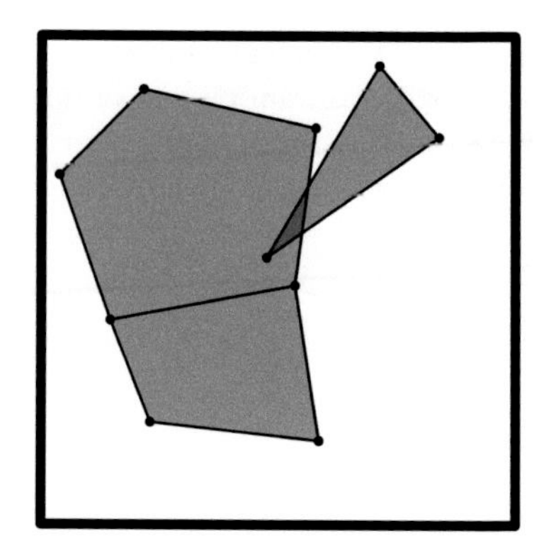

Fig. 1 *Left* a valid polyhedral complex in $\mathbb{R}^2$. *Right* an invalid polyhedral complex in $\mathbb{R}^2$. There are invalid intersections where the *triangle* and the *pentagon* overlap

we work with vector spaces over $\mathbb{Z}_2$ these coefficients reduce to 0 and 1 and the boundary of v_σ is

$$\partial v_\sigma = \sum_{\tau \text{ facet of } \sigma} v_\tau$$

These coefficients need to ensure that $\partial_n \partial_{n+1} = 0$.

The cell complex can be represented in full abstraction as the boundary maps, abstracting away the geometry of this definition completely.

Most commonly, we use simplicial complexes—complexes where the polyhedra are all simplices. Simplices are the same shapes that show up in the simplex method in optimization. Geometrically, an n-dimensional simplex is the convex hull of $n + 1$ points in general position: where no $k + 1$ points lie on the same k-dimensional plane. More interesting for our applications is the idea of an *abstract simplicial complex*.

Definition 2 An *abstract simplicial complex* Σ on a set of (totally ordered) vertices V is a collection of subsets of vertices (simplices) such that whenever some set $\{v_0, \dots, v_k\}$ is in Σ, so is every subset of that set.

We usually represent a simplex as a sorted list of its constituent vertices.

The boundary map assigns to the facet $[v_0, \dots, v_{i-1}, v_{i+1}, \dots, v_k]$ the coefficient $(-1)^i$ so that the full expression of the boundary map is

$$\partial[v_0, \dots, v_k] = \sum_{i=0}^{k} (-1)^i [v_0, \dots, \hat{v}_i, \dots, v_k]$$

where $\hat{v}$ means to leave v out of the simplex.

We illustrate this definition in Fig. 2.

Now consider a closed chain of edges, such as $a - b - c - d - a$ in Fig. 3. The boundary of the sum of these edges includes each vertex twice: once from each edge

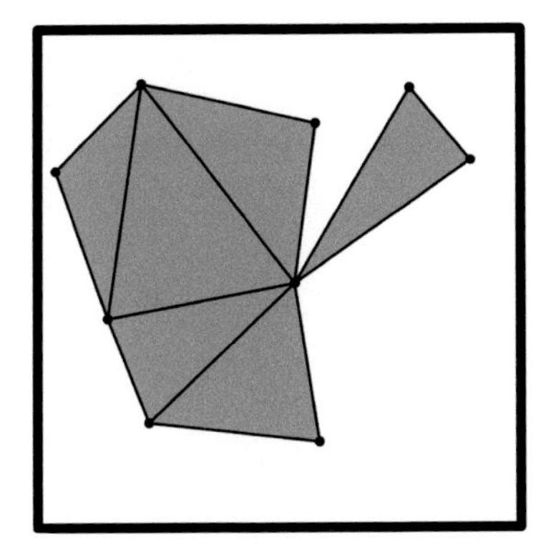

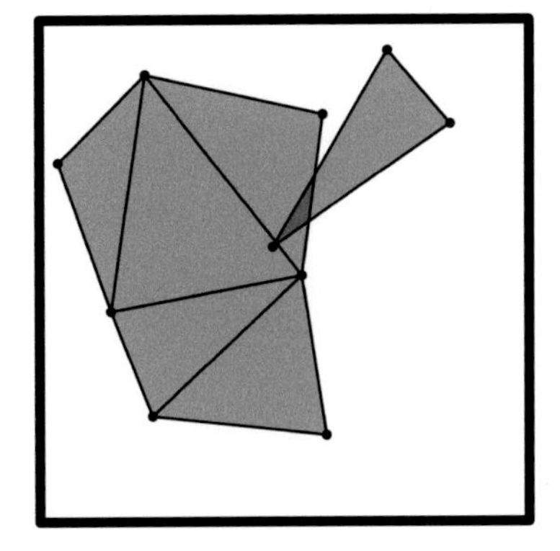

Fig. 2 *Left* a valid simplicial complex in $\mathbb{R}^2$. *Right* an invalid simplicial complex in $\mathbb{R}^2$. There are invalid intersections where two *triangles* overlap

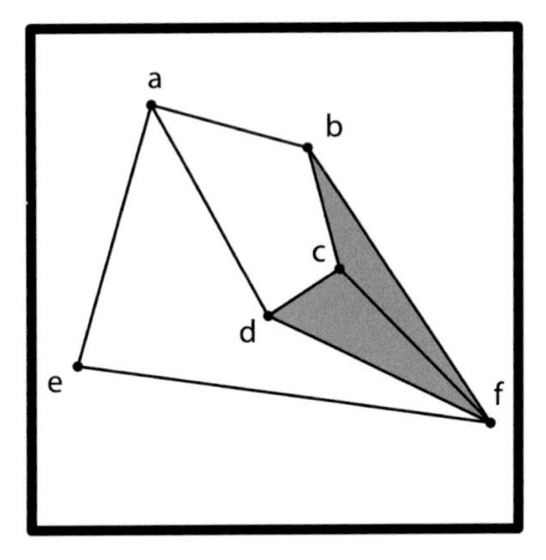

Fig. 3 An illustration of cycles and boundaries. $a - b - c - d - a$ is an essential cycle, while $b - c - d - f - b$ is a non-essential cycle, filled in by higher-dimensional cells

that includes the vertex. The coefficients of these vertices cancel out to 0, so that the closed chain is in $\ker \partial_1$. This generalizes: an element of $\ker \partial_n$ is a collection of n-cells that enclose an $n + 1$-dimensional hypervolume of some sort, in the same way that a closed chain of edges can be seen as enclosing some surface.

Some of these closed chains in a given complex end up being filled in, such as the sequence $b - c - d - f - b$ in Fig. 3, while others have an empty void enclosed. The cells that fill in a closed cycle are part of $C_{n+1}\Sigma$, and the boundary map applied to those cells precisely hits the enclosing cell collection. Thus, $\operatorname{img} \partial_{n+1}$ is the collection of closed cycles that are filled in. This means that the vector space quotient $\ker \partial_n / \operatorname{img} \partial_{n+1}$ is precisely the *essential* enclosures: those that detect a void of some sort.

Definition 3 The n-dimensional *homology* $H_n(\Sigma)$ of a cell complex Σ is the vector space $\ker \partial_n / \operatorname{img} \partial_{n+1}$.

Later in the text we will also need the concept of *cohomology*. This is the homology of the vector space dual of the chain complex: we write $C^n\Sigma = C_n\Sigma$, and $\delta^n = \partial_n^T$; the transposed matrix. Elements of $C^n\Sigma$ correspond to $\mathbb{R}$-valued maps defined on the n-dimensional cells. The n-dimensional cohomology $H^n(\Sigma)$ is $\ker \delta^n / \mathrm{img}\, \delta^{n-1}$.

3 Persistence

The tools and definitions in Sect. 2 all are most relevant when we have a detailed description of the topological shape under study. In any data-driven situation, such as when facing big or complex data, the data accessible tends to take the shape of a discrete point cloud: observations with some similarity measure, but no intrinsic connection between them.

Persistence is the toolbox, introduced in [51] and developed as the foundation of topological data analysis that connects discrete data to topological tools acting on combinatorial or continuous shapes.

At the heart of persistence is the idea of sweeping a parameter across a range of values and studying the ways that a shape derived from data changes with the parameter change. For most applications, the shape is constructed by "decreasing focus": each data point is smeared out over a larger and larger part of the ambient space until the smears start intersecting. We can sometimes define these shapes using only dissimilarity between points, removing the role of an ambient space completely so that data studied can have arbitrary representations as long as a dissimilarity measure is available. These intersection patterns can be used to build cell complexes that then can be studied using homology, cohomology, and other topological tools.

The most commonly used construction for this smearing process is the *Vietoris-Rips complex*. For a data set $\mathbb{X}$, the vertex set is the set of data points. We introduce a simplex $[x_0, \dots, x_k]$ to the complex $\mathrm{VR}_\varepsilon(\mathbb{X})$ precisely when all pairwise dissimilarities are small enough: $d(x_i, x_j) < \varepsilon$. An illustration can be found in Fig. 4.

At each parameter value ε, there is a simplicial complex $\mathrm{VR}_\varepsilon(\mathbb{X})$. As the parameter grows, no intersections vanish—so no existing simplices vanish with a growing parameter. By functoriality—a feature of the homology construction—there is a kind of continuity for topological features: the inclusion maps of simplicial complexes generate linear maps between the corresponding homology (or cohomology) vector spaces. For a growing sequence $\varepsilon_0 < \varepsilon_1 < \varepsilon_2 < \varepsilon_3$, there are maps

$$\begin{array}{ccccccc} \mathrm{VR}_{\varepsilon_0}(\mathbb{X}) & \hookrightarrow & \mathrm{VR}_{\varepsilon_1}(\mathbb{X}) & \hookrightarrow & \mathrm{VR}_{\varepsilon_2}(\mathbb{X}) & \hookrightarrow & \mathrm{VR}_{\varepsilon_3}(\mathbb{X}) \\ H_k\mathrm{VR}_{\varepsilon_0}(\mathbb{X}) & \to & H_k\mathrm{VR}_{\varepsilon_1}(\mathbb{X}) & \to & H_k\mathrm{VR}_{\varepsilon_2}(\mathbb{X}) & \to & H_k\mathrm{VR}_{\varepsilon_3}(\mathbb{X}) \\ H^k\mathrm{VR}_{\varepsilon_0}(\mathbb{X}) & \leftarrow & H^k\mathrm{VR}_{\varepsilon_1}(\mathbb{X}) & \leftarrow & H^k\mathrm{VR}_{\varepsilon_2}(\mathbb{X}) & \leftarrow & H^k\mathrm{VR}_{\varepsilon_3}(\mathbb{X}) \end{array}$$

For a diagram of vector spaces like these, there is a consistent basis choice across the entire diagram. This basis choice is, dependent on the precise argument made, either a direct consequence of the structure theorem for modules over a Principal Ideal Domain (result available in most commutative algebra textbooks, e.g. [53])

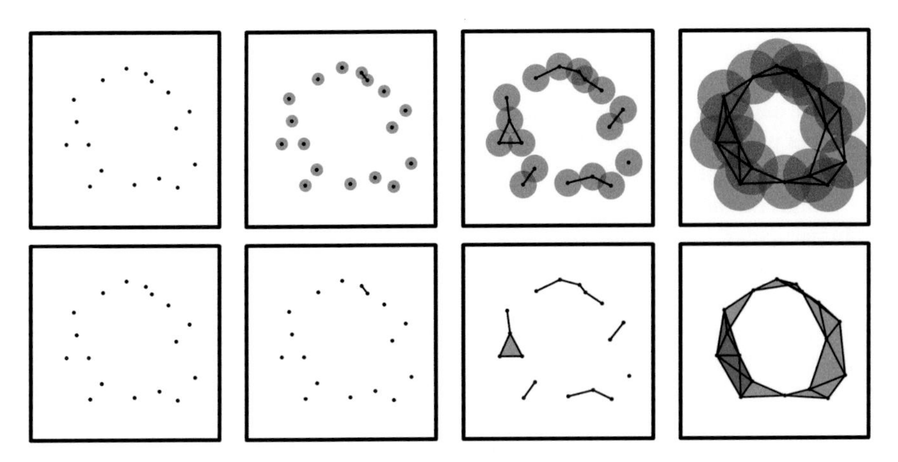

Fig. 4 The growth of a Vietoris-Rips complex as points are smeared out. *Top row* is the view of the data with each data point surrounded by a $\varepsilon/2$ radius ball, while the *bottom row* shows the corresponding abstract complex as it grows. At the very end, the *circle*-like nature of the point cloud can be detected in the Vietoris-Rips complex. This will stick around until ε is large enough that the hole in the *middle* of the *top right figure* is filled in

or a direct consequence of Gabriel's theorem [56] on decomposing modules over tame quivers. The whole diagram splits into components of one-dimensional vector spaces with a well defined start and endpoint along the diagram. These components correspond precisely to topological features, and tell us at what parameter value a particular feature shows up, and at what value it is filled in and vanishes. The components are often visualized as a *barcode*, as can be seen in Fig. 5.

Features that exist only along a very short range of parameter values can be considered noisy: probably the result of sampling errors or inherent noise in the production of the data. These show up along the diagonal of the persistence diagram. Features that exist along a longer range of parameter values are more likely to be

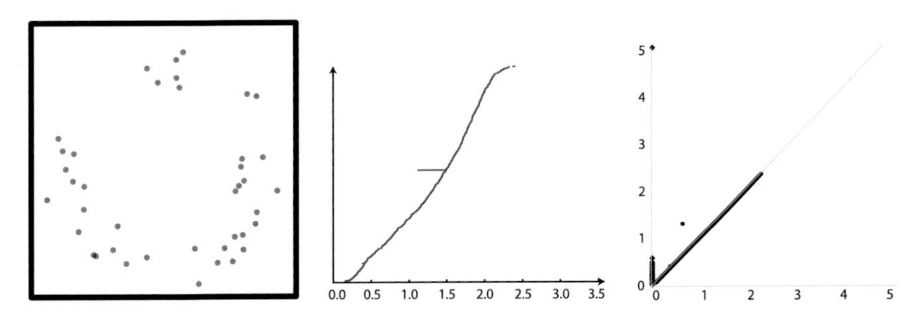

Fig. 5 To the *left*, a point cloud. In the *middle*, the corresponding persistence barcode for dimension 1 homology. To the *right*, the persistence diagram for dimension 0 (*diamonds* along the y-axis) and dimension 1. We see a large amount of very short intervals, and then one significantly larger interval corresponding to the *circle*-like shape of the data

features inherent in the source of the data—and the length of the corresponding bar is a measure of the size of the feature.

These barcode descriptors are stable in the sense that a bound on a perturbation of a data set produces a bound on the difference between barcodes. This stability goes further: data points can vanish or appear in addition to moving around, and there still are bounds on the difference of barcodes.

In particular, this means that with guarantees on sampling density and noise levels, large enough bars form a certificate for the existence of particular topological features in the source of a data set.

Additional expositions of these structures can be found in the survey articles by Carlsson [24] and by Ghrist [57]. For the algebraic and algorithmic focused aspects, there are good surveys available from the first author [113], and by Edelsbrunner and Harer [47, 48].

3.1 Persistence Diagrams as Features

Homology serves as a rough descriptor of a space. It is naturally invariant to many different different types of transformations and deformations. Unfortunately, homology groups of a single space (for example, data viewed at a single scale) are highly unstable and lose too much of the underlying geometry. This is where persistence enters the picture. Persistence captures information in a stable way through the filtration. For example, the Vietoris-Rips filtration encodes information about the underlying metric space.

Therefore, by choosing an appropriate filtration, we can encode information about the space. The first such instance was referred to as *topological inference*. The intensity levels of brain activity in fMRI scans was investigated using Euler characteristic curves [6].

These curves have a long history in the probabilistic literature [5, 116–118], are topological in nature and can be inferred from persistence diagrams. The Euler characteristic can be computed by taking the alternating sum of the ranks of homology groups (or equivalently Betti numbers),

$$\chi(X) = (-1)^k \mathrm{rk}(H_k(X))$$

If X is parameterized by t, we obtain an Euler characteristic curve. Surprisingly, the expectation of this quantity can be computed analytically in a wide range of settings. This makes it amenable for machine learning applications. Another notable application of this approach can be found in distinguishing stone tools from different archaeological sites [93].

These methods work best in the functional setting where the underlying space is fixed (usually some triangulated low dimensional manifold).

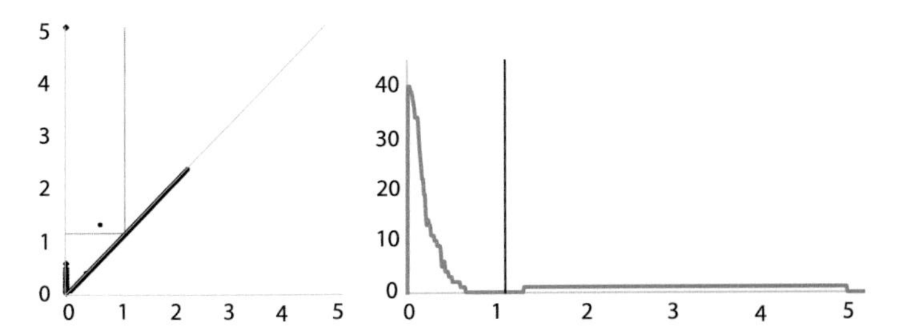

Fig. 6 The relation between an Euler characteristic curve and the corresponding persistence diagram. To the *left*, a persistence diagram, with the quadrant anchored at some (t, t) marked out. To the *right*, the Euler characteristic curve from the corresponding data set, with the corresponding t marked

The persistence diagram encodes more information—the Euler characteristic curve can easily be computed from a persistence diagram by taking the alternating sum over different dimensions for each quadrant anchored at (t, t) as in Fig. 6.

There are several different approaches to using persistence diagrams as features. Initially, it was observed that the space of persistence diagrams can be transformed into a metric space [112]. A natural metric for persistence diagrams is the *bottleneck matching distance*. Given two diagrams Dgm(F) and Dgm(G) (corresponding to two filtrations F and G), the bottleneck distance is defined as

$$d_B(\mathrm{Dgm}(F), \mathrm{Dgm}(G) = \inf_{\phi \in bijections} \sup_{p \in F} d_\infty(p, \phi(p))$$

This has the benefit of always being well-defined, but also has been shown to be not as informative as other distances—namely, Wasserstein distances.

The most commonly used Wasserstein distances used are:

1. 1-Wasserstein distance—W_1
2. 2-Wasserstein distance—W_2

Under some reasonable conditions, persistence diagrams satisfy stability under these metric as well [40] (albeit with a worse constant in front).

While first order moments exist in this space in the form of Frechet means, this space is generally quite complicated. For example, while means exist, there are no guarantees they are unique [112]. Below, we have an example of this phenomenon. This presents numerous algorithmic challenges both for computing the means themselves, as well as for interpretation. This can be made Hölder continuous by considering the *distribution* of persistence diagrams [81].

Ultimately, the problem with viewing the space of persistence diagrams as a metric space is that the space is insufficiently nice to allow for standard machine learning techniques. Furthermore, the standard algorithmic solution for computing bottleneck distance is the Hungarian algorithm for computing the maximum weight bipartite matching between the two diagrams. This computes an explicit matching between points and has at worst an $O(n^3)$ complexity where n is the number of points in the

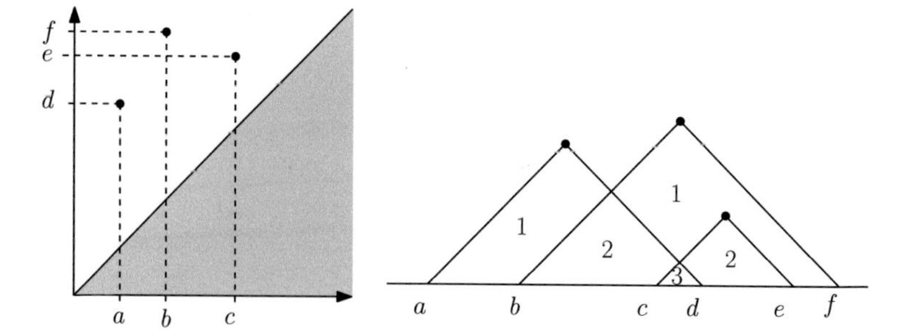

Fig. 7 The construction of a persistence landscape from a persistence diagram. We rotate the diagram by 45°. Each point in the persistence diagram creates a region in the persistent landscape by considering the 45° lines from the point. These correspond to *vertical* and *horizontal lines* from the point to diagonal. To each region in the landscape, we assign the number which corresponds to how many times it is covered. Distances between landscapes are computed by integrating the absolute difference point-wise

persistence diagrams. This is often expensive, which led to the development of the following algorithms.

The key insight came with the development of the *persistence landscape*. This is a functional on the space of the persistence diagrams in line with the kernel trick in machine learning. The main idea is to raise the diagram into a functional space (usually a Hilbert space), where the space behaves fundamentally like Euclidean space, making techniques like support vector machines feasible.

There have been several approaches to constructing functionals on persistence diagrams. The most developed is the *persistence landscape* [19]. This assigns to each point in the plane a support on how many points lie above it. We illustrate the process in Fig. 7, but it assigns to each point in the plane a number which corresponds to how many points in the persistence diagram cover it. In addition to being useful for machine learning algorithms, it is also much faster to compute that distances directly on persistence diagrams (which are based on bipartite matching problems).

The algebraic structure connecting persistence diagrams to functionals was partially addressed in [3]. In this work, Adcock et al. show that the algebraic structure in persistence diagram has a family of functionals which can be used to parameterize the family. This was used to train a SVM classifier on the MINST handwriting dataset. The performance of the classifier is near state-of-the-art, where it is important to mention this the case for generic features rather than the specially chosen ones in current state-of-the-art techniques for handwriting recognition. The same techniques were also used to classify hepatic lesions [4].

3.1.1 Applications

Here we recount some successful applications of the above techniques to real-world data.

The first is on a study of the effects of psilocybin (e.g. magic mushrooms) on the brain using fMRI [86]. In this case, persistence diagram based features are shown to clearly divide the brain activity under the effects of psilocybin from normal brain activity. The authors found that while normal brain activity is highly structured, brain activity under psilocybin is much more chaotic, connecting parts of the brain which are usually not connected.

The second application we highlight the use of topological features to distinguish stone tools coming from different archaeological sites [93]. In this work, the authors began with three dimensional models of the tools obtained from scans. Then they computed the Euler characteristic curves given by curvature, e.g. they used curvature as the filtering function. They found that training a classifier using these curves, they were able to obtain a high classification accuracy ($\sim$80 %).

The final application we highlight is for detecting and classifying periodicity in gene expression time series [84]. Gene expressions are a product of the cell cycle and in this work, the authors recognize that in sufficiently high dimensional space, periodicity is characterized by closed one forms (i.e. circles). The work in the following section makes a similar observation, but parametrizes the circle rather than compute a feature. Circles are characterized by one-dimensional homology and so the authors use the 1-dimensional persistence diagram in order to compare the periodicity of different gene expressions. To obtain, the persistence diagram, the authors embed the time series in high dimension using a sliding window embedding (also known as a Takens' embedding or a delay embedding). The idea is, given a time series $x(1), x(2), \ldots$, take a sliding window over a time series and map each point to the vector of the window. For example, for a window size of three, a data point at time 1, $x(1)$ would be mapped to the vector $[x(1), x(2), x(3)]$ which is in $\mathbb{R}^3$. After some normalization, the authors computed the persistence diagram of the embedded time series which they used to compare different gene expressions.

3.2 *Cohomology and Circular Coordinates*

One particular derived technique from the persistent homology described here is using persistent cohomology to compute coordinate functions with values on the circle. We have already mentioned cohomology, and it plays a strong role in the development of fast algorithms. For the applications to coordinatization, we use results from homotopy theory—another and far less computable part of algebraic topology. This approach was developed by the first author joint with de Silva and Morozov [78, 101, 102].

An equivalence class element in $H^1(X, \mathbb{Z})$—an equivalence class of functions $X \to \mathbb{Z}$—corresponds to an equivalence class of functions $X \to S^1$ to the circle. The correspondence is algorithmic in nature, and efficient to compute. In particular, for any specific function $X \to \mathbb{Z}$ in a cohomology equivalence class The drawback at this stage is that applied to complexes like the Vietoris-Rips complex produces maps that send all data points to a single point on the circle.

We can work around this particular problem by changing the target domain of the function from $\mathbb{Z}$ to $\mathbb{R}$. As long as we started out with a $\mathbb{Z}$-valued function, and stay in the same equivalence class of functions, the translation to circle-valued coefficient maps remains valid. So we can optimize for as smooth as possible a circle-valued map. This turns out to be, essentially, a LSQR optimization problem: the circle valued function related to a cocycle ζ with the coboundary matrix B is

$$\arg\min_{z} \|\zeta - Bz\|_2$$

reduced modulo 1.0. The cocycle is a 1-dimensional cocycle, and B is the coboundary map from 0-cochains to 1-cochains. This makes the correspondingly computed z a 0-cochain—a circle-valued function on the vertices and hence on the data set itself.

4 Mapper

Mapper is a different approach to topological data analysis. Proposed in 2008, it is much faster than persistent homology, and produces an intrinsic shape of an arbitrary data set as a small simplicial complex. This complex can be used for visualization or for further analysis. Applications of this method have been widespread: from medical research through financial applications to politics and sports analyses. This section is based on several articles by Singh et al. [71, 104].

At the core of the Mapper algorithm is the idea that data can be viewed through "lenses"—coordinate functions displaying interesting characteristics of the data set. For any such lens, the data can be stratified according to values of that lens, and local summaries within each stratum can be related to each other to form a global picture of the data set. We see the process illustrated in Fig. 8.

To be precise, given a dataset $\mathbb{X}$ and some function $\ell : \mathbb{X} \to \mathbb{R}^k$ and a cover of $\mathbb{R}^k$ by overlapping open subsets U_i (for instance open balls or open axis-aligned hypercubes), we compute all inverse images $\ell^{-1}(U_i)$. Each such inverse image might contain data points separated from each other—using a clustering algorithm of the user's choice, each inverse image is broken down into its component clusters. Finally, since the sets U_i cover $\mathbb{R}^n$, some of them will overlap. These overlaps may contain data points: when they do, a data point contained in clusters from several inverse images $\ell^{-1}(U_{i_0}), \ell^{-1}(U_{i_1}), \dots, \ell^{-1}(U_{i_k})$ gives rise to a k-simplex spanned by the corresponding clusters. The collection of clusters from the various layers with these connecting simplices forms a simplicial complex describing the inherent shape of the data set. For any given data point, its corresponding location in the simplicial complex can easily be found.

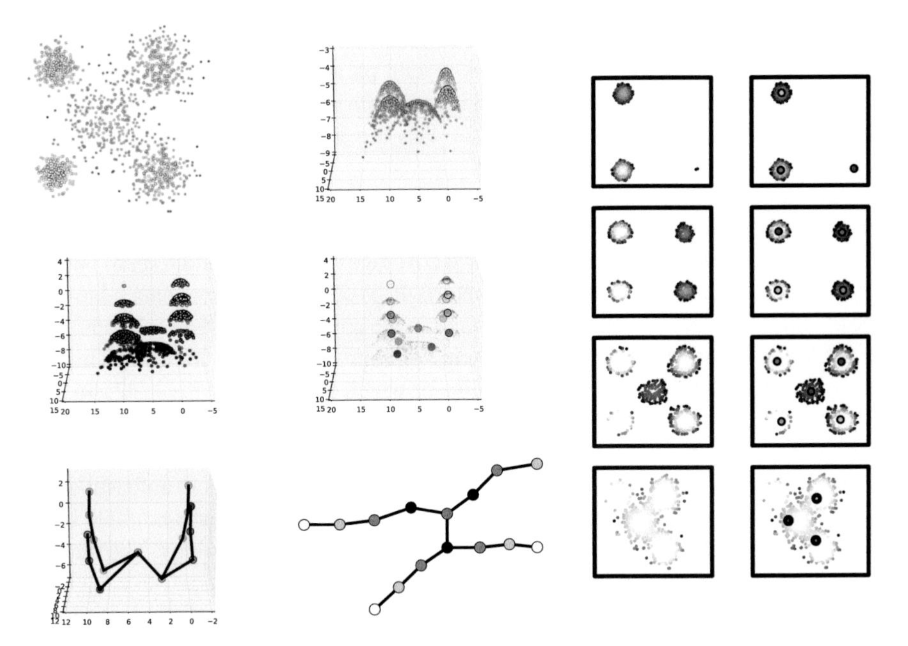

Fig. 8 The entire Mapper pipeline applied to random samples from a mixture of Gaussians, as viewed through the lens of Gaussian density estimation. In the *left two columns* are: the original data set; a graph of the density function on the data set; this same graph split up according to the Mapper method; computed clusters within each section of the split; the corresponding mapper graph in 3D; the resulting Mapper graph in 2D. In the *right two columns*, we see the split and cluster process in more detail: the *left* of these two has the points as split into sections while the *right* has these same points as well as their cluster centers

5 Optimization

Topological tools group objects by qualitative behaviors, in ways that can be deformed to each other within each group. Finding good representatives for qualitative features often turn out to be a case of searching within such a class for an optimal member.

Computing a representative circle-valued coordinate from a cohomology class $[\zeta]$ is a matter of computing $\arg\min_x \phi(\zeta - Bx)$ for some penalty function ϕ defining the optimality of the coordinate function, where B is the coboundary matrix of the triangulation. In [102], the penalty function chosen was $\phi(w) = \|w\|_2$, whereas for other applications, other penalty functions can be used.

The work of computing optimal homology cycles has gotten a lot of attention in the field, using growing neighborhoods, total unimodularity or computational geometry and matroid theory [21, 34, 35, 43, 54]. Unimodularity in particular turns out to have a concrete geometric interpretation: simplifying the optimization significantly, it requires all subspaces to be torsion free. An interesting current direction of research is the identification of problems which become tractable when the equivalence class is fixed. There are many examples of fixed-parameter tractable algorithms—where there is an exponential dependence on a certain parameter (such as dimension). In such instances, it would be beneficial to identify the global structure of the data and optimize within each (co)homology class. This has been used indirectly in network and other shortest path routing [23, 62].

Another area where homology shows up as a tool for optimization is in evaluating coverage for sensor agents—such as ensembles of robots, or antenna configurations. Here, for a collection of agents with known coverage radii and a known set of boundary agents, degree 2 homology of the Vietoris-Rips complex of the agents relative to the boundary reveals whether there are holes in the coverage, and degree 1 homology of the Vietoris-Rips complex reveals where holes are located [59, 99, 100]. This has given rise to a wealth of applications, some of which can be found in [2, 42, 46, 80, 108, 115].

In other parts of topological data analysis, optimization formulations or criteria form the foundations of results or constructions—in ways that turn out unfeasible and require approximations or simplifications for practical use. The main example is for the various stability results that have shown up for persistent homology. The metrics we use for persistence diagrams, bottleneck and Wasserstein distances, take the form of optimization problems over spaces of bijections between potentially large finite sets [20, 27, 28, 31, 32, 38, 39, 41, 44, 69].

6 Applications

6.1 Local to Global

Topological techniques are designed to extract global structure from local information. This local information may be in the form of a metric or more generally a similarity function. Often a topological viewpoint can yield new insights into existing techniques. An example of this *persistence-based clustering* [30]. This work is closely related with *mode-seeking clustering* techniques [36]. This class of methods assumes the points are sampled from some underlying density function and defines the clusters as the modes of the density function (e.g. the basins of attraction of the peaks of the density function). There are generally two steps involved:

1. Estimation of the underlying density function
2. Estimation of the peaks

These techniques have the advantage that the number of clusters is not required as input. Rather the main problem is to determine which peaks are "real" versus which peaks are noise. For example, in *mean-shift clustering*, points are flowed to local maxima incrementally, but require a stopping parameter (as we never exactly hit the peak). There are many other criteria which have been proposed—however, it turns out that persistence provides an important insight. Due to the stability of the

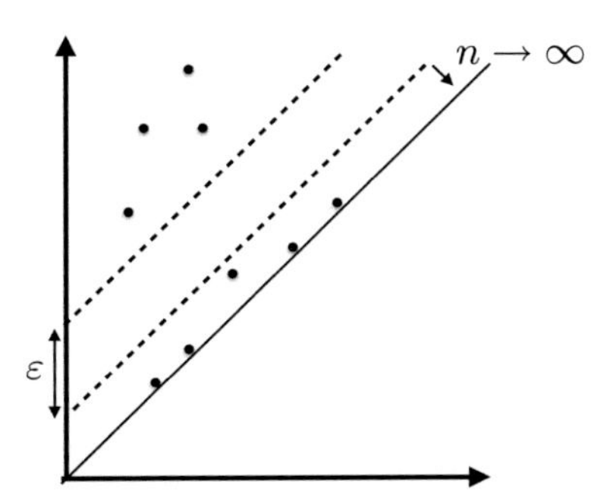

Fig. 9 A persistence diagram with a gap of ε. The topological noise and the "true" features are separated by an empty region of size ε. Note that under mild assumptions on the underlying sample, the noise goes to 0 as the number of points goes to infinity, therefore the gap increases with increasing amounts of data

persistence diagram, if we the first step (i.e. estimation of the density function) is done correctly, then the persistence diagram is provably close. Furthermore, if there is a separation of noise and the peaks (i.e. a gap as shown in Fig. 9), then we can estimate the number of clusters as well. It can also be shown that the noise goes to zero as the number of points increases, ensuring that the gap exists if we have sufficiently many points.

This approach also allows for the identification of stable and unstable parts of the clustering. The main idea is that since the persistence diagram is stable, the number of clusters is also stable. Furthermore, persistent clusters can be uniquely identified in the presence of resampling, added noise, etc. The idea is illustrated in Fig. 10. This can be useful when determining unstable regions for tasks such as segmentation [105]. Here unstable regions are themselves considered separate segments.

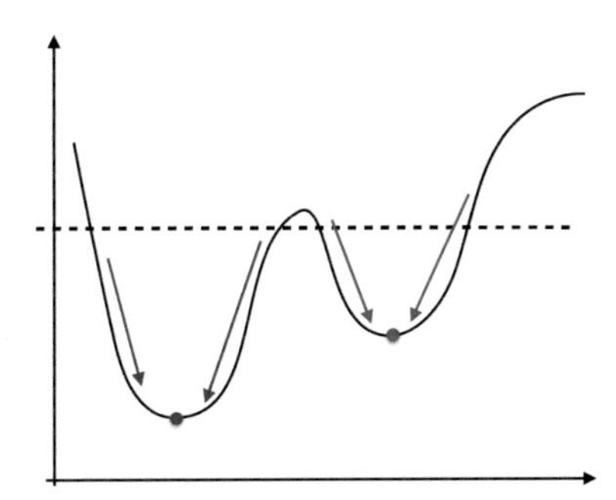

Fig. 10 Persistence based clustering is based on the idea that the basins of attraction of the peaks of a density function are clusters. Here we show the negative of a density function (so we look for valleys rather than peaks), with two clusters. For clustering, there also exists a spatial stability for persistent clusters, since if we consider a point before two clusters meet, they are disjoint—shown here by the *dashed line*

In addition to clustering based on density (or a metric—in which case we obtain single-linkage clustering), topological methods can find clusters which have similar local structure. Often we consider data as living in Euclidean space or a Riemannian manifold. While this may be *extrinsically* true, i.e. the data is embedded in such a space, the *intrinsic* structure of data is rarely this nice. A natural generalization of a manifold is the notion of a *stratified space*. This can be thought of as a mixture of manifolds (potentially of different dimensions) which are glued together in a nice way. Specifically, the intersection of two manifold pieces is itself be a manifold. The collection of manifolds of a given dimension is called a *stratum*. We omit the technical definition, but refer the reader to the excellent technical notes [74].

The problem of *stratified manifold learning* is to identify the manifold pieces directly from the data. The one dimensional version of this problem is the graph construction problem, which has been considered for reconstruction of road networks from GPS traces [1]. In this setting, zero dimensional strata are intersections, forks and merges (i.e. vertices in the graph) while one dimensional strata are the connecting roads (i.e. edges in the graph). Some examples are shown in Fig. 11.

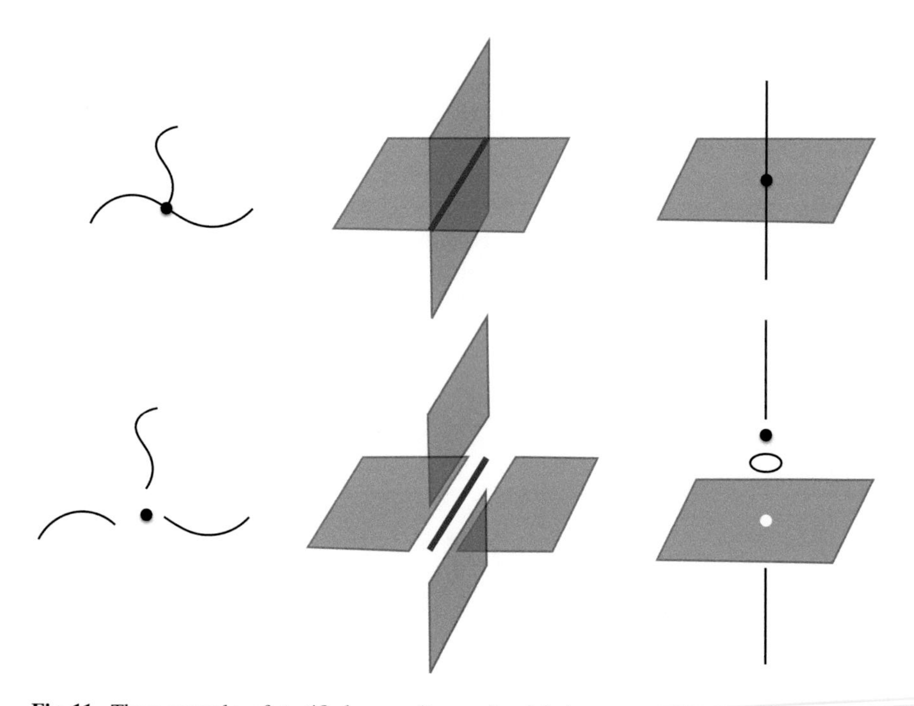

Fig. 11 Three examples of stratified spaces (*top row*) and their corresponding strata (*bottom row*). On the *left*, we have part of a graph with three edges (1-strata) coming from a vertex (0-strata). The intersection of two 1-strata, must be a 0-strata. In the *middle*, we have two planes intersecting along a *line*. This gives four 2 strata, which meet together at a 1-strata. On the *right* we have a line through a plane. The plane is a 2-strata, the *line* is divided into two pieces and the intersection is a 1 and 0-strata. The intersection is a point, however to ensure it is a valid stratification we must consider the loop around the point

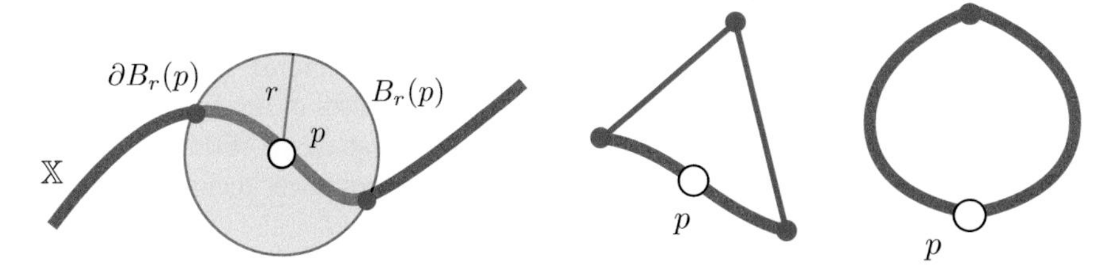

Fig. 12 The intuition behind local homology. We consider the local neighborhood around a point, in this case an intersection between the *ball* of radius r at point p and the space $\mathbb{X}$. The boundary is two points which are collapsed together to make a *circle*

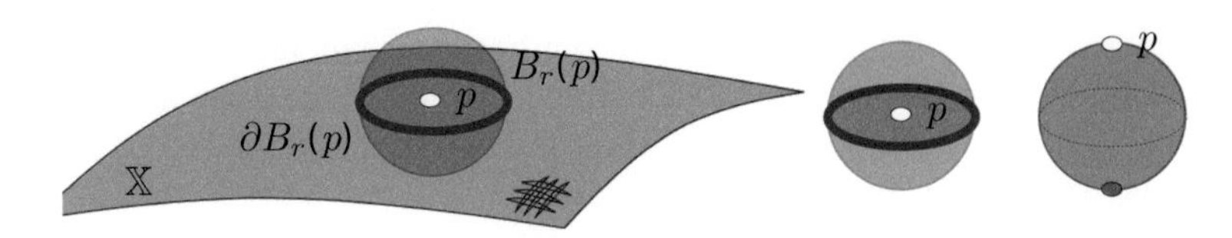

Fig. 13 A higher dimensional example for local homology. In this case an intersection between the *ball* of radius r at point p and the space $\mathbb{X}$ is a disc and the boundary is a *circle*. When we collapse the boundary (i.e. *circle*) to a point we obtain a sphere

The problem has been considered in more generality in [14]. Intuitively, the problem can be thought of as determining the local structure of a space and then clustering together points which share the same structure. The topological tool for determining the local structure is called *local homology*. The idea is to consider the structure of the local neighborhood. An important technical point is that we consider the quotient of the neighborhood modulo its boundary. By collapsing the boundary to a point (as shown in Figs. 12 and 13), we can distinguish different dimensions. In the case of a k-dimensional manifold, each point will have the same structure—that of a k-dimensional sphere. In the cases shown in Fig. 11, we obtain a different answer. On the left we obtain a wedge of two circles, in the middle a wedge of three spheres and on right we obtain a wedge of a sphere and two circles.

While preliminary results have been promising, this is currently an active area of research.

6.2 *Nonlinear Dimensionality Reduction*

Dimensionality reduction is well rooted in data analysis, as a way to reduce an unmanageably wide data set to a far more narrow and thus more easily analyzed derived data set. Classical techniques often work by linear projections, as is done by principal component analysis or by random projection methods. While some techniques for non-linear dimensionality reduction have been known since the 1960s [63–65], a more significant boost in the development of new reduction methods showed up in the late 1990s.

van der Maaten et al. [72] distinguish between three types of nonlinear dimensionality reduction techniques: those that try to preserve global features; those that try to preserve local features; and those that globally align mixtures of linear techniques.

Where some methods are focused on retaining most or all distances in the dataset—such as multidimensional scaling [63], many nonlinear techniques focus on retaining *closeness*.

Isomap [111] generates approximations to the geodesic distance of a dataset as computed on a neighborhood graph. As compared to MDS, it puts more emphasis on retaining closeness by first focusing the scaling on local connectivity before generating a coordinate set globally approximating these geodesic distances.

Local techniques fit a linear local dimensionality reduction to small neighborhoods in the data and then gluing the local coordinates to a global description. First out was Local Linear Embeddings (LLE): fitting a local tangent plane through each data points, and then minimizes the distortion of these local tangents [95]. This minimization reduces to an eigenvector computation.

Several improvements on LLE have been constructed as eigenvector computations of the Laplacian operator, or through enriched representations of the local tangent descriptions [13, 45, 70]. The Laplacian or Laplace operator is a classical operator in algebraic topology. The coboundary operator δ has a dual operator δ^*—represented by the same matrix as the boundary operator. The Laplacian is defined as the composition $\delta^*\delta$. The operator smooths out a function along the connectivity of the underlying space, and its eigenmaps form smooth—in the sense of keeping nearby points close together—and produces globally defined functions that retain closeness of data points.

Isomap and LLE both suffer from weaknesses when constructing coordinate functions on data sets with holes. One possible solution was offered by Lee and Verleysen [68], who give a graph algorithm approach to cutting the data set to remove the non-trivial topology. They give a complexity of $O(n \log_2 n)$ for their cutting procedure, based on using Dijkstra's algorithm for spanning trees. Such a cut can also be produced based on persistent cohomology, with a representative cocycle demonstrating a required cut to reduce topological complexity [9, 26, 58]. While the worst case complexity for this computation is matrix-multiplication time, for many data sets, linear complexity has been observed [12, 121].

Some shapes require more linear coordinates to represent accurately than the intrinsical dimension would indicate. A first example is the circle: while a one-dimensional curve, any one-dimensional projection will have to collapse distant points to similar representations. With the techniques we describe in Sect. 3.2, we can generate circle-valued coordinates for the data points. This has been used in finding cyclic structures [10, 37] and for analyzing quasiperiodic or noisily recurrent signals in arbitrary dimension [94, 114].

Mapper provides an approach to dimensionality reduction with intrinsic coordinate spaces: instead of providing features on a line or a circle, the Mapper output is a small, finite model space capturing the intrinsic shape of the original data set.

The often large reduction in representation size with a Mapper reduction enables speedups in large classes of problems. Classic dimensionality reduction such as done by MDS, Isomap or LLE can be done on the Mapper model, and with coordinate values pulled back and interpolated onto the original data points themselves, while optimization problems could be solved on the Mapper model to produce seed values or approximate solutions when pulled up to the original data points. As long as all functions involved are continuous, and the Mapper analysis sufficiently fine grained, each vertex of the Mapper model corresponds to a compact set of data points with trivial topology and each higher dimensional simplex corresponds to a connection between sets of data points.

6.3 Dynamics

Topological methods have a long history of use in simplifying, approximating and analyzing dynamical systems. For this approach, the Conley index—a homology group computed in a small neighborhood in a dynamical system—gives a measure of the local behavior of the dynamical system, stable and useful for nonlinear and multiparameter cases. This approach has found extensive use [22, 75, 79].

Computing persistence on point clouds from dynamical systems, and then using clustering to extract features from the resulting invariants has found some use. In [15], bifurcation detection for dynamical systems using persistent cohomology was explored, while in [66] clustered persistence diagrams helped classify gait patterns to detect whether and what people were carrying from video sequences.

The idea of using the Takens delay embedding [109] to create point clouds representative of dynamic behavior from timeseries data has emerged simultaneous from several groups of researchers in topological data analysis. Harer and Perea [85] used 1-dimensional persistent homology to pick out appropriate parameters for a delay embedding to improve accuracy for the embedded representation of the original dynamical system. The same idea of picking parameters for a delay embedding, but with different approaches for subsequent analyses were described by Skraba et al. [103], and later used as the conceptual basis for the analysis of the dynamics of motion capture generated gait traces by Vejdemo-Johansson et al. [114]. The work in [114] uses persistent cohomology to detect intrinsic phase angle coordinates, and then use these either to create an average gait cycle from a sequence of samples, or to generate gait cycle classifiers functions, indicating similarity of a new sequence of gait samples to the sequences already seen.

From the same group of researchers, persistent homology and cohomology has been used for motion planning in robotics. Moduli spaces for grasping procedures give geometric and topological ways of analyzing and optimizing potential grasp plans [87, 88, 90], and 1-dimensional persistent homology provides suggestions for grasp sites for arbitrary classes of objects with handles [92, 106, 107]. Topology also generates constraints for motion planning optimization schemes, and produces approaches for caging grasps of wide classes of objects [89, 91, 119, 120].

6.4 Visualization

Topological techniques are common in visualization, particularly so called scientific visualization. Perhaps the most prominent of these applications are topological skeleta and the visualization and simplification of two or three dimensional scalar and vector fields.

6.4.1 Topological Skeleta

There has been a large amount of work on topological skeleta extraction. Here we highlight two types of constructions (without exhaustively describing all related work)

- Reeb graphs
- Contour trees

Though there are many variations the main idea behind both constructions is that given a space $\mathbb{X}$ and a real-valued function, i.e.

$$f : \mathbb{X} \to \mathbb{R}$$

a topological summary can be computed by taking every possible function value $a \in \mathbb{R}$, and considering its preimage, $f^{-1}(a) \in \mathbb{X}$. For each preimage, we can count the number of connected components.[1] If we consider very small intervals rather than just points, we see that we can connect these components if they overlap. Connecting these connected components together using this criteria, we obtain a graph (again under reasonable assumptions). The resulting graph is called a *Reeb graph.*

By only considering the connected components, a potentially high-dimensional structure can be visualized as a graph. However, the input need not be high-dimensional, as these constructions have are useful as shape descriptors for 2-dimensional shapes. In addition, they are a crucial part of understanding 3-dimensional data sets, where direct visualization is impossible.

When the underlying space is contractible, there is additional structure, which allows for more efficient computation of the structure, interestingly in any dimension [25]. This is mainly due to the observation that if the underlying space is contractible (such as on a convex subset of Euclidean space), then the Reeb graph has the structure of a *tree*, and is therefore called a *contour tree.*

Mapper can be thought of as a "fuzzy" Reeb graph, where connectivity is scale-dependent and is computed via clustering rather than as an intrinsic property of the space.

---

[1]Technically, these are path-connected components. However, this distinction is a mathematical formality, as the two are indistinguishable in any form of sampled data.

6.4.2 Features and Simplification

In addition to visualizing a scalar function, the presence of noise, or more generally multi-scale structure, makes the ability to perform simplification desirable. By simplifying a function, larger structures become clearer as they are no longer overwhelmed by large numbers of smaller features.

There has been a substantial amount of work done on simplifying functions on topological spaces. Initially, the work was done on two dimensional manifolds (surfaces) using the Morse-Smale complex [50] and has been extended to three dimensions [49, 60]. Morse theory connects the topological and geometric properties of a space in terms of the critical points of a function on that space. For example, consider the height function on a sphere. This has two critical points, the global minimum (the bottom of the sphere) and maximum (the top of the sphere). If we distort the function to add another maxima, we also add a saddle. Simplification in this setting proceeds by considering the reverse of this process. By combining minima and saddles (or maxima and saddles), it simplifies the underlying function. The order of simplification can be done in a number of different ways, such as distance based (i.e. distance between critical points). The persistence ordering is given if it is done by relative heights (e.g. small relative heights first) and the methods are closely tied to Morse theory [17].

The natural extension from scalar fields is to vector fields, that is each point is assigned a vector. These are often used to model flows in simulations of fluids or combustion. Simplifying these is much more difficult than simplifying vector fields. However, the notion of fixed point naturally generalizes critical points of a scalar function. These are studied in *topological dynamics*. We highlight two topological approaches which are based on Conley index theory and degree theory respectively. The Conley index [96] is a topological invariant based on homology, which is an extension of Morse theory. The main problem in this approach is the requirement to find a neighborhood which isolates the fixed points from the rest of the flow. This neighborhood (called an *isolating neighborhood*), must be nice with the respect to the flow, in particular, the flow must not be internally tangent to the boundary of the neighborhood.

The second approach is based on a variant of persistence called *robustness* [52]. A vector field in Euclidean space can be thought of as a map from $\mathbb{R}^n \to \mathbb{R}^n$ and we can compute its robustness diagram [33]. This carries similar information and and shares similar stability properties as the persistence diagram. Furthermore, this can be connected to degree theory, which is yet another invariant developed to describe maps. Famously, the existence of Nash equilibrium in game theory is the consequence of a fixed point theorem (i.e. Brouwer fixed point theorem). Using a classical result from differential topology, which states that if a part of the flow has has degree zero, then is can be deformed to a fixed point free vector field, a general vector field can be simplified using the robustness order in the same way as persistence order gives an ordering in the case of scalar fields.

This is only a brief glimpse at topology and visualization as this is a large field of research. The main motivation for using topological methods for large complex data sets is that they ensure consistency, whereas ad-hoc methods may introduce artifacts during simplification.

7 Software and Limitations

These methods are implemented in a range of software packages. There is a good survey of the current state of computation and computational timings written by Otter et al. [83]. We will walk through a selection of these packages and their strengths and weaknesses here.

The available software can be roughly divided by expected platform. Some packages are specifically adapted to work with the statistics and data analysis platform R, some for interacting well with Matlab, and some for standalone use or to work as libraries for software development.

In R, the two main packages are pHom and R-TDA. pHom is currently abandoned, but still can be found and included. R-TDA [55] is active, with a userbase and maintenance. Both are built specifically to be easy to use from R and integrate into an R data analysis workflow.

When working in Matlab, or in any other java-based computational platform—such as Maple or Mathematica—the main software choice is JavaPlex [110] or JPlex [97]. JPlex is the predecessor to JavaPlex, built specifically for maximal computational efficiency on a Java platform, while JavaPlex was built specifically to make extension of functionality easy. Both of them are also built to make the user experience as transparent and accessible as possible: requiring minimal knowledge in topological data analysis to be usable. While less efficient than many of the more specialized libraries, JavaPlex has one of the most accessible computational pipelines. The survey by Otter et al. [83] writes “However, for small data sets (less than a million simplices) the software Perseus and javaPlex are best suited because they are the easiest to use[…]”.

Several other packages have been constructed that are not tied to any one host platform: either as completely standalone processing software packages, or as libraries with example applications that perform many significant topological data analysis tasks. Oldest among these is ChomP [76]. ChomP contains a C++ library and a couple of command line applications to compute persistent homology of cubical sets, and has been used in dynamical systems research [8]. Perseus [82] works on Vietoris-Rips complexes generated from data sets, as well as from cubical and simplicial complexes. Perseus needs its input data on a particular format, with meta data about the data points at the head of the input file, which means many use cases may need to adjust input data to fit with Perseus expectations. DIPHA [11] is the first topological data analysis program with built in support for distributed computing: building on the library PHAT [12], DIPHA works with MPI for parallelization or distribution of computation tasks. DIPHA takes in data, and produces a persistence diagram, both in their own file format—the software distribution includes Matlab functions to convert to and from the internal file format. Last among the standalone applications, Dionysus [77] is a library for topological data analysis algorithm development in C++ and Python. The package comes with example application for persistent homology and cohomology, construction of Vietoris-Rips complexes and a range of further techniques from computational topology.

Two more libraries are focused on use for software developers. Gudhi [73] is a library that focuses on the exploration of different data structures for efficient computations of persistent homology or with simplicial complexes. CTL[2] is a very recent library maintained by Ryan Lewis. The library is still under development, and currently supports persistent homology and complex construction, and has plans to support persistent cohomology, visualizations and bindings to other languages and platforms.

Complex construction the current computational bottleneck in topological data analysis. Simplicial complexes are built up dimension by dimension and in higher dimensions, a small number of points can result in a large number of simplices. For example in 2000 points in dimension 6 can easily yield overall billion simplicies. We do have the option of limiting our analysis to low dimension (e.g. clustering only requires the graph to be built), and there are techniques which yield an approximate filtration while maintaining a linear size [98]. Current research is finding further speedups as well as modifying this to a streaming model. The second problem is that although the volume of data is getting larger, the data itself does not cover the entire space uniformly and preforming a global analysis where we have insufficient data in some regions is impossible. One approach that is currently being explored is how to construct "likely" analysis to fill in regions where data is sparse (e.g. anomalies).

8 Conclusions

At the state of the field today, topological data analysis has proven itself to produce descriptors and invariants for topological and geometric features of data sets. These descriptors are

COORDINATE-FREE so that the descriptors are ultimately dependent only on a measure of similarity or dissimilarity between observations. Ambient space, even data representation and their features are not components of the analysis methods, leading to a set of tools with very general applicability.

STABLE UNDER PERTURBATIONS making the descriptors stable against noise. This stability forms the basis for a topological inference.

COMPRESSED so that even large data sets can be reduced to small representations while retaining topological and geometric features in the data.

[2]http://ctl.appliedtopology.org/.

Looking ahead, the adaptation and introduction of classical statistical and inferential techniques into topological data analysis is underway [16, 18, 29, 112].

The problem of efficient constructions of simplicial complexes encoding data geometry remains both under-explored and one of the most significant bottlenecks for topological data analysis.

Over the last few years, big data techniques have been developed which perform well for specific tasks: building classifiers, linear approaches or high speed computations of simple invariants of large volume data sets. As data and complexity grows, the need emerges for methods that support interpretation and transparency—where the data is made accessible and generalizable without getting held back by the simplicity of the chosen models. These more qualitative approaches need to include both visualization and structure discovery: nonlinear parametrization makes comparison and correlation with existing models easier. The problems we encounter both in non-standard optimization problems and in high complexity and large volume data analysis are often NP-hard in generality. Often, however, restricting the problem to a single equivalence class under some equivalence relation—often the kinds found in topological methods—transforms the problem to a tractable one: examples are maximizing a function over only one persistent cluster, or finding optimal cuts using cohomology classes to isolate qualitatively different potential cuts. The entire area around these directions is unexplored, wide open for research. We have begun to see duality, statistical approaches and geometric features of specific optimization problems show up, but there is a wealth of future directions for research.

As for data, the current state of software has problems both with handling streaming data sources and data of varying quality. The representations available are dependent on all seen data points, which means that in a streaming or online setting, the computational problem is constantly growing with the data stream. Data quality has a direct impact on the computational results. Like with many other techniques, topological data analysis cannot describe what is not present in the data but rather will produce a description of the density indicated by the data points themselves. If the data quality suffers from variations in the sampling density, the current software is not equipped to deal with the variations. There is research [30] into how to modify the Vietoris-Rips construction to handle well-described sampling density variations, but most of the major software packages have yet to include these modifications.

All in all, topological data analysis creates features and descriptors capturing topological and geometric aspects of complex and wide data.

References

1. Aanjaneya, M., Chazal, F., Chen, D., Glisse, M., Guibas, L., Morozov, D.: Metric graph reconstruction from noisy data. Int. J. Comput. Geom. Appl. **22**(04), 305–325 (2012)
2. Adams, H., Carlsson, G.: Evasion paths in mobile sensor networks. Int. J. Robot. Res. **34**(1), 90–104 (2015)
3. Adcock, A., Carlsson, E., Carlsson, G.: The ring of algebraic functions on persistence bar codes. http://comptop.stanford.edu/u/preprints/multitwo (2012)

4. Adcock, A., Rubin, D., Carlsson, G.: Classification of hepatic lesions using the matching metric. Comput. Vis. Image Underst. **121**, 36–42 (2014)
5. Adler, R.J.: The Geometry of Random Fields, vol. 62. Siam (1981)
6. Adler, R.J.: Some new random field tools for spatial analysis. Stochast. Environ. Res. Risk Assess. **22**(6), 809–822 (2008)
7. Amari, S.I., Nagaoka, H.: Methods of Information Geometry, vol. 191. American Mathematical Society (2007)
8. Arai, Z., Kalies, W., Kokubu, H., Mischaikow, K., Oka, H., Pilarczyk, P.: A database schema for the analysis of global dynamics of multiparameter systems. SIAM J. Appl. Dyn. Syst. **8**(3), 757–789 (2009)
9. Babson, E., Benjamini, I.: Cut sets and normed cohomology with applications to percolation. Proc. Am. Math. Soc. **127**(2), 589–597 (1999)
10. Bajardi, P., Delfino, M., Panisson, A., Petri, G., Tizzoni, M.: Unveiling patterns of international communities in a global city using mobile phone data. EPJ Data Sci. **4**(1), 1–17 (2015)
11. Bauer, U., Kerber, M., Reininghaus, J.: Distributed computation of persistent homology. In: ALENEX, pp. 31–38. SIAM (2014)
12. Bauer, U., Kerber, M., Reininghaus, J., Wagner, H.: PHAT-persistent homology algorithms toolbox. In: Mathematical Software-ICMS 2014, pp. 137–143. Springer (2014)
13. Belkin, M., Niyogi, P.: Laplacian eigenmaps for dimensionality reduction and data representation. Neural Comput. **15**(6), 1373–1396 (2003)
14. Bendich, P., Wang, B., Mukherjee, S.: Local homology transfer and stratification learning. In: Proceedings of the Twenty-Third Annual ACM-SIAM Symposium on Discrete Algorithms, pp. 1355–1370. SIAM (2012)
15. Berwald, J., Gidea, M., Vejdemo-Johansson, M.: Automatic recognition and tagging of topologically different regimes in dynamical systems. Discontinuity Non-linearity Complex. **3**(4), 413–426 (2015)
16. Blumberg, A.J., Gal, I., Mandell, M.A., Pancia, M.: Robust statistics, hypothesis testing, and confidence intervals for persistent homology on metric measure spaces. Found. Comput. Math. **14**(4), 745–789 (2014)
17. Bremer, P.T., Edelsbrunner, H., Hamann, B., Pascucci, V.: A multi-resolution data structure for two-dimensional morse-smale functions. In: Proceedings of the 14th IEEE Visualization 2003 (VIS'03), p. 19. IEEE Computer Society (2003)
18. Bubenik, P.: Statistical Topology Using Persistence Landscapes (2012)
19. Bubenik, P.: Statistical topological data analysis using persistence landscapes. J. Mach. Learn. Res. **16**, 77–102 (2015)
20. Bubenik, P., Scott, J.A.: Categorification of persistent homology. arXiv:1205.3669 (2012)
21. Busaryev, O., Cabello, S., Chen, C., Dey, T.K., Wang, Y.: Annotating simplices with a homology basis and its applications. In: Algorithm Theory-SWAT 2012, pp. 189–200. Springer (2012)
22. Bush, J., Gameiro, M., Harker, S., Kokubu, H., Mischaikow, K., Obayashi, I., Pilarczyk, P.: Combinatorial-topological framework for the analysis of global dynamics. Chaos: Interdiscip. J. Nonlinear Sci. **22**(4), 047,508 (2012)
23. Cabello, S., Giannopoulos, P.: The complexity of separating points in the plane. In: Proceedings of the Twenty-Ninth Annual Symposium on Computational Geometry, pp. 379–386. ACM (2013)
24. Carlsson, G.: Topology and data. Am. Math. Soc. **46**(2), 255–308 (2009)
25. Carr, H., Snoeyink, J., Axen, U.: Computing contour trees in all dimensions. Comput. Geom. **24**(2), 75–94 (2003)
26. Chambers, E.W., Erickson, J., Nayyeri, A.: Homology flows, cohomology cuts. SIAM J. Comput. **41**(6), 1605–1634 (2012)
27. Chazal, F., Cohen-Steiner, D., Glisse, M., Guibas, L.J., Oudot, S.Y.: Proximity of persistence modules and their diagrams. In: Proceedings of the 25th Annual Symposium on Computational Geometry, SCG'09, pp. 237–246. ACM, New York, NY, USA (2009). doi:10.1145/1542362.1542407

28. Chazal, F., Cohen-Steiner, D., Guibas, L.J., Oudot, S.Y.: The Stability of Persistence Diagrams Revisited (2008)
29. Chazal, F., Fasy, B.T., Lecci, F., Rinaldo, A., Wasserman, L.: Stochastic convergence of persistence landscapes and silhouettes. In: Proceedings of the Thirtieth Annual Symposium on Computational Geometry, p. 474. ACM (2014)
30. Chazal, F., Guibas, L.J., Oudot, S.Y., Skraba, P.: Persistence-based clustering in riemannian manifolds. J. ACM (JACM) **60**(6), 41 (2013)
31. Chazal, F., de Silva, V., Glisse, M., Oudot, S.: The structure and stability of persistence modules. arXiv:1207.3674 (2012)
32. Chazal, F., de Silva, V., Oudot, S.: Persistence stability for geometric complexes. arXiv:1207.3885 (2012)
33. Chazal, F., Skraba, P., Patel, A.: Computing well diagrams for vector fields on $\mathbb{R}^n$. Appl. Math. Lett. **25**(11), 1725–1728 (2012)
34. Chen, C., Freedman, D.: Quantifying homology classes. arXiv:0802.2865 (2008)
35. Chen, C., Freedman, D.: Hardness results for homology localization. Discrete Comput. Geom. **45**(3), 425–448 (2011)
36. Cheng, Y.: Mean shift, mode seeking, and clustering. IEEE Trans. Pattern Anal. Mach. Intell. **17**(8), 790–799 (1995)
37. Choudhury, A.I., Wang, B., Rosen, P., Pascucci, V.: Topological analysis and visualization of cyclical behavior in memory reference traces. In: Pacific Visualization Symposium (PacificVis), 2012 IEEE, pp. 9–16. IEEE (2012)
38. Cohen-Steiner, D., Edelsbrunner, H., Harer, J.: Stability of persistence diagrams. Discrete Comput. Geom. **37**(1), 103–120 (2007)
39. Cohen-Steiner, D., Edelsbrunner, H., Harer, J.: Extending persistence using Poinca and Lefschetz duality. Found. Comput. Math. **9**(1), 79–103 (2009). doi:10.1007/s10208-008-9027-z
40. Cohen-Steiner, D., Edelsbrunner, H., Harer, J., Mileyko, Y.: Lipschitz functions have L_p-stable persistence. Found. Comput. Math. **10**(2), 127–139 (2010)
41. Cohen-Steiner, D., Edelsbrunner, H., Morozov, D.: Vines and vineyards by updating persistence in linear time. In: Proceedings of the Twenty-Second Annual Symposium on Computational Geometry, SCG'06, pp. 119–126. ACM, New York, NY, USA (2006). doi:10.1145/1137856.1137877
42. de Silva, V., Ghrist, R., Muhammad, A.: Blind swarms for coverage in 2-D. In: Robotics: Science and Systems, pp. 335–342 (2005)
43. Dey, T.K., Hirani, A.N., Krishnamoorthy, B.: Optimal homologous cycles, total unimodularity, and linear programming. SIAM J. Comput. **40**(4), 1026–1044 (2011)
44. Dey, T.K., Wenger, R.: Stability of critical points with interval persistence. Discrete Comput. Geom. **38**(3), 479–512 (2007)
45. Donoho, D.L., Grimes, C.: Hessian eigenmaps: Locally linear embedding techniques for high-dimensional data. Proc. Natl. Acad. Sci. **100**(10), 5591–5596 (2003)
46. Dłotko, P., Ghrist, R., Juda, M., Mrozek, M.: Distributed computation of coverage in sensor networks by homological methods. Appl. Algebra Eng. Commun. Comput. **23**(1), 29–58 (2012). doi:10.1007/s00200-012-0167-7
47. Edelsbrunner, H., Harer, J.: Persistent homology—a survey. In: Goodman, J.E., Pach, J., Pollack, R. (eds.) Surveys on Discrete and Computational Geometry: Twenty Years Later, Contemporary Mathematics, vol. 453, pp. 257–282. American Mathematical Society (2008)
48. Edelsbrunner, H., Harer, J.: Computational Topology: An Introduction. AMS Press (2009)
49. Edelsbrunner, H., Harer, J., Natarajan, V., Pascucci, V.: Morse-smale complexes for piecewise linear 3-manifolds. In: Proceedings of the Nineteenth Annual Symposium on Computational Geometry, pp. 361–370. ACM (2003)
50. Edelsbrunner, H., Harer, J., Zomorodian, A.: Hierarchical morse complexes for piecewise linear 2-manifolds. In: Proceedings of the Seventeenth Annual Symposium on Computational Geometry, pp. 70–79. ACM (2001)

51. Edelsbrunner, H., Letscher, D., Zomorodian, A.: Topological persistence and simplification. In: 41st Annual Symposium on Foundations of Computer Science, 2000. Proceedings, pp. 454–463 (2000)
52. Edelsbrunner, H., Morozov, D., Patel, A.: Quantifying transversality by measuring the robustness of intersections. Found. Comput. Math. **11**(3), 345–361 (2011)
53. Eisenbud, D.: Commutative Algebra with a View Toward Algebraic Geometry, vol. 150. Springer (1995)
54. Erickson, J., Whittlesey, K.: Greedy optimal homotopy and homology generators. In: Proceedings of the Sixteenth Annual ACM-SIAM Symposium on Discrete Algorithms, pp. 1038–1046. Society for Industrial and Applied Mathematics (2005)
55. Fasy, B.T., Kim, J., Lecci, F., Maria, C.: Introduction to the R package TDA. arXiv:1411.1830 (2014)
56. Gabriel, P.: Unzerlegbare Darstellungen I. Manuscripta Mathematica **6**(1), 71–103 (1972). doi:10.1007/BF01298413
57. Ghrist, R.: Barcodes: the persistent topology of data. Bull. Am. Math. Soc. **45**(1), 61–75 (2008)
58. Ghrist, R., Krishnan, S.: A topological max-flow-min-cut theorem. In: Proceedings of Global Signal Inference (2013)
59. Ghrist, R., Muhammad, A.: Coverage and hole-detection in sensor networks via homology. In: Proceedings of the 4th International Symposium on Information Processing in Sensor Networks, p. 34. IEEE Press (2005)
60. Gyulassy, A., Natarajan, V., Pascucci, V., Hamann, B.: Efficient computation of morse-smale complexes for three-dimensional scalar functions. IEEE Trans. Vis. Comput. Graph. **13**(6), 1440–1447 (2007)
61. Hatcher, A.: Algebraic Topology. Cambridge University Press (2002)
62. Huang, K., Ni, C.C., Sarkar, R., Gao, J., Mitchell, J.S.: Bounded stretch geographic homotopic routing in sensor networks. In: INFOCOM, 2014 Proceedings IEEE, pp. 979–987. IEEE (2014)
63. Kruskal, J.B.: Multidimensional scaling by optimizing goodness of fit to a nonmetric hypothesis. Psychometrika **29**(1), 1–27 (1964)
64. Kruskal, J.B.: Nonmetric multidimensional scaling: a numerical method. Psychometrika **29**(2), 115–129 (1964)
65. Kruskal, J.B., Wish, M.: Multidimensional Scaling, vol. 11. Sage (1978)
66. Lamar-Leon, J., Baryolo, R.A., Garcia-Reyes, E., Gonzalez-Diaz, R.: Gait-based carried object detection using persistent homology. In: Bayro-Corrochano, E., Hancock, E. (eds.) Progress in Pattern Recognition, Image Analysis, Computer Vision, and Applications, no. 8827 in Lecture Notes in Computer Science, pp. 836–843. Springer International Publishing (2014)
67. Le Roux, B., Rouanet, H.: Geometric Data Analysis. Springer, Netherlands, Dordrecht (2005)
68. Lee, J.A., Verleysen, M.: Nonlinear dimensionality reduction of data manifolds with essential loops. Neurocomputing **67**, 29–53 (2005). doi:10.1016/j.neucom.2004.11.042
69. Lesnick, M.: The Optimality of the Interleaving Distance on Multidimensional Persistence Modules. arXiv:1106.5305 (2011)
70. Li, X., Lin, S., Yan, S., Xu, D.: Discriminant locally linear embedding with high-order tensor data. IEEE Trans. Syst. Man Cybern. Part B: Cybern. **38**(2), 342–352 (2008)
71. Lum, P.Y., Singh, G., Lehman, A., Ishkanov, T., Vejdemo-Johansson, M., Alagappan, M., Carlsson, J., Carlsson, G.: Extracting insights from the shape of complex data using topology. Sci. Rep. **3** (2013). doi:10.1038/srep01236
72. van der Maaten, L.J., Postma, E.O., van den Herik, H.J.: Dimensionality reduction: a comparative review. J. Mach. Learn. Res. **10**(1–41), 66–71 (2009)
73. Maria, C., Boissonnat, J.D., Glisse, M., Yvinec, M.: The Gudhi library: simplicial complexes and persistent homology. In: Mathematical Software-ICMS 2014, pp. 167–174. Springer (2014)
74. Mather, J.: Notes on Topological Stability. Harvard University Cambridge (1970)

75. Mischaikow, K.: Databases for the global dynamics of multiparameter nonlinear systems. Technical report, DTIC Document (2014)
76. Mischaikow, K., Kokubu, H., Mrozek, M., Pilarczyk, P., Gedeon, T., Lessard, J.P., Gameiro, M.: Chomp: Computational homology project. http://chomp.rutgers.edu
77. Morozov, D.: Dionysus. http://www.mrzv.org/software/dionysus/ (2011)
78. Morozov, D., de Silva, V., Vejdemo-Johansson, M.: Persistent cohomology and circular coordinates. Discrete Comput. Geom. **45**(4), 737–759 (2011). doi:10.1007/s00454-011-9344-x
79. Mrozek, M.: Topological dynamics: rigorous numerics via cubical homology. In: Advances in Applied and Computational Topology: Proceedings Symposium, vol. 70, pp. 41–73. American Mathematical Society (2012)
80. Muhammad, A., Jadbabaie, A.: Decentralized computation of homology groups in networks by gossip. In: American Control Conference, ACC 2007, pp. 3438–3443. IEEE (2007)
81. Munch, E., Turner, K., Bendich, P., Mukherjee, S., Mattingly, J., Harer, J.: Probabilistic fréchet means for time varying persistence diagrams. Electron. J. Statist. **9**(1), 1173–1204 (2015). doi:10.1214/15-EJS1030. http://dx.doi.org/10.1214/15-EJS1030
82. Nanda, V.: Perseus: The Persistent Homology Software (2012)
83. Otter, N., Porter, M.A., Tillmann, U., Grindrod, P., Harrington, H.A.: A roadmap for the computation of persistent homology. arXiv:1506.08903 [physics, q-bio] (2015)
84. Perea, J.A., Deckard, A., Haase, S.B., Harer, J.: Sw1pers: Sliding windows and 1-persistence scoring; discovering periodicity in gene expression time series data. BMC Bioinf. (Accepted July 2015)
85. Perea, J.A., Harer, J.: Sliding windows and persistence: an application of topological methods to signal analysis. Found. Comput. Math. **15**(3), 799–838 (2013)
86. Petri, G., Expert, P., Turkheimer, F., Carhart-Harris, R., Nutt, D., Hellyer, P.J., Vaccarino, F.: Homological scaffolds of brain functional networks. J. R. Soc. Interface **11**(101) (2014). doi:10.1098/rsif.2014.0873
87. Pokorny, F.T., Bekiroglu, Y., Exner, J., Björkman, M.A., Kragic, D.: Grasp Moduli spaces, Gaussian processes, and multimodal sensor data. In: RSS 2014 Workshop: Information-based Grasp and Manipulation Planning (2014)
88. Pokorny, F.T., Bekiroglu, Y., Kragic, D.: Grasp moduli spaces and spherical harmonics. In: Robotics and Automation (ICRA), 2014 IEEE International Conference on, pp. 389–396. IEEE (2014)
89. Pokorny, F.T., Ek, C.H., Kjellström, H., Kragic, D.: Topological constraints and kernel-based density estimation. In: Advances in Neural Information Processing Systems 25, Workshop on Algebraic Topology and Machine Learning, 8 Dec, Nevada, USA (2012)
90. Pokorny, F.T., Hang, K., Kragic, D.: Grasp moduli spaces. In: Robotics: Science and Systems (2013)
91. Pokorny, F.T., Kjellström, H., Kragic, D., Ek, C.: Persistent homology for learning densities with bounded support. In: Advances in Neural Information Processing Systems, pp. 1817–1825 (2012)
92. Pokorny, F.T., Stork, J., Kragic, D., others: Grasping objects with holes: A topological approach. In: 2013 IEEE International Conference on Robotics and Automation (ICRA), pp. 1100–1107. IEEE (2013)
93. Richardson, E., Werman, M.: Efficient classification using the Euler characteristic. Pattern Recogn. Lett. **49**, 99–106 (2014)
94. Robinson, M.: Universal factorizations of quasiperiodic functions. arXiv:1501.06190 [math] (2015)
95. Roweis, S.T., Saul, L.K.: Nonlinear dimensionality reduction by locally linear embedding. Science **290**(5500), 2323–2326 (2000)
96. Salamon, D.: Morse theory, the conley index and floer homology. Bull. London Math. Soc **22**(2), 113–140 (1990)
97. Sexton, H., Vejdemo-Johansson, M.: jPlex. https://github.com/appliedtopology/jplex/ (2008)
98. Sheehy, D.R.: Linear-size approximations to the vietoris-rips filtration. Discrete Comput. Geom. **49**(4), 778–796 (2013)

99. de Silva, V., Ghrist, R.: Coordinate-free coverage in sensor networks with controlled boundaries via homology. Int. J. Robot. Res. **25**(12), 1205–1222 (2006). doi:10.1177/0278364906072252
100. de Silva, V., Ghrist, R.: Coverage in sensor networks via persistent homology. Algebraic Geom. Topol. **7**, 339–358 (2007)
101. de Silva, V., Morozov, D., Vejdemo-Johansson, M.: Dualities in persistent (co)homology. Inverse Prob. **27**(12), 124,003 (2011). doi:10.1088/0266-5611/27/12/124003
102. de Silva, V., Vejdemo-Johansson, M.: Persistent cohomology and circular coordinates. In: Hershberger, J., Fogel, E. (eds.) Proceedings of the 25th Annual Symposium on Computational Geometry, pp. 227–236. Aarhus (2009)
103. de Silva, V., Škraba, P., Vejdemo-Johansson, M.: Topological analysis of recurrent systems. In: NIPS 2012 Workshop on Algebraic Topology and Machine Learning, 8 Dec, Lake Tahoe, Nevada, pp. 1–5 (2012)
104. Singh, G., Mémoli, F., Carlsson, G.E.: Topological methods for the analysis of high dimensional data sets and 3D object recognition. In: SPBG, pp. 91–100 (2007)
105. Skraba, P., Ovsjanikov, M., Chazal, F., Guibas, L.: Persistence-based segmentation of deformable shapes. In: 2010 IEEE Computer Society Conference on Computer Vision and Pattern Recognition Workshops (CVPRW), pp. 45–52. IEEE (2010)
106. Stork, J., Pokorny, F.T., Kragic, D., others: Integrated motion and clasp planning with virtual linking. In: 2013 IEEE/RSJ International Conference on Intelligent Robots and Systems (IROS), pp. 3007–3014. IEEE (2013)
107. Stork, J., Pokorny, F.T., Kragic, D., others: A topology-based object representation for clasping, latching and hooking. In: 2013 13th IEEE-RAS International Conference on Humanoid Robots (Humanoids), pp. 138–145. IEEE (2013)
108. Tahbaz-Salehi, A., Jadbabaie, A.: Distributed coverage verification in sensor networks without location information. IEEE Trans. Autom. Control **55**(8), 1837–1849 (2010)
109. Takens, F.: Detecting strange attractors in turbulence. Dyn. Syst. Turbul. Warwick **1980**, 366–381 (1981)
110. Tausz, A., Vejdemo-Johansson, M., Adams, H.: javaPlex: a research platform for persistent homology. In: Book of Abstracts Minisymposium on Publicly Available Geometric/Topological Software, p. 7 (2012)
111. Tenenbaum, J.B., De Silva, V., Langford, J.C.: A global geometric framework for nonlinear dimensionality reduction. Science **290**(5500), 2319–2323 (2000)
112. Turner, K., Mileyko, Y., Mukherjee, S., Harer, J.: Fréchet means for distributions of persistence diagrams. Discrete Comput. Geom. **52**(1), 44–70 (2014)
113. Vejdemo-Johansson, M.: Sketches of a platypus: persistent homology and its algebraic foundations. Algebraic Topol.: Appl. New Dir. **620**, 295–320 (2014)
114. Vejdemo-Johansson, M., Pokorny, F.T., Skraba, P., Kragic, D.: Cohomological learning of periodic motion. Appl. Algebra Eng. Commun. Comput. **26**(1–2), 5–26 (2015)
115. Vergne, A., Flint, I., Decreusefond, L., Martins, P.: Homology based algorithm for disaster recovery in wireless networks. In: 2014 12th International Symposium on Modeling and Optimization in Mobile, Ad Hoc, and Wireless Networks (WiOpt), pp. 685–692. IEEE (2014)
116. Worsley, K.J.: Local maxima and the expected Euler characteristic of excursion sets of χ^2, F and t fields. Adv. Appl. Probab. 13–42 (1994)
117. Worsley, K.J.: Boundary corrections for the expected Euler characteristic of excursion sets of random fields, with an application to astrophysics. Adv. Appl. Probab. 943–959 (1995)
118. Worsley, K.J.: Estimating the number of peaks in a random field using the Hadwiger characteristic of excursion sets, with applications to medical images. Ann. Stat. 640–669 (1995)
119. Zarubin, D., Pokorny, F.T., Song, D., Toussaint, M., Kragic, D.: Topological synergies for grasp transfer. In: Hand Synergies—How to Tame the Complexity of Grapsing, Workshop, IEEE International Conference on Robotics and Automation (ICRA 2013), Karlsruhe, Germany. Citeseer (2013)
120. Zarubin, D., Pokorny, F.T., Toussaint, M., Kragic, D.: Caging complex objects with geodesic balls. In: 2013 IEEE/RSJ International Conference on Intelligent Robots and Systems (IROS), pp. 2999–3006. IEEE (2013)

121. Zomorodian, A., Carlsson, G.: Computing persistent homology. Discrete Comput. Geom. **33**(2), 249–274 (2005)

Author Biographies

Mikael Vejdemo-Johansson received his Ph.D. in Mathematics—Computational Homological Algebra—from the Friedrich-Schiller University in Jena, Germany in 2008. Since then he has worked on research into Topological Data Analysis in research positions at Stanford, St Andrews, KTH Royal institute of Technology and the Jozef Stefan Institute. The bulk of his research is into applied and computational topology, especially persistent cohomology and applications, topological software and applications of the Mapper algorithm. In addition to these topics, he has a wide spread of further interests: from statistical methods in linguistics, through network security and the enumeration of necktie knots, to category theory applied to programming and computer science. He has worked with data from the World Color Survey, from motion capture and from political voting patterns, and published in a spread of journals and conferences, including Discrete and Computational Geometry, IEEE Transactions on Visualization and Computer Graphics and Nature Scientific Reports.

Primoz Skraba received his Ph.D. in Electrical Engineering from Stanford University in 2009. He is currently a Senior Researcher at the Jozef Stefan Institute and an assistant professor of CS at the University of Primorska in Slovenia. He also spent two years as a visiting researcher at INRIA in France. His main research interests are in applied and computational topology as well as its applications to computer science including data analysis, machine learning, sensor networks, and visualization. His work has worked with a variety of types of data. Some examples including cross-lingual text analysis, wireless network link data as well as the study of personal mobility patterns. He has published in numerous journals and conferences including the Journal of the ACM, Discrete and Computational Geometry, IEEE Transactions on Visualization and Computer Graphics, the Symposium of Computational Geometry and the Symposium of Discrete Algorithms.

Applications of Big Data Analytics Tools for Data Management

Mo Jamshidi, Barney Tannahill, Maryam Ezell, Yunus Yetis and Halid Kaplan

Abstract Data, at a very large scale, has been accumulating in all aspects of our lives for a long time. Advances in sensor technology, the Internet, social networks, wireless communication, and inexpensive memory have all contributed to an explosion of "Big Data". Our interconnected world of today and the advent of cyber-physical or system of systems (SoS) are also a key source of data accumulation- be it numerical, image, text or texture, etc. SoS is basically defined as an integration of independently operating, non-homogeneous systems for certain duration to achieve a higher goal than the sum of the parts. Recent efforts have developed a promising approach, called *"Data Analytics"*, which uses statistical and computational intelligence (CI) tools such as principal component analysis (PCA), clustering, fuzzy logic, neuro-computing, evolutionary computation, Bayesian networks, data mining, pattern recognition, deep learning, etc. to reduce the size of "Big Data" to a manageable size and apply these tools to (a) extract information, (b) build a knowledge base using the derived data, (c) optimize validation of clustered knowledge through evolutionary computing and eventually develop a non-parametric model for the "Big Data", and (d) Test and verify the model. This chapter attempts to construct a bridge between SoS and Data Analytics to develop reliable models for such systems. Four applications of big data analytics will be presented, i.e. solar, wind, financial and biological data.

Keywords Machine learning · Data analytics · Computational intelligence · Statistical techniques · Finances · Solar energy · Biology

---

M. Jamshidi (✉) · B. Tannahill · M. Ezell · Y. Yetis · H. Kaplan
ACE Laboratory, The University of Texas, San Antonio, TX, USA
e-mail: moj@wacong.org

A. Emrouznejad (ed.), *Big Data Optimization: Recent Developments and Challenges*, Studies in Big Data 18, DOI 10.1007/978-3-319-30265-2_8

1 Introduction

System of Systems (SoS) are integrated, independently operating systems working in a cooperative mode to achieve a higher performance. A detailed literature survey on definitions of SoS and many applications can be found in texts by Jamshidi [1, 2]. Application areas of SoS are vast. They span from software systems like the Internet to cloud computing, health care, and cyber-physical systems all the way to such hardware dominated cases like military missions, smart grid of electric energy, intelligent transportation, etc. Data analytics and its statistical and intelligent tools including clustering, fuzzy logic, neuro-computing, data mining, pattern recognition, principle component analysis (PCA), Bayesian networks, independent component analysis (ICA), regression analysis and post-processing such as evolutionary computation have their own applications in forecasting, marketing, politics, and all domains of SoS. SoS's are generating "Big Data" which makes modeling of such complex systems a big challenge.

A typical example of SoS is the future smart grid, destined to replace the conventional electric grid. The small-scale version of the smart grid is known as a micro-grid, designed to provide electric power to a home, an office complex or a small local community. A micro-grid is an aggregation of multiple distributed generators (DGs) such as renewable energy sources, conventional generators, and energy storage units which work together as a power supply networked in order to provide both electric power and thermal energy for small communities which may vary from one common building to a smart house or even a set of complicated loads consisting of a mixture of different structures such as buildings, factories, etc. [2]. Typically, a micro-grid operates synchronously in parallel with the main grid. However, there are cases in which a micro-grid operates in islanded mode, or in a disconnected state [3, 4]. Accurate predictions of received solar power can reduce operating costs by influencing decisions regarding buying or selling power from the main grid or utilizing non-renewable energy generation sources.

In the new era of smart grid and distributed power generation big data is accumulated in a very large scale.. Another important example is the financial markets. Nowadays, artificial neural networks (ANNs), as a data analytics tool, have been applied in order to predict the stock exchange stock market index. ANNs are one of the data mining techniques that have the learning capability of the human brain. Due to stochastic nature of financial data several research efforts have been made to improve computational efficiency of share values [5, 6]. One of the first financial market prediction projects was by Kimoto et al. [7] have used an ANN for the prediction of Tokyo stock exchange index. Mizuno et al. [8] applied an ANN to the Tokyo stock exchange to predict buying and selling signals with an overall prediction rate of 63 %. Sexton et al. [9] have determined that use of momentum and start of training in neural networks may solve the problems that may occur in training process. Langdell [10] has utilized neural networks and decision trees to model behaviour of financial stock and currency exchange rates data.

The object of this chapter is to use Big Data Analytics approaches to predict or forecast the behaviour of three important aspects of our times—Renewable energy availability, biological white cell behaviour, and stock market prediction. In each case, large amounts of data are used to achieve these goals.

The remainder of this chapter is as follows: Sect. 2 briefly describes, big data analytics and five statistical and artificial intelligence-based tools. These are PCA, fuzzy logic, fuzzy C-Means, Artificial neural networks and genetic algorithms. Four applications of big data analytics—Solar energy, wind energy, Stock market index prediction and biological white blood cells behavior are presented. Section 4 provides some conclusions.

2 Big Data and Big Data Analytics

In this section a brief description of big data and data analytics are first given. Then five tools of statistics and AI will follow.

2.1 Big Data

Big data is a popular term used to describe the exponential growth and availability of structured and/or unstructured data. Big data may be as important to business—and society—as the Internet has become. Big data is defined as a collection of data so large and complex that it becomes difficult to process using on-hand database management tools or traditional data processing techniques. These data sets have several attributes such as *volume* which is due to social media, advanced sensors, inexpensive memory, system of systems, (Cyber-physical systems), etc. [2, 11]. Big data comes with different *frequencies* such as RFID tags, electric power smart meters, etc. It has different *varieties* such as structured and unstructured text, video, audio, stock ticker data, financial transactions, etc. Big data changes due to special events and hence has variability such as daily, seasonal, and event-triggered peaks such as world sports events, etc. [12, 13] They can have multiple sources, relationships, and linkages, hence they possess *complexity*. For a more detailed coverage big data, refer to Chapter 1 in this volume.

2.2 Big Data Analytics

Data analytics represents a set of statistical and artificial intelligence (AI) tools that can be applied to the data on hand to reduce its size, mine it, seek patterns and

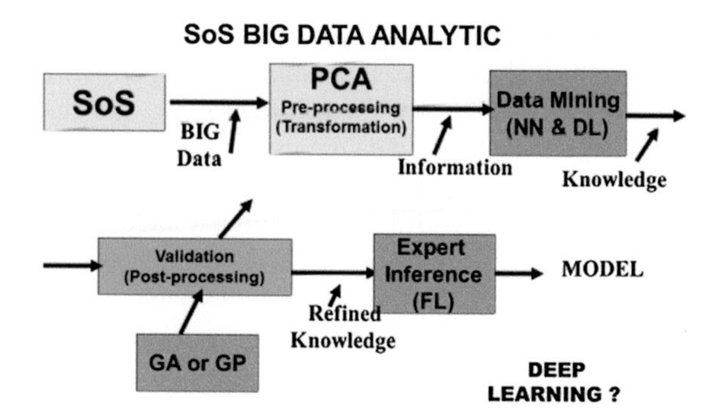

Fig. 1 A possible paradigm for big data analytics

eventually deliver a non-parametric model. These tools are best to be used in a hybrid mode, where the big data is first pre-processed, mined for patterns, and a knowledge base is developed based on attained information, and eventually a model is constructed and evaluated. Figure 1 shows the process just described [14].

2.2.1 Principal Component Analysis

Principal Components Analysis (PCA) is a statistical scheme which uses an orthogonal transformation to identify patterns in data sets so that its similarities and differences are highlighted as a set of values of linearly uncorrelated values called *Principal Components*. Since patterns in data can be hard to find in data of high dimensions, PCA can often help reduce the dimension of the data while bringing up the principal meaning of the information in the data. In other words, PCA can first find the pattern and then compress the data. PCA can work both with numerical as well as image data. Principal components will be independent if the data set is jointly normally distributed. PCA is sensitive to the relative scaling of the original variables [15, 16]. The following steps cans summarize simple steps to perform PCA [16].

Algorithm 1 Standard PCA

Step 1 Get a data set
Step 2 Subtract the mean from each data value
Step 3 Calculate the covariance matrix
Step 4 Calculate the eigenvectors and eigenvalues of the covariance matrix
Step 5 Choosing components and forming a feature vector

The principal component of the data will be near to the eigenvector of the covariance matrix with the largest eigenvalue. It is noted that this algorithm is not necessarily applicable fro a truly “Big Data” scenarios. In such cases one may utilize deep belief networks [17].

2.2.2 Fuzzy Logic

Fuzzy logic can be defined in two terms. On one hand it refers to multi-level logic based on fuzzy sets, which was first introduced by Zadeh in 1965 [18]. Fuzzy sets is the foundation of any logic, regardless of the number of truth levels it assumes. Fuzzy sets represent a continuum of logical values between 0 (completely false) and 1 (completely true). Hence, fuzzy logic treats many possibilities in reasoning in truth, i.e. human reasoning. Therefore, the theory of fuzzy logic deals with two problems (1) the fuzzy set theory, which deals with the vagueness found in semantics, and (2) the fuzzy measure theory, which deals with the ambiguous nature of judgments and evaluations [18].

On the other hand, fuzzy logic refers to a collection of techniques based on approximate reasoning called *fuzzy systems*, which include fuzzy control, fuzzy mathematics, fuzzy operations research, fuzzy clustering, etc. The primary motivation and "banner" of fuzzy logic is to exploit tolerance of costly exact precision., so if a problem does not require precision, one should not have to pay for it. The traditional calculus of fuzzy logic is based on fuzzy IF-THEN rules like: IF pressure is low and temperature is high then throttle is medium.

In this chapter fuzzy clustering, also called C-Means (next section), and fuzzy reasoning for building a non-parametric fuzzy expert system will be utilized for Data Analytics. However, its application to "Big Data" is subject to future research and development in such areas as deep architectures and deep learning.

2.2.3 Fuzzy C-Means Clustering

Cluster analysis, or clustering is a process of observation where the same clusters share some similar features. This is an unsupervised learning approach that has been used in various fields including machine learning, data mining, bio-informatics, and pattern recognition, as well as applications such as medical imaging and image segmentation.

Fuzzy C-means clustering algorithm is based on minimization of the following objective function:

$$J_m = \sum_{i=1}^{N} \sum_{j=1}^{C} u_{ij}^m \left\| x_i - c_j \right\|^2, 1 \leq m < \infty$$

where u_{ij} is the degree of membership of x_i being in cluster j, x_i is the ith of the d-dimensional measured data, c_j is the d-dimension center of the cluster, and $\|*\|$ is any norm expressing the similarity between any measured data and the center. Fuzzy partition is carried out by iterative optimization of the objective function shown above, with an update in membership u_{ij} and cluster centers c_j by:

$$u_{ij} = \frac{1}{\sum_{k=1}^{C} \frac{\|x_i - c_j\|}{\|x_i - c_k\|}^{\frac{2}{m-1}}}$$

where $c_j = \frac{\sum_{i=1}^{N} u_{ij}^m x_i}{\sum_{i=1}^{N} u_{ij}^m}$

This iteration will stop when $max_{ij}\left\{u_{ij}^{k+1} - u_{ij}^{k}\right\} < \varepsilon$ where ε is a terminator criterion between 0 and 1 and k is the iteration step. This procedure converges to a local minimum or a saddle point of J_m. Application of fuzzy C-Means to biological white cells will be given in Sect. 3.3.

2.2.4 Traditional Artificial Neural Networks

Traditional artificial neural networks (ANNs) are information processing system that was first inspired by generalizations of mathematical model of human brain of human neuron (see Fig. 2).

Each neuron receives signals from other neurons or from outside (input layer). The Multi-Layer Perceptron (MLP), shown in Fig. 2 has three layers of neurons, where one input layer is present. Every neuron employs an activation function that fires when the total input is more than a given threshold. In this chapter, we focus on MLP networks that are layered feed-forward networks, typically trained with static *backpropagation*. These networks are used for application static pattern classification [19, 20].

One of the learning methods in MLP neural networks selects an example of training, make a forward and a backward pass. The primary advantage of MLP

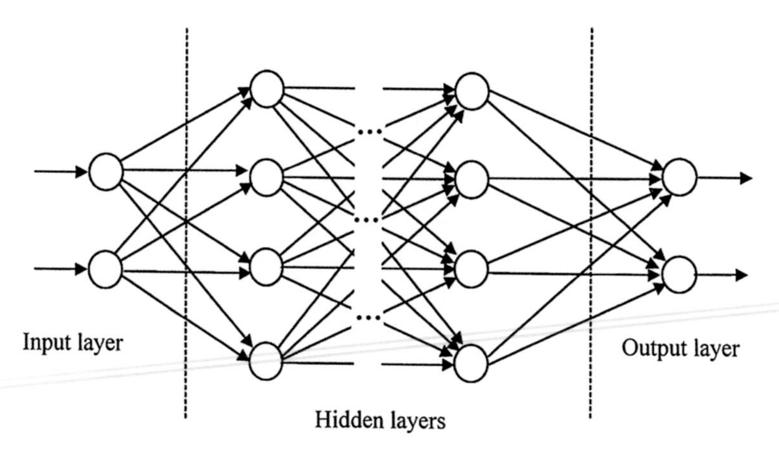

Fig. 2 Architecture of a feed forward multi-layer perceptron

networks is their ease of use and approximation of any input or output map. The primary disadvantage is that they train very slowly and require a lot of training data. It should be said that the learning speed will dramatically decrease according to the increase of the number of neurons and layers of the networks.

However, while traditional ANN have been around for over 30 years, they are not suitable for big data and deep architectures need to be considered [17]. However, these algorithm have to be used for real "big" data. But, the full efficiency of these algorithms have yet to be proven to the best of our knowledge.

2.2.5 Traditional Genetic Algorithms

Genetic algorithm (GA) is a heuristic search approach that mimics the process of natural selection and survival of the fittest. It belongs to a larger class of computational techniques called *evolutionary computations* which also include *genetic programming*, where symbolic and text data is optimized. Heuristically, solutions to optimization problems using techniques inspired by natural evolution, such as inheritance, mutation, selection, and crossover will be obtained after several generations.

In GA, an initial population is formed (see Fig. 3), possibly by chromosomes randomly represented by 0 and 1 bits. Then each pair of the chromosomes are crossed over or mutated (single bit only) after selection. Once new extended population is obtained, through comparison of fitness function extra members are eliminated and the resulting new fitter population replaces the old one. Figure 3 shows the lifecycle of the GA algorithm [21]. GA has been used in Sect. 3.1 to enhance solar energy forecasting via ANN.

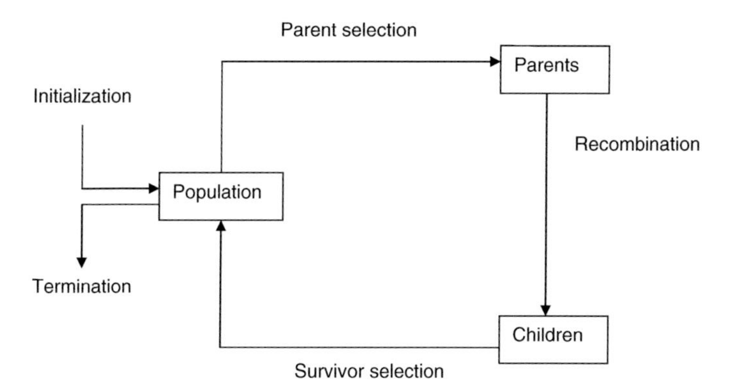

Fig. 3 GA algorithm's cycle [21]

3 Applications of Data Analytics

In this section four different applications of physical, financial and biological data analytics will be presented.

3.1 *Solar Energy Forecasting*

This section provides an example showing how Data Analytics can be used to generate models to forecast produced photovoltaic (or solar) energy to assist in the optimization of a micro-grid SoS. Tools like fuzzy interference, neural networks, PCA, and genetic algorithms are used.

The object of this section and next is to use a massive amount of environmental data in order to derive an unconventional model capable of performing solar and wind energy predictions, which could be leveraged to control energy trading, power generation, and energy storage strategies to optimize operating costs. Recent studies primarily produced by Tannahill [22] have been performed that start looking into different aspects of this problem.

To ensure that the environmental input data for the different data analytics tools is comprehensive, data from different sources was combined to form the full dataset. This was possible because of the solar research projects occurring in Golden, CO, where the National Renewable Energy Laboratory (NREL) is conducting long term research and data recording to support the growing renewable energy industry. Data from the Solar Radiation Research Laboratory (SRRL), SOLPOS data made available by the Measurement and Instrumentation Data Center (MIDC), and data from the Iowa Environmental Mesonet (IEM) Automated Surface Observing System (ASOS) station near the Golden, CO site was also included to have current weather data in the set [22].

Data from the month of October 2012 was combined from the different sources of data. This final set includes one sample for each minute of the month and incorporates measured values for approximately 250 different variables at each data point. The data set was sanitized to only include data points containing valid sensor data prior to the analysis.

Once the viability of this approach was established, a year's worth of data from 2013 was retrieved from the online resources and was similarly sanitized in order to serve as a data set against which the effectiveness of the data analytics techniques could be evaluated for an entire year.

Since the micro-grid would benefit from predicted values of solar irradiance, it was decided that the output of the data analytics should be predicted values of three key irradiance parameters (Global Horizontal Irradiance (GHI), Direct Horizontal Irradiance (DHI), and Direct Normal Irradiance (DNI)). These values were shifted by 60 min so that they would serve as output datasets for the training of the fuzzy

Inference System and Neural Network fitting tools that ultimately provided the means of non-parametric model generation in this exercise.

Once the objective of the data analytics was determined, relevant inputs to the data analytics tools needed to be identified. The full dataset contains approximately 250 different variables. Unfortunately, due to the curse of dimensionality, including all these variables in the data analytics was not practical due to memory and execution time constraints. If this exercise was to be conducted using distributed *cloud computing*, the number of variables to be considered might not need to be down-selected; however, since this effort took place on a single PC, the number of variables needed to be reduced. Ideally, a subject matter expert would be available to optimally identify the best variables to include in the evaluated dataset, or an adaptive training algorithm could be used to automatically perform the selection process. For the purposes of this section, several variables were selected based on intuition, including cloud levels, humidity, temperature, wind speed, and current irradiance levels.

Next, cleanup of the reduced dimension dataset was started by removing all data points containing invalid values from the data set. For instance, during night hours, many solar irradiance parameters contained negative values. Once these invalid data points were removed, the data set was further reduced by removing data points in which GHI, DHI, and DNI levels were very low. The primary reason for this second step was to reduce the amount of time and memory necessary for analysis. Figure 4 is contains the measurements of GHI, DHI, and DNI over one day in the cleaned dataset.

After cleaning took place, the data could be fed into either of the two non-parametric model generating tools, the Fuzzy Inference System Generator and Back-Propagation Neural Network training tools included in the Matlab's fuzzy logic toolbox and the ANN Toolbox.

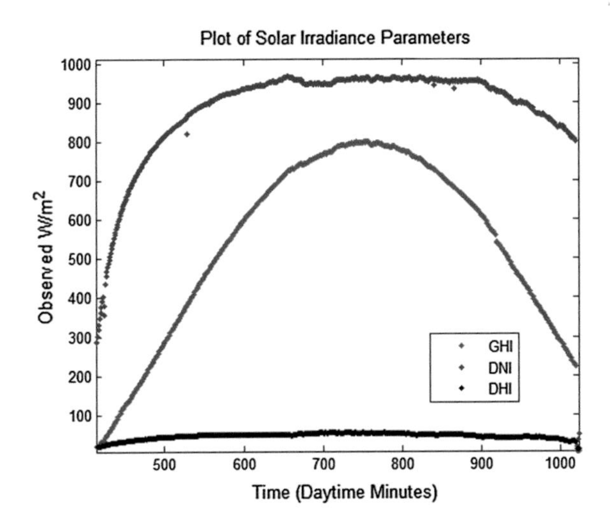

Fig. 4 Three key irradiance parameter plot for a clear day [22]

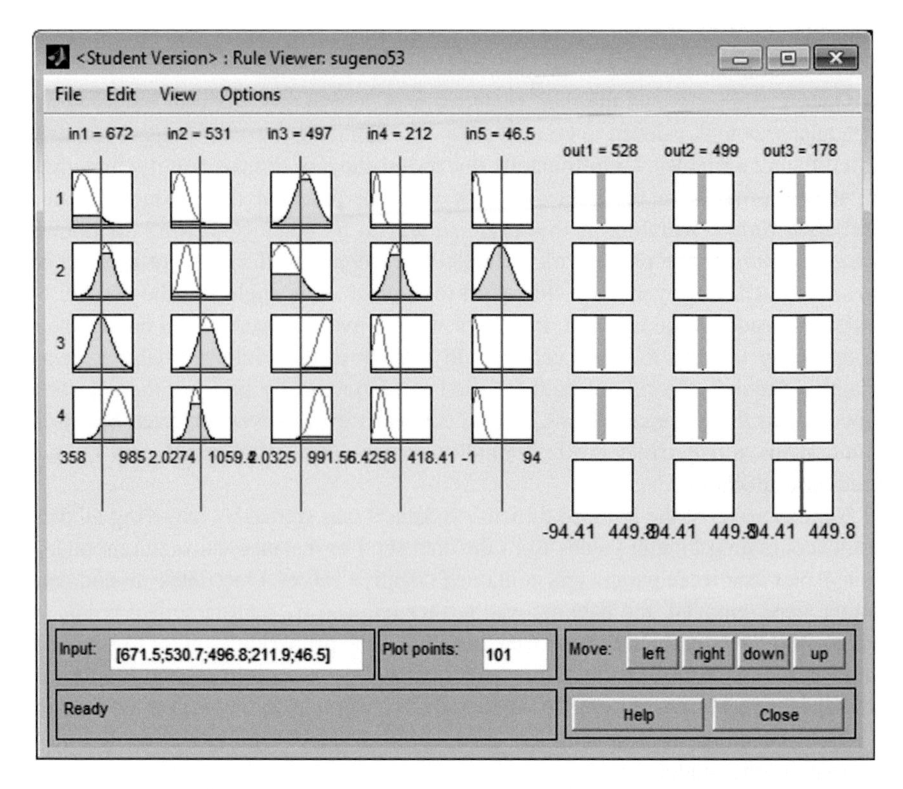

Fig. 5 Rule view of generated FIS membership functions [22]

The Matlab Fuzzy Logic Toolbox function used in this step, *genfis3* uses Fuzzy C-Means clustering (Sect. 2.2.3) to cluster values for each variable which produces fuzzy membership functions for each of the variables in the input matrix and output matrix. It then determines the rules necessary to map each of the fuzzy inputs to the outputs to best match the training data set. These membership functions and rules can be viewed using the Matlab's FIS GUI tools such as *ruleview*. Figure 5 shows the results of running *genfis3* on only four different variables in the dataset.

When comparing the predicted values to the actual values of GHI, DHI, and DNI, differences in the observed and predicted data points could corresponds to the presence of clouds or other anomalies that could not be predicted an hour in advance using the variables input to the function. In addition to unpredictable weather phenomena, such anomalies could include sensor accuracy error, data acquisition noise or malfunctions, missing data, and other issues associated with the acquisition of the environmental data.

The second model generating method used was the Matlab's ANN training toolbox. By default, this tool uses the Levenberg-Marquardt back propagation method to train the network to minimize its mean squared error performance. Figure 6 shows a representation of the feed-forward neural network generated when training using 13 inputs variables and one hidden layer comprised of 10 neurons.

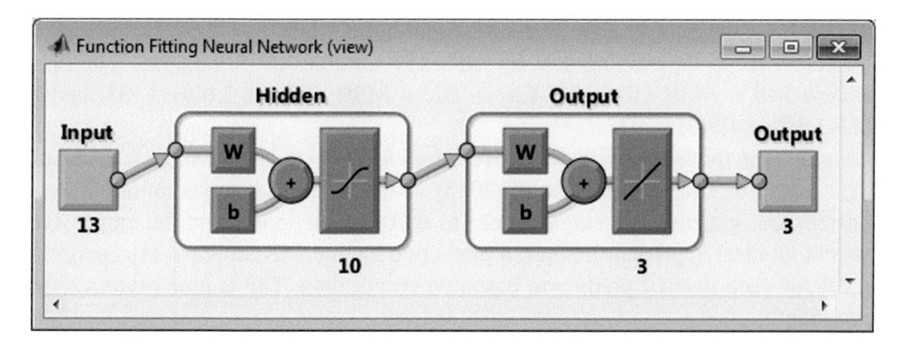

Fig. 6 Trained feed-forward neural network representation

In order to improve the performance of the model generation methods, the following additional data derived points were included in the training sets: $x(t)^2$, sin (x(t)), cos(x(t)), slope(x(t−1):x(t)), slope(x(t−60):x(t)), mean(x(t−60):x(t)), and stdev(x(t−60):x(t)). These functions in some cases made the data more useable and added memory to the training algorithms, which greatly improved performance. Where necessary, PCA was used to reduce the dimension of the data while minimizing the amount of useful data lost.

The results of this analysis showed that the best results were achieved with the Neural Network training tool. The results (mean Square Errors) are discussed in Table 1. NN10 refers to a standard feed forward neural network consisting of 10 neurons in one hidden layer.

These models performed significantly better than a simple predictor based on the average solar parameter values observed at different points during the day. The *nftool* model training time was noted to be linear with respect to the number of variables in the training data set, but *genfis3*'s training time was observed to be much longer with higher dimensions. Its training time appears to be a function of the training set dimension squared. The training time with both functions was linearly related to the number of variable instances in the training dataset, but the slope of this line was an order of magnitude smaller when using the *nftool*.

Next, Genetic Algorithms (Sect. 2.2.4) were used to improve the accuracy of the best performing generated model while minimizing the number of inputs necessary to implement the system. Each population element was a vector of binary variables indicating which of the available variables in the training set would be used for training the model. This is useful because it would reduce the amount of sensors necessary to implement a system.

Table 1 Performance comparison of *GENFIS3* and *NFTOOL* prediction models

Model type	Input params.	PCA?	Final dim	MSE GHI	MSE DNI	MSE DHI	R
FIS	244	Y	50	2.96E + 03	2.08E + 04	1.05E + 03	0.967
NN10	244	Y	150	9.97E + 02	2.95E + 03	4.57E + 02	0.994

Over time, the genetic algorithm solver within the Matlab's Global Optimization Toolbox reduced the training data set from 244 variables to 74 variables. The final solution had a MSE GHI of 7.42E + 02, a MSE DNI of 2.60E + 03, and a MSE GHI of 1.78E + 02.

Finally, the methods discussed above were used to evaluate the models against a larger data set (the entire year of 2013). After some experimentation, it was apparent that generating a single model to make predictions about the entire year was not an ideal approach. Instead, a prediction engine was designed to generate a model for each desired prediction based on recent data. The engine performs the following:

1. Searches the pre-processed data set in order to find the prediction time index of interest
2. Fetches the appropriate training data set (10 day window starting an hour before the prediction time)
3. Performs any additional pre-processing to prepare the training data set for model generation training.
4. Trains two solar prediction models using the training data.
5. Selects the models with the highest R values and uses them to make the requested prediction

The prediction engine was used to make a prediction every 6 h throughout the year. The resulting error statistics are shown in Table 2 below.

Next, the prediction model was used to generate a high time resolution prediction data set over a short period (1 day) in order to illustrate the effectiveness of the engine with a more easily displayed data set (see Figs. 7, 8 and 9).

3.2 *Wind Energy Forecasting*

Many of the data analytics techniques discussed in Sect. 3.1 were also evaluated in their ability to generate models capable of predicting available wind power capacity. Similar to the solar predictions, knowing the available wind power ahead of time could be useful in energy trading or control algorithms.

It was decided that the output of the data analytics for this exercise should be predicted values of wind speed at three different altitudes (19 ft, 22 ft, and 42 ft in

Table 2 Solar prediction engine error statistics

	RMSE GHI	RMSE DHI	RMSE DNI	Solar model R value
Mean	4.250	1.491	−0.9052	0.981739
Min	−651.4	−707.8	−234.9	0.941555
Max	778.6	809.0	198.3	0.998421
STDEV	90.55	111.3	40.88	0.009399
RMSE	90.62	111.3	40.88	N/A

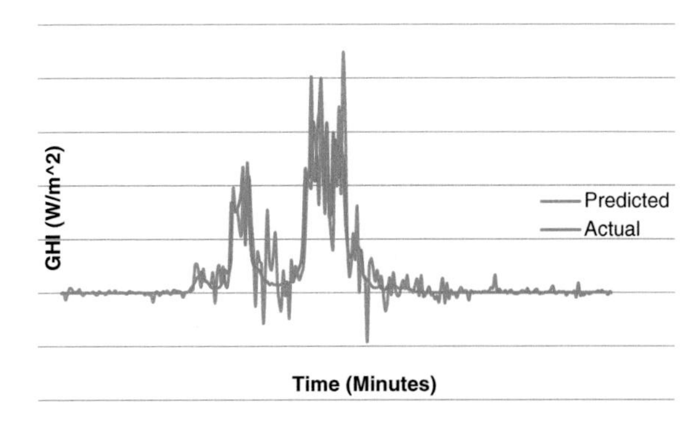

Fig. 7 Prediction engine GHI results (5 min resolution)

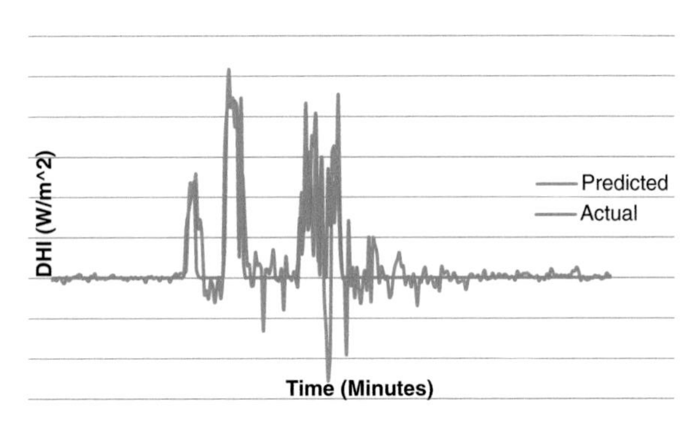

Fig. 8 Prediction engine DHI results (5 min resolution)

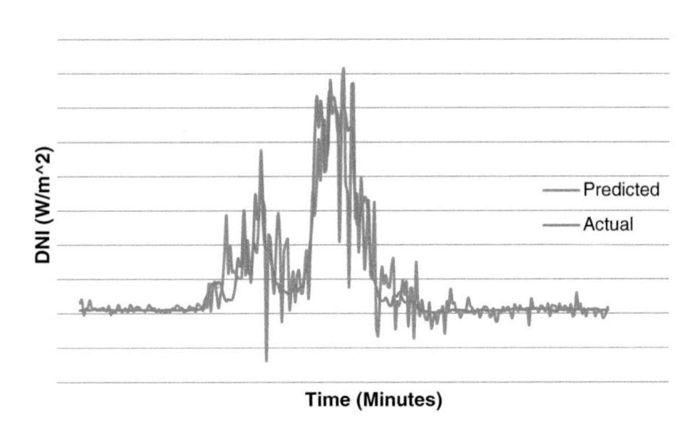

Fig. 9 Prediction engine DNI results (5 min resolution)

altitude). These values were shifted by 60 min so that they would serve as output datasets for the training of the neural networks investigated in the following neural network types:

- Standard Feed-Forward Neural Network
- Time Delay Network
- Nonlinear Autoregressive Network with Exogenous Inputs (NARXNET) Neural Network
- Layer Recurrent Neural Network

Two different input data sets were used for this investigation. The first was merely reduced to a dimension of 21 using PCA [16]. The second was first expanded to include derived, preprocessed values including slope and average values from the past parameter values. Then, this second dataset was also reduced to a dimension of 21 using PCA.

A variety of configurations were tested with each network type, but surprisingly, the best performing neural networks were those using a pre-expanded (via nonlinear expansion) data set fed into a conventional feed forward neural network with ten neurons in the hidden layer. The figures below show the results and error generated using this network to predict wind speed an hour in advance.

With this information in mind, year-long predictions were made using the model prediction engine discussed in Sect. 2.2. For the purposes of this work, predicted wind power availability was calculated assuming the wind speed across the entire wind turbine blade area was the same. It was also assumed that the air density was 1.23 kg/m^3 throughout the year, and that the power coefficient Cp was 0.4. These assumptions were used in conjunction with the wind turbine equation found [23] to calculate power density availability (W/m^2). This quantity can be multiplied by the cumulative sweep area of a wind turbine farm to calculate total available wind power; however, this step was left out of this exercise to keep the results more generalized.

The statistics of the resulting data from the year-long wind power prediction are included in Table 3.

Figure 10 shows a graph of the regression analysis's index values for both the solar and wind prediction model performance over the year-long data set.

A higher resolution prediction loop based is shown in Fig. 11.

3.3 *Financial Data Analytics*

In this section artificial neural networks (ANN) have been used to predict stock market index. ANN has long been used as a data mining tool. This section presents a forecasting scheme of NASDAQ's stock values using ANN. For that purpose, actual values for the exchange rate value of the NASDAQ Stock Market index were used. The generalized feed forward network, used here, was trained with stock market prices data between 2012 and 2013. Prediction of stock market price is an

Table 3 Wind prediction engine error statistics

	RMSE DNI	Wind model R value
Mean	−0.01645	0.9113
Min	−10.15	0.6670
Max	9.617	0.9700
STDEV	1.614	0.03700
RMSE	1.614	N/A

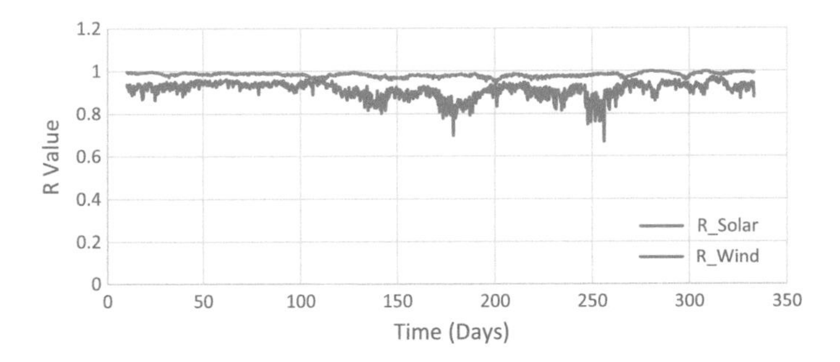

Fig. 10 Prediction engine model R values (6 h resolution)

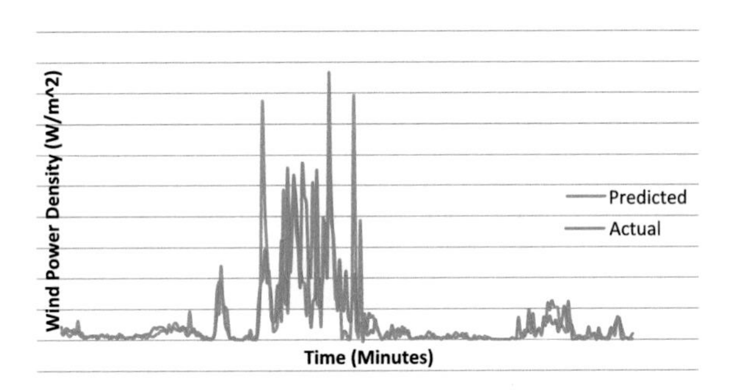

Fig. 11 Prediction engine average wind power density results (5 min resolution)

important issues in national economies. Many researchers have proposed forecasting market price [24, 25].

3.3.1 Training Process

Training process of ANN is through a gradient-based optimization approach (similar to conjugate gradient approach) through adjusting of inter-layer links weights between neurons. Celebrated "back propagation" was used here.

The initial values of weights are determined through a random generation. The network is adjusted based on a comparison of the output and the target during the training (Fig. 12).

The training process requires a set of examples of proper network behaviour and target outputs. During training, the network of the weights and biases are repeated to minimize the network performance function. Mean square error (MSE) performance index during training of feed-forward neural network. MSE is the average squared error between the network outputs and the target outputs [27].

Training is the process of propagating errors back though the system from the output layer towards. Backpropagation is a common paradigm for the training process, utilizing errors between layers of the network. Output is the only layer

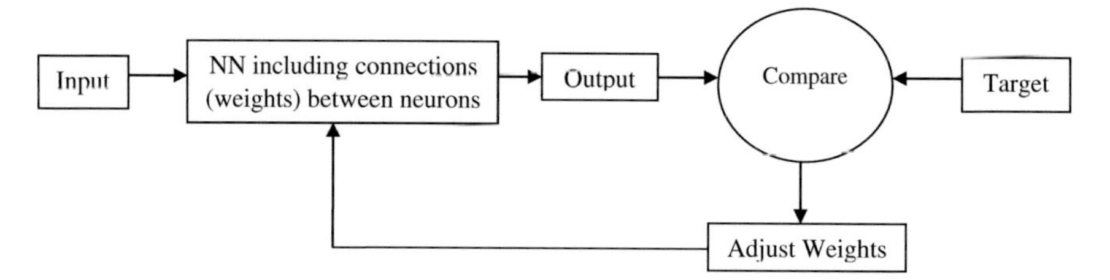

Fig. 12 A network based on a desired target and output comparison [26]

which has a target value. Training occurs until the errors in the weights reduces to a pre-set minimum value. Training has been kept on until its iterative process reaches such that the errors in weights are below a threshold. MLP is the most common feed-forward networks.

With these settings, the input and target vectors will be randomly divided into three sets as follows: 70 % will be used for training and 15 % will be used to validate the network in generalizing and to stopping training before overfitting. The last 15 % will be used as a completely independent test of network generalization.

For simulation purpose, the NASDAQ dataset of daily stock prices has been used [28, 29]. We used five input variables for ANN such as opening price, the highest price, the lowest price, volume of stock, and adjusted daily closing price of the day. Moreover, the architecture and pre-set parameters for ANN were: 10 hidden neurons, 0.4 as learning rate, 0.75 as momentum constant and 1000 was chosen as maximum epochs. Mean squared error (MSE) is the average squared of error between outputs and targets. If the test curve had increased significantly before the validation curve increased, it means it is possible that some over fitting might have occurred. The result of the simulation were acceptable and reasonable.

Error histogram provided additional verification of network performance. It can be clearly seen that errors are between −120 and +100 (Fig. 13). Data set represented hundreds of thousands, so these errors were found negligible considering that the error was smaller than about 0.02 % of targets.

Each input variable of ANN was preprocessed. Mean value, average of the training set was small as compared to its standard deviation. Index rage was between −1 and +1 [30]. We were able to use simple formula which is *Index (x) = (Index(x)—Min (Index))/(Max (Index)—Min (Index))* [30]. It can be clearly seen the regression plot of the training set (see Figs. 14 and 15). Each of the figures corresponds to the target from the output array. Regression values (correlation coefficients) are very close to 1. It indicates that the correlation between the forecasted figures and the target is very high.

Regression was used to validate the network performance. For a perfect fit, the data fell along a 45° degree line, where the network outputs are equal to the targets. For this problem, the fit is reasonably good for all data sets, with R values in each case of 0.99 or above (Figs. 13 and 14).

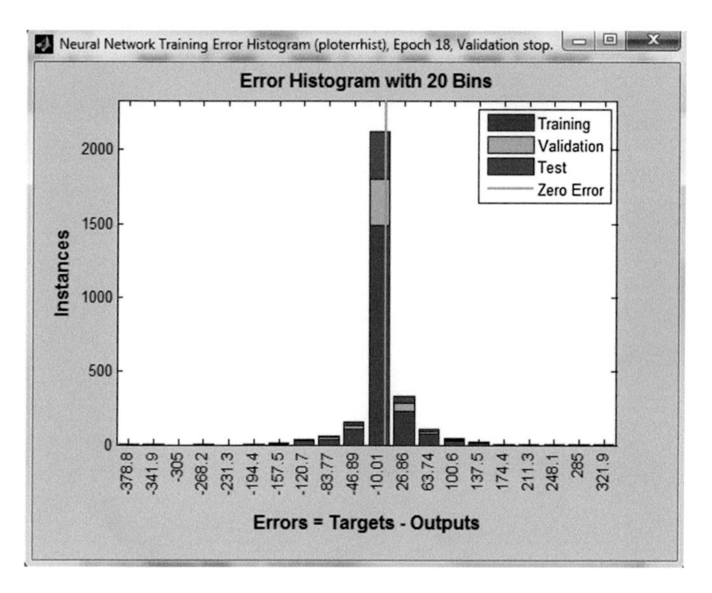

Fig. 13 Error histogram

3.3.2 Biological Data Analytics

This brief section describes the creation of a server on the University of Texas, San Antonio (UTSA) Open Cloud Institute's private cloud and how it was used to run code on some data that is typically too large to be handled by a personal computer.

A virtual machine on the UTSA private cloud running Ubuntu 12.04 as its operating system was created. The code and images were securely copied onto the server. The code was opened in the operating system's *vi editor* and was edited to be used on a batch of images before execution began.

The Fuzzy C-Means algorithm (Sect. 2.2.3) was utilized and the code was modified to be used on our dataset. The data set was a batch of grey-scale images and Fuzzy C-Means algorithm has been used to form clusters based on the pixel values. The code was originally used for clustering one image but it was embedded in a loop to read 312 images, one at a time, find and print the centroids and membership values and then show the clustered images.

Each image is the image of a white blood cell on a blood film and we have chosen to divide the image data (the pixel values) into 4 clusters which are represented by colors white, black, light gray and dark gray. Each image was first converted into the form of a vector of pixel values.

It took the same amount of time to run the code for one image on the virtual machine of the personal laptop as it took on the cloud, but it kept running the code on the cloud for one image after another without crashing the server, whereas running it on the laptop would heat up the CPU and crash it. Figure 16 shows the results, while the centroids of R G B for all 4 clusters are summarized in Table 4.

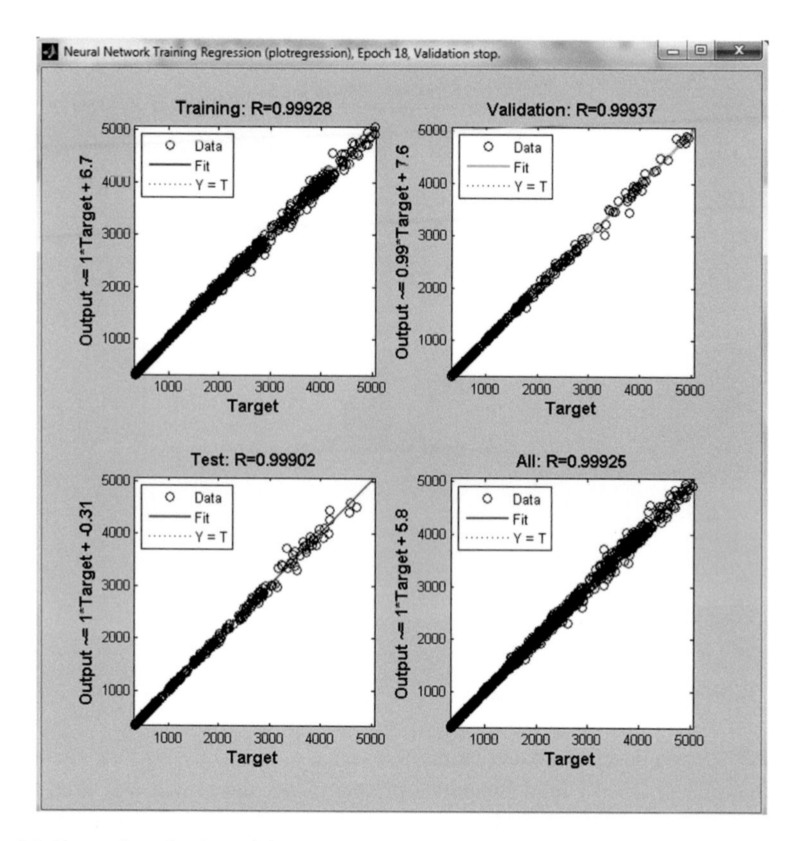

Fig. 14 Regression plot for training

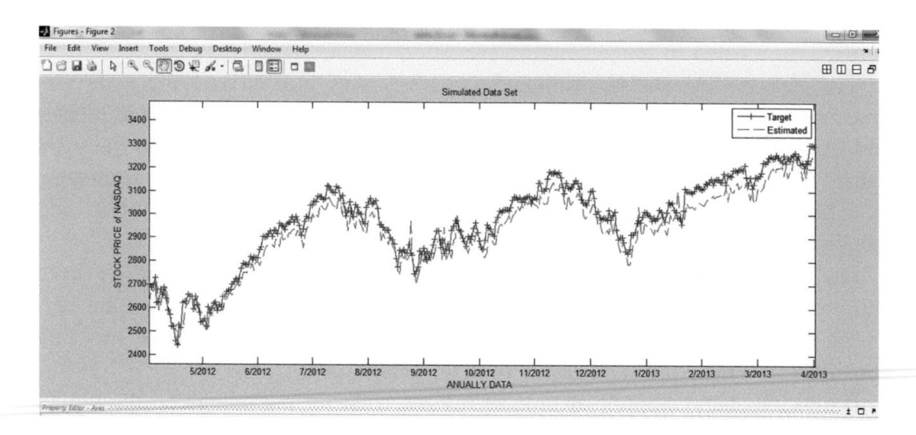

Fig. 15 Target and estimated data of annually NASDAQ data

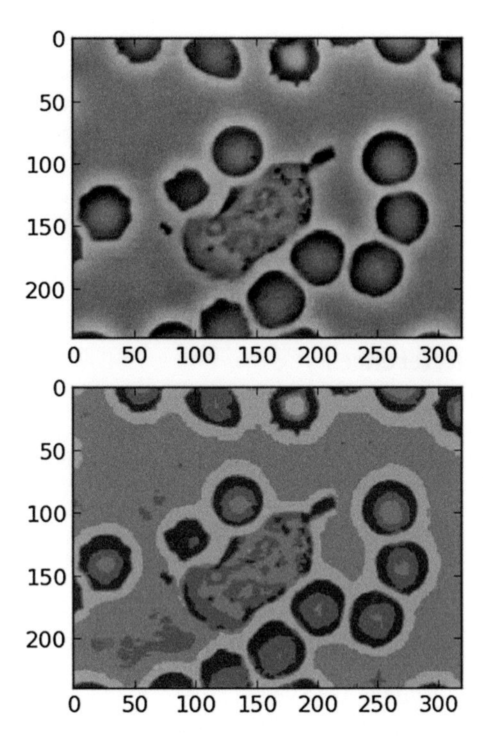

Fig. 16 The results of a fuzzy C-mean clustering for a white blood cell

Table 4 Clustering centroids results for biological data

	R	G	B
Cluster 1	0.72771727	0.65201056	0.67813281
Cluster 2	0.47189973	0.40160155	0.42703744
Cluster 3	0.58333669	0.50191924	0.53275565
Cluster 4	0.29300364	0.22106555	0.24503933

4 Conclusions

This chapter describes the use of a few machine learning tools of data analytics paradigms to assess future estimates of physical, economical and biological variables. Physical phenomena like the intermittency of renewable energy sources, the volatility of stock markets or the inconclusiveness of health related issues all cry out for accurate estimation. Solar and wind energy data were used to forecast availability of solar and wind energy up to 6 h and can be extended to 24 h.

ANN was used to predict future behavior of stock market indices such as NASDAQ index in the United States. The stock market index prediction using artificial neural networks has produced a very efficient results for over a year. Regression index R was within 2 % of actual value. The final application of data analytics was the clustering of biological white blood cells utilizing UTSA's Open Cloud Institute's private cloud for extensive computing.

Future work will include testing much larger data for analytics work and utilization of new tools like "Deep Learning" [31, 32]. Deep learning can also be combined with tools like Bayesian networks, PCA, GA, GP (genetic programming g) and fuzzy logic to enhance the final forecast.

Acknowledgments This work was supported, in part, by the Lutcher Brown Distinguished Chair, the University of Texas at San Antonio, USA. With Southwest Research Institute, San Antonio, TX, USA

References

1. Jamshidi, M. (ed.) Systems of Systems Engineering—Principles and Applications, CRC—Taylor & Francis Publishers, London, UK (2008)
2. Jamshidi, M. (ed.): System of Systems Engineering—Innovations for the 21st Century. Wiley, New York (2009)
3. Manjili, Y.S., Rajaee, A., Jamshidi, M., Kelley, B.: Fuzzy control of electricity storage unit for energy management of micro-grids. In: Proceedings of World Automation Congress, Puerto Vallarta Mexico, pp. 1–6 (2012)
4. Tannahill, B.K., Maute, C., Yetis, Y., Ezell, M.N., Jaimes, A., Rosas, R. Motaghi, A., Kaplan, H., Jamshidi, M.: Modeling of system of systems via data analytics—case for "big data" in SoS. In: Proceedings of 8th IEEE SoSE, Maui, HI, 2–4 June 2013, EDAS # 1569744777 (2013)
5. Sergiu, C.: Financial Predictor via Neural Network (2011). http://www.codeproject.com
6. Jamshidi, M., Tannahill, B., Yetis, Y., Kaplan, H.: Big data analytics via soft computing paradigms. In: Sadeghian, A., Tahayori, H. (eds.) Frontiers of Higher Order Fuzzy Sets, pp. 229–258. Springer, New York, Germany (2015)
7. Kimoto, T., Asawaka, K., Yoda, M., Takeoka, M.: Stock market prediction system with modular neural network. In: Processing of International Joint Conference on Neural Networks, pp. 1–6 (1990)
8. Mizono, H., Kosaka, M., Yajma, H., Komoda, N.: Application of neural network to technical analysis of stock market prediction. Stud. Inf. Control **7**(3), 111–120 (1998)
9. Sexton, S.R., Dorsey, R.E., Johnson, J.D.: Toward global optimization of neural network: a Comparison of the genetic algorithm and backpropagation. Decis. Support Syst. **22**, 117–185 (1998)
10. Langdell, S.: Examples of the use of data mining in financial applications. http://www.nag.com/IndustryArticles/DMinFinancialApps.pdf. Accessed 15 Jul 2015
11. Jamshidi, M., Tannahill, B.K.: Big data analytics paradigms—from PCA to deep learning. In: Proceedings of AAAI Workshop on Big Data, Stanford, CA (2014)
12. http://www.sas.com, 20 Feb 2014
13. http://en.wikipedia.org/wiki/Big_Data, 25 Jul 2015
14. http://en.wikipedia.org/wiki/Data_analytics, 25 Jul 2015
15. Shlens, J.: A Tutorial on Principal Component Analysis. http://www.snl.salk.edu/~shlens/pca.pdf, 22 Apr 2009
16. Smith, L.I.: A Tutorial on Principal Components Analysis, 26 Feb 2002
17. Hinton, G., Osindero, S., Teh, Y.-W.: A fast learning algorithm for deep belief nets. Neural Comput. **18**(7), 1527–1554
18. Zilouchian, A., Jamshidi, M. (eds.): Intelligent Control Systems with Soft Computing Methodologies. CRC Publishers, Boca Raton, FL (2001). Chap. 8–10

19. Rumelhart, D.E., Hinton, G.E., Wiliams, R.J.: Learning internal representation by error propagation. In: McClelland, J.L. (eds.) Parallel Distributed Processing Explorations in the Microstructures of Cognition, vol. 1, pp. 318–362. MIT Press, Cambridge (1986)
20. Zilouchian, A., Jamshidi, M. (eds.) Intelligent Control Systems with Soft Computing Methodologies. CRC Publishers, Boca Raton, FL (2001) Chap. 2
21. www.cse.hcmut.edu.vn/~dtanh/download/Genetic_Algorithm.doc. Accessed 15 Jul 2015
22. Tannahill, B.: Big Data Analytics Techniques: Predicting Renewable Energy Capacity to Facilitate The Optimization of Power Plant Energy Trading And Control Algorithms, M.S. thesis, ACE Laboratory, ECE Department, The University of Texas at San Antonio (2013)
23. The Royal Academy of Engineering. Wind Turbine Power Calculations (2014). http://www.raeng.org.uk/education/diploma/maths/pdf/exemplars_advanced/
24. Philip, M.T., Poul, K., Choy, S.O., Reggie, K., Ng, S.C., Mark, J., Jonathan, T., Kai, K., Tak-Lam, W.: Design and Implementation of NN5 for hong stock price forecasting. J. Eng. Appl. Artif. Intell. **20**, 453–461 (2007)
25. Wu, X., Fund, M., Flitman, A.: Forecasting stock performance using intelligent hybrid systems. Springer, Berlin (2001), pp. 447–456
26. Ball, R., Tissot, P.: Demonstration of Artificial Neural Network in Matlab. Division of Near shore Research, Texas A&M University (2006)
27. Neenwi, S., Asagba, P.O., Kabari, L.G.: Predicting the Nigerian stock market using artificial neural network. Eur. J. Comput. Sci. Inf. **1**(1), 30–39 (2013)
28. Yahoo Historical Prices of NASDAQ http://finance.yahoo.com/q/hp?s=%5EIXIC&a=00&b=2&c=2013&d=03&e=15&f=2013&g=d, Accessed 5 May 2013
29. Mathworks Website. http://www.mathworks.com/help/nnet/ug/analyze-neural-network-performance-after-training.html, Nov 2013
30. Sergiu, C.: Financial Predictor via Neural Network. http://www.codeproject.com, Accessed 25 May 2011
31. http://www.deeplearning.net/tutorial/, 20 Jul 2015
32. Moussavi, A., Jamshidi, M.: Cascaded and Partially Cascaded Stacked Autoencoders Applied to Traffic Flow Prediction. Neural Comput. Appl. J. (2015), Submitted

Author Biographies

Mo M. Jamshidi (F-IEEE, F-ASME, AF-AIAA, F-AAAS, F-TWAS, F-NYAS) He holds honorary doctorate degrees from University of Waterloo, 2004 and Technical University of Crete, Greece, 2004 and Odlar Yourdu University, Baku, Azerbaijan, 1999. Currently, he is the Lutcher Brown Endowed Distinguished Chaired Professor at the University of Texas, San Antonio, TX, USA. He has advised over 70 MS and nearly 55 Ph.D. students. He has over 740 technical publications including 68 books (11 text books), research volumes, and edited volumes. and near 7000 science citations. He is the Founding Editor or co-founding editor or Editor-in-Chief of 5 journals including *IEEE Control Systems Magazine* and the *IEEE Systems Journal*. He has received numerous honours and awards, including five from the IEEE, among others. He is currently involved in research on system of systems engineering, cyber-physical systems, with emphasis on open cloud computing, robotics, UAVs, autonomous vehicles, biological and sustainable energy systems, including smart grids and big data analytic.

Barnabas K. Tannahill He holds an MS EE degree from the University of Texas at San Antonio, 2014. While studying at UTSA, he focused his studies on control systems and data analytics. Currently, he is employed as a Senior Research Engineer at Southwest Research Institute, San Antonio, Texas, USA, where he has been a full-time employee for 10 years.

Maryam Nezamabadi Ezell received her MSEE student at UTSA in the Autonomous Control Engineering (ACE) Laboratory in 2015. She has recently completed her graduate project titled: *"Investigating the Use of an Evolutionary Algorithm and Geometric Shapes in Approximating a Target Image for Compression in Python"*. She has 3 joint publications including a book chapter. She has presented papers in 2013 World Conference on Soft Computing (WCSC) and 2013 UTSA's 7th Architecture, Business, Engineering and Science Student Conference (ABES). She won the third place for best joint paper presentation in the 2013 ABES Conference and was part of the organizing committee in 2013 WCSC Conference. Currently she is working for IBM Corp., Austin, TX, USA.

Yunus Yetis received the B.S. in Electronic and Electrical Engineering from Kirikkale University, Turkey in 2005, the M. S. degree from The University of Texas at San Antonio, USA in 2013. He is currently a Ph.D. student in the Autonomous Control Engineering (ACE) Laboratory in the electrical and computer engineering program at UTSA. where he expects to graduate in 2017. His current research interests include complex energy systems modelling and control, input—output model, and prediction algorithms, big data analytic. He has 4 technical publications including 1 book's chapter. Mr. Yetis has presented papers in multiple conferences, and has won best project at the Kirikkale, Turkey in 2011 and International scholarship from Turkish Ministry of Education. He has been involved in the organization of the 2013 World Conference on Soft Computing (WCSC), 2014 World Automation Congress (WAC 2014), and student conferences at UTSA.

Halid Kaplan received a B.S. in Electrical and Electronics Engineering from the Kahramanmaras Sutcu Imam University, Turkey in 2011. He completed the M.S.E.E from The University of Texas at San Antonio in 2014. He is currently a Ph.D. student in Electrical and Computer Engineering Department at the University of Texas at San Antonio. He expects to graduate in May 2018. He is member of Autonomous Control Engineering Laboratory (ACE). His current research interests are: renewable energy forecasting systems, artificial intelligent control applications, data analytics and cloud computing. He had been teaching assistantships at Introduction to the Electrical Engineering Profession course. He has 3 publications including a book chapter. He has co-authored conference papers at the WAC 2014 and the SoSE 2013 conferences. He was a member of conference organization committee of Third Annual Word Conference On Soft Computing (WCSC) 2013 and (ABES 2014) student conferences at UTSA.

Optimizing Access Policies for Big Data Repositories: Latency Variables and the Genome Commons

Jorge L. Contreras

Abstract The design of access policies for large aggregations of scientific data has become increasingly important in today's data-rich research environment. Planners routinely consider and weigh different policy variables when deciding how and when to release data to the public. This chapter proposes a methodology in which the timing of data release can be used to balance policy variables and thereby optimize data release policies. The global aggregation of publicly-available genomic data, or the "genome commons" is used as an illustration of this methodology.

Keywords Commons • Genome • Data sharing • Latency

1 Introduction

Since its beginnings in the 1980s, genomic science has produced an expanding volume of data about the genetic makeup of humans and other organisms. Yet, unlike many other forms of scientific data, genomic data is housed primarily in publicly-funded and managed repositories that are broadly available to the public (which I have collectively termed the "genome commons" [1]). The availability of this public data resource is due, in large part, to the data release policies developed during the Human Genome Project (HGP), which have been carried forward, in modified form, to the present. These policies developed in response to numerous design considerations ranging from practical concerns over project coordination to broader policy goals.

One innovative approach developed by policy designers for the genome commons was the use of timing mechanisms to achieve desired policy outcomes. That is, by regulating the rate at which information is released to the public ("knowledge

J.L. Contreras (✉)
University of Utah, S.J. Quinney College of Law, 383 South University St., Salt Lake City, UT 84112, USA
e-mail: jorge.contreras@law.utah.edu

A. Emrouznejad (ed.), *Big Data Optimization: Recent Developments and Challenges*, Studies in Big Data 18, DOI 10.1007/978-3-319-30265-2_9

latency"), and then by imposing time-based restrictions on its use ("rights latency"), policy designers have addressed the concerns of multiple stakeholders while at the same time offering significant benefits to the broader research enterprise and to the public [1, 2].

The lessons learned from the genomc commons and its utilization of latency techniques for achieving policy goals is informative for all repositories of scientific data. In this chapter, I describe the use of latency variables in designing access and usage policies for data repositories (also known as "information commons" [3] with specific reference to the development of the genome commons over the past three decades.

2 Information Commons and Latency Variables

1. The Dynamic Nature of Information Commons

Information commons, by their nature, are dynamic: the pool of data constituting the commons may expand and contract over time. The total pool of data constituting an information commons at any given time its "knowledge base". The magnitude of the knowledge base at a given time (K_t), and the rate at which it changes over time (K(t)) can be adjusted and optimized via policy mechanisms.

Just as the pool of *data* within an information commons may expand and contract over time, so may the set of *rights* applicable to the data within the commons. That is, for a given commons, the nature and duration of the usage restrictions on each data element may evolve over time, and the aggregate pool of usable data within the commons will likewise change. For purposes of discussion, I will term the portion of the knowledge base of an information commons, the use of which is materially encumbered,[1] as its "encumbered knowledge base" (K_e) and the portion of its knowledge base, the use of which is not materially encumbered, or which is generally accessible and usable by a relevant public community, as its "unencumbered knowledge base" (K_u). The total pool of data within the commons at any given time (T) constitutes the sum of its encumbered and unencumbered knowledge bases ($K_T = K_e + K_u$).

There may be large bodies of publicly-accessible information that are almost entirely encumbered ($K_e \gg K_u$). One such example is the database of currently-issued U.S. patents. It contains much information, but little freedom to use it, at least in the near-term. Other information commons may contain less information, but few limitations on use ($K_u > K_e$). Thus, each information commons may have a different combination of knowledge and rights latency. And, by

[1]By "materially encumbered" I mean that one or more material restrictions on the use of the data exist. These might include a contractual or policy embargo on presentation or publication of further results based on that data. At the extreme end of the spectrum, patent rights that wholly prevent use of the data can be viewed as another variety of encumbrance.

extension, as the knowledge and rights aspects of an information commons change over time, the relationship between these two variables also changes, providing the basis for the latency analysis described below and the opportunity to optimize commons design based on adjustment of latency variables as described below.

2. Latency Variables

a. *Knowledge Latency*. For any given data element, there will be a period of time between its creation/discovery and the point at which it becomes accessible to the relevant community through the commons. I term this period "knowledge latency" (L_k). Knowledge latency defers the introduction of data into the commons, thus reducing the amount of data that would otherwise reside within the commons at a given time.[2]

b. *Rights Latency*. Just as there may be a delay between the generation of data and its deposit into an information commons, there may be a delay between the appearance of data in the commons and its free usability. I term this delay "rights latency" (L_r).[3] Thus, just as knowledge latency (L_k) affects the total quantity of knowledge (K_T) within an information commons at a given time, rights latency affects the amount of unencumbered knowledge (K_{uT}) within the commons at a given time.

   Rights latency for a particular information commons may reflect a variety of factors including policy-imposed embargos on the use of data and, in the extreme case, patent rights. True public domain commons such as a compendium of Dante's sonnets, in which no copyright or contractual encumbrances exist, would have a rights latency of zero ($L_r = 0$). Commons that include data covered by patents would have a rights latency equal to the remaining patent term ($L_r = P(t)$).[4] Most information commons would fall somewhere between these two extremes.

c. *Latency Variables as Policy Design Tools*. Knowledge latency and rights latency can be modulated by policy designers in order to optimize policy outcomes. For example, policy designers who wish to disseminate information to

[2]Knowledge latency in a given information commons may be expressed either as a *mandated* value (derived from policy requirements), or as an *actual* value. It goes without saying that the *actual* value for knowledge latency may deviate from the *mandated* value for a number of reasons, including technical variations in data deposit practices and intentional or inadvertent non-compliance by data generators. As with any set of policy-imposed timing requirements (e.g., time periods for making filings with governmental agencies), it is important to consider the *mandated* time delay for the deposit of data to an information commons. Because a mandated value is also, theoretically, the *maximum* amount of time that should elapse before a datum is deposited in the commons, knowledge latency is expressed in this chapter in terms of its *maximum* value.

[3]As with knowledge latency, this term may be applied to an individual datum (i.e., representing the time before a particular datum becomes freely usable) or to the commons as a whole (i.e., representing the *maximum* time that it will take for data within the commons to become freely usable).

[4]In the U.S. and many other countries, the patent term lasts for twenty years from the date of filing.

the public as quickly as possible, but to limit the ability of users to exploit that information commercially or otherwise, will likely choose designs that embody low knowledge latency but high rights latency ($L_r > L_k$). On the other hand, those who wish to enable "insiders" to benefit from the use of data for some period before it is released to the public, but who then wish for the data to be freely usable, would opt for designs that embody high knowledge latency and low rights latency ($L_k > L_r$). In addition, as discussed in Contreras [2], the effective interplay of latency variables can mediate between the requirements of competing stakeholder interests and enable the creation of commons where disagreement might otherwise preclude it.

The following case study of the genome commons illustrates the ways in which policy designers have utilized latency variables to achieve desired outcomes when constructing an information commons.

3 Evolution and Design of the Genome Commons

1. *The Expanding Genomic Data Landscape*

The principal databases for the deposit of genomic sequence data are GenBank, which is administered by the National Center for Biotechnology Information (NCBI) a division of the National Library of Medicine at the U.S. National Institutes of Health (NIH), the European Molecular Biology Library (EMBL), and the DNA Data Bank of Japan (DDBJ). These publicly-funded and managed repositories are synchronized on a daily basis and offer open public access to their contents [4]. In addition to sequence data, these databases accept expressed sequence tags (ESTs), protein sequences, third party annotations and other data. NCBI also maintains the RefSeq database, which consolidates and annotates much of the sequence data found in GenBank.

In addition to DNA sequence data, genomic studies generate data relating to the association between particular genetic markers and disease risk and other physiological traits. This type of data, which is more complex to record, search and correlate than the raw sequence data deposited in GenBank, is housed in databases such as the Database of Genotypes and Phenotypes (dbGaP), also operated by the National Library of Medicine. dbGaP can accommodate phenotypic data, which includes elements such as de-identified subject age, ethnicity, weight, demographics, drug exposure, disease state, and behavioral factors, as well as study documentation and statistical results. Given potential privacy and regulatory concerns regarding phenotypic data, dbGaP allows access to data on two levels: open and controlled. Open data access is available to the general public via the Internet and includes non-sensitive summary data, generally in aggregated form. Data from the controlled portion of the database may be accessed only under conditions specified by the data supplier, often requiring certification of the user's identity and research purpose.

The sheer size of the genome commons is matched only by the breathtaking rate at which it is expanding. The genome of simple organisms such as the *E. coli* bacterium contains approximately five million base pairs, that of the fruit fly *Drosophila melanogaster* contains approximately 160 million, and that of *Homo sapiens* contains approximately 3.2 billion base pairs. Over its decade-long existence, the HGP mapped the 3.2 billion base pairs comprising the human genome. To do so, it sequenced tens of billions of DNA bases (gigabases), creating what was then an unprecedented accumulation of genomic data. According to NCBI, between 1982 and 2015 the amount of data in GenBank has doubled every eighteen months [5]. As of June 2015, the database contained approximately 1.2 trillion nucleotide bases from more than 425 million different genomic sequences (Ibid.). The rapid growth of GenBank and other genomic databases is attributable, in part, to the technological advancement of gene sequencing equipment, computational power and analytical techniques. According to one report, a single DNA sequencer in 2011 was capable of generating in one day what the HGP took ten years to produce [6].

2. *Bermuda and the Origins of Rapid Genomic Data Release*

The HGP began as a joint project of NIH and the U.S. Department of Energy (DOE), with additional support from international partners such as the Wellcome Trust in the UK and the Japanese government. From the outset it was anticipated that the HGP would generate large quantities of valuable data regarding the genetic make-up of humans and other organisms. Thus, in 1988 the U.S. National Research Council recommended that all data generated by the project "be provided in an accessible form to the general research community worldwide" [7]. In 1992, shortly after the project was launched, NIH and DOE developed formal guidelines for the sharing of HGP data among project participants [8]. These guidelines required that all DNA sequence data generated by HGP researchers be deposited in GenBank, making it available to researchers worldwide. This public release of data was viewed as necessary to avoid duplication of effort, to coordinate among multiple research centers across the globe, and to speed the use of DNA data for other beneficial purposes [9].

At the time, most U.S. federal agencies that funded large-scale scientific research (e.g., NASA and the DOE) required that data generated using federal funds be released to the public at the completion of the relevant project, usually after the principal researchers published their analyses in the scientific literature. This typically resulted in a release of data one to two years following the completion of research [10, 11]. Seeking to accelerate this time frame, NIH and DOE agreed that DNA sequence data arising from the HGP should be released six months after it was generated, whether or not the principal researchers had yet published their analyses. HGP's 6-month data release requirement was considered to be aggressive and viewed as a victory for open science.

This perception changed, however, in 1996. Early that year, prior to commencing sequencing the human genome,[5] the HGP leadership met in Bermuda to plan the next phase of the project. Among the things they considered was the rate at which HGP data should be released to the public, and whether the 6-month "holding period" approved in 1992 should continue. Several arguments for eliminating the holding period were presented. From a pragmatic standpoint, some argued that sequencing centers working on the HGP required regularly-updated data sets in order to avoid duplication of effort and to optimize coordination of the massive, multi-site project. Waiting six months to obtain data was simply not practical if the project were to function effectively. But perhaps more importantly, the concept of rapid data release became endowed with an ideological character: the early release of data was viewed as necessary to accelerate the progress of science [12].

After substantial debate, it was agreed that a change was needed, and a document known as the "Bermuda Principles" emerged [13]. This short document established that the DNA sequence information generated by the HGP and other large-scale human genomic sequencing projects should be "freely available and in the public domain in order to encourage research and development and to maximize its benefit to society." Most importantly, it went on to require that all such DNA sequences be released to the public a mere *twenty-four hours* after assembly.

The Bermuda Principles were revolutionary in that they established, for the first time, that data from public genomic projects should be released to the public almost immediately after their generation, with no opportunity for researchers generating the data to analyze it in private. This policy was heralded as a major victory for open science, and soon became the norm for large-scale genomic and related biomedical research [9, 14].

3. *Second Generation Policies and the Ft. Lauderdale Accord*

An initial draft of the human genome sequence was published by the HGP in 2001 to much fanfare. In 2003, the Wellcome Trust, a major UK-based funder of biomedical research, convened a meeting in Ft. Lauderdale, Florida to revisit rapid data release practices in the "post-genome" world. While the Ft. Lauderdale participants "enthusiastically reaffirmed" the 1996 Bermuda Principles, they also expressed concern over the inability of data generating researchers to study their results and publish analyses prior to the public release of data [15]. The most significant outcome of the Ft. Lauderdale meeting was a consensus that the Bermuda Principles should apply to each "community resource project" (CRP), meaning "a research project specifically devised and implemented to create a set of data, reagents or other material whose primary utility will be as a resource for the broad scientific community." Under this definition, the 24-h rapid release rules of Bermuda would be applicable to large-scale projects generating non-human sequence data, other basic genomic data maps, and other collections of complex

[5] Prior work had focused on simple model organisms and technology development.

biological data such as protein structures and gene expression information. In order to effectuate this data release requirement, funding agencies were urged to designate appropriate efforts as CRPs and to require, as a condition of funding, that rapid, pre-publication data release be required for such projects.

Despite this support, the Ft. Lauderdale participants acknowledged that rapid, pre-publication data release might not be feasible or desirable in all situations, particularly for projects other than CRPs. In particular, the notion of a CRP, the primary goal of which is to generate a particular data set for general scientific use, is often distinguished from "hypothesis-driven" research, in which the investigators' primary goal is to solve a particular scientific question, such as the function of a specific gene or the cause of a disease or condition. In hypothesis-driven research, success is often measured by the degree to which a scientific question is answered rather than the assembly of a quantifiable data set. Thus, the early release of data generated by such projects would generally be resisted by the data generating scientists who carefully selected their experiments to test as yet unpublished theories. Releasing this data before publication might allow a competing group to "scoop" the data generating researchers, a persistent fear among highly competitive scientists.

In the years following the Ft. Lauderdale summit, numerous large-scale genomic research projects were launched with increasingly sophisticated requirements regarding data release. Some of these policies utilized contractual mechanisms that are more tailored and comprehensive than the broad policy statements of the HGP era [9]. Moreover, increasingly sophisticated database technologies have enabled the provision of differentiated levels of data access, the screening of user applications for data access, and improved tracking of data access and users.

4. *Third Generation Data Release Policies*

Between 2003 and 2006, the technologies available for genomic research continued to improve in quality and decrease in cost, resulting in the advent of so-called genome-wide association studies (GWAS). These studies differ from pure sequencing projects in that their goal is not the generation of large data sets (such as the genomic sequence of a particular organism), but the discovery of disease markers or associations hidden within the genome. They thus have greater potential clinical utility and lie further along the path to ultimate commercialization than raw sequence data. Moreover, as the types of data involved in large-scale genomics projects expanded, the community of researchers participating in these projects has become more diverse. Today, many scientists with backgrounds outside of genomics, including medical researchers, medical geneticists, clinicians and epidemiologists, actively lead and participate in GWAS projects. Yet these researchers do not necessarily share the norms of rapid pre-publication data release embraced by the model organism and human genomics communities since the early days of the HGP. In many cases, particularly when patient data are involved, these researchers are accustomed to an environment in which data is tightly guarded and released only after publication of results, and then only in a limited, controlled manner.

Accordingly, when the federally-backed Genetic Association Information Network (GAIN) was established in 2006 to conduct GWA studies of six common diseases, its data release policies reflected a compromise among data generators and data users. Data generators agreed to "immediate" release of data generated by the project, but for the first time a temporal restriction was placed on *users* of the data. That is, in order to secure a period of exclusive use and publication priority for the data generators, data users were prohibited from submitting abstracts and publications based on GAIN data for a specified "embargo" period (rights latency), generally fixed at nine months.

Shortly thereafter, a similar embargo-based approach was adopted by NIH in its institute-wide policy regarding the generation, protection and sharing of data generated by federally-funded GWA studies, as well as subsequent genomic studies [9, 16]. The NIH GWAS Policy states that users of GWAS data should refrain from submitting their analyses for publication, or otherwise presenting them publicly, during an "exclusivity" period of up to twelve months from the date that the data set is first made available. While the agency expresses a "hope" that "genotype-phenotype associations identified through NIH-supported and maintained GWAS datasets and their obvious implications will remain available to all investigators, unencumbered by intellectual property claims," it stops short of prohibiting the patenting of resulting discoveries.

5. *Private Sector Initiatives*

A noteworthy parallel to the government-sponsored projects discussed above is that of private-sector initiatives in the genome sciences. The first of these was organized by pharmaceutical giant Merck in 1994, which established a large public database of short DNA segments known as expressed sequence tags (ESTs). The stated purpose of the so-called Merck Gene Index was to increase the availability of basic knowledge and the likelihood of discovery in support of proprietary therapeutic innovations and product development [17]. Another important, but less publicized, motivation for placing the EST data into the public domain was reputedly to pre-empt the patenting of these genetic sequences by private biotechnology companies [9, 18].

A similar effort known as the SNP Consortium was formed in 1999 by a group of private firms and the Wellcome Trust to identify and map genetic markers referred to as "single nucleotide polymorphisms" (SNPs) and to release the resulting data to the public, unencumbered by patents. The consortium accomplished this goal by filing U.S. patent applications covering the SNPs it discovered and mapped, and then ensuring that these applications were contributed to the public domain prior to issuance [19]. This approach ensured that the consortium's discoveries would act as prior art defeating subsequent third party patent applications, with a priority date extending back to the initial filings. The SNP Consortium's innovative "protective" patenting strategy has been cited as a model of private industry's potential to contribute to the public genome commons [18].

Since the successful conclusion of the SNP Consortium project, other privately-funded research collaborations have adopted similar data release models.

In recent years, however, these efforts have implemented timing mechanisms into their data release policies. For example, the International SAE Consortium (SAEC) was formed in 2007 to fund the identification of DNA markers for drug-induced serious adverse events. SAEC adopted a "defensive" patent filing strategy similar to that of the SNP Consortium, but secures for data-generating scientists a period of exclusivity during which they have the sole ability to analyze data and prepare papers for publication [20]. Like the policies adopted by some government-funded projects, SAEC imposes a nine-month embargo on publication or presentation of publicly-released data. But in addition SAEC utilizes a delayed-release principle, allowing data generating researchers to retain data internally for a period of up to twelve months while they analyze and prepare publications derived from the data.

6. *The Public Domain Genome*

In contrast to the governmental and private sector projects described above is the Harvard-led Personal Genome Project. The PGP, launched in 2008 to significant press coverage, solicits volunteers to submit tissue samples and accompanying phenotypic data [21]. Researchers are then authorized to analyze the submitted samples and publish any resulting genomic information on the PGP web site. All such data is released without restriction under a "CC0" Creative Commons copyright waiver.[6] The PGP approach differs markedly from that of the projects described above in that it dispenses entirely with any attempt to restrict the use of the genomic data. PGP requires its contributors to waive all privacy-related rights when contributing their tissue samples to the project, and gives no preference to use of the data by researchers of any kind. As explained by the PGP, "Privacy, confidentiality and anonymity are impossible to guarantee in a context like the PGP where public sharing of genetic data is an explicit goal. Therefore, the PGP collaborates with participants willing to waive expectations of privacy. This waiver is not for everyone, but the volunteers who join make a valuable and lasting contribution to science."

7. *NIH's 2014 Genomic Data Sharing (GDS) Policy*

In late 2014, after five years of deliberation, NIH adopted a new institute-wide Genomic Data Sharing (GDS) policy governing the release of genomic data generated by NIH-funded studies [22]. Under the GDS policy, human genomic data must be submitted to NIH promptly following cleaning and quality control (generally within three months after generation). Once submitted, this data may be retained by NIH for up to six months prior to public release. Non-human and model organism data, on the other hand, may be retained by data producers until their initial analyses of the data are published, representing a much longer lead time (at least 12–24 months). In both cases, once released, data is not subject to further

[6]Creative Commons is a non-profit organization that makes available a suite of open access licenses intended to facilitate the contribution of content and data to the public. See creativecommons.org.

embargoes or restrictions on analysis or publication. The GDS policy thus diverges from the GWAS and similar federal policies in that it (a) permits the withholding of data from the public for a fixed period of time, and (b) does not utilize embargoes on data usage following its release [23]. In these respects, the GDS policy resembles private sector policies such as those adopted by iSAEC more than prior federal policies.

4 Latency Analysis and the Genome Commons

While data release policies are typically drafted by funding agencies, NIH in particular has given substantial weight to the views of the scientific community when developing policy. Thus, the role and influence of other stakeholder groups is not to be underestimated: the development of data release policies in the genome sciences has been a process of negotiation and compromise. The evolution of the genome commons illuminates three principal policy considerations: (1) promoting the advancement of science by making genomic data as widely available as possible (scientific advancement, typically espoused by funders and public-interest advocates); (2) addressing the tension between publication priority of data generators and data users (publication priority, typically espoused by data generators), and (3) minimizing patent-related encumbrances on genomic data sets (minimizing encumbrances, espoused by both funders and data users). The interplay of these design considerations, and the latency-based compromises that were effected to satisfy competing requirements of relevant stakeholders, resulted in the policies that are in effect today.

In those NIH genomic data release policies adopted after the Bermuda Principles and prior to the 2014 GDS policy, data must be released rapidly to public databases. The motivations underlying this requirement have been discussed above: there is the explicit desire to accelerate the progress of science, and a less explicit, but strongly implied, desire to limit patent encumbrances on genomic data. However, once concerns regarding publication priority between data generators and data users emerged, a need for policy change became evident.

Most other federal agencies, as well as private initiatives in the genome sciences, have addressed this conflict by permitting data generators to withhold their data from the public for a specified time period, generally 9–12 months, after which it is released without encumbrance (like the "retention" strategy adopted by private sector initiatives in genomics described in Sect. 5 above). Until the GDS policy, however, NIH policy makers took the opposite approach. Instead of allowing data generators to retain their data for a protected period and then releasing it unencumbered, these pre-2014 NIH policies continued to require rapid data release, while imposing a publication embargo on users (an "embargo strategy"). The GDS policy, however, signals a change in NIH's approach. Under GDS, either NIH or the data generators may withhold data from the public for a period of either

Table 1 Latency analysis of genomic data release policies

	Knowledge latency (L_k)	Rights latency (L_r)
Bermuda	0	0
Embargo (GAIN, GWAS)	0	9–12 months
Retention (SAEC)	9–12 months	0+
Public Domain (PGP)	0	0
GDS (Human)	6 months	0
GDS (Non-human)	Publication (12–24 months)	0

6 months (for human genomic data) or until the time of publication (for non-human data). These different policy approaches are compared in Table 1.

Table 1 highlights the differences and similarities among the latency-based approaches to genomic data release. At one extreme are Bermuda-based policies (including PGP), in which data must be released without restriction immediately. Embargo and retention policies attempt to address competing policy considerations in a manner that outwardly appears similar. That is, under both an embargo and a retention strategy the data generator has a period of 9–12 months during which it retains exclusive rights to analyze and publish papers concerning the data. But in practice there are material differences between the retention strategy ($L_k = 12/L_r = 0$) and the embargo strategy ($L_k = 0/L_r = 12$). These differences are driven by material externalities that distinguish government-funded projects from privately-funded projects.

A retention strategy lengthens knowledge latency and, by definition, extends the time before data is released to the public. NIH has repeatedly stated its position that genomic data should be released as rapidly as possible for the advancement of science and the public good. The embargo approach accomplishes this goal by minimizing knowledge latency while still protecting the data generators' publication interests. However, the embargo strategy involves a significant tradeoff in terms of enforceability. Usage embargos in NIH's recent data release policies are embodied in click-wrap agreements[7] or online certifications that must be acknowledged upon making a data request. The enforceability of these mechanisms is uncertain [24, 25]. However, even the *most* robust contractual embargo provides the data generator with *less* protection than withholding data from the public (i.e., if a user has no data, it cannot breach its obligation to refrain from publishing). Moreover, a retention strategy gives the data generator a true "head start" with respect to the data, during which time no third party may analyze or build upon it, whereas an embargo strategy enables third parties to analyze and build upon data *during* the embargo period, putting them in a position to publish their results the moment the embargo expires, even if they strictly comply with its terms during the embargo period.

[7]A "click-wrap" agreement (alternatively referred to as a "click-through" or "click-to-accept" agreement or license) is "an electronic form agreement to which [a] party may assent by clicking an icon or a button or by typing in a set of specified words" [24].

With all of these comparative disadvantages to the data generator, why did the NIH policies adopted between 2006 and 2014 adopt an embargo strategy rather than a retention strategy? The answer may lie in regulatory constraints on NIH's ability to control the development of intellectual property. Unlike private sector groups, NIH must operate within the bounds of the federal Bayh-Dole Act, which prohibits federal agencies from preventing federally-funded researchers from patenting their results.[8] Thus, while NIH's post-Bermuda data release policies acknowledge the requirements of the Bayh-Dole Act, they discourage the patenting of genomic data. The enforceability, however, of policy provisions that merely "urge" or "encourage" data generators and users not to seek patents is questionable [2, 26]. Lacking a strong policy tool with which to prohibit outright the patenting of genomic data, NIH policy makers employed rapid pre-publication data release as a surrogate to reach the same result. The Bermuda Principles, in particular, ensured both that data produced by the HGP and other large-scale sequencing projects would be made publicly-available before data generators could seek to patent "inventions" arising from that data, and in a manner that would also make the data available as prior art against third party patent filings.

In contrast, private sector groups such as SAEC adopted lengthier knowledge latency periods to protect the publication priority of their researchers, but did so in conjunction with explicit patent-defeating strategies. These groups, unlike federal agencies, have the freedom to impose express contractual limitations on patenting without running afoul of the requirements of the federal Bayh-Dole Act. These policies may thus be optimized with respect to the policy goals of minimizing encumbrances and protecting data generators' publication priority, but are less optimal than the government-led policies in terms of broad disclosure of knowledge and scientific advancement.

Why, then, did NIH change course in 2014 by adopting a retention strategy over an embargo strategy in its GDS policy? In terms of human data, the GDS policy offers modest protection of producer lead time, while delaying data release by only six months. In the case of non-human data, however, the GDS policy retreats further, allowing potentially large data release delays tied to publication of results. There are several possible reasons that NIH may have softened its rapid data release requirements as to non-human data, ranging from different norms within the relevant scientific communities to the public perception that non-human data may be less crucial for human health research. As for the patent-deterring effects that are promoted by rapid release policies, NIH appears to have concluded that the issue is less pressing given recent U.S. judicial decisions making it more difficult to obtain patents on human DNA [23].[9] It is not clear, however whether this conclusion is well-justified [23].

[8]The Bayh-Dole Act of 1980, P.L. 96-517, *codified at* 35 U.S.C. §§200-12, rationalized the previously chaotic rules governing federally-sponsored inventions and strongly encourages researchers to obtain patents on inventions arising from federally-funded research.

[9]The GDS policy refers specifically to the U.S. Supreme Court's decision in *Assn. for Molecular Pathology v. Myriad Genetics*, 133 S.Ct. 2107 (2013).

In summary, policy makers over the years have sought to optimize the principal policy goals pursued in the genome commons through the modulation of latency variables:

(1) *Scientific Advancement* is highest under policies that minimize knowledge latency. While data is retained by data generators in private, overall scientific advancement cannot occur. Thus, under a retention strategy, scientific advancement is lower than it would be under an embargo or public domain policy.
(2) *Minimizing Encumbrances* (e.g., patent protection) is achieved by two different means. Private sector groups implementing retention strategies employed contractual anti-patenting policies coupled with "protective" patent filings, generally resulting in reliable freedom from patent encumbrances. Government-funded projects, which cannot avail themselves of these techniques, must rely on early disclosure of information as prior art (low knowledge latency). While the effectiveness of these measures is debatable, they are likely not as strong as those under a retention strategy. NIH recently cast doubt on the need for strong patent-deterrence policy measures in light of recent U.S. judicial decisions.
(3) *Publication priority* for data generators was explicitly sacrificed under public projects such as the HGP in the service of scientific advancement and other policy goals. While embargo policies attempted to improve priority for data generators, the enforceability of contractual embargo provisions is less certain than simple withholding of data under a retention policy. Thus, retention policies, such as those now employed in the NIH GDS policy, yield the highest priority for data generators.

5 Conclusion

Big data repositories are typically subject to policies regarding the contribution and use of data. These policies, far from being unimportant legal "boilerplate," are key determinants of the scope and value of data within the repository. The designers of the repositories of genomic data that have been growing since the HGP used a variety of latency-related techniques to optimize the achievement of policy goals. In particular, their modulation of knowledge latency and rights latency with respect to data inputs and outputs achieved, in varying degrees, policy goals directed at scientific advancement, minimization of intellectual property encumbrances and securing publication priority for data generators. Understanding and utilizing these techniques can be useful for the designers of other data repositories in fields as diverse as earth science, climatology and astronomy who wish to optimize the achievement of competing policy goals.

References

1. Contreras, J.L.: Prepublication data release, latency and genome commons. Science **329**, 393–94 (2010a)
2. Contreras, J.L.: Data Sharing, latency variables and science commons. Berkeley Tech. L.J. **25**, 1601–1672 (2010b)
3. Ostrom, E., Hess, C.: A framework for analyzing the knowledge commons. In: Hess, C., Ostrom, E. (eds.) Understanding Knowledge as a Commons: From Theory to Practice. MIT Press, Cambridge, Mass (2007)
4. Benson, B.A., Clark, K., Karsch-Mizrachi, I., Lipman, D.J., Ostell, J., Sayers, E.W.: GenBank. Nucleic Acids Res. **42**, D32–D37 (2014). doi:10.1093/nar/gkt1030
5. Natl. Ctr. Biotechnology Info. (NCBI): Growth of GenBank and WGS. http://www.ncbi.nlm.nih.gov/genbank/statistics (2015). Accessed 14 June 2015
6. Pennisi, E.: Will computers crash genomics? Science **331**, 666–667 (2011)
7. Natl. Res. Council (NRC): Mapping and Sequencing the Human Genome. Natl. Acad. Press, Washington (1988)
8. Oak Ridge Natl. Lab. (ORNL): NIH, DOE guidelines encourage sharing of data, resources. Hum. Genome News **4**, 4. http://www.ornl.gov/sci/techresources/Human_Genome/publicat/hgn/pdfs/Vol4No5.pdf (1993)
9. Contreras, J.L.: Bermuda's legacy: policy, patents, and the design of the genome commons. Minn. J.L. Sci. Tech. **12**, 61–125 (2011)
10. Natl. Res. Council (NRC): Bits of Power—Issues in Global Access to Scientific Data. Natl. Acad. Press, Washington (1997)
11. Reichman, J.H., Uhlir, P.F.: A contractually reconstructed research commons for scientific data in a highly protectionist intellectual property environment. Law Contemp. Probs. **66**, 315–462 (2003)
12. Intl. Human Genome Sequencing Consortium (IHGSC): Initial sequencing and analysis of the human genome. Nature **409**, 860–914 (2001)
13. Bermuda Principles: Summary of principles agreed at the first international strategy meeting on human genome sequencing. http://www.ornl.gov/sci/techresources/Human_Genome/research/bermuda.shtml (2006)
14. Kaye, J., et al.: Data sharing in genomics—re-shaping scientific practice. Nat. Rev. Genet. **10**, 331–335 (2009)
15. Wellcome Trust: Sharing Data from Large-Scale Biological Research Projects: A System of Tripartite Responsibility: Report of meeting organized by the Wellcome Trust and held on 14–15 January 2003 at Fort Lauderdale, USA. http://www.genome.gov/Pages/Research/WellcomeReport0303.pdf (2003)
16. Natl. Inst. Health (NIH): Policy for sharing of data obtained in NIH supported or conducted Genome-Wide Association Studies (GWAS). Fed. Reg. **72,** 49,290 (2007)
17. Merck & Co., Inc.: First installment of merck gene index data released to public databases: cooperative effort promises to speed scientific understanding of the human genome. http://www.bio.net/bionet/mm/bionews/1995-February/001794.html (1995)
18. Marshall, E.: Bermuda rules: community spirit, with teeth. Science **291**, 1192–1193 (2001)
19. Holden, A.L.: The SNP consortium: summary of a private consortium effort to develop an applied map of the human genome. Biotechniques **32**, 22–26 (2002)
20. Contreras, J.L., Floratos, A., Holden, A.L.: The international serious adverse events consortium's data sharing model. Nat. Biotech. **31**, 17–19 (2013)
21. Personal Genome Project (PGP): About the PGP. http://www.personalgenomes.org (2014). Accessed 25 June 2014
22. Natl. Inst. Health (NIH): Final NIH genomic data sharing policy. Fed. Reg. **79**, 51345–51354 (2014)
23. Contreras, J.L.: NIH's genomic data sharing policy: timing and tradeoffs. Trends Genet. **31**, 55–57 (2015)

24. Kunz, C.L., et al.: Click-through agreements: strategies for avoiding disputes on validity of assent. Bus. Lawyer **57**, 401–429 (2001)
25. Delta, G.B., Matsuura, J.H. (eds.): Law of the Internet, 3rd edn. Aspen, New York (2014)
26. Rai, A.K., Eisenberg, R.S.: Bayh-Dole reform and the progress of biomedicine. Law Contemp. Probs. **66**, 289–314 (2003)
27. GAIN Collaborative Research Group: New models of collaboration in genome-wide association studies: the genetic association information network. Nat. Genet. **39**, 1045–1051 (2007)

Author Biography

Jorge L. Contreras (JD Harvard; BSEE, BA Rice) is an Associate Professor at the University of Utah College of Law, holding an adjunct appointment in the Department of Human Genetics. He previously directed the Intellectual Property program at Washington University in St. Louis. Professor Contreras has written and spoken extensively on the institutional structures of intellectual property, technical standardization and biomedical research. He currently serves as a member of the Advisory Council of the National Center for the Advancement of Translational Sciences (NCATS) at the U.S. National Institutes of Health (NIH), and previously served as Co-Chair of the National Conference of Lawyers and Scientists, and a member of the National Advisory Council for Human Genome Research. He co-edited the first text addressing the intersection of bioinformatics and the law (ABA Publishing 2013), and is the recipient of the 2014 Elizabeth Payne Cubberly Faculty Scholarship Award from American University.

Big Data Optimization via Next Generation Data Center Architecture

Jian Li

Abstract The use of Big Data underpins critical activities in all sectors of our society. Achieving the full transformative potential of Big Data in this increasingly digital and interconnected world requires both new data analysis algorithms and a new class of systems to handle the dramatic data growth, the demand to integrate structured and unstructured data analytics, and the increasing computing needs of massive-scale analytics. As a result, massive-scale data analytics of all forms have started to operate in data centers (DC) across the world. On the other hand, data center technology has evolved from DC 1.0 (tightly-coupled silos) to DC 2.0 (computer virtualization) in order to enhance data processing capability. In the era of big data, highly diversified analytics applications continue to stress data center capacity. The mounting requirements on throughput, resource utilization, manageability, and energy efficiency demand seamless integration of heterogeneous system resources to adapt to varied big data applications. Unfortunately, DC 2.0 does not suffice in this context. By rethinking of the challenges of big data applications, researchers and engineers at Huawei propose the High Throughput Computing Data Center architecture (HTC-DC) toward the design of DC 3.0. HTC-DC features resource disaggregation via unified interconnection. It offers Peta Byte (PB) level data processing capability, intelligent manageability, high scalability and high energy efficiency, hence a promising candidate for DC 3.0. This chapter discusses the hardware and software features HTC-DC for Big Data optimization.

Keywords Big data · Data center · System architecture

J. Li (✉)
Futurewei Technologies Inc., Huawei Technologies Co. Ltd, 2330 Central Expressway, Santa Clara, CA 95050, USA
e-mail: jian.li1@huawei.com

A. Emrouznejad (ed.), *Big Data Optimization: Recent Developments and Challenges*, Studies in Big Data 18, DOI 10.1007/978-3-319-30265-2_10

1 Introduction

1.1 Challenges of Big Data Processing

During the past few years, applications that are based on big data analysis have emerged, enriching human life with more real-time and intelligent interactions. Such applications have proven themselves to become the next wave of mainstream of online services. At the dawn of the big data era, higher and higher demand on data processing capability has been raised. Given industry trend and being the major facilities to support highly varied big data processing tasks, future data centers (DCs) are expected to meet the following big data requirements (Fig. 1)[1]:

- **PB/s-level data processing capability** ensuring aggregated high-throughput computing, storage and networking;
- **Adaptability** to highly-varied run-time resource demands;
- **Continuous availability** providing 24 × 7 large-scaled service coverage, and supporting high-concurrency access;
- **Rapid deployment** allowing quick deployment and resource configuration for emerging applications.

1.2 DC Evolution: Limitations and Strategies

DC technologies in the last decade have been evolved (Fig. 2) from DC 1.0 (with tightly-coupled silos) to current DC 2.0 (with computer virtualization). Although data processing capability of DCs have been significantly enhanced, due to the limitations on throughput, resource utilization, manageability and energy efficiency, current DC 2.0 shows its incompetence to meet the demands of the future:

- **Throughput**: Compared with technological improvement in computational capability of processors, improvement in I/O access performance has long been lagged behind. With the fact that computing within conventional DC architecture largely involves data movement between storage and CPU/memory via I/O ports, it is challenging for current DC architecture to provide PB-level high throughput for big data applications. The problem of I/O gap is resulted from low-speed characteristics of conventional transmission and storage mediums, and also from inefficient architecture design and data access mechanisms.
  To meet the requirement of future high throughput data processing capability, adopting new transmission technology (e.g. optical interconnects) and new storage medium can be feasible solutions. But a more fundamental approach is

[1]This chapter is based on “High Throughput Computing Data Center Architecture—Thinking of Data Center 3.0”, white paper, Huawei Technologies Co. Ltd., http://www.huawei.com.

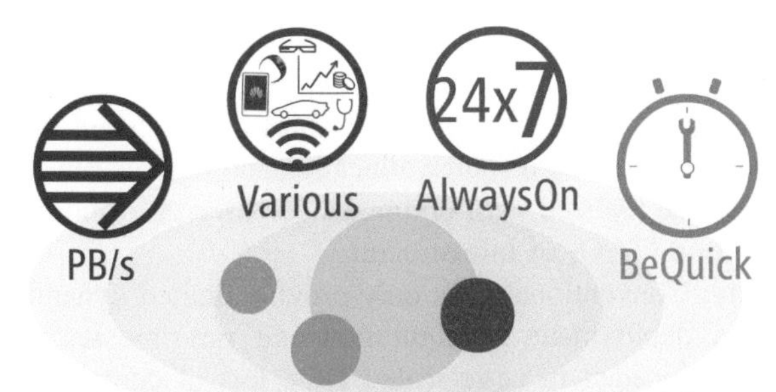

Fig. 1 Needs brought by big data

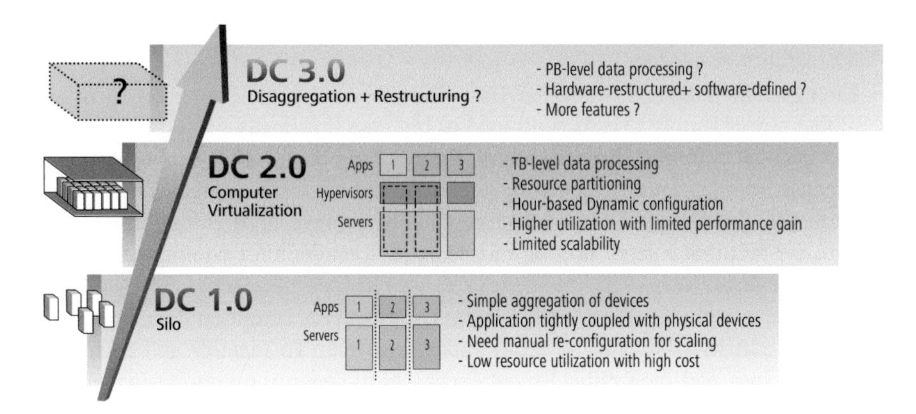

Fig. 2 DC evolution

to re-design DC architecture as well as data access mechanisms for computing. If data access in computing process can avoid using conventional I/O mechanism, but use ultra-high-bandwidth network to serve as the new I/O functionality, DC throughput can be significantly improved.

- **Resource Utilization**: Conventional DCs typically consist of individual servers which are specifically designed for individual applications with various pre-determined combinations of processors, memories and peripherals. Such design makes DC infrastructure very hard to adapt to emergence of various new applications, so computer virtualization technologies are introduced accordingly. Although virtualization in current DCs help improve hardware utilization, it cannot make use of the over-fractionalized resource, and thus making the improvement limited and typically under 30 % [1, 2]. As a cost, high overhead exists with hypervisor which is used as an essential element when implementing computer virtualization. In addition, in current DC architecture, logical pooling of resources is still restricted by the physical coupling of in-rack hardware devices. Thus, current DC with limited resource utilization cannot support big data applications in an effective and economical manner.

One of the keystones to cope with such low utilization problem is to introduce resource disaggregation, i.e., decoupling processor, memory, and I/O from its original arrangements and organizing resources into shared pools. Based on disaggregation, on-demand resource allocation and flexible run-time application deployment can be realized with optimized resource utilization, reducing Total Cost of Operation (TCO) of infrastructure.

- **Manageability**: Conventional DCs only provide limited dynamic management for application deployment, configuration and run-time resource allocation. When scaling is needed in large-scaled DCs, lots of complex operations still need to be completed manually.
  To avoid complex manual re-structuring and re-configuration, intelligent self-management with higher level of automation is needed in future DC. Furthermore, to speed up the application deployment, software defined approaches to monitor and allocate resources with higher flexibility and adaptability is needed.
- **Energy Efficiency**: Nowadays DCs collectively consume about 1.3 % of all global power supply [3]. As workload of big data drastically grows, future DCs will become extremely power-hungry. Energy has become a top-line operational expense, making energy efficiency become a critical issue in green DC design. However, the current DC architecture fails to achieve high energy efficiency, with the fact that a large portion of energy is consumed for cooling other than for IT devices.
  With deep insight into the composition of DC power consumption (Fig. 3), design of each part in a DC can be more energy-efficient. To identify and eliminate inefficiencies and then radically cut energy costs, energy-saving design of DC should be top-to-bottom, not only at the system level but also at the level of individual components, servers and applications.

Fig. 3 DC power consumption

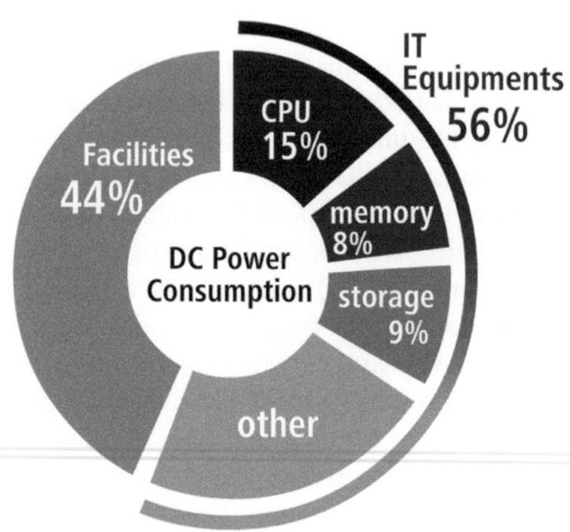

1.3 Vision on Future DC

Future DCs should be enabled with the following features to support future big data applications:

- **Big-Data-Oriented**: Different from conventional computing-centric DCs, data-centric should be the key design concept of DC 3.0. Big data analysis based applications have highly varied characteristics, based on which DC 3.0 should provide optimized mechanisms for rapid transmission, highly concurrent processing of massive data, and also for application-diversified acceleration.
- **Adaptation for Task Variation**: Big data analysis brings a booming of new applications, raising different resource demands that vary with time. In addition, applications have different need for resource usage priority. To meet such demand variation with high adaptability and efficiency, disaggregation of hardware devices to eliminate the in-rack coupling can be a key stone. Such a method enables flexible run-time configuration on resource allocation, ensuring the satisfactory of varied resource demand of different applications.
- **Intelligent Management**: DC 3.0 involves massive hardware resource and high density run-time computation, requiring higher intelligent management with less need for manual operations. Application deployment and resource partitioning/allocation, even system diagnosis need to be conducted in automated approaches based on run-time monitoring and self-learning. Further, Service Level Agreement (SLA) guaranteeing in complex DC computing also requires a low-overhead run-time self-manageable solution.
- **High Scalability**: Big data applications require high throughput low-latency data access within DCs. At the same time, extremely high concentration of data will be brought into DC facilities, driving DCs to grow into super-large-scaled with sufficient processing capability. It is essential to enable DCs to maintain acceptable performance level when ultra-large-scaling is conducted. Therefore, high scalability should be a critical feature that makes a DC design competitive for the big data era.
- **Open, Standard based and Flexible Service Layer**: With the fact that there exists no unified enterprise design for dynamical resource management at different architecture or protocol layers, from IO, storage to UI. Resources cannot be dynamically allocated based on the time and location sensitive characteristics of the application or tenant workloads. Based on the common principles of abstraction and layering, open and standard based service-oriented architecture (SOA) has been proven effective and efficient and has enabled enterprises of all sizes to design and develop enterprise applications that can be easily integrated and orchestrated to match their ever-growing business and continuous process improvement needs, while software defined networking (SDN) has also been proven in helping industry giants such as Google to improve its DC network resource utilization with decoupling of control and data forwarding, and centralized resource optimization and scheduling. To provide competitive big data related service, an open, standard based service layer should be enabled in future

DC to perform application driven optimization and dynamic scheduling of the pooled resources across various platforms.

- **Green**: For future large-scale DC application in a green and environment friendly approach, energy efficient components, architectures and intelligent power management should be included in DC 3.0. The use of new mediums for computing, memory, storage and interconnects with intelligent on-demand power supply based on resource disaggregation help achieving fine-grained energy saving. In addition, essential intelligent energy management strategies should be included: (1) Tracking the operational energy costs associated with individual application-related transactions; (2) Figuring out key factors leading to energy costs and conduct energy-saving scheduling; (3) Tuning energy allocation according to actual demands; (4) Allowing DCs to dynamically adjust the power state of servers, and etc.

2 DC3.0: HTC-DC

2.1 HTC-DC Overview

To meet the demands of high throughput in the big data era, current DC architecture suffers from critical bottlenecks, one of which is the difficulty to bridge the I/O performance gap between processor and memory/peripherals. To overcome such problem and enable DCs with full big-data processing capability, we propose a new high throughput computing DC architecture (HTC-DC), which avoids using conventional I/O mechanism, but uses ultra-high-bandwidth network to serve as the new I/O functionality. HTC-DC integrates newly-designed infrastructures based on resource disaggregation, interface-unified interconnects and a top-to-bottom optimized software stack. Big data oriented computing is supported by series of top-to-bottom accelerated data operations, light weighted management actions and the separation of data and management.

Figure 4 shows the architecture overview of HTC-DC. Hardware resources are organized into different pools, which are links up together via interconnects. Management plane provides DC-level monitoring and coordination via DC Operating System (OS), while business-related data access operations are mainly conducted in data plane. In the management plane, a centralized Resource Management Center (RMC) conducts global resource partitioning/allocation and coordination/scheduling of the related tasks, with intelligent management functionalities such as load balancing, SLA guaranteeing, etc. Light-hypervisor provides abstract of pooled resources, and performs lightweight management that focuses on execution of hardware partitioning and resource allocation but not get involved in data access. Different from conventional hypervisor which includes data access functions in virtualization, light-hypervisor focuses on resource management, reducing complexity and overhead significantly. As a systematical DC 3.0 design, HTC-DC also

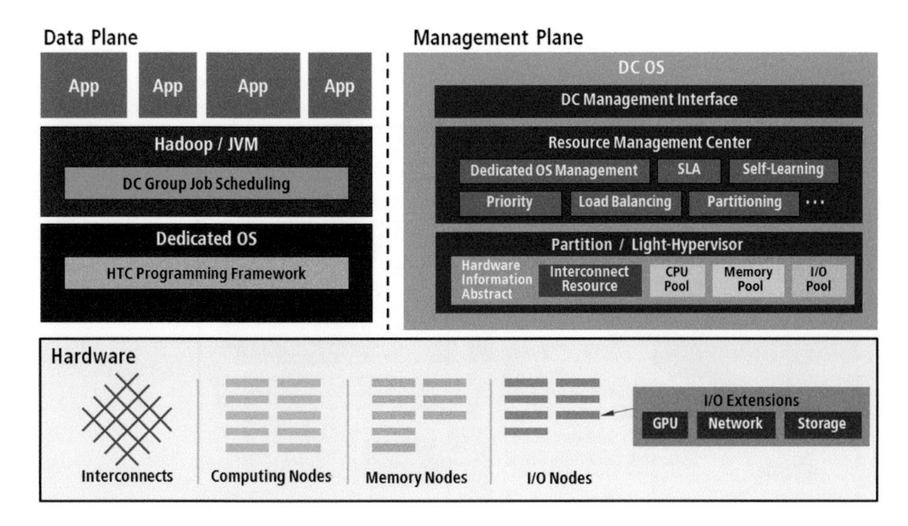

Fig. 4 HTC-DC architecture

provides a complete software stack to support various DC applications. A programming framework with abundant APIs is designed to enable intelligent run-time self-management.

2.2 *Key Features*

Figure 5 illustrates the hardware architecture of HTC-DC, which is based on completely-disaggregated resource pooling. The computing pool is designed with heterogeneity. Each computing node (i.e. a board) carries multiple processors (e.g., x86, Atom, Power and ARM, etc.) for application- diversified data processing. Nodes in memory pool adopt hybrid memory such as DRAM and non-volatile memory (NVM) for optimized high- throughput access. In I/O pool, general-purposed extension (GPU, massive storage, external networking, etc.) can be supported via different types of ports on each I/O node. Each node in the three pools is equipped with a cloud controller which can conduct diversified on-board management for different types of nodes.

2.3 *Pooled Resource Access Protocol (PRAP)*

To form a complete DC, all nodes in the three pools are interconnected via a network based on a new designed Pooled Resource Access Protocol (PRAP). To reduce the complexity of DC computing, HTC-DC introduces PRAP which has

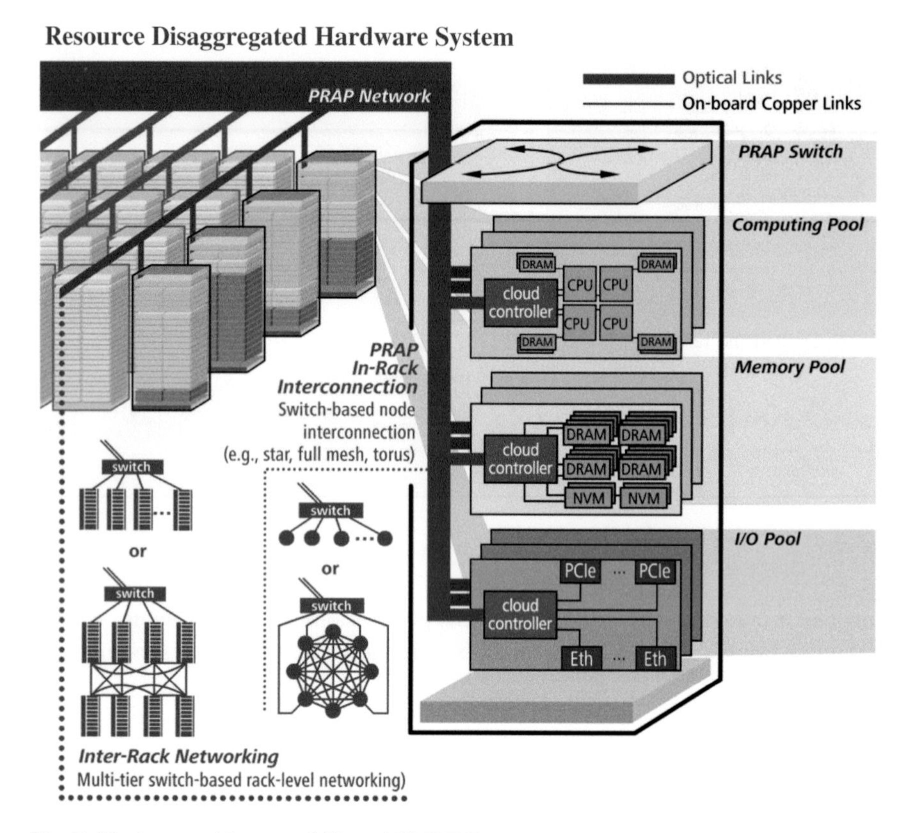

Fig. 5 Hardware architecture of Huawei HTC-DC

low-overhead packet format, RDMA-enabled simplified protocol stack, unifying the different interfaces among processor, memory and I/O. PRAP is implemented in the cloud controller of each node to provide interface-unified interconnects. PRAP supports hybrid flow/packet switching for inter-pool transmission acceleration, with near-to-ns latency. QoS can be guaranteed via run-time bandwidth allocation and priority-based scheduling. With simplified sequencing and data restoring mechanisms, light-weight lossless node-to-node transmission can be achieved.

With resource disaggregation and unified interconnects, on-demand resource allocation can be supported by hardware with fine-granularity, and intelligent management can be conducted to achieve high resource utilization (Fig. 6). RMC in the management plane provides per-minute based monitoring, on-demand coordination and allocation over hardware resources. Required resources from the pools can be appropriately allocated according to the characteristics of applications (e.g. Hadoop). Optimized algorithm assigns and schedules tasks on specific resource partitions where customized OSs are hosted. Thus, accessibility and bandwidth of remote memory and peripherals can be ensured within the partition, and hence end-to-end SLA can be guaranteed. Enabled with self-learning mechanisms,

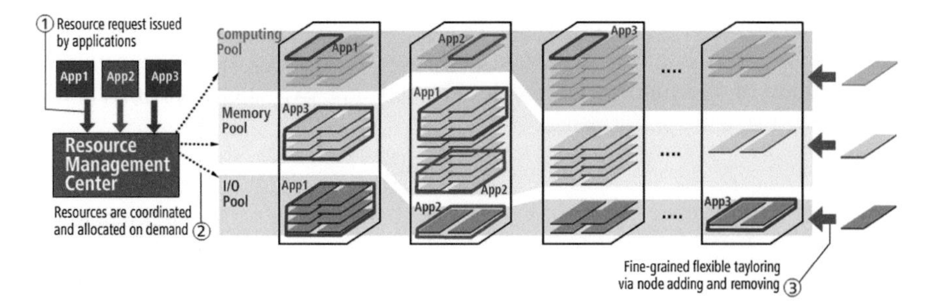

Fig. 6 On-demand resource allocation based on disaggregation

resource allocation and management in HTC-DC requires minimal manual operation, bringing intelligence and efficiency.

2.4 Many-Core Data Processing Unit

To increase computing density, uplift data throughput and reduce communication latency, Data Processing Unit (DPU, Fig. 7) is proposed to adopt lightweight-core based many-core architecture, heterogeneous 3D stacking and Through-Silicon Vias (TSV) technologies. In HTC-DC, DPU can be used as the main computing component. The basic element of DPU is Processor-On-Die (POD), which consists of NoC, embedded NVM, clusters with heavy/light cores, and computing accelerators. With software-defined technologies, DPU supports resource partitioning and QoS-guaranteed local/remote resource sharing that allow application to directly access resources within its assigned partition. With decoupled multi-threading support, DPU executes speculative tasks off the critical path, resulting in enhanced

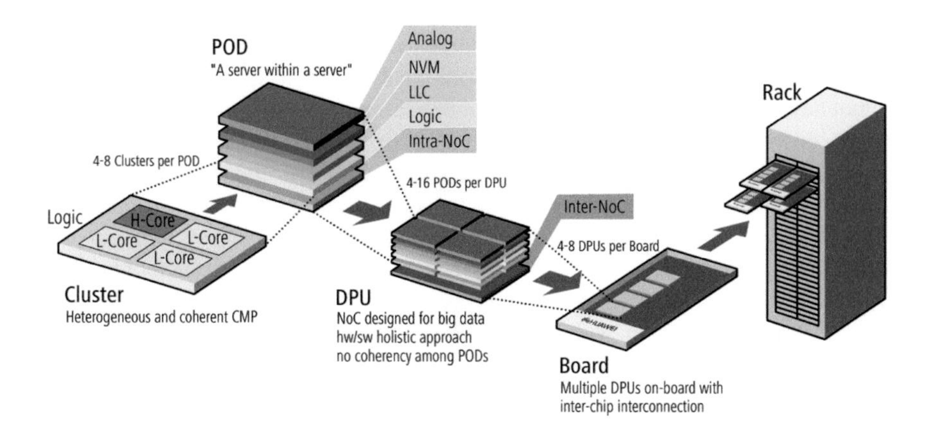

Fig. 7 Many-core processor

overall performance. Therefore static power consumptions can be significantly reduced. Especially, some of the silicon chip area can be saved by using the optimal combinations of the number of synchronization and execution pipelines, while maintaining the same performance.

2.5 *NVM Based Storage*

Emerging NVM (including MRAM or STT-RAM, RRAM and PCM, etc.) has been demonstrated with superior performance over flash memories. Compared to conventional storage mediums (hard-disk, SSD, etc.), NVM provides more flattened data hierarchy with simplified layers, being essential to provide sufficient I/O bandwidth. In HTC-DC, NVMs are employed both as memory and storage. NVM is a promising candidate for DRAM replacement with competitive performance but lower power consumption. When used as storage, NVM provides 10 times higher IOPS than SSD [4], bringing higher data processing capability with enhanced I/O performance.

Being less hindered by leakage problems with technology scaling and meanwhile having a lower cost of area, NVM is being explored extensively to be the complementary medium for the conventional SDRAM memory, even in L1 caches. Appropriately tuning of selective architecture parameters can reduce the performance penalty introduced by the NVM to extremely tolerable levels while obtaining over 30 % of energy gains [5].

2.6 *Optical Interconnects*

To meet the demand brought by big data applications, DCs are driven to increase the data rate on links (>10 Gbps) while enlarging the scale of interconnects (>1 m) to host high-density components with low latency. However due to non-linear power consumption and signal attenuation, conventional copper based DC interconnects cannot have competitive performance with optical interconnects on signal integrity, power consumption, form factor and cost [6]. In particular, optical interconnect has the advantage of offering large bandwidth density with low attenuation and crosstalk. Therefore a re-design of DC architecture is needed to fully utilize advantages of optical interconnects. HTC-DC enables high-throughput low-latency transmission with the support of interface-unified optical interconnects. The interconnection network of HTC-DC employs low-cost Tb/s-level throughput optical transceiver and co-packaged ASIC module, with tens of pJ/bit energy consumption and low bit error rate for hundred-meter transmission. In addition, with using intra/inter-chip optical interconnects and balanced space-time-wavelength design, physical layer scalability and the overall power consumption can be enhanced. Using optical transmission that needs no signal synchronization, PRAP-based interconnects provide higher degree of freedom on topology choosing, and is enabled to host ultra-large-scale nodes.

2.7 *DC-Level Efficient Programming Framework*

To fully exploit the architectural advantages and provide flexible interface for service layer to facilitate better utilization of underlying hardware resource, HTC-DC provides a new programming framework at DC-level. Such a framework includes abundant APIs, bringing new programming methodologies. Via these APIs, applications can issue requests for hardware resource based on their demands. Through this, optimized OS interactions and self-learning-based run-time resource allocation/scheduling are enabled. In addition, the framework supports automatically moving computing operations to near-data nodes while keeping data transmission locality. DC overhead is minimized by introducing topology-aware resource scheduler and limiting massive data movement within the memory pool.

As a synergistic part of the framework, Domain Specific Language (HDSL) is proposed to reduce the complexity of parallel programming in HTC-DC. HDSL includes a set of optimized data structures with operations (such as Parray, parallel processing of data array) and a parallel processing library. One of the typical applications of HDSL is for graph computing. HDSL can enable efficient programming with demonstrated competitive performance. Automated generation of distributed code is also supported.

3 Optimization of Big Data

Optimizing Big Data workloads differ from workloads typically run on more traditional transactional and data-warehousing systems in fundamental ways. Therefore, a system optimized for Big Data can be expected to differ from these other systems. Rather than only studying the performance of representative computational kernels, and focusing on central-processing-unit performance, practitioners instead focus on the system as a whole. In a nutshell, one should identify the major phases in a typical Big Data workload, and these phases typical apply to the data center in a distributed fashion. Each of these phases should be represented in a distributed Big Data systems benchmark to guide system optimization.

For example, the MapReduce Terasort benchmark is popular a workload that can be a “stress test” for multiple dimensions of system performance. Infrastructure tuning can result in significant performance improvement for such benchmarks. Further improvements are expected as we continue full-stack optimizations on both distributed software and hardware across computation, storage and network layers. Indeed, workloads like Terasort can be very IO (Input-Output) intensive. That said, it requires drastically higher throughput in data centers to achieve better performance. Therefore, HTC-DC, our high-throughput computing data center architecture works perfectly with such big data workloads.

Finally, we plan to combine this work with a broader perspective on Big Data workloads and suggest a direction for a future benchmark definition effort. A number of methods to further improve system performance look promising.

4 Conclusions

With the increasing growth of data consumption, the age of big data brings new opportunities as well as great challenges for future DCs. DC technology has evolved from DC 1.0 (tightly-coupled server) to DC 2.0 (software virtualization) with enhanced data processing capability. However, the limited I/O throughput, energy inefficiency, low resource utilization and hindered scalability of DC 2.0 have become the bottlenecks to meet the demand of big data applications. As a result, a new, green and intelligent DC 3.0 architecture capable to adapt to diversified resource demands from various big-data applications is in need.

With the design of ultra-high-bandwidth network to serve as the new I/O functionality instead of conventional schemes, HTC-DC is promising to serve as a new generation of DC design for future big data applications. HTC-DC architecture enables high throughput computing in data centers. With its resource disaggregation architecture and unified PRAP network interface, HTC-DC is currently under development to integrate many-core processor, NVM, optical interconnects and DC-level efficient programming framework. Such a DC will ensure PB-level data processing capability, support intelligent management, be easy and efficient to scale, and significantly save energy cost. We believe HTC-DC can be a promising candidate design for future DCs in the Big Data era.

Acknowledgments Many Huawei employees have contributed significantly to this work, among others are, Zhulin (Zane) Wei, Shujie Zhang, Yuangang (Eric) Wang, Qinfen (Jeff) Hao, Guanyu Zhu, Junfeng Zhao, Haibin (Benjamin) Wang, Xi Tan (Jake Tam), Youliang Yan.

References

1. http://www.energystar.gov/index.cfm?c=power_mgt.datacenter_efficiency_consolidation
2. http://www.smartercomputingblog.com/system-optimization/a-data-center-conundrum/
3. http://www.google.com/green/bigpicture/#/datacenters/infographics
4. http://www.samsung.com/global/business/semiconductor/news-events/press-releases/detail?newsId=12961
5. Komalan, M., et.al.: Feasibility exploration of NVM based I-cache through MSHR enhancements. In: Proceeding in DATE'14
6. Silicon Photonics Market & Technologies 2011–2017: Big Investments, Small Business, Yole Development (2012)

Author Biography

Jian Li is a research program director and technology strategy leader at Huawei Technologies. He was the chief architect of Huawei's FusionInsights big data platform and is currently leading strategy and R&D efforts in IT technologies, working with global teams around the world. Before joining Huawei, he was with IBM where he worked on advanced R&D, multi-site product development and global customer engagements on computer systems and analytics solutions with significant revenue growth. A frequent presenter at major industry and academic conferences around the world, he holds over 20 patents and has published over 30 peer-reviewed papers. He earned a Ph. D. in electrical and computer engineering from Cornell University. He also holds an adjunct position at Texas A&M University. In this capacity, he continues to collaborate with leading academic researchers and industry experts.

Big Data Optimization Within Real World Monitoring Constraints

Kristian Helmholt and Bram van der Waaij

Abstract Large scale monitoring systems can provide information to decision makers. As the available measurement data grows, the need for available and reliable interpretation also grows. To this, as decision makers require the timely arrival of information, the need for high performance interpretation of measurement data also grows. Big Data optimization techniques can enable designers and engineers to realize large scale monitoring systems in real life, by allowing these systems to comply to real world constrains in the area of performance, reliability and reliability. Using several examples of real world monitoring systems this chapter discusses different approaches in optimization: data, analysis, system architecture and goal oriented optimization.

Keywords Measurement · Decision · Monitoring · Constraint-based optimization

1 Introduction

Monitoring systems enable people to respond adequately to changes in their environment. Big Data optimization techniques can play an important role in realization of large scale monitoring systems. This chapter describes the relationship between Big Data optimization techniques and real world monitoring in terms of optimization approaches that enable monitoring systems to satisfy real world deployment constraints.

K. Helmholt (✉) · B. van der Waaij
Netherlands Organisation for Applied Scientific Research (TNO), Groningen, The Netherlands
e-mail: kristian.helmholt@tno.nl

B. van der Waaij
e-mail: bram.vanderwaaij@tno.nl

A. Emrouznejad (ed.), *Big Data Optimization: Recent Developments and Challenges*, Studies in Big Data 18, DOI 10.1007/978-3-319-30265-2_11

This chapter starts with a high level overview of the concept of monitoring in Sect. 2. After that the relationship between large scale monitoring systems and Big Data is described in Sect. 3. This is done using constraints on monitoring systems with respect to performance, availability and reliability, which are critical success factors for monitoring systems in general and also have a strong relationship with Big Data. Then, in Sect. 4, several solution approaches from the field of Big Data optimization are presented for designers and engineers that need to stay within constraints as set forward by the context of the deployment of a monitoring system. Also, the impact on several other constraints is taken into account. This chapter ends with Sect. 5 presenting conclusions.

2 Monitoring

A thorough understanding of the relationship between Big Data optimization and monitoring, starts with a global understanding of the concept of monitoring. This understanding provides a means to comprehend the Big Data related constraints put upon monitoring by the real world, which will be described in the next section. In this section an understanding of monitoring systems will be provided by describing a real world example.

2.1 General Definition

The word 'monitor' is supposed to be derived from the Latin monitor ("warner"), related to the Latin verb monere ("to warn, admonish, remind"). A monitoring system could therefore be defined as a collection of components that interact in order to provide people and/or other systems with a warning with respect to the state of another object or system. A (very) small scale example of a monitoring system is a smoke detector: it continuously measures the visibility of the surrounding air. Once a threshold level with respect to that visibility has been crossed, an alarm is sounded. In this chapter, the definition of a monitoring system is extended to systems that are not primarily targeted at warning, but possibly also at learning. This is because modern day monitoring systems can and need to adapt to changes in the environment they monitor. This means that modern monitoring systems can—for example—also be used to find the relationship between behavior of different objects and/or parameters in the system they observe. So, in turn the definition of a monitoring system in this chapter becomes:

> *"a collection of components that interact in order to provide people and/or other systems with information with respect to the state of a system of other objects, based on (real-time) measurement data on that system (of objects)."*

In this chapter the scope is limited to large scale monitoring systems. In the remainder of this chapter the abbreviation LSMS is used. These systems collect vast amounts of (measurement) data—using sensors—on the real world and process it to information, which could then be used to decide upon or to learn from. As will become clear in the next paragraphs, LSMSs can be used for many different systems. For example, a dike LSMS can be used to monitor dikes for failure (which would result in flooding). Another example is an LSMS that monitors the health and wellbeing of hundreds of thousands of cows throughout their lifetime.

In order to structure the material in this chapter, a simple but effective monitoring framework is used to position functions of a monitoring system. This framework—depicted in Fig. 1—enables describing the relationship with Big Data optimization later on in this chapter. The basic idea of this framework is that in all monitoring systems three generic steps can be distinguished. Using the relatively simple example of a radar the following three steps can be described:

1. **Measurement**: *mapping a certain aspect of reality to a unit of measurement of a physical magnitude (or quantity) using a sensor.* For example a radar that sends out radio waves and receives reflections which tells something about distances from the object to the radar antenna.
2. **Interpretation**: *interpretation of sensor data into information—using expert models—about an* ***object*** *or system to be monitored.* For example interpretation of radar data into information about an object that is flying towards a radar.
3. **Decision to inform**: *applying some kind of rule set to determine if a human or other system should be informed (e.g. warned).* For example, based on the speed of an object flying towards the radar, a warning could be produced if an object is coming into fast or too close.

Note that there are other generic and more elaborate descriptions of monitoring (& control) systems, like the famous Observe, Orient, Decide, and Act (often abbreviated as OODA) loop [1] or the 'Knowledge Discovery' as described in [2]. These shall not be discussed in this chapter, since this simple decomposition in Measurement, Interpretation and Decision (MID) is only needed as a means to position monitoring constraints from a Big Data optimization perspective later on. Also, control systems are beyond the scope of this chapter.

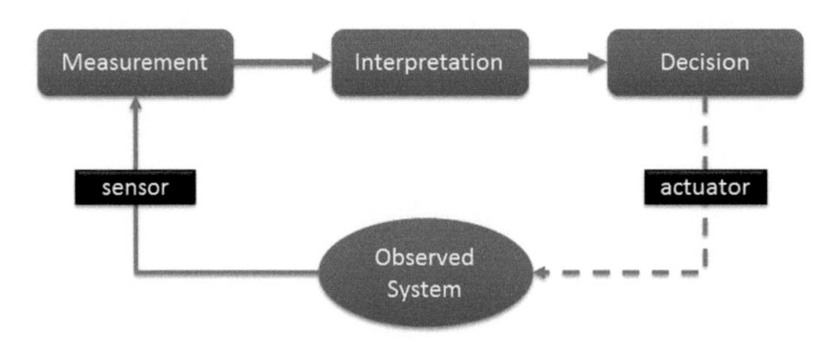

Fig. 1 Measurement, interpretation, decision framework

This concludes the very high level description of monitoring in general. With this in mind, an example of an LSMS is presented in the next section. This example will be used to describe specific Big Data related aspects in the section on Big Data related constraints.

2.2 Dike Monitoring Example

In the Netherlands dikes prevent more than half of the population of 17 million people from losing their houses, losing their livelihood and ultimately drowning. Construction and maintenance of dikes is relatively expensive, because other than safety, dikes provide no direct financial output, like a factory that produces products that can be sold. Note that the Netherlands there is more than 3500 km of primary dikes and more than 10,000 km of secondary dikes. From both a safety as well as economic point of view, it is valuable to both know how and when a dike could fail. From a socio-economic point of view dikes should be safe enough, but too much over dimensioning is a waste of money. Since this could be spent elsewhere, for example in healthcare where lives are also at stake.

Several years ago, it seemed appropriate to a consortium of water boards (i.e. local/regional governments targeted at water management), research organizations and industrial partners to start working on an LSMS for dikes [3, 4]. Advances in computer science, geotechnical sciences and the decline of costs of computation, communication and storage hardware, made it seem as if there could be a business case for dike monitoring. In other words the expected costs of the LSMS seemed to be less than the avoided costs in case of inefficient dike maintenance (too little or

Fig. 2 Dike being monitored with an large scale monitoring system

too much). The result of the collaboration was the creation of a test facility for the monitoring of dikes (Fig. 2). Several experimental dikes have been constructed there, which contained many different types of sensors [3].

Measurement. A type of sensor used that produced large amounts of data is fiberglass cloth, which can be used to measure the shape of a dike, by wrapping it onto the dike. When the dike changes shape, the fiberglass cloth bends resulting in different measurement data. Another type of sensor is the ShapeAccelArray/Field (SAAF), a string of sensors that provides (relative) position, acceleration and orientation information. Like fiberglass sensors, a SAAF can provide a vast amount of data in a short amount of time. For example a 3 m SAAF can contain 10 measurement points, that each can provide 9 measurement values. Providing 10 * 9 values in total every 5 s. A final example for determining the shape of a dike is to use remote sensing such as satellite data [5].

Interpretation. The end-users of the dike LSMS are interested in the stability of the dike: what is the chance that it will fail. There is no physical instrument that measures this abstract 'unit of measurement', it has to be calculated according to a computational model that all LSMS involved parties agree upon. This model must take into account significantly contributing to the dike stability. For example the geotechnical make-up of the dike and the (expected) forces acting upon the dike, like water levels on both sides. This can be done without using sensor data and by assuming possible (extreme) parameter values for the models involved and calculate the likelihood of failure. However, while progress has been made in the last decennia with theoretical models for dike behavior, uncertainty remained. For this test facility dike stability models have therefor been adapted to use measurements from sensors inside or targeted at the dike. During the experiments at the facility, sensor developers could find out if the data from their (new) sensors contributed to reducing uncertainty about dike failure. Geotechnical model builders could find out if their models were using data from sensor efficiently. Hence the name of the test facility, which was 'IJkdijk', as 'ijk' means 'to calibrate' in Dutch.

Decision. A dike LSMS is an excellent example of the need for a monitoring system that produces information on which people can rely. When an LSMS warns that a dike will fail within the next two days, people are almost forced to start evacuating if there is no possibility to strength the dike anymore. If it turns out the LSMS was wrong (i.e. the dike would not have failed), the impact of this false alarm is huge from a socio-economic point of view. When an LSMS does not warn, while the dike is about to fail, resulting in a flood nobody was warned about, the socio-economic impact is also very huge. So in short: if society wants to trust LSMSs for dikes there must be no error or failure in the chain of components that transform sensor data into decision information.

As shown by the dike LSMS, there can be strong requirements to an LSMS system that act as heavy constraints, which are also related to Big Data optimization, as this can influence the extent to which an LSMS can meet the requirements. These will be covered in the next section.

3 Big Data Related Constraints to Monitoring

In order to be economically sustainably deployed in the real world there are certain requirements a LSMS must meet. Designers and engineers of an LSMS that have to come up with a technical solution, consider these requirements as constraints to their solution. An obvious constraint is that the financial value of the information produced by the LSMS should not exceed the costs of the LSMS itself. This includes costs of hardware, energy, labor, etc. In this chapter the focus is on three types of constraints (performance, availability and reliability) that are both strongly related to the added value of a monitoring system to society, as well as Big Data optimization. It could be argued that there are other such constraints, but for reasons of scope and size of this chapter it has been decided not to include those.

In order to describe the relationship between Big Data and LSMSs from the viewpoint of constraints in an efficient way, this relationship will first be described from the viewpoint of data collection and interpretation. After describing the relationship from constraints, solution approaches for keeping within the constraints will be presented in the next section.

3.1 The Big Data in Monitoring

The relevance of Big Data optimization to LSMS depends—to a large extent—on the way data is collected and interpreted. In this subsection this relationship will be explored by looking at different aspects of data collection and interpretation in LSMSs.

3.1.1 Abstraction Level of Interpreted Information

In the dike LSMS example the LSMS has to provide information on 'dike stability'. This is an abstract concept and cannot directly be measured and has to be interpreted. The value of this interpreted 'unit of measurement' can only be produced by using a computational model that computes it, based on sensor measurements and a model of the dike. This can result in much computational effort, even if there are relatively few sensors installed.

The higher the abstraction level of a interpreted 'unit of measurement' is the more (sub)models tend to be required. This is illustrated by an example of an LSMSs for underground pipelines illustrates this [6]. This LSMS determines the chance of failure of a segment of underground pipeline and is depicted in Fig. 3. The interpretation uses based on:

1. Actual soil movements, measured by underground position sensors.
2. Expected soil movements, based on soil behavior models, where (expected) external forces onto the soil are also taken into account.

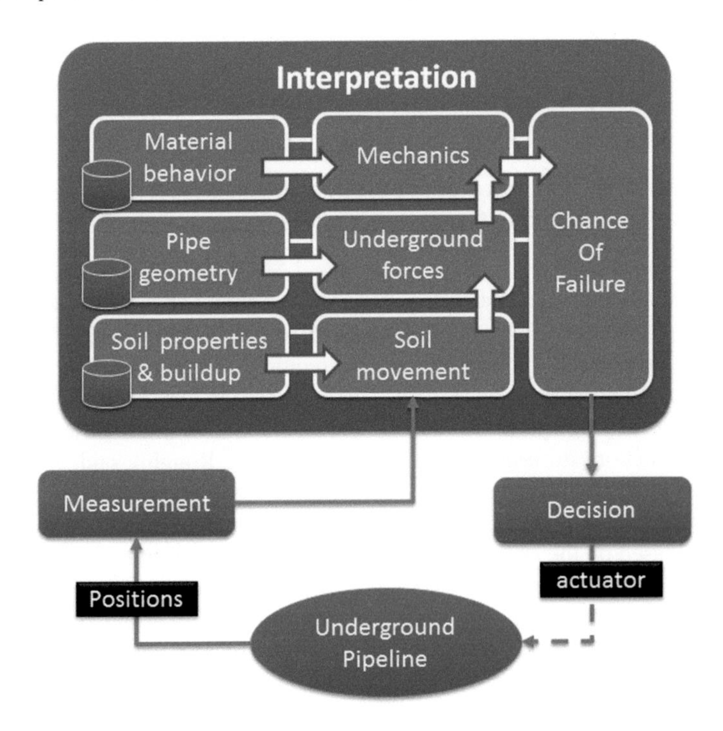

Fig. 3 Example of interpretation using model with a high level of abstraction

3. Estimated underground geometry of pipeline structures, based on construction and maintenance plans.
4. Estimated mechanical properties of pipeline segments and structures, based on physics and analysis of similar pipeline segments elsewhere.

The LSMS uses to the actual ground movements and the assumed model of the underground to create a better estimate of the actual underground. It then computes forces on the pipe structure and uses those in combination with estimated mechanical properties to determine an estimate of the structural reliability. Note that it does also take into account possible variations of parameters involved, due to the uncertainty.

Each of these different aspects (e.g. soil movement, structural behavior of pipe geometry based on ground forces) requires a model of its own. Next to sensor data, it also requires data on the ground buildup and geometries of pipelines. Because the ground buildup tends to vary, just as the pipeline geometries, estimating the chance of failure for all segments of—for example a country like—the Netherlands results in an explosion of computations to be carried out on a large dataset.

3.1.2 Temporal Issues

The way data has to be collected and interpreted (also) depends on what LSMS (end-)users expect the LSMS to do, from a temporal point of view:

1. assessment of the **current state** of the system under observation
2. an estimation of **possible future states** of that system

As was shown above in the case of the LSMS for underground pipelines, assessing the current state of a system under observation can already require many computations because of the need to evaluate different models involved. When information on possible future states is also required, there is need for even more computation and possibly extra storage. This is due to the fact that the interpretation part of the monitoring system needs a model of the system under observation, that allows to predict or estimate future states of the system. Roughly said, two types of models can be distinguished.

Descriptive model: the inner workings of a system are understood to a certain extent, and based on the application of the model to the measurement data it can be predicted how the system will behave next (within a certain band of uncertainty). For example: if you measure the time, you can use models of the earth's rotation around the sun to determine your position in space, relative to the sun.

Phenomenological model: the inner workings of a system are not known, but based on trends and/or other relationships between data segments in past observed measurements, it can be predicted—within a (known) band of uncertainty—how the system under observation will behave in the nearby future. Such a model can be built on observing the system and looking at the relationship between different measurement values throughout time. See [7] for an example of a monitoring system that observes (patterns in behavior) of elderly people.

The establishment of phenomenological prediction models requires learning or 'knowledge discovery', a process that is described in [2]. This can require analyzing large amounts of historical sensor data. This is because—especially when little to knowledge is available on the system to be monitored—many sensor data needs to be taken into account, as well as long historical time series of sensor data. Also these models tend to be intertwined during their development phase with humans. During the learning phase, researchers and engineers will tend to look at different parameters by adding more (or less) sensors, apply different ways of interpretation (change algorithms), etc. In the case of dike LSMSs it turned out during research that it is sometimes very difficult to create a descriptive model, since dikes sometimes are from medieval times and no construction schematics are present. Little is known about the inner structure of these dikes, in this case the use of phenomenological models might be needed to determine structural behavior in the future, based on passed measurements.

While the establishment of a phenomenological model might require analysis of large amounts of data, once it has been established and the behavior of an observed system does no longer change, this is analysis is no longer needed. However, in practice the issue remains how to be sure that the behavior of the system under observation does not change. Not surprisingly, (end-)users of an LSMS tend to prefer to have a understanding of the systems for which they are responsible and thus prefer a **descriptive** model. A phenomenological model can sometimes help to deduce a descriptive model, by trying to build such a model that explains cause and effect relationships found in measured sensor data during the establishment of a phenomenological model. A detailed description of this kind of system modelling and learning is beyond the scope of this chapter.

3.1.3 Growth of Available Data

Next to the abstraction level of information and temporal issues, there is another Big Data aspect to monitoring, which is the growth of available data [8, 9]. Due to the advances in microelectronics, sensors can produce more data in a faster way. Data can be transported using more bandwidth and be stored in much larger quantities. While processing power also has increased, the growth of data remains an issue. Especially in the case of the establishment of phenomenological models—as described above—it is, in the beginning, often unknown what data should be measured and processed by the LSMS and what data is not relevant for producing the desired information. This results in the acquirement of a large amount of data, which all has to be processed.

Based on this general high level overview of several aspects of the relationship between Big Data and monitoring, it is possible to look at the relationship from three different constraint-based points of view.

3.2 Performance

The ability to act upon information (e.g. a warning) is key in the successful deployment of a LSMS. In this chapter the speed at which an LSMS can produce the desired information is defined as its performance. The amount of time between the moment a warning is issued and the last point in time the warning can be acted upon is defined in this chapter as the 'time window'. The larger the time window, the better the performance of an LSMS. For example, if it will take a year to carry out maintenance, than the monitoring system will have to warn at least a year ahead when it predicts failure 'in about a year'.

The size of a time window can be influenced by several factors. The MID decomposition is used to describe the relationship with Big Data from the performance constraint point of view.

Measurement. The amount of time involved in measurement is roughly determined by three things. First, the amount of time needed for measuring a physical aspect of the real world. Secondly, the amount of time needed for transporting the measurement to a location where interpretation of the data can take place. And finally, the amount of time needed for storing the information at a location, so it can be accessed by a data processing component. The total time involved largely depends on the amount of data measured and available bandwidth.

Note that the needed bandwidth and type of sensor are often closely intertwined. This relationship between a (visual) sensor and transport bandwidth is clearly shown by the example of a camera. Does it—for example—provide an HD movie stream of 25 images per second or does it act as an hourly snapshot camera? Or does the camera only send a new snapshot when something changes in the image?

Interpretation. The amount of time involved in interpretation roughly depends on three things. First, the amount of time needed for retrieving data from storage for processing. Then, the amount of time needed for processing data into information using (a) model(s) and finally the amount of time needed for storing information after processing. Note that if the data is processed in a streaming fashion instead of batch wise, less time might be needed.

Decision. The amount of time needed for making a decision to inform/warn (end-)users of an LSMS is roughly determined by two factors. First, the amount of time it takes for a rule set to be applied on the produced information. The amount of time it takes for a issued warning to reach the people or systems that can act on the information. Simply stated: the more time is needed for each of these steps, the less time remains for the warning time window.

3.3 Availability

Besides performance constraints, there are also availability constraints. A LSMS that is not available and does not provide a warning when it is needed is rather useless. It does not matter if the unavailability is due to expected monitoring systems maintenance or due to unexpected other causes. Note that the level of availability depends on the specific context of an LSMS. For example, if a system under observation can quickly change its behavior, the LSMS obviously needs a high level of availability, since else this change would be missed. However, in the case of slow changing systems, temporal non availability is allowed, because the time window (as described in the performance constraint viewpoint above) is relatively large.

The availability of an LSMS can be influenced by several factors. The MID decomposition is used to describe the relationship with Big Data from the availability constraint point of view.

Measurement. The availability of an LSMS with respect to measurement depends on the availability of the sensors, transport and storage. Especially storage is important from a Big Data point of view: this is where techniques for fast storage and retrieval of data come into play, as will be shown in the solution approach section later on.

Interpretation and Decision. The availability of the interpretation and decision parts depends on the availability of storage and computational facilities. Designers and engineers of an LSMS must take into account that these facilities can fail and be temporarily unavailable. This is where Big Data techniques can come into play with respect to redundancy, as will be shown later on.

3.4 Reliability

Even if the performance and availability of an LSMS are within the constraints set forward by the end-users in their context, an LSMS might still not be deployable because of reliability constraints. As described in the dike LSMS example in the monitoring section: not issuing a warning on structural dike failure is not acceptable. Also, sending out false alarms is not acceptable as people will no longer respond to alarms, even if this system is right at the time. In practice, reliability is closely related to availability, because the temporarily unavailability of components of a monitoring system can make the entire system less reliable.

The reliability of an LSMS can be influenced by several factors. The MID decomposition is used to describe the relationship with Big Data from the reliability constraint point of view.

Measurement. Reliability can be influenced by measurement errors, transport and storage errors. This means that data might get corrupted somehow.

Interpretation and decision. The reliability of the information produced by the LSMS directly relies on the data processing algorithms used. At the same time, the reliability is also influenced by the reliability of the computational and storage facilities (i.e. protection against corruption of data and information).

4 Solutions Within Constraints

In the previous sections monitoring systems have been generically described from a Big Data optimization point of view. Specific Big Data related requirements have been identified that have to be met in order for a monitoring system to be useful. In this section different solution approaches for meeting these constraints will be described.

For reasons of scope and size of this chapter, solution approaches concerning faster and bigger hardware will not be described, even though they help in certain situations. The focus will be on the data, algorithms and system architecture/design.

4.1 Approaches in Optimization

In the previous sections three types of constraints have been listed with respect to Big Data (optimization). First, the maximum size of the warning time window. Secondly, the availability of a monitoring system and finally the reliability of the warning itself. The relation with Big Data lies within the fact that the amount or availability of data involved can cause constraint violation. Optimization techniques can help avoid these violations.

In this section optimization techniques are categorized using several approaches in design and implementation of monitoring systems:

1. **Data oriented**. Deal with the amount of data by somehow reducing the volume, before it is processed by the data interpretation algorithms.
2. **Analysis oriented**. Enhance data interpretation algorithms, enabling them to deal with large amounts of data.
3. **System/architecture oriented**. Instead of reducing the amount of data or altering the data to information algorithms, the system as a whole is set up in such a way it can deal with large(r) amounts of data and/or in less time.

Note that designers can also try to resort to a fourth approach:

4. **Goal oriented**. Loosen the constraints. Only possible if there is still a business case for the system after loosening constraints.

Often, design and/or engineering solutions for keeping a monitoring system within constraints, cannot be positioned on a specific axis of orientation. For example, a solution might be data and algorithm oriented at the same time. In this chapter however, the distinction is made in order to position different types of solutions. Note that solution approaches will be described. Not the solutions themselves, since they require more detailed explanation.

4.1.1 Data Oriented Optimization

In general the amount of data to be processed can prevent a LSMS from having a small enough time window. A relatively simple optimization step is to reduce the volume of the (monitoring) data, resulting in a reduction of time needed in all process steps. In the case of the IJkdijk LSMS, it was learned that well maintained dikes—not being in a critical structural health condition—'change' relatively slow. A warning dike LSMS could therefor suffice with a sample rate of 5 min. Reduction of data involved can be performed in a number of different ways. Since this book is about Big Data optimization, increasing the speed of transport, retrieval and storage is not covered.

Reducing the size of the data often results in a reduction of processing time, since less data for each analysis step tends to take less time for analysis. An often used approach to reduce data is aggregation. Instead of having a sample each minute, 60 samples are aggregate—using an average—into a sample each hour. Note that this may come at a price with respect to the reliability constraint: if it is important that 'sudden spikes' in measurement data are to be analyzed too, using averaging might result in a loss of spikes in the data set.

Another way of reducing data for analysis is to convert data to another domain that can be processed just as efficiently. For example cyclic signals in actual measurements can sometimes be converted from the time domain into the frequency domain using Fourier transformations. There are also techniques to convert measurement data graphs automatically into a sequence of symbols. For example: segmenting the graph into sections of horizontal (A), riser (B) and falling (C) slopes, the graph can be converted into a sequence of A, B, C symbols. With non-cyclic signals this can result in a heavy data reduction. Note that again, this optimization technique can only be applied, if the LSMS as a whole stays within the reliability constraint: users must still be able to rely on the LSMS. An example of data reduction technique is called SAX and is described in [10].

A third data oriented optimization technique is to reorder the measurement data in such a way, that the analysis takes less time. An example case is the design of a cow LSMS in which the health and wellbeing of hundreds of thousands of cows is monitored throughout their lifetime [11]. This LSMS provides information on dairy production, which is needed by farmers to maximize the return on investment in milk production. At farms weight data is collected device centric, as depicted in Fig. 4. This means for example that there are a few scales (at each milking robot) that determine weight and many cows. Every time a scale is tread upon, the weight is digitally stored with a timestamp. The milking robot knows which cow (i.e. RFID tag) was being milked (and weighed) at a certain point in time. By combining the data from the milking robot with the scales, the weight of a cow at a certain point in time can be deduced. The interpretation part of the cow LSMS is cow-centric. It provides information per cow through time. By (also) storing weight data in a cow-centric way, the algorithm for interpretation has to carry out far less data retrieval, speeding up the analysis.

In general data oriented optimization techniques do not increase availability or reliability. In fact, it might even decrease reliability because information is lost. This is where another design and/or engineering degree of freedom comes into play: optimize the analysis.

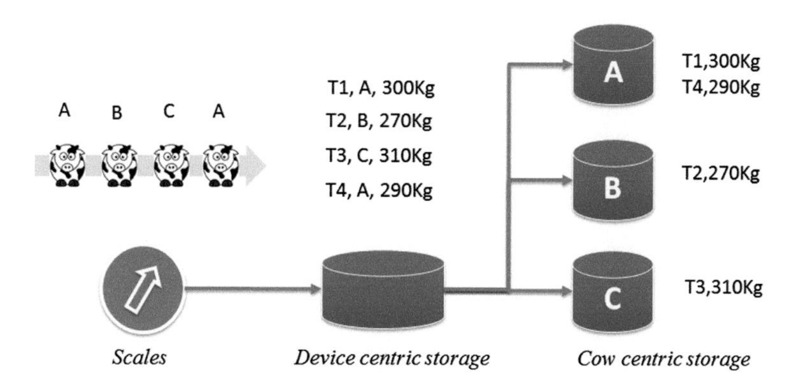

Fig. 4 Device centric and object centric storage

4.1.2 Analysis Oriented Optimization

During the analysis phase, data is converted into information using processing algorithms. The type of algorithms determines the reliability of the information and also the processing speed. So in order to meet time and reliability requirements, an approach is to focus on the algorithm at hand. Roughly stated, there are two kinds of analysis oriented approaches. The first approach is to optimize (or change) a specific algorithm (run on a single computer). This can influence the size of the time window or the level of reliability. A simple example of this approach—with respect to time constraints—is sorting a long list of measurement data. When using a 'bubblesort' algorithm, it takes for more time in general than using a 'quicksort' algorithm. Optimizing and/or designing improved data interpretation algorithms is very domain specific and outside the scope of this chapter.

Another well-known approach—with respect to the size of the time window—is to 'divide-and-conquer': carry out calculations for an algorithm in parallel. This can be done by creating a distributed algorithm or by separating data and run the same algorithm in parallel. Much Big Data optimization techniques revolve around some kind of 'Map Reduce' approach, where a larger problem is reduced into smaller one that can be handled separately [8, 9, 12, 13]. As stated before, this chapter is not about explaining these kind of solutions. They are multiple and require detailed explaining on their own.

4.1.3 System Architecture Oriented Optimization

Data oriented optimization and analysis oriented optimization are targeted at separate parts of the MID steps. It is also possible to optimize the monitoring system as a whole, across the 'Measurement' and 'Interpretation' part. In this section a number of them are reviewed.

More performance by distributing interpretation

By bringing the analysis towards the collection of data, it is sometimes possible to reduce the amount of transport, storage, retrieval and/or computation. An example can be found in [2]. This can be achieved by doing part of the analysis closer to the measurement points, reducing the need for transportation and central processing power (and time). Also, it is possible to carry out intelligent forwarding of information [7], where data or information is only forwarded when needed, by doing a small amount of interpretation close to the sensor, removing the need for a central point of interpretation to carry out all interpretation.

More performance with enough reliability by combined processing

Combining streaming and batch processing of data can also reduce the amount of processing time, while keeping reliability at acceptable levels. The idea is to have two separate and 'parallel' lines of processing data. The first processing line is targeted at speed: data that arrives is immediately processed and calculations are finished before the next data arrives. In this chapter this is defined as streaming data processing. Algorithms used for this type of processing often deal with a small portion of a time series, close to the most recent point in time ('sliding window'). Warnings can be quickly issued, because processing takes place all the time. If the LSMS for some reason crashed, it can be 'rebooted' quickly again, thus increasing availability. However, because only a small portion of the historical behavior is taken into account, the monitoring system might miss certain long term trends. This might reduce the amount of reliability. This is where the second processing line comes into play. Next to the stream processing line there is also a batch processing line. It is targeted at processing more (historical) data in a single batch, but it requires more time. By using the batch processing line as a means to calibrate the stream processing line (e.g. 'long term trends'), it is possible to get 'the best of both worlds'.

The lambda architecture [14] is an example of combining processing lines. It states that the batch processing should be performed each time over all the data. To be able to produce also analyses over the latest data, a streaming system must be set up next to the batch which can processes the data in a streaming manner as long as the batch processing takes. At the moment the batch processing is finished, the results of the streaming analyses are overwritten by the batch results. The streaming analyses starts again, initialized with the newly batch results and the batch also is starting again.

Affordable performance through elasticity of resources

The analysis-oriented divide and conquer approach can require so much computational resources that the costs of storage and computation exceed the value of the information that the monitoring system produces. An approach to reduce the costs is to make use of resources in an elastic way.

The idea is to collect data at a relative low sample rate, which requires less storage and computational power for analysis. Once an unknown anomaly in the measurements is detected, the sample rate is increased and the anomaly can be analyzed using more sophisticated data processing algorithms, that are (temporarily) hosted on an elastic computing cloud. This is depicted in Fig. 5. If variation on this theme is doing a pre-analysis at the source of the data collections, instead of in the interpretation phase. An example can be found in an LSMS that monitors the impact of earthquakes in the Netherlands, which is currently under development. It is constructed to develop a deeper understanding of how vibrations of the underground impact houses. The quakes are caused by compacting of the subsurface at a depth of approximately 3 km, which in turn is the result of the decreasing of pore pressure in an underlying gas reservoir [15]. The LSMS design consists of hundreds of vibration sensors at several hundred houses. The structural health of the building is logged (e.g. 'cracks in walls') and by analyzing the combination of earthquake data, vibrations in houses and the structural health, researchers try to establish quantifiable cause and effect relationships between quakes and the resulting damage to a house. Since these types of quakes and impact on the houses are new, it is not known what for example the sample rate should be. Research and engineers currently thus want to collect data at a high sample rate. In order to keep the amount of processing as low as possible, the idea to reduce is to only forward high sample rate data if it is quake related. Only then computing power is needed for processing. This could be supplied by a

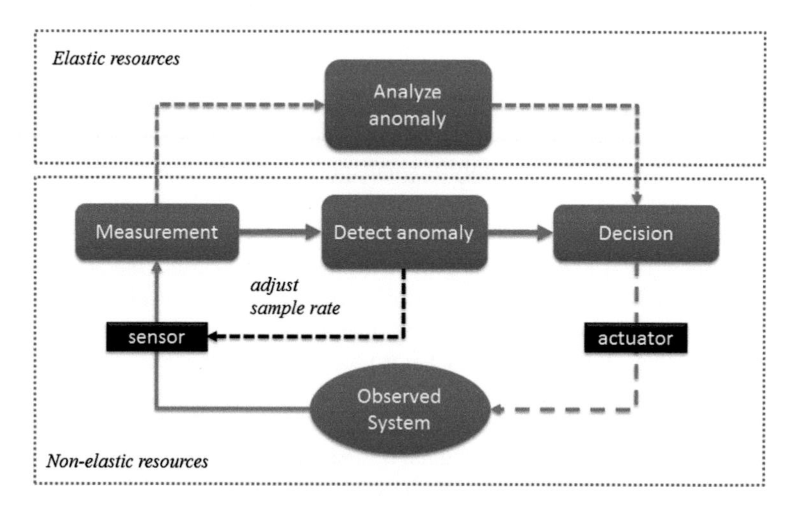

Fig. 5 Using elastic resources in monitoring

computational elastic cloud. After all quake data has been processed, the sample rate of forwarded data can be reduced again, which results in less computational power. More information on Big Data and elasticity can be found—for example—in [16].

Increasing availability through redundancy

As a final system architecture oriented optimization approach, the concept of adding redundancy is discussed. The basic idea is that failure of one of the components does not result in failure of the entire system ('no single point of failure'). By adding multiple components with the same functionality, some could fail, without a total failure of the entire LSMS. Describing general techniques for designing high availability (or sometimes called fault-tolerant systems) by using redundancy is beyond the scope of this chapter. Big Data computational platforms like Hadoop [17] and Project Storm [18] can provide—if configured properly—resilience against failing computational nodes. Also, NoSQL databases like Cassandra can store data with multiple copies of that data distributed over racks and even data centers if necessary.

An important aspect of increasing availability through redundancy is assuming everything can (and will) fail. Also, designers should create a system that tries to continue functioning, even if several components are not available anymore. The loss of a component should not result in a stop of the entire data to information processing chain. Instead, data and/or information should be rerouted to redundant components that still work. If loss of data is unacceptable (e.g. in a learning LSMS) data should only be allowed to be deleted in the monitoring chain, as long as there are at least two copies remaining elsewhere in the chain.

Increasing reliability through human error resilience

No matter how well tested data processing algorithms are, it is always possible a mistake has been made. Resulting in the production of unreliable information. One of the design goals of the Lambda architecture [14] discussed earlier is to be more robust against these kinds of human errors. LSMSs that are based on the Lambda architecture should preserve all original data and that it should be possible to reproduce al information using that data. Next to the streaming processing, there is the possibility to start a batch process that could use the entire dataset to reproduce the information. In this way the effects of a discovered error in a data processing algorithm can be mitigated to some extent.

4.1.4 Goal Oriented Optimization

As mentioned before, as a means of last resort, designers and engineers might loosen constraints. In a way, this means slightly altering the goal of the system. In practice designers and/or engineers sometimes discover not all of the original constraints from the specification do not have to be as tight as originally specified. By loosening constraints just a little, it is sometimes still possible to come up with a LSMS that still provides more value through information, than costs needed for realization and operation of an LSMS. For example, there are techniques—like the Early Accurate Result Library (EARL) system—which help determine the sample rate or the size of data samples, while staying with a specific error bound [19].

4.2 Summary of Optimizations and Constraints

As mentioned earlier on, there is no one-to-one mapping of constraints (time window, availability and reliability) and optimization techniques. In practice applying a specific technique often influences the achievement of several constraints. In the table below these relationships are once more illustrated and summarized, based on the descriptions above. A '+' means that the chance of staying within a constraint is increased, a '–' means that the chance is decreased. For example, aggregation of data can increase the chance of meeting a time window requirement, while decreasing the chance of meeting a reliability constraint. A '0' means little to no effect. Note that these are very generic scores, provided to show that improvements in the area of one constraint might result in issues with other constraints. In practice, the specific case and optimization technique at hand might lead to other values in the table (Table 1).

4.2.1 Impact on Other Constraints

Applying one or more Big Data optimization techniques as discussed in the previous paragraphs, can come at a price with respect to other constraints. The relationship between Big Data related constraints (performance, availability and reliability) have already been discussed. However, other constraints might be impacted too. This

Table 1 Relationship between optimations and constraints

	Performance	Reliability	Availability
Change data			
Aggregation	+	–	0
Concept conversion	+	0/–	0
Reordering	+	0	0
Change algorithms			
More efficient algorithm	+	0	0
Parallelization	++	0	0
Change architecture			
Faster: bring the analysis to the data	++	0	
Faster: combining streaming and batch	++	0	
Deal with many different sensor connections		0	++
Adapt resources according to need	0	0	–
Add redundant components	–	+	++
Create resilience against human error	0/–	+	0/+
Change goal			
Less perfect answers are acceptable	++	–	0

chapter would not be complete without mentioning several constraints that are important in modern day LSMS deployment too. This will be done in these last subsections.

Energy consumption

LSMS can consume large amounts of energy. Optimization techniques that require more hardware because of parallelization or redundancy, obviously tend to use more energy. Depending on the value of the information produced by the LSMS or the energy available in the deployment area of an LSMS, the energy usage might be more than allowed. The link between Big Data analytics and energy usage is covered in more detail in [20].

Maintainability

Intricate Big Data optimization techniques might enable designers and engineers of an LSMS to stay within the constrains as specified, but they could make it far more difficult for (other) designers and engineers to understand the LSMS as a whole. The ability to understand a system is also a constraint in everyday practice: without people that understand how the LSMS actually works it is difficult to maintain or improve the system. This understanding is needed, if for example unexpected errors take place and the LSMS needs some 'debugging' or a (partial) redesign. Even more so, if other people, which have not designed or implemented the system, have doubts about the reliability of the LSMS, it is very important that other experts—not involved in the realization of the LSMS—can get an understanding of the LSMS and provide the non-experts with a satisfactory answer to their doubts. In other words: maintainability through auditability is also a constraint, which can be influenced by Big Data optimization techniques in a negative way.

5 Conclusion

In this chapter the relationship between Big Data optimization and monitoring in the real world has been explored from different view-points. Several aspects of data collection and interpretation have been used as a viewpoint, as well as three types of constraints for large scale monitoring systems (abbreviated as LSMSs) that are related to Big Data: performance, availability and reliability constraints. Finally, several solution approaches involving Big Data optimization techniques have been provided that enable designers and engineers of LSMSs to provide a solution that stays within performance, availability and reliability constraints.

References

1. McCauley-Bell, P., Freeman, R.: Quantification of belief and knowledge systems in information warfare. In: Proceedings of the Fifth IEEE International Conference on Fuzzy Systems (1996). doi:10.1109/FUZZY.1996.552568
2. Chen, C.L.P., Zhang, C.-Y.: Data-intensive applications, challenges, techniques and technologies: a survey on big data. Inf. Sci. (2014). doi:10.1016/j.ins.2014.01.015
3. FloodControl IJkdijk. http://www.ijkdijk.nl/en/. Accessed 15 June 2015
4. Pals, N., De Vries, A., De Jong, A., Boertjes, E.: Remote and in situ sensing for dike monitoring: the Ijkdijk Experience. In: de Amicis, R., et al. (eds.) GeoSpatial Visual Analytics, pp 465–475. Springer, Netherlands (2009)
5. Roedelsperger, S., Coccia, A., Vicente, D., Trampuz, C., Meta, A.: The novel FastGBSAR sensor: deformation monitoring for dike failure prediction. In: 2013 Asia-Pacific Conference on Synthetic Aperture Radar (APSAR), pp. 420−423 (2013)
6. Helmholt, K., Courage, W.: Risk management in large scale underground infrastructures. In: 2013 IEEE International Systems Conference (SysCon) (2013). doi:10.1109/SysCon.2013.6549991
7. Jiang, P., Winkley, J., Zhao, C., Munnoch, R., Min, G., Yang, L.T.: An intelligent information forwarder for healthcare big data systems with distrib-uted wearable sensor. IEEE Syst. J. (2014). doi:10.1109/JSYST.2014.2308324
8. Kejariwal, A.: Big data challenges: a program optimization perspective. In: Second International Conference on Cloud and Green Computing (CGC) (2012). doi:10.1109/CGC.2012.17
9. Gu, L., Zeng, D., Li, P., Guo, S.: Cost Minimization for big data processing in geo-distributed data centers. IEEE Trans. Emerg. Top. Comput. (2014). doi:10.1109/TETC.2014.2310456
10. Lin, J., Keogh, E., Lonardi, S., Chiu, B.: A Symbolic representation of time series, with implications for streaming algorithms. In: DMKD '03 Proceedings of the 8th ACM SIGMOD Workshop on Research Issues in Data Mining and Knowledge Discovery, pp. 2–11 (2003)
11. Vonder, M.R., Donker, G.: Cow-centric data made available in real-time for model development (2014). http://www.smartagrimatics.eu/Portals/4/SAM2014/1%20Farming%20in%20the%20Cloud/1.3%20Smart%20Livestock%20Farming/1.3.5%2020140618%20Agrimatics%202014%20-%20SDF%20%20InfoBroker%20v7.pdf. Accessed 15 June 2015
12. Ma, Y., Wang, L., Liu, D., Liu, P., Wang, J., Tao, J.: Generic parallel programming for massive remote sensing. In: 2012 IEEE International Conference on Data Processing (2012). doi:10.1109/CLUSTER.2012.51
13. Nita, M.-C., Chilipirea, Dobre, C., Pop, F.: A SLA-based method for big-data transfers with multi-criteria optimization constraints for IaaS. In: 2013 11th Roedunet International Conference (RoEduNet) (2013). doi:10.1109/RoEduNet.2013.6511740

14. Marz, N.: Lambda architecture. http://lambda-architecture.net. Accessed 15 June 2015
15. Ketelaar, G., van Leijen, F., Marinkovic, P., Hanssen, R.: Multi-track PS-InSAR datum connection. In: 2007 IEEE International Geoscience and Remote Sensing Symposium (IGARSS) (2007). doi:10.1109/IGARSS.2007.4423346
16. Han, R., Nie, L., Ghanem, M.M., Guo, Y.: Elastic algorithms for guaranteeing quality monotonicity in big data mining. In: 2013 IEEE International Conference on Big Data (2013). doi:10.1109/BigData.2013.6691553
17. Bicer, T., Wei Jiang, Agrawal, G.: Supporting fault tolerance in a data-intensive computing middleware. In: 2010 IEEE International Symposium on Parallel and Distributed Processing (IPDPS) (2010). doi:10.1109/IPDPS.2010.5470462
18. Chandarana, P., Vijayalakshmi, M.: Big data analytics frameworks. In: 2014 International Conference on Circuits, Systems, Communication and Information Technology Applications (CSCITA) (2014). doi:10.1109/CSCITA.2014.6839299
19. Laptev, N., Zeng, K., Zaniolo, C.: Very fast estimation for result and ac-curacy of big data analytics: the EARL system. In: 2013 IEEE 29th International Conference on Data Engineering (ICDE) (2013). doi:10.1109/ICDE.2013.6544928
20. Kambatla, K., Kollias, G., Kumar, V., Grama, A.: Trends in big data analytics. J. Parallel Distributed Comput. (2014). doi:10.1016/j.jpdc.2014.01.003

Author Biographies

Kristian Helmholt is a Senior Consultant at TNO. After obtaining a master's degree in applied computing science, he worked on the creation of applications and services Internet Service Providers at the turn of the millennium. After that he started working on using sensor networks for data collection in different application domains like logistics and agriculture. By now he has specialized in the application of ICT for monitoring and control in asset management of large scale physical infrastructures, for example underground gas and electricity networks. A recurring theme in his work is the affordable and reliable acquisition of data which can be combined and refined into information for decision makers.

Bram van der Waaij is Senior Research Scientist at TNO. His department focuses on the monitoring and control of large scale infrastructures and how the many related organizations can use this insight in an optimal manner. He specializes in cloud based sensor data storage, Bigdata streaming analysis and multi-party collaboration. Research topics include cloud based analysis, column based databases and generic anomaly detection algorithms. All with a focus on inter organizational collaboration and multi-use of sensor systems. He applies his knowledge in various industry sectors, among others: dike management, smart grids and agriculture.

Smart Sampling and Optimal Dimensionality Reduction of Big Data Using Compressed Sensing

Anastasios Maronidis, Elisavet Chatzilari, Spiros Nikolopoulos and Ioannis Kompatsiaris

Abstract Handling big data poses as a huge challenge in the computer science community. Some of the most appealing research domains such as machine learning, computational biology and social networks are now overwhelmed with large-scale databases that need computationally demanding manipulation. Several techniques have been proposed for dealing with big data processing challenges including computational efficient implementations, like parallel and distributed architectures, but most approaches benefit from a dimensionality reduction and smart sampling step of the data. In this context, through a series of groundbreaking works, Compressed Sensing (CS) has emerged as a powerful mathematical framework providing a suite of conditions and methods that allow for an almost lossless and efficient data compression. The most surprising outcome of CS is the proof that random projections qualify as a close to optimal selection for transforming high-dimensional data into a low-dimensional space in a way that allows for their almost perfect reconstruction. The compression power along with the usage simplicity render CS an appealing method for optimal dimensionality reduction of big data. Although CS is renowned for its capability of providing succinct representations of the data, in this chapter we investigate its potential as a dimensionality reduction technique in the domain of image annotation. More specifically, our aim is to initially present the challenges stemming from the nature of big data problems, explain the basic principles, advantages and disadvantages of CS and identify potential ways of exploiting this theory in the domain of large-scale image annotation. Towards this end, a novel Hierarchical Compressed Sensing (HCS) method is proposed. The new method dramatically decreases the computational complexity, while displays robustness equal to

A. Maronidis (✉) · E. Chatzilari · S. Nikolopoulos · I. Kompatsiaris
Information Technologies Institute, Centre for Research and Technology Hellas, Thessaloniki, Greece
e-mail: amaronidis@iti.gr

E. Chatzilari
e-mail: ehatzi@iti.gr

S. Nikolopoulos
e-mail: nikolopo@iti.gr

I. Kompatsiaris
e-mail: ikom@iti.gr

A. Emrouznejad (ed.), *Big Data Optimization: Recent Developments and Challenges*, Studies in Big Data 18, DOI 10.1007/978-3-319-30265-2_12

the typical CS method. Besides, the connection between the sparsity level of the original dataset and the effectiveness of HCS is established through a series of artificial experiments. Finally, the proposed method is compared with the state-of-the-art dimensionality reduction technique of Principal Component Analysis. The performance results are encouraging, indicating a promising potential of the new method in large-scale image annotation.

Keywords Smart sampling · Optimal dimensionality reduction · Compressed Sensing · Sparse representation · Scalable image annotation

1 Introduction

Recently, the pattern recognition community has been struggling to develop algorithms for improving the classification accuracy at the expense of additional computational cost or potential need for extra data storage. Nowadays though, as enormous volumes of data—originating mainly from the internet and the social networks—are accumulated, great interest has been placed on the theoretical and practical aspects of extracting knowledge from massive data sets [2, 21]. In addition, the dimensionality of the data descriptors is always increasing, since it leads to better classification results in large-scale datasets [42]. However, as the volume (i.e., either the number of dimensions or the number of samples) increases, it becomes more difficult for classification schemes to handle the data. It is therefore clear that the need to establish a fair compromise among the three primitive factors, i.e., classification accuracy, computational efficiency and data storage capacity proves of utmost importance.

Although the analysis intuitions behind big data are pretty much the same as in small data, having bigger data consequently requires new methods and tools for solving new problems, or solving the old problems in a more efficient way. Big data are characterized by their variety (i.e., multimodal nature of data ranging from very structured ones like ontologies to unstructured ones like sensor signals), their velocity (i.e., real-time and dynamic aspect of the data) and of course their volume. In order to deal with these new requirements, a set of technological paradigms that were not particularly characteristic for small datasets have been brought into the forefront of interest, including: smart sampling, incremental updating, distributed programming, storage and indexing, etc.

Among these paradigms, of particular interest we consider the case of smart sampling, where massive datasets are sampled and the analysis algorithms are applied only on the sampled data. However, in order for smart sampling to be effective we need to be able to extract conclusions from small parts of the dataset as if we were working on the entire dataset. Although there are works in the literature claiming that building a model on sampled data can be as accurate as building a model on the entire dataset [16], there are still questions that remain unsolved as: What is the right sample size? How do we increase the quality of samples? How do we ensure that the selected samples are representative of the entire dataset? Although there have been well-known techniques for sampling and dimensionality reduction, including Principal Component Analysis (PCA) [32], the recently proposed mathematical theory of Compressed Sensing (CS) has the potential to answer some of these questions.

Compressed Sensing is a signal processing technique for dimensionality reduction that can be used as a means to succinctly capture the most important information of high-dimensional data. It offers a reversible scheme for compressing and reconstructing pieces of digital data. The effectiveness of CS is based on two conditions: (a) the *sparsity* of the initial data and (b) the *incoherence* between the spaces of compressed and uncompressed data. Roughly speaking, the sparsity expresses the extend to which the data contain non-zero values, while incoherence is the property to transform sparse data into dense and vice versa. The functionality of CS relies upon some quite interesting mathematical findings, which constitute the outcome of many scientific fields including Linear Algebra, Optimization Theory, Random Matrix Theory, Basis Expansions, Inverse Problems and Compression [3].

Within the framework of CS, data are considered as vectors lying in a high-dimensional space. This space is endowed with a set of basis vectors so that an arbitrary vector can be represented as a linear combination of these basis vectors. The goal of CS is firstly to design a transformation matrix that can be used to project any initial high-dimensional vector to a new space of a much smaller dimensionality [1] and secondly to ensure all those conditions that the transformation matrix must obey so as for the compressed vector to be perfectly reconstructable in the initial high-dimensional space [22].

Regarding the first objective, the great leap of CS is the rather surprising finding that even a random matrix, i.e., whose values at each position have been generated for instance by a Gaussian distribution, can serve as an appropriate transformation matrix under certain circumstances, and that these circumstances are usually met in real-world situations [13]. The ability to use a random, rather than a complicated and difficult to generate, transformation matrix for compressing the data, offers a huge advantage in the context of big data and specifically with respect to velocity- and memory-related challenges, e.g., real-time processing, live-streaming, etc.

With respect to the second objective, CS achieves solving an ill-posed underdetermined linear system of equations. The reconstruction is achieved through the optimization of a dedicated objective energy function encoding the sparsity of the reconstructed vector, subject to the above linear system constraints. The fact that the use of the l_1 norm for encoding the energy of an arbitrary vector reduces the optimization problem to a standard linear programming problem, makes the complexity properties of CS particularly suited for handling huge volumes of data [3].

The use of CS has already proven beneficial for a multitude of applications such as Image Processing (Annotation, Denoising, Restoration, etc.), Data Compression, Data Acquisition, Inverse Problems, Biology, Compressive Radar, Analog-to-Information Converters and Compressive Imaging [25, 27, 35, 53]. It is this success that has triggered the scientific interest about its potential to alleviate some of the big data related problems.

Our aim in this chapter is to present the challenges stemming from the nature of big data problems, explain the basic principles, advantages and disadvantages of CS and identify potential ways for exploiting this theory in the large-scale image annotation domain. More specifically, the contribution of this chapter is fourfold: First, we investigate the sensitivity of CS to the randomness of the projections. Second, we study the performance of CS as a function of the number of the reduced dimensions. Third, we propose a novel Hierarchical Compressed Sensing (HCS) dimensionality reduction method and we show that the proposed method displays performance equivalent to the typical CS, while exponentially reduces the computational load required during the dimensionality reduction process. Fourth, we practically establish a connection between the performance of CS-like methods and the level of the data sparsity, through a series of artificial experimental sessions. Finally, apart from the above contributions, we identify the advantages and disadvantages of the new method through a comparison with the widely used and always state-of-the-art dimensionality reduction technique of Principal Component Analysis [32].

The remainder of this chapter is organized as follows. A background knowledge of the CS framework along with the prerequisite notation is provided in Sect. 2. The proposed methodology for compressing the volume of big data is analytically presented in Sect. 3. Experimental results are provided in Sect. 4. A number of related works on data compression, dimensionality reduction as well as a variety of CS applications are presented in Sect. 5. Finally, conclusions are drawn in Sect. 6.

2 Background

The inception of the CS theory dates back to the valuable works of Candès [13], Romberg [9], Tao [10] and Donoho [22]. Until that time, the computer science community was governed by the Shannon–Nyquist sampling theorem and its implications [38, 43]. This theorem dictates that when acquiring a signal, the sampling frequency must be at least twice the highest frequency contained within the signal in order to permit a perfect reconstruction of the sampled signal to the initial space. In fact, this theorem imposes an inherent limitation to the compression capacity of any sampling process. However, it has been proven that under a set of certain circumstances, it is feasible to use significantly lower sampling rates than the Nyquist rate and at the same time allow for the signal's almost perfect reconstruction by exploiting its sparsity [12].

From that moment on, a huge leap has been realized in digital signal processing, through a plethora of important works and publications. The principal outcome of the research in this field is a suite of conditions and methods that allow for effectively compressing and perfectly reconstructing data under the assumption of the sparsity of the data and the incoherence of the sampling means, mathematical notions that will be analytically explained later in this section. Before we proceed further into the CS methodology, let us first provide the necessary notation along with the fundamental knowledge that will be used throughout this manuscript.

2.1 *Notation and Fundamental Knowledge*

Two principal domains, the *representation* and the *sensing* domain, and three main mathematical entities, a *raw vector*, a *sensing matrix* and a *compressed vector* are involved in the bi-directional compression-reconstruction scheme, each playing a particular role. The two domains are vector-spaces, each endowed with a specific set of basis vectors. The former offers as a basis for representing raw uncompressed vectors, while the latter offers as a basis for representing compressed vectors. Finally, the sensing matrix constitutes an intermediary between the two domains by compressing and reconstructing vectors from the one domain to the other. In the remainder of this chapter, vectors and matrices will be denoted by bold, while scalars by plain type letters.

2.1.1 Representation Domain

Consider an n-length raw uncompressed vector $\mathbf{x} \in \mathbb{R}^{n\times 1}$ and let $\{\mathbf{r}_i\}_{i=1}^{n}$ with $\mathbf{r}_i \in \mathbb{R}^{n\times 1}$, for all i, be an orthonormal basis, which is referred to as the representation basis, so that

$$\mathbf{x} = \sum_{i=1}^{n} t_i \mathbf{r}_i , \tag{1}$$

where $\{t_i\}_{i=1}^{n}$ is the set of coefficients of $\mathbf{x}$ in the representation basis given by $t_i = \langle \mathbf{x}, \mathbf{r}_i \rangle = \mathbf{r}_i^T \mathbf{x}$. Equation 1 can be also formulated in matrix notation as

$$\mathbf{x} = \mathbf{R}\mathbf{t} , \tag{2}$$

where the *representation matrix* $\mathbf{R} = [\mathbf{r}_1, \mathbf{r}_2, \ldots, \mathbf{r}_n] \in \mathbb{R}^{n\times n}$ is constructed by concatenating the basis vectors of $\{\mathbf{r}_i\}_{i=1}^{n}$ column-wise and $\mathbf{t} \in \mathbb{R}^{n\times 1}$ is a $n \times 1$ vector containing the coefficients $\{t_i\}_{i=1}^{n}$.

2.1.2 Sensing Domain

Let also $\mathbf{y} \in \mathbb{R}^{m\times 1}$ denote an m-length vector, with $m \ll n$ and $\{\mathbf{s}_j\}_{j=1}^{m}$ with $\mathbf{s}_j \in \mathbb{R}^{n\times 1}$, for all j, be a collection of *sensing* vectors, which from now on will be referred to as the sensing basis, so that

$$y_j = \mathbf{s}_j^T \mathbf{x} , \tag{3}$$

where $\{y_j\}_{j=1}^{m}$ comprises a set of m dimensions in the reduced space, which are also known in the CS literature as *measurements*. From now onwards, the terms reduced dimensions and measurements will be used interchangeably throughout the manuscript. In similar terms, Eq. 3 can be formulated in matrix notation as

$$\mathbf{y} = \mathbf{S}\mathbf{x} , \tag{4}$$

where the *sensing matrix* $\mathbf{S} = [\mathbf{s}_1, \mathbf{s}_2, \ldots, \mathbf{s}_m]^T \in \mathbb{R}^{m \times n}$ is constructed by concatenating the basis vectors $\{\mathbf{s}_j\}_{j=1}^m$ row-wise and $\mathbf{x} \in \mathbb{R}^{n \times 1}$ is a $n \times 1$ vector containing the coefficients $\{x_j\}_{j=1}^n$. At this point, it is worth emphasizing that the vector $\mathbf{x}$ used in Eq. 4 is the same as the one used in Eq. 2. Also note that

$$\mathbf{y} = \sum_{j=1}^{n} x_j \mathbf{S}^{(j)}, \tag{5}$$

where $\mathbf{S}^{(j)}$ is the jth column of matrix $\mathbf{S}$. Since $m \ll n$, the set of vectors $\{\mathbf{S}^{(j)}\}_{j=1}^n$ acts as an over-complete basis of the sensing domain. Setting $\mathbf{D} = \mathbf{SR}$ and combining Eqs. 2 and 4 we have

$$\mathbf{y} = \mathbf{Sx} = \mathbf{SRt} = \mathbf{Dt}. \tag{6}$$

The matrix $\mathbf{D} \in \mathbb{R}^{m \times n}$ is a transformation matrix, which reduces the initial dimensionality of vector $\mathbf{x}$ to a much smaller number of dimensions providing a compressed vector $\mathbf{y}$. It is worth mentioning that in the particular case where the representation basis is the "spikes" basis, i.e., $\mathbf{R} = \mathbf{I}$, then $\mathbf{D}$ becomes equal to the sensing matrix $\mathbf{S}$. For this reason, conventionally, $\mathbf{D}$ is also called the sensing matrix.

2.2 Compressed Sensing and Sparse Reconstruction

Compressed Sensing addresses the bi-directional problem of transforming a signal to a compressed version and reversely perfectly reconstructing the uncompressed signal from its compressed version. The reverse process is also referred to as Sparse Reconstruction. Towards the forward direction, CS addresses the problem of compressing a raw signal through the use of a sensing matrix $\mathbf{D}$. Formally speaking, given a sparse vector $\mathbf{t}$ lying in the representation domain, the aim is to measure a compressed version $\mathbf{y} = \mathbf{Dt}$ of that vector lying in the sensing domain. Towards this end, it is of great importance to appropriately design the sensing matrix $\mathbf{D}$, so that it enables a subsequent reconstruction of the uncompressed signal. Qualitatively, this means that the information maintained during the compression process must be enough to enable the almost lossless recovery of the signal.

On the other hand, towards the reverse direction, Sparse Reconstruction addresses the problem of recovering the raw signal from the knowledge of its compressed version. The recovery is performed using again the matrix $\mathbf{D}$, which in the Sparse Reconstruction context is often called *dictionary*. From now on, for the sake of simplicity, we will consistently use the term dictionary, for both forward and reverse direction. There is a variety of algorithms that guarantee the perfect reconstruction of the raw signal, provided that during the compression process all the appropriate conditions are satisfied by the dictionary and the raw vector. With respect to the former, the principal property that a dictionary must satisfy is the so-called Restricted Isometry Property (RIP) [11], as described later in the manuscript. Regarding the latter, two properties, namely sparsity and compressibility [19], play a key role in both directions. These properties are analytically described in the next subsection.

2.2.1 Sparsity and Compressibility

A vector $\mathbf{t} = (t_i)_{i=1}^n \in \mathbb{R}^n$ is called k-sparse, if

$$\|\mathbf{t}\|_0 := card\{i : t_i \neq 0\} \leq k\,. \tag{7}$$

In a few words, sparsity counts the number of non-zero components in a vector. From now on, we will denote as Σ_k the set of all k-sparse vectors. As already mentioned, in reality we rarely meet proper sparse vectors. Usually the case is that the majority of the vector components are close to zero and only a few of them have significantly larger values. In this case, another relaxed metric, the compressibility, is preferred to approximate the vector's sparsity [19].

Let $1 \leq p \leq \infty$ and $r > 0$. A vector $\mathbf{t} = (t_i)_{i=1}^n \in \mathbb{R}^n$ is called p-compressible with constant C and rate r, if

$$\sigma_k(\mathbf{t})_p := \min_{\tilde{\mathbf{t}} \in \Sigma_k} \|\mathbf{t} - \tilde{\mathbf{t}}\|_p \leq C \cdot k^{-r} \text{ for any } k \in \{1, 2, \ldots, n\}\,. \tag{8}$$

Actually, this formula measures the extent to which a vector $\mathbf{t}$ can be adequately approximated by a sparse vector. In practice, as long as the above formula is difficult to calculate, a variety of heuristics have been proposed for estimating vector compressibility. For instance, in our work, inspired by [19], we propose an approach based on the growth rate of the vector components. A more detailed description of our approach is provided in Sect. 3.2.1.

2.2.2 Signal Reconstruction

Given a compressed vector lying in the sensing domain, the aim is to reconstruct it in the representation domain. Formally speaking, the problem can be stated as recover $\mathbf{t}$ from the knowledge of $\mathbf{y} = \mathbf{Dt}$ or equivalently recover $\mathbf{x}$ from the knowledge of $\mathbf{y} = \mathbf{Sx}$. Solving $\mathbf{y} = \mathbf{Dt}$ (or $\mathbf{y} = \mathbf{Sx}$) for $\mathbf{t}$ (or $\mathbf{x}$, respectively) is an ill-posed problem, since $m < n$, hence there are infinite many solutions of the linear system. These solutions lie on a $(n - m)$-dimensional hyperplane: $\mathcal{H} = \mathcal{N}(\mathbf{D}) + \mathbf{t}'$, where $\mathcal{N}(\mathbf{D})$ constitutes the null space of $\mathbf{D}$ and $\mathbf{t}'$ is an arbitrary solution.

Despite the ill-posedness of the reconstruction problem, various algorithms have been devised in order to effectively resolve it exploiting the assumption of sparsity of the signal in the representation domain. In this context, several l_p-norm based objective criteria have been employed using the formulation in Eq. 9.

$$(l_p): \quad \mathbf{t}' = \operatorname{argmin}\|\mathbf{t}\|_p \text{ s.t. } \mathbf{y} = \mathbf{Dt}\,. \tag{9}$$

Along these lines, the l_0-norm (i.e., $p = 0$) minimization approach has been initially proposed, which reflects signal sparsity. A severe shortcoming of this approach though is that trying to solve the l_0-norm problem proves to be NP-hard. Nevertheless, several greedy algorithms like Matching Pursuit (MP) or Orthogonal Matching Pursuit (OMP) have been proposed for efficiently solving this optimization problem.

Alternatively, the l_2-norm has also been adopted instead in the above formulation (Eq. 9), for overcoming the limitations that stem from using the l_0-norm, leading to the Least Squares problem. The big advantage using the l_2-norm instead of the l_0-norm is that it provides a closed-form solution $\mathbf{t}' = \mathbf{D}^T \left(\mathbf{D}\mathbf{D}^T\right)^{-1} \mathbf{y}$. However, the l_2-norm barely reflects the signal sparsity.

A good compromise between signal sparsity reflexivity and problem solvability is provided by substituting the l_2-norm with the l_1-norm in the above formulation (Eq. 9), leading to the Basis Pursuit (BP) optimization problem. Adopting the l_1-norm reduces the optimization problem to a linear programming one, which after the advances made by introducing the interior point method [18] proves to be a desirable approach, in terms of computational efficiency. Although a variety of alternative approaches has also been proposed for performing signal reconstruction [17], it is worth mentioning that the BP and the OMP are the most widely used approaches in practice. Furthermore, there is no strong evidence that the one is consistently better than the other. For instance, despite the big advantage of BP in terms of computational efficiency in most of the cases, in [47] through a direct comparison between the two, it is clearly asserted that OMP proves faster and easier to implement than BP when dealing with highly sparse data.

2.2.3 Dictionary Design

In parallel to the development of algorithms that perform sparse signal reconstruction, much effort has been allocated in designing a dictionary that allows for effectively compressing and perfectly recovering a raw vector. This dictionary could be either adaptive or non-adaptive depending on different motivations stemming from either classification or representation problems. The key property that such a dictionary must satisfy is the exact recovery property. In mathematical terms, a dictionary **D** has the exact recovery property of order k, which qualitatively means that it allows the perfect reconstruction of k-sparse vectors, if for every index set $\mathcal{I}$, with $|\mathcal{I}| \leq k$ there exists a unique solution $\mathbf{t}'$ with support $\mathcal{I}$ to the minimization problem:

$$\mathbf{t}' = \operatorname{argmin} \|\mathbf{t}\|_1 \ \text{ s.t. } \mathbf{D}\mathbf{t} = \mathbf{y}\,. \tag{10}$$

The exact recovery property is equivalent to another important mathematical property which is well-known in the literature as the Restricted Isometry Property (RIP). A dictionary **D** satisfies the RIP of order k if there exists a $\delta_k \in (0, 1)$ such that

$$(1 - \delta_k)\|\mathbf{t}\|_2^2 \leq \|\mathbf{D}\mathbf{t}\|_2^2 \leq (1 + \delta_k)\|\mathbf{t}\|_2^2 \ \text{ for all } \mathbf{t} \in \Sigma_k\,. \tag{11}$$

In a few words, this condition dictates that $\mathbf{D}$ must be almost an isometry, i.e., approximately preserve the distance between any pair of k-sparse vectors in the representation and the sensing vector spaces. Apart from the RIP, other equivalent properties like the Null Space Property (NSP) [19] have also been presented in the literature. Although all these properties constitute important theoretical weapons, in practice it is difficult to prove that a dictionary has the RIP or the NSP. Alternatively, another more tractable metric, the Mutual Coherence (MC) has been introduced instead. The MC of a Dictionary $\mathbf{D}$ is defined as:

$$\mu(\mathbf{D}) = \max_{i \neq j} \frac{|\mathbf{d}_i^T \mathbf{d}_j|}{\|\mathbf{d}_i\|_2 \|\mathbf{d}_j\|_2}, \tag{12}$$

where $\mathbf{d}_i$ is the ith column of $\mathbf{D}$. Intuitively, the MC estimates the affinity among the basis vectors of the dictionary. A low-rank approximation method, which uses a column sampling process, has been presented in [37] for estimating the coherence of a matrix. It has been proven in the literature that if $\mathbf{t} \in \mathbb{R}^n \backslash \{\mathbf{0}\}$ is a solution of the (l_0) problem (see Eq. 9) satisfying

$$\|\mathbf{t}\|_0 < \frac{1}{2}\left(1 + \frac{1}{\mu(\mathbf{D})}\right), \tag{13}$$

then $\mathbf{s}$ is the unique solution of (l_0) and (l_1) problems, which implies that MP, OMP and BP are guaranteed to solve both of these problems [24]. This finding infers that the smaller the mutual coherence of a dictionary is, the better the recovery of the signal will be.

Along these lines, it has been proven that a matrix containing random values generated for instance by independent and identically distributed (i.i.d.) Gaussians with mean zero and variance $1/n$, satisfies with extremely high probability the mutual coherence criteria emanating from the CS theory and qualifies as an appropriate dictionary [12, 13]. This rather counter-intuitive finding actually renders random projections suitable for perfectly reconstructing compressed data. The huge advantage of using random projections is the computational ease of constructing a random matrix. In practice, the rows of the resulting matrix, constituting the dictionary, are orthonormalized through a Singular Value Decomposition (SVD) step, since working with orthonormal bases is supposed to provide both practical and intuitive benefits to the whole process. Another advantage of using random projections is that it offers a non-adaptive way for constructing dictionaries independent of the specific problem domain and parameters.

3 Methodology

In this chapter, based on the outcomes of the CS theory, we propose the use of random projections for reducing the dimensionality of large-scale datasets for performing data classification. Our methodology has been substantiated and validated within the case study of image annotation. In this case study, a number of images annotated with a predefined set of c high-level concepts c_k, for $k = 1, 2, \ldots, c$ is used as the training set. Given a previously unseen test image, the goal is to recognize which of these concepts are depicted and hence classify accordingly the image. Towards this end, given a test image, a set of visual features are extracted providing a corresponding feature vector. The resulting feature vector is subsequently subject to dimensionality reduction yielding a compressed vector, which is finally used for classification. The steps of the above pipeline approach are analytically described in the following subsections.

3.1 Feature Extraction

A variety of techniques has been proposed for extracting a representative set of features from data, including SIFT [33], SURF [5] and MPEG7 [36]. Among these techniques, for the purpose of our study we choose SIFT, which has been proven quite robust in classification problems [33]. Given a raw image, 128-dimensional gray SIFT features are extracted at densely selected key-points at four scales, using the vl-feat library [49]. Principal Component Analysis (PCA) is then applied on the SIFT features, decreasing their dimensionality from 128 to 80. The parameters of a Gaussian Mixture model with $K = 256$ components are learned by Expectation Maximization from a set of descriptors, which are randomly selected from the entire set of descriptors extracted by an independent set of images. The descriptors are encoded in a single feature vector using the Fisher vector encoding [40]. Moreover, each image is divided in 1×1, 3×1 and 2×2 regions, resulting in 8 total regions, also known as *spatial pyramids*. A feature vector is extracted for each pyramid by the Fisher vector encoding and the feature vector corresponding to the whole image (1×1) is calculated using sum pooling [15]. Finally, the feature vectors of all 8 spatial pyramids are concatenated to a single feature vector of 327680 components, which comprise the overall data dimensionality that is to be reduced by CS in the experiments.

3.2 Smart Sampling

As we have seen in the previous subsection, the feature extraction phase provides us with data lying in a very high-dimensional space. In an attempt to make these data more tractable we perform smart sampling by considerably reducing their dimensionality, while retaining the most important information by exploiting the theoretical findings of the CS framework. However, in order to guarantee the correct functionality of CS, we need to know the optimal number of reduced dimensions. It has been theoretically proven that this number is strongly associated with the data sparsity/compressibility [19]. For this purpose, we propose a heuristic approach to estimate the compressibility of the data.

3.2.1 Optimal Number of Reduced Dimensions and Data Compressibility

From the presentation of the core theory, the main contribution of CS is the effective and efficient reduction of the data dimensionality using random projections. As we have seen, random projections consist of calculating random measurements, where each measurement is a linear combination of the initial vector components with random scalar component-coefficients. Intuitively, by using random projections, every measurement is given equal significance, an attribute that in [20] has been characterized as "democratic", in the sense that the signal information is compressed and uniformly distributed to the random measurements. However, due to this democratic attribute of random projections, theoretical bounds on the number m of measurements required have been derived [19]. More specifically, it has been proven that the least number m required to guarantee the perfect reconstruction of k-sparse vectors using the l_0 or l_1 optimization approaches must be on the order of $O(k \cdot log(\frac{n}{k}))$, where n is the initial vector dimensionality and k is the number of non-zero features of the vector.

Apparently, from the above discussion, in order to find the optimal number of reduced dimensions, we need to estimate k. Towards this direction, for a given vector, per each feature we calculate its absolute value, i.e., its distance from zero, and we sort the obtained absolute values in ascending order. Using the sorted values, we calculate their first order differences and we find the index with the maximum first order difference. This index, essentially conveys information about the data compressibility, since it is indicating a large step of the features from smaller to larger absolute values. Hence, the features with indices larger than the above critical index have noticeably larger absolute values than the remaining features. Therefore, they could be considered as non-zero's providing a reasonable estimation of k.

3.2.2 Dimensionality Reduction Using Compressed Sensing

Based on the capability of using random projections, an $m \times n$ ($m \ll n$) transformation matrix (i.e., dictionary) $\mathbf{D}$, whose entries contain random values generated by a Gaussian distribution with mean zero and variance $1/n$ is constructed. The rows of this matrix are subsequently orthonormalized through an SVD process, so that $\mathbf{D}\mathbf{D}^T = \mathbf{I}$. Using the resulting matrix, the initial high-dimensional data are projected onto a much lower dimensional space through a matrix multiplication and in this target space the new data representations are used for classification.

Despite its power, in some practical cases when the number m of the measurements is too high, CS proves inefficient in handling big data, since due to insufficient memory resources it fails to perform SVD and big matrix multiplications. In this chapter, in order to alleviate this shortcoming, we propose a novel method based on a hierarchical implementation of CS as described in the next subsection.

3.2.3 Hierarchical Compressed Sensing

An innovative way to apply CS in big data with computational efficiency, called Hierarchical Compressed Sensing (HCS) is proposed. HCS is an iterative method that reduces the dimensionality of an initial n-dimensional feature vector to half at each iteration. An initialization step, a depth parameter d and an iteration parameter j in the range $1, 2, \ldots, d$ are used during the process.

At the initialization step, given an n-dimensional vector $\mathbf{x}_0 = [x_1, \ldots, x_n]^T$, its components are arbitrarily permuted. To keep the notation as simple as possible, let us consider the above $\mathbf{x}_0$ as the permuted vector. Subsequently, $\mathbf{x}_0$ is subdivided into a set $\{\mathbf{x}_0^i\}_{i=1}^{\omega}$ of ω vector-blocks of size $q = \frac{n}{\omega}$ each, with $\mathbf{x}_0^i = [x_{(i-1)\cdot q+1}, x_{(i-1)\cdot q+2}, \ldots, x_{(i-1)\cdot q+q}]^T$, for $i = 1, 2, \ldots, \omega$. The previous permutation step is used for the sake of generality, since no underlying sparsity structure is known in the vector decomposition phase.

At the first iteration, i.e., $j = 1$, a random transformation matrix $\mathbf{D}_1$ of size $\frac{q}{2} \times q$ is firstly orthonormalized through an SVD step and then used for reducing the data of each block $\mathbf{x}_0^i$ to half its initial dimensionality, producing $\mathbf{x}_1^i$, for $i = 1, 2, \ldots, \omega$, through $\mathbf{x}_1^i = \mathbf{D}_1 \mathbf{x}_0^i$. The process is then iterated to the obtained blocks, for $j = 2, 3, \ldots, d$ using Eq. 14:

$$\mathbf{x}_j^i = \mathbf{D}_j \mathbf{x}_{j-1}^i \in \mathbb{D}^{\frac{q}{2^j}}, \tag{14}$$

where $\mathbf{x}_j^i$ is the ith block obtained after iteration j and $\mathbf{D}_j \in \mathbb{D}^{\frac{q}{2^j} \times \frac{q}{2^{j-1}}}$ is the random transformation matrix corresponding to the jth iteration. Hence, the new blocks at iteration j are of half the dimensionality of the blocks at iteration $j - 1$. The finally reduced vector consists of the blocks obtained at iteration d concatenated and therefore is of $\frac{n}{2^d}$ dimensions. At this point, it should be added that the random matrices at each iteration of HCS are not constructed from scratch, but sampled as submatrices from $\mathbf{D}_1$, avoiding the considerable extra computational effort that would be allocated by consecutively calling a random function.

The functionality of HCS relies upon the assumption of the data sparsity. In fact, when subdividing a sparse vector we expect that, on average, the vector segments are also supposed to inherit the sparsity of the main vector, therefore permitting the use of random projections per each segment in a hierarchical manner. However, it must be emphasized that the disadvantage of HCS is that using a large depth parameter (i.e., significantly reducing the dimensionality), there is no guarantee that at each iteration the data are adequately sparse.

From the above discussion, the computational effort using HCS consists of performing SVD on a random matrix along with a number of matrix multiplications, per each iteration, during the dimensionality reduction process. The computational complexity of performing SVD on a $m \times n$ matrix is on the order of $\mathcal{O}(4m^2n + 8mn^2 + 9n^3)$. Similarly, the computational complexity of multiplying a $m \times n$ matrix by a $n \times 1$ vector is on the order of $\mathcal{O}(mn)$. In order to compare HCS with CS, for the sake of simplicity, let us investigate the case where the number of dimensions in the reduced space is $\frac{n}{2}$ or equivalently, in HCS terms, the depth d is one. Under these circumstances, using CS requires SVD on a $n \times \frac{n}{2}$ random matrix with computational complexity $\mathcal{O}(5n^3)$, and a multiplication between the transpose of the resulting matrix by the initial $n \times 1$ vector, which is $\mathcal{O}(\frac{n^2}{2})$. On the other hand, using HCS with ω blocks, requires SVD calculation of a matrix of size $\frac{n}{\omega} \times \frac{n}{2\omega}$, which is $\mathcal{O}(\frac{1}{\omega^3}5n^3)$, that is $\mathcal{O}(\omega^3)$ times less than using CS. Subsequently, the resulting transformation matrix is multiplied by the ω vector-blocks, which is $\mathcal{O}(\frac{1}{\omega^2}\frac{n^2}{2})$, that is $\mathcal{O}(\omega^2)$ times less than using CS. This result, clearly shows that the computational benefit using HCS is exponentially associated with the number of blocks used.

3.3 Classification and Evaluation

The vectors obtained after the dimensionality reduction methodology described in Sect. 3.2 are used in the classification phase. For classifying the data, we use Support Vector Machines (SVM) [48], because of its popularity and effectiveness. For each concept c_k, $k = 1, 2, \ldots, c$, a binary linear SVM classifier $(\mathbf{w}_k, \mathbf{b}_k)$, where $\mathbf{w}_k$ is the normal vector to the hyperplane and $\mathbf{b}_k$ the bias term, is trained using the labelled training set. The images labelled with c_k are chosen as positive examples, while all the rest are used as negative examples in an one-versus-all fashion. For each test signal $\mathbf{x}_i$, the distance from the hyperplane $\mathcal{V}(\mathbf{x}_i, c_k)$ is extracted by applying the SVM classifier (see Eq. 15).

$$\mathcal{V}(\mathbf{x}_i, c_k) = <\mathbf{w}_k, \mathbf{x}_i> + \mathbf{b}_k . \tag{15}$$

The values $\mathcal{V}(\mathbf{x}_i, c_k)$ obtained from Eq. 15 are used as prediction scores, which indicate the likelihood that a sample $\mathbf{x}_i$ depicts the concept c_k.

For assessing the ability of the presented methods to correctly predict the set of concepts depicted on a test sample, the mean Average Precision (*mAP*) performance measure has been used. Average Precision (*AP*) measures the rank quality of the retrieved images per each concept. For each test image $\mathbf{x}_i$ among a collection of N test samples, using the values $\mathcal{V}(\mathbf{x}_i, c_k)$, we produce a $c \times 1$ score vector. By concatenating these score vectors for all N test samples column-wise, a $c \times N$ score matrix is obtained. For each concept k, the kth row of this matrix is sorted according to the scores in descending order providing a ranked list of samples. This ranked list is used for calculating the Average Precision $AP(c_k)$, per each concept c_k, with the help of Eq. 16:

$$AP(c_k) = \frac{\sum_{i=1}^{N} \left(P_{c_k}(i) \times \delta_{c_k}(i)\right)}{N_{c_k}}, \tag{16}$$

where $P_{c_k}(i)$ is the precision in the first i retrieved items in the ranked list, $\delta_{c_k}(i)$ equals 1 if the ith item is relevant to the c_kth concept and zero otherwise, and N_{c_k} is the total number of the items relevant to the c_kth concept. The mAP is the mean of the average precisions over all concepts (see Eq. 17).

$$mAP = \frac{\sum_{k=1}^{C} AP(c_k)}{C}. \tag{17}$$

3.4 Dependance of Random Projections to Data Sparsity in Classification Problems

A reasonable question that might arise from the above presentation is why are random projections supposed to work in classification problems. One could plausibly ask why not use a random feature selection (RFS) approach instead of calculating random linear combinations of the initial features. Furthermore, another question could be how does the data sparsity affect the performance of random projections. In an attempt to answer these questions, we propose a methodology for comparing the performance of HCS with RFS as a function of the sparsity level of the data. The methodology is based on "sparsifying" the original dataset using different threshold values and investigating the robustness of the two above methods in the resulting artificial datasets. From a practical point of view, sparsifying the data is supposed to deteriorate the classification performance of both methods, as long as it leads to considerable loss of information. However, from this artificial sparse data construction, we expect that a number of important findings regarding the effect of the sparsity level of the data on random projections can be derived. The above presented methodology has been substantiated on a real-world dataset and the results are presented in Sect. 4.

4 Experiments

We conducted a series of experiments in order to investigate the potential of CS as a dimensionality reduction technique in the problem of image annotation in large-scale datasets. More specifically, the goal of the experiments was fivefold: First, to investigate the sensitivity of CS to the specific random values of the dictionary used. Second, to study the relation between the classification accuracy using CS and the number of reduced dimensions. Third, to demonstrate the gain in computational efficiency using HCS instead of CS and in parallel to show that HCS exhibits robustness equivalent to CS. Fourth, to establish a connection between the performance of CS and the level of the data sparsity and to provide credibility to CS as a smart random approach over other naive random schemes. Finally, to compare the performance of HCS with the state-of-the-art dimensionality reduction method of PCA in the image classification problem. All the experiments were run on a 12 core Intel® Xeon (R) CPU ES-2620 v2 @ 2.10 GHz with 128 GB memory.

4.1 Dataset Description

For the experiments we have used the benchmarking dataset of the PASCAL VOC 2012 competition [26]. The dataset consists of 5717 training and 5823 test images collected from flickr. The images are annotated with 20 concepts c_k, for $k = 1, 2, \ldots, 20$ in a multi-label manner (person, bird, cat, cow, dog, horse, sheep, airplane, bicycle, boat, bus, car, motorbike, train, bottle, chair, dining table, potted plant, sofa and TV monitor). Performing the feature extraction procedure presented in Sect. 3.1 on the above dataset, we came up with a set of 327680-dimensional feature vectors. The dimensionality of the obtained feature vectors was subsequently reduced by utilizing CS as well as the novel HCS method.

Motivated by the interesting findings presented in [40], we exhaustively tested a variety of normalization schemes involved in diverse parts of the whole dimensionality reduction procedure, i.e., prior or posterior to the application of CS, in order to check how the accuracy is affected. We examined two specific normalization types, the square root and the l_2 normalization as in [40]. For each configuration, we performed classification and after evaluating the accuracy results, we concluded that the optimal configuration is square-rooting followed by l_2 normalization followed by CS. Actually, applying any normalization scheme after CS deteriorates the accuracy performance of SVM. So, the above configuration was set and used throughout the remaining experiments.

4.2 Sensitivity of Compressed Sensing to Randomness

For the remainder of our experimental study, as long as different random matrices were involved, we needed to ensure that it is not the specific set of random values used for producing the random matrices that makes it work, but rather the nature of the randomness per se. Although the above claim constitutes a theoretical finding regarding representation problems, the purpose of this section is to provide the analogous guarantee that this finding is valid in terms of classification, too. Towards this direction, we performed an experiment for investigating the variation of the classification accuracy as this is derived by using a number of different matrices. More specifically, we ran 100 sessions using different random values in the transformation matrix. These values were generated by a Gaussian distribution with mean zero and standard deviation $1/n$, where n is the data dimensionality (i.e., $n = 327680$). At each session, we fixed the number of measurements m to 2000 and after reducing the initial data to m dimensions, we subsequently performed classification and calculated the corresponding *mAP*. The mean *mAP* across the series of 100 sessions was 0.4321 and the standard deviation was 0.0034. This standard deviation is quite small verifying the above claim about the insensitivity to the specific set of random values used in the transformation matrix.

4.3 Compressibility Estimation and Optimal Dimensionality Investigation

For estimating the compressibility of the data, we applied the methodology described in Sect. 3.2.1 to each specific sample of the dataset. For visualization purposes, we provide an example of the process applied to an arbitrary sample from the dataset. The sorted distances of the vector features from zero are plotted in Fig. 1. From this figure, it is clear that the majority of the distances are close to zero, while only a small portion of them considerably differs. The first order differences of these distances are illustrated in Fig. 2. The maximum value (peak) of the latter figure is realized at 324,480. According to the proposed methodology, the number k of non-zero

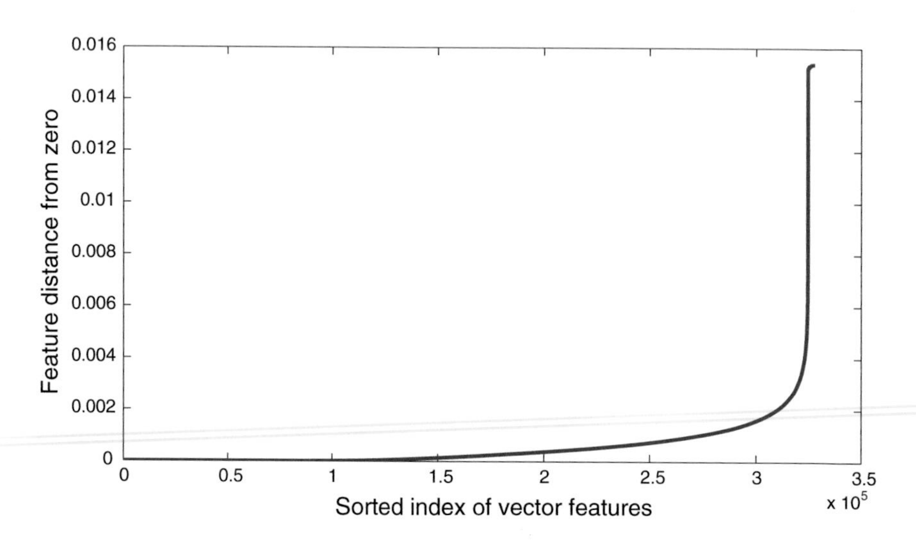

Fig. 1 Feature distances from zero sorted in ascending order

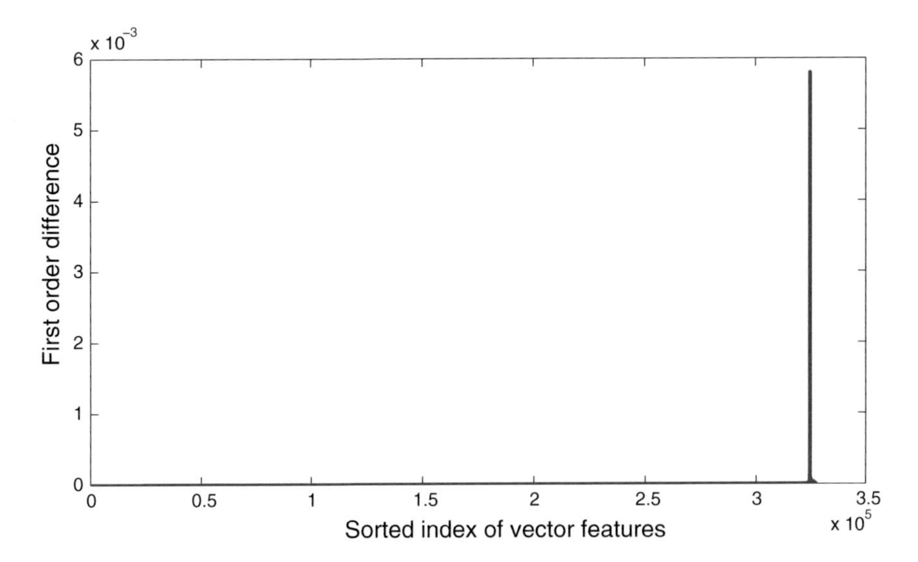

Fig. 2 First order differences of feature distances from zero

features of this particular sample can be approximated by counting the number of feature indices lying on the right side of this peak, which in this case is 3200. Adopting the above methodology to all samples of the dataset we achieved estimating k for every sample. As a k-value representative of the whole dataset we set 3220, which is the maximum k across the samples. As a consequence, on average, only around 1 % of the vector features have non-zero values, hence the data could be considered as compressible and eligible for the application of CS. In addition, recalling from Sect. 3.2.1 that the number m of measurements must be on the order of $k \cdot log(\frac{n}{k})$, in our case this value is approximately 6464, which means that theoretically, at least on the order of 6464 dimensions are required in order to guarantee the stable performance of CS.

Our next concern was to investigate how the classification accuracy varies as a function of the number m of measurements. For this purpose, a series of consecutive experiments was carried out, where at each iteration we varied m in the range from 1 to 20,000, while maintaining the remaining settings unchanged. In this context, using different integers for m, we constructed random matrices of different sizes. Using these matrices we subsequently projected the initial high-dimensional data onto the corresponding m-dimensional space.

The *mAP* results obtained using the SVM classifier are illustrated in Fig. 3. The reduced number of dimensions is depicted in the horizontal axis, while the *mAP* is depicted in the vertical axis. The *mAP* obtained as a function of the number of dimensions using CS is depicted with the solid blue curve. For comparison reasons, the baseline *mAP* using directly SVM on the initial data, with no dimensionality reduction is also depicted with the dashed red line. A couple of important remarks could be drawn from Fig. 3. First, it is clear that the *mAP* is ever increasing as a function

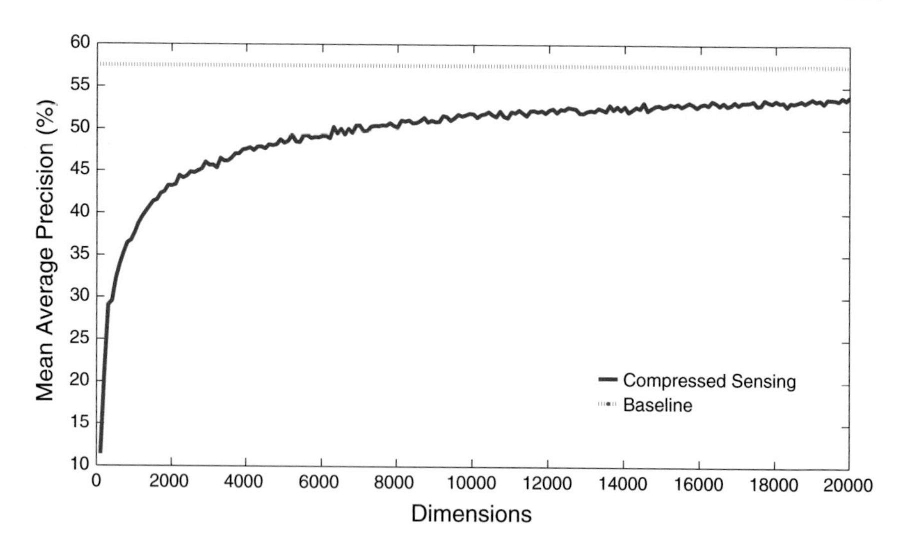

Fig. 3 Mean average precision using compressed sensing in a range of different dimensions

of the number of the reduced dimensions and asymptotically converges to a certain value around 55 %. Moreover, a closer inspection of Fig. 3 interestingly shows that although the *mAP* sharply increases in low dimensions, the growth rate becomes smoother in the neighborhood of the theoretically required number of dimensions (i.e., 6464), which can be associated with the above compressibility estimation analysis.

4.4 *Investigating the Robustness of Hierarchical Compressed Sensing*

From the presentation of Sect. 3.2.3, although it has been theoretically shown that increasing ω benefits the HCS process, however the question how can ω affect the classification performance using HCS is still open. In an attempt to answer this question, we set the depth $d = 1$ and varied the number of blocks ω in the range $1, 2, 2^2, \ldots, 2^9$. For each setting, we counted the time elapsed during the dimensionality reduction process and we calculated the classification performance using HCS. The results are collectively illustrated in Fig. 4. The horizontal axis depicts the number of blocks used. The left vertical axis depicts the *mAP*, while the right vertical axis depicts the computational time required for reducing the dimensionality of the training data using HCS. The latter includes the time required for both the SVD calculation and the matrix multiplication processes.

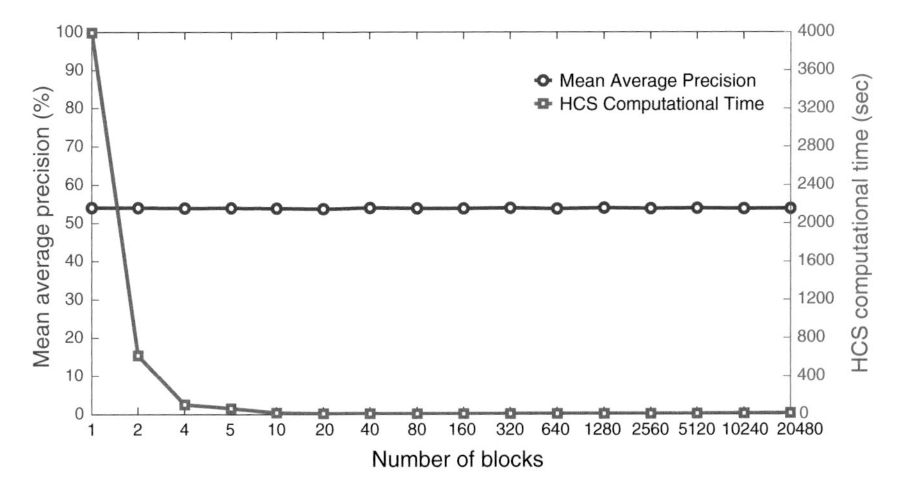

Fig. 4 Mean average precision and training computational time versus number of blocks used in Hierarchical Compressed Sensing

Observing the "HCS Computational Time" curve in Fig. 4, it is clear that as the number of blocks increases, the computational complexity decreases exponentially. Furthermore, interestingly, the robustness of HCS using different numbers of blocks is evident (see Fig. 4, "Mean Average Precision" curve). These results can be associated with the theoretical findings, verifying the analysis presented in Sect. 3.2.3. In addition, Fig. 4 provides an implicit comparison between HCS and CS. More specifically, when using one block (i.e., $\omega = 1$), HCS actually collapses to CS. It is clear that HCS displays classification performance equal to CS, which combined with the computational complexity advantage of HCS shows the superiority of HCS over CS.

4.5 Dependance of Random Projections to Data Sparsity

In this section, we experimentally investigate the relation between the data sparsity and the performance of CS-like methods. Since CS and HCS display equivalent performance, for this experiment we used only the latter. Moreover, for computational simplicity, for each sample, we used only the spatial pyramid corresponding to the whole image out of the eight pyramids obtained through the feature extraction process. Following the methodology presented in Sect. 3.4, we used six different threshold values leading to six corresponding sparsity levels of the data. Figure 5 depicts the average number of non-zero features of the data per each sparsity level.

For each specific sparsity level and dimensionality we calculated the difference between the *mAP*'s obtained by using HCS and RFS and we estimated the percentage gain in classification performance obtained by HCS over RFS. The results are

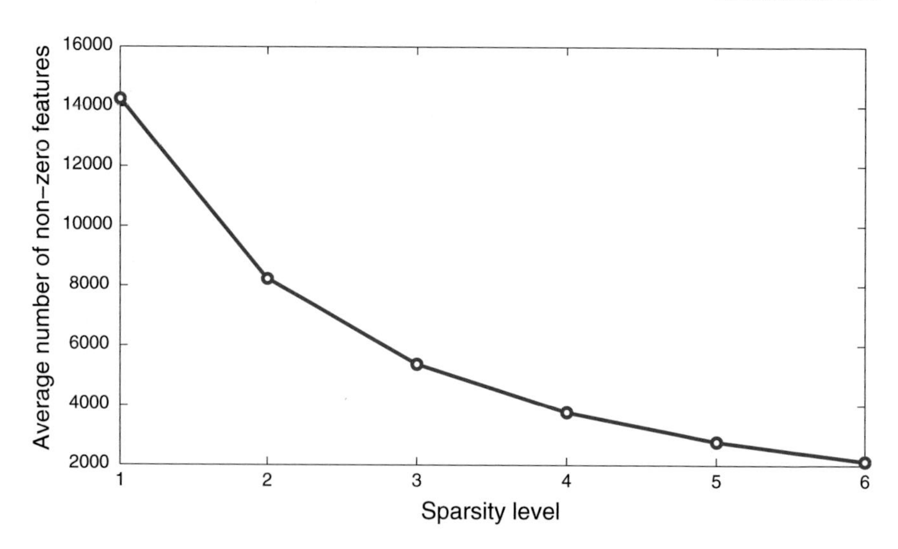

Fig. 5 Average number of non-zero's per each sparsity level

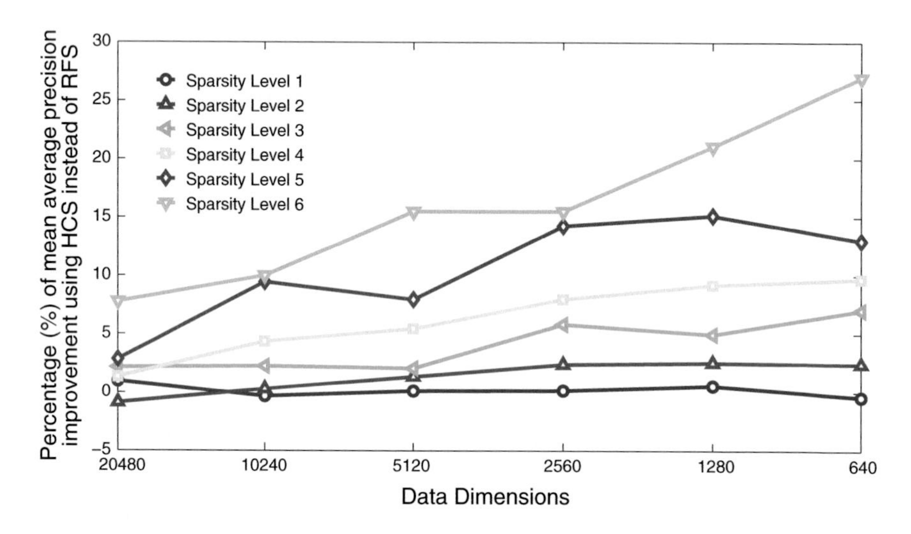

Fig. 6 Percentage (%) of mean average precision improvement using HCS instead of RFS

illustrated in Fig. 6. From Fig. 6, it is evident that as the sparsity level increases, the percentage of *mAP* improvement using HCS instead of RFS increases, too. Moreover, independently of the sparsity level, this percentage increases (up to $\simeq$27 %) as the dimensionality decays highlighting the robustness of HCS in contrast to RFS.

Intuitively, at a first glance, seemingly there is nothing special about random projections (e.g., HCS) against RFS, due to the random nature of both. However, random projections clearly take into account all the initial data features, while in contrast, selecting a number of specific features inevitably avoids the rest leading to considerable loss of information. This advantage in conjunction with its democratic nature, provide credibility to HCS as a smart dimensionality reduction method over other naive random feature selection schemes under the data sparsity assumption.

4.6 Comparison with Principal Component Analysis

In this section, we compare the performance of HCS with the PCA method [32]. For computational reasons, that will be explained later in this section, in this experiment we used only the spatial pyramid corresponding to the whole image. That is, the dimensionality of the data was 40,960. The *mAP* results are plotted in Fig. 7. From Fig. 7, we observe that although in low dimensions, the superiority of PCA over HCS is evident, by increasing the number of dimensions, HCS exhibits classification performance competitive to PCA. The deterioration in performance of HCS when reducing the number of dimensions can be attributed to the democratic nature of the CS measurements, which postulates a representative number of dimensions in order to maintain the important information. On the contrary, PCA shows impressive robustness in low dimensions, which is well justified by the fact that PCA by definition attempts to encode the data information into the least possible eigenvectors.

The advantage of PCA in terms of robustness comes at a cost of excessive computational and memory requirements. More specifically, PCA requires the computation and eigen-analysis of the data covariance matrix, which is of size n^2, where n is the dimensionality of the original data. This is a considerable computational complexity, which in some cases prohibits the application of PCA at all. On the other hand, CS requires the construction of an $m \times n$ random matrix, where m is much smaller

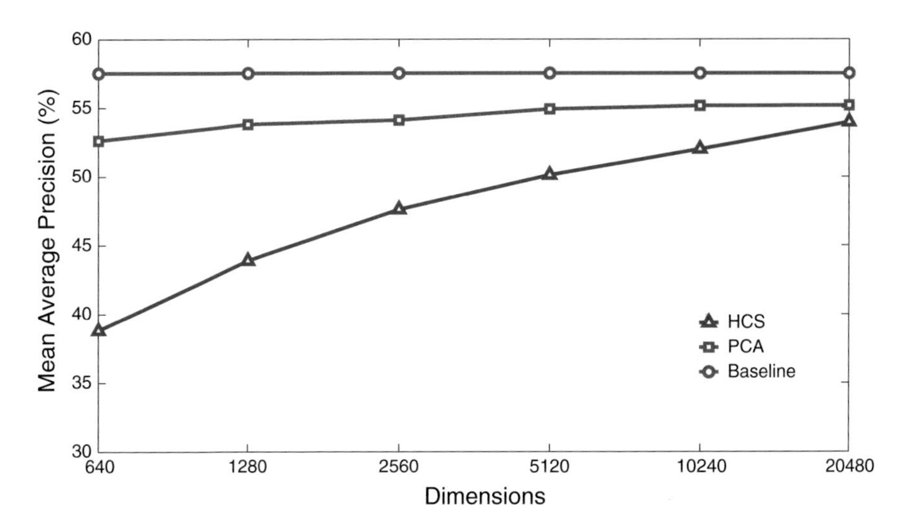

Fig. 7 Comparison between HCS and PCA

than n, and the SVD of this matrix during the basis orthonormalization step. Furthermore, using HCS, the SVD is decomposed into a number of smaller SVD steps, leading to a computational load orders of magnitude less than the corresponding of PCA. In summary, despite the weak performance of HCS in low dimensions, it must be emphasized that the profit in terms of computational complexity using HCS may prove of utmost importance in real-world problems. As a matter of fact, it is worth mentioning that in this particular experiment, it was infeasible to apply PCA on the data consisting of all eight spatial pyramids. For this reason, as stated above, we used only the one out of the eight spatial pyramids. This limitation accentuates the disadvantage of PCA in terms of computational complexity.

5 Related Work

In this section, a brief review of the bibliography related to our work is presented. The review is divided into two subsections. The first subsection includes related works on dimensionality reduction and data compression, while the second subsection presents some interesting applications of CS in pattern analysis.

5.1 Dimensionality Reduction and Data Compression

A plethora of methodologies have been proposed for reducing the dimensionality of big data. In [45], dimensionality reduction methods have been classified into three main categories. The first category consists of methods based on statistics and information theory and among others includes vector quantization methods and Principal Component Analysis [32]. In the vector quantization context, recently, effort has been allocated on the optimization of the kernel K-Means. For instance, a clustering algorithm, which models multiple information sources as kernel matrices is proposed in [57].

The second category includes methods based on Dictionaries, where a vector is represented as a linear combination of the dictionary atoms. For instance, an overcomplete dictionary design method that combines both the representation and the class discrimination power is proposed in [58] for face recognition. The presented method is an extension of the well-known K-SVD algorithm [1] and is referred to as the D-KSVD, since a discriminative term is added into the main objective function of the regular K-SVD.

The third category consists of methods that seek for "interesting" projections leading to learning transformation matrices. For instance, within this category fall methods based on random projections. Moreover, advances have been made in manifold learning through the development of adaptive techniques that address the selection of the neighborhood size as well as the local geometric structure of the manifold

[59]. Finally, Linear Discriminant Projections (LDP) have been proposed as a useful tool for large-scale image recognition and retrieval [8].

The notion of sparsity seems to play a key role in recent advances in the computer science community. In [7], the authors learn an over-complete dictionary using the K-SVD algorithm along with a set of predefined image patches. Using this dictionary they compress facial images by representing them as sparse linear combinations of the dictionary atoms. The potential of fusing index compression with binary bag of features representation has been investigated in [29]. The proposed method is based on an inverted file structure, which is merely concerned with the distances of the non-zero positions of sparse vectors. A "sparsified" version of Principal Component Analysis has also been proposed [60]. Using elastic net methods, the principal components are represented as sparse linear combinations of the original variables, as opposed to the typical PCA algorithm. The discriminative nature of sparse representations to perform classification has been experimentally proven in [55]. Solving a dedicated linear programming sparsity-encoding problem, the authors propose a face recognition algorithm robust to expressions, illumination and occlusions.

Product quantization techniques that decompose a vector space as a Cartesian product of quantized subspaces have been proposed for constructing short codes representing high-dimensional vectors [30]. In this vein, transform coding and product quantization have been used for approximating nearest neighbor search in large-scale retrieval, content similarity, feature matching and scene classification [6]. Similarly, in [31] using the so-called Vector of Locally Aggregated Descriptors (VLAD), the authors use product quantization for performing nearest neighbor search during the indexing process for searching the most similar images in a very large image database.

Hashing has also been proven a computationally attractive technique, which allows one to efficiently approximate kernels for very high-dimensional settings by means of a sparse projection into a lower dimensional space. For instance, in [44] hashing has been implemented for handling thousands of classes on large amounts of data and features. In the same fashion, specialized hash functions with unbiased inner-products that are directly applicable to a large variety of kernel methods have also been introduced. Exponential tail bounds that help explain why hash feature vectors have repeatedly led to strong empirical results are provided in [51]. The authors demonstrate that the interference between independently hashed subspaces is negligible with high probability, which allows large-scale multi-task learning in a very compressed space.

Considerable effort has been allocated to associate the Hamming distance between codewords with semantic similarity leading to hashing techniques. Towards this end, a spectral graph based method, which uses the Laplacian eigenvectors is proposed in [52]. Similarly, an unsupervised random projections-based method with computational requirements equal to spectral hashing is proposed in [41]. The Hamming distance between the binary codes of two vectors are related to the value of a shift-invariant kernel between the vectors. Along the same lines, in [46] the authors learn compact binary codes by employing boosting techniques in conjunction with locality sensitive hashing and Restricted Boltzmann Machines.

5.2 Pattern Analysis Using Compressed Sensing

Recently, considerable effort has been allocated in bridging the semantic gap, i.e., to assign semantic information to sparse representations, in the fields of computer vision and pattern recognition [54]. It has been shown through a variety of experimental settings that sparse representations convey important semantic information. This assertion has been corroborated within diverse challenging application domains such as face recognition, image reconstruction, image classification, etc.

In this context, semantic issues have been tackled using a sparse image reconstruction based approach in the challenging object recognition problem [50]. The images are encoded as super-vectors consisting of patch-specific Mixture Gaussian Models and reconstructed as linear combinations of a number of training images containing the objects in question. The reconstruction coefficients are inherited by the corresponding labels providing annotation to the query image. Towards the same direction, the problem of accelerating sparse coding based scalable image annotation has also been addressed [27]. The pursuit of an accurate solution is based on an iterative bilevel method, which achieves reducing the large-scale sparse coding problem to a series of smaller sub-problems.

The contribution of sparse representations, CS and Dictionary learning to the object recognition problem has been reviewed in [39]. It is emphasized that recent studies have shown that the specific choice of features is no longer crucial rather than the dimensionality and the sparsity of the features. Combining these methods with the appropriate discriminant criteria may lead to performance superior to the performance of the state of the art discriminative algorithms in classification problems.

Unlike most feature extraction algorithms that have been proposed for dimensionality reduction, in [56] the authors propose a CS based method, which combines sparse representation with feature selection. An objective function involving a matrix that weights the features according to their discriminant power is minimized in a sparse reconstruction fashion by adopting an OMP based algorithm. In the same work, a two-stage hierarchical architecture using directional and adaptive filters for feature detection has been implemented. The proposed methodology has been applied and evaluated in the pedestrian detection problem proving its efficiency in real-world problems.

Sparsity, Reconstruction error and Discrimination power are combined in a common objective function in [28]. A hybrid approach that combines the discrimination power of discriminative methods with the reconstruction capacity of sparse representation has been proposed in an attempt to handle corrupted signals. However, there is a trade-off between the two aforementioned aspects, which stems from the potential noise, the completeness of the data, the number of outliers, etc., that must be effectively handled. Even in the case where the data are noise-free and the discriminative methods are more robust, it has been experimentally shown that by combining the two methods superior performance is obtained.

Compressed Sensing has also been utilized for recovering the sparse foreground of a scene as well as the silhouettes of the foreground objects. In [14], based on the CS representation of the background image, the authors propose a method devoid of any foreground reconstruction for performing background subtraction and object detection, using convex optimization and greedy methods. The authors also recover the silhouettes of foreground objects by learning a low-dimensional CS representation of the background image, robust against background variations.

Sparse representations with over-complete dictionaries have also been applied on image de-noising [25]. Utilizing the K-SVD technique, the authors train dictionaries based on a set of high-quality image patches or based on the patches of the noisy image itself. The image is iteratively de-noised through a sparse coding and a dictionary update stage. Based on a similar approach, K-SVD has been utilized for the restoration of images and video [35]. In this work, the sparse representations are obtained via a multi-scale dictionary learned using an example based approach. Finally, the above de-noising algorithm has also been extended for color image restoration, demosaicing and inpainting [34].

From the above presentation, CS has already been utilized in a variety of domains pertaining to semantic pattern analysis. In our work though, we investigate the connection between the CS performance as a dimensionality reduction technique and a number of factors, including the data sparsity, the number of dimensions and the randomness of the projections in the domain of image annotation. The results obtained are quite interesting advocating the potential application of our methodology to other domains as well.

6 Conclusions

Concluding this chapter, we are now at the position to answer some of the questions posed in the Introduction. Although CS has been initially proposed as a powerful representational tool, in this chapter, it has been given credibility in the image classification problem, too. In our study, it has been shown that undoubtedly CS offers a way to increase the quality of samples obtained by naive random feature selection schemes. The power of CS relies upon its democratic attribute in handling the data features, while random feature selection schemes completely avoid some features leading to considerable loss of information.

Several normalization approaches, proposed in the literature, have been investigated showing that no normalization scheme should be involved in the classification process posterior to the application of CS. Furthermore, the effectiveness of CS has been proven independent of the specific set of random values used for constructing the dictionary and instead has been attributed to the nature of the randomness per se. The connection between the theoretical and practical boundaries on the optimal number of reduced dimensions has also been established. The CS performance has been proven strongly dependent to the data sparsity verifying the corresponding theoretical findings. More specifically, it has been verified that the sparser the data are, the more robust the performance of CS becomes.

The main contribution of this chapter is the novel hierarchical parametric HCS method for efficiently and effectively implementing CS in big data. The proposed method dramatically decreases the computational complexity of CS, while displays robustness equivalent to the typical CS. Through a comparison with PCA, it has been shown that HCS displays performance competitive to PCA provided that an appropriate number of dimensions is kept. This fact in conjunction with the superiority of HCS over PCA in terms of computational complexity, provide credibility to HCS as a smart sampling method in the domain of image classification.

This chapter comprises a preliminary study in smart sampling of big data using CS-like methods. The early results obtained are very encouraging proving the potential of this kind of methods in the image annotation case study. In the future, we intend to extend this study in more datasets. Moreover, apart from the quantitative estimation of the data sparsity, a qualitative investigation of the nature of sparsity and its impact on the effectiveness of CS is encompassed in our future plans. In this context, based on the interesting works presented in [4] and [23], we envisage that the sparsity structure also plays some crucial role and should be equally considered as a factor affecting CS in classification problems. Potential findings in this direction could also be exploited by the proposed HCS method. For instance, in the vector decomposition phase, there is plenty of space for investigating more sophisticated ways to divide a vector, based on sparsity patterns contained in the vector rather than using arbitrary vector-components per each segment.

Acknowledgments This work was supported by the European Commission Seventh Framework Programme under Grant Agreement Number FP7-601138 PERICLES.

References

1. Aharon, M., Elad, M., Bruckstein, A.: SVD: an algorithm for designing overcomplete dictionaries for sparse representation. IEEE Trans. Signal Process. **54**(11), 4311–4322 (2006)
2. Bacardit, J., Llorà, X.: Large-scale data mining using genetics-based machine learning. Wiley Interdiscip. Rev.: Data Mining Knowl. Discov. **3**(1), 37–61 (2013)
3. Baraniuk, R.: Compressive sensing. IEEE Signal Process. Mag. **24**(4) (2007)
4. Baraniuk, R.G., Cevher, V., Duarte, M.F., Hegde, C.: Model-based compressive sensing. IEEE Trans. Inf. Theory **56**(4), 1982–2001 (2010)
5. Bay, H., Ess, A., Tuytelaars, T., Van Gool, L.: Speeded-up robust features (SURF). Comput. Vis. Image Underst. **110**(3), 346–359 (2008)
6. Brandt, J.: Transform coding for fast approximate nearest neighbor search in high dimensions. In: 2010 IEEE Conference on Computer Vision and Pattern Recognition (CVPR), pp. 1815–1822. IEEE (2010)
7. Bryt, O., Elad, M.: Compression of facial images using the K-SVD algorithm. J. Vis. Commun. Image Represent. **19**(4), 270–282 (2008)
8. Cai, H., Mikolajczyk, K., Matas, J.: Learning linear discriminant projections for dimensionality reduction of image descriptors. IEEE Trans. Pattern Anal. Mach. Intell. **33**(2), 338–352 (2011)
9. Candes, E.J., Romberg, J.: Quantitative robust uncertainty principles and optimally sparse decompositions. Found. Comput. Math. **6**(2), 227–254 (2006)
10. Candès, E.J., Romberg, J., Tao, T.: Robust uncertainty principles: Exact signal reconstruction from highly incomplete frequency information. IEEE Trans. Inf. Theory **52**(2), 489–509 (2006)
11. Candes, E.J., Tao, T.: Decoding by linear programming. IEEE Trans. Inf. Theory **51**(12), 4203–4215 (2005)

12. Candès, E.J., Wakin, M.B.: An introduction to compressive sampling. IEEE Signal Process. Mag. **25**(2), 21–30 (2008)
13. Candès, E.J., et al.: Compressive sampling. In: Proceedings of the International Congress of Mathematicians, vol. 3, pp. 1433–1452. Madrid, Spain (2006)
14. Cevher, V., Sankaranarayanan, A., Duarte, M.F., Reddy, D., Baraniuk, R.G., Chellappa, R.: Compressive sensing for background subtraction. In: Computer Vision-ECCV 2008, pp. 155–168. Springer (2008)
15. Chatfield, K., Lempitsky, V., Vedaldi, A., Zisserman, A.: The Devil is in the Details: An Evaluation of Recent Feature Encoding Methods (2011)
16. Chawla, N.V., Hall, L.O., Bowyer, K.W., Kegelmeyer, W.P.: Learning ensembles from bites: a scalable and accurate approach. J. Mach. Learn. Res. **5**, 421–451 (2004)
17. Chen, S., Donoho, D.: Basis pursuit. In: 1994 Conference Record of the Twenty-Eighth Asilomar Conference on Signals, Systems and Computers, 1994, vol. 1, pp. 41–44. IEEE (1994)
18. Dantzig, G.B.: Linear Programming and Extensions. Princeton University Press (1998)
19. Davenport, M.A., Duarte, M.F., Eldar, Y.C., Kutyniok, G.: Introduction to Compressed Sensing. Preprint 93 (2011)
20. Davenport, M.A., Laska, J.N., Boufounos, P.T., Baraniuk, R.G.: A simple proof that random matrices are democratic. arXiv:0911.0736 (2009)
21. Dean, J., Ghemawat, S.: Mapreduce: simplified data processing on large clusters. Commun. ACM **51**(1), 107–113 (2008)
22. Donoho, D.L.: Compressed sensing. IEEE Trans. Inf. Theory **52**(4), 1289–1306 (2006)
23. Duarte, M.F., Eldar, Y.C.: Structured compressed sensing: from theory to applications. IEEE Trans. Signal Process. **59**(9), 4053–4085 (2011)
24. Elad, M.: Sparse and Redundant Representations: From Theory to Applications in Signal and Image Processing. Springer (2010)
25. Elad, M., Aharon, M.: Image denoising via sparse and redundant representations over learned dictionaries. IEEE Trans. Image Process. **15**(12), 3736–3745 (2006)
26. Everingham, M., Van Gool, L., Williams, C., Winn, J., Zisserman, A.: The Pascal Visual Object Classes Challenge 2012 (2012)
27. Huang, J., Liu, H., Shen, J., Yan, S.: Towards efficient sparse coding for scalable image annotation. In: Proceedings of the 21st ACM international conference on Multimedia, pp. 947–956. ACM (2013)
28. Huang, K., Aviyente, S.: Sparse representation for signal classification. In: NIPS, pp. 609–616 (2006)
29. Jégou, H., Douze, M., Schmid, C.: Packing bag-of-features. In: 2009 IEEE 12th International Conference on Computer Vision, pp. 2357–2364. IEEE (2009)
30. Jegou, H., Douze, M., Schmid, C.: Product quantization for nearest neighbor search. IEEE Trans. Pattern Anal. Machine Intell. **33**(1), 117–128 (2011)
31. Jégou, H., Douze, M., Schmid, C., Pérez, P.: Aggregating local descriptors into a compact image representation. In: 2010 IEEE Conference on Computer Vision and Pattern Recognition (CVPR), pp. 3304–3311. IEEE (2010)
32. Jolliffe, I.: Principal Component Analysis. Wiley Online Library (2005)
33. Lowe, D.G.: Object recognition from local scale-invariant features. In: The Proceedings of the Seventh IEEE International Conference on Computer vision, 1999, vol. 2, pp. 1150–1157. IEEE (1999)
34. Mairal, J., Elad, M., Sapiro, G.: Sparse representation for color image restoration. IEEE Trans. Image Process. **17**(1), 53–69 (2008)
35. Mairal, J., Sapiro, G., Elad, M.: Learning multiscale sparse representations for image and video restoration. Technical report, DTIC Document (2007)
36. Manjunath, B.S., Ohm, J.R., Vasudevan, V.V., Yamada, A.: Color and texture descriptors. IEEE Trans. Circ. Syst. Video Technol. **11**(6), 703–715 (2001)
37. Mohri, M., Talwalkar, A.: Can matrix coherence be efficiently and accurately estimated? In: International Conference on Artificial Intelligence and Statistics, pp. 534–542 (2011)

38. Nyquist, H.: Certain topics in telegraph transmission theory. Trans. Am. Inst. Electr. Eng. **47**(2), 617–644 (1928)
39. Patel, V.M., Chellappa, R.: Sparse representations, compressive sensing and dictionaries for pattern recognition. In: 2011 First Asian Conference on Pattern Recognition (ACPR), pp. 325–329. IEEE (2011)
40. Perronnin, F., Sánchez, J., Mensink, T.: Improving the Fisher kernel for large-scale image classification. In: Computer Vision-ECCV 2010, pp. 143–156. Springer (2010)
41. Raginsky, M., Lazebnik, S.: Locality-sensitive binary codes from shift-invariant kernels. In: Advances in Neural Information Processing Systems, pp. 1509–1517 (2009)
42. Sánchez, J., Perronnin, F.: High-dimensional signature compression for large-scale image classification. In: 2011 IEEE Conference on Computer Vision and Pattern Recognition (CVPR), pp. 1665–1672. IEEE (2011)
43. Shannon, C.E.: Communication in the presence of noise. Proc. IRE **37**(1), 10–21 (1949)
44. Shi, Q., Petterson, J., Dror, G., Langford, J., Strehl, A.L., Smola, A.J., Vishwanathan, S.: Hash kernels. In: International Conference on Artificial Intelligence and Statistics, pp. 496–503 (2009)
45. Sorzano, C.O.S., Vargas, J., Montano, A.P.: A survey of dimensionality reduction techniques. arXiv:1403.2877 (2014)
46. Torralba, A., Fergus, R., Weiss, Y.: Small codes and large image databases for recognition. In: IEEE Conference on Computer Vision and Pattern Recognition, 2008. CVPR 2008, pp. 1–8. IEEE (2008)
47. Tropp, J.A., Gilbert, A.C.: Signal recovery from random measurements via orthogonal matching pursuit. IEEE Trans. Inf. Theory **53**(12), 4655–4666 (2007)
48. Vapnik, V.: The Nature of Statistical Learning Theory. Springer (2000)
49. Vedaldi, A., Fulkerson, B.: Vlfeat: An open and portable library of computer vision algorithms. In: Proceedings of the International Conference on Multimedia, pp. 1469–1472. ACM (2010)
50. Wang, C., Yan, S., Zhang, L., Zhang, H.J.: Multi-label sparse coding for automatic image annotation. In: IEEE Conference on Computer Vision and Pattern Recognition, 2009. CVPR 2009, pp. 1643–1650. IEEE (2009)
51. Weinberger, K., Dasgupta, A., Langford, J., Smola, A., Attenberg, J.: Feature hashing for large scale multitask learning. In: Proceedings of the 26th Annual International Conference on Machine Learning, pp. 1113–1120. ACM (2009)
52. Weiss, Y., Torralba, A., Fergus, R.: Spectral hashing. In: Advances in Neural Information Processing Systems, pp. 1753–1760 (2009)
53. Willett, R.M., Marcia, R.F., Nichols, J.M.: Compressed sensing for practical optical imaging systems: a tutorial. Opt. Eng. **50**(7), 072,601–072,601 (2011)
54. Wright, J., Ma, Y., Mairal, J., Sapiro, G., Huang, T.S., Yan, S.: Sparse representation for computer vision and pattern recognition. Proc. IEEE **98**(6), 1031–1044 (2010)
55. Wright, J., Yang, A.Y., Ganesh, A., Sastry, S.S., Ma, Y.: Robust face recognition via sparse representation. IEEE Trans. Pattern Anal. Mach. Intell. **31**(2), 210–227 (2009)
56. Yang, J., Bouzerdoum, A., Tivive, F.H.C., Phung, S.L.: Dimensionality reduction using compressed sensing and its application to a large-scale visual recognition task. In: The 2010 International Joint Conference on Neural Networks (IJCNN), pp. 1–8. IEEE (2010)
57. Yu, S., Tranchevent, L.C., Liu, X., Glanzel, W., Suykens, J.A., De Moor, B., Moreau, Y.: Optimized data fusion for kernel k-means clustering. IEEE Trans. Pattern Anal. Mach. Intell. **34**(5), 1031–1039 (2012)
58. Zhang, Q., Li, B.: Discriminative k-svd for dictionary learning in face recognition. In: 2010 IEEE Conference on Computer Vision and Pattern Recognition (CVPR), pp. 2691–2698. IEEE (2010)
59. Zhang, Z., Wang, J., Zha, H.: Adaptive manifold learning. IEEE Trans. Pattern Anal. Mach. Intell. **34**(2), 253–265 (2012)
60. Zou, H., Hastie, T., Tibshirani, R.: Sparse principal component analysis. J. Comput. Graph. Stat. **15**(2), 265–286 (2006)

Author Biographies

Mr. Anastasios Maronidis (M) received his diploma degree in Mathematics in 2008 and his Master degree on Digital Media in Computer Science in 2010, both from the Aristotle University of Thessaloniki (AUTH), Greece. He is currently a research associate in Information Technologies Institute (ITI) at the Centre for Research and Technology Hellas (CERTH). He has participated in several EU funded ICT research projects (e.g., MOBISERV, i3Dpost, RESTORE, Live+Gov, PERICLES) and his research interests include digital signal and image processing, pattern recognition, machine learning and computer vision. His scientific work has been published in peer-reviewed journals, international conferences and book chapters.

Dr. Elisavet Chatzilari (F) received her diploma degree in Electronics and Computer Engineering from Aristotle university of Thessaloniki, Greece in 2008. She also holds a Ph.D. degree on social media based scalable concept detection, University of Surrey (2014). She is currently a post-doctoral research fellow in Information Technologies Institute (ITI) at the Centre for Research & Technology Hellas (CERTH). She has participated in a number of EC-funded ICT projects (e.g., We-knowIT, X-Media, GLOCAL, CHORUS+, SocialSensor, Live+Gov, i-Treasures) and her research interests include image processing, natural language processing, multimodal analysis using social media and machine learning (supervised learning, active learning, cross domain learning). Her scientific work has been published in peer-reviewed journals, international conferences and book chapters.

Dr. Spiros Nikolopoulos (M) received his diploma degree in Computer Engineering and Informatics and the MSc degree in Computer Science & Technology from university of Patras, Greece in 2002 and 2004 respectively. He also holds a Ph.D. degree on Semantic multimedia analysis using knowledge and context, Queen Mary University of London (2012). He is currently a post-doctoral research fellow in Information Technologies Institute (ITI) at the Centre for Research & Technology Hellas (CERTH). He has participated in a number of EC-funded ICT projects (e.g., MinervaPlus, ECRYPT, X-Media, GLOCAL, CHORUS+, Live+Gov, i-Treasures, Pericles) and his research interests include image analysis, indexing and retrieval, multimodal analysis, multimedia analysis using semantic information and integration of context and content for multimedia interpretation. His scientific work has been published in peer-reviewed journals, international conferences and book chapters.

Dr. Ioannis Kompatsiaris (M) received the Diploma degree in electrical engineering and the Ph.D. degree in 3-D model based image sequence coding from Aristotle University of Thessaloniki in 1996 and 2001, respectively. He is a Senior Researcher (Researcher A') with CERTH-ITI and head of the Multimedia Knowledge & Social Media Analytics lab. His research interests include semantic multimedia analysis, indexing and retrieval, social media and big data analysis, knowledge structures, reasoning and personalization for multimedia applications, eHealth and environmental applications. He is the co-author of 83 papers in refereed journals, 35 book chapters, 8 patents and more than 250 papers in international conferences. Since 2001, Dr. Kompatsiaris has participated in 43 National and European research programs, in 11 of which he has been the Project Coordinator and in 29 the Principal Investigator. He has been the co-organizer of various international conferences and workshops and has served as a regular reviewer for a number of journals and conferences. He is a Senior Member of IEEE and member of ACM.

Optimized Management of BIG Data Produced in Brain Disorder Rehabilitation

Peter Brezany, Olga Štěpánková, Markéta Janatová, Miroslav Uller and Marek Lenart

Abstract Brain disorders resulting from injury, disease, or health conditions can influence function of most parts of human body. Necessary medical care and rehabilitation is often impossible without close cooperation of several diverse medical specialists who must work jointly to choose methods that improve and support healing processes as well as to discover underlying principles. The key to their decisions are data resulting from careful observation or examination of the patient. We introduce the concept of scientific dataspace that involves and stores numerous and often complex types of data, e.g., the primary data captured from the application, data derived by curation and analytic processes, background data including ontology and workflow specifications, semantic relationships between dataspace items based on ontologies, and available published data. Our contribution applies big data and cloud technologies to ensure efficient exploitation of this dataspace, namely, novel software architectures, algorithms and methodology for its optimized management and utilization. We present its service-oriented architecture using a running case study and results of its data processing that involves mining and visualization of selected patterns optimized towards big and complex data we are dealing with.

Keywords Optimized dataspace management · Data mining · Brain rehabilitation

---

P. Brezany (✉) · M. Lenart
Faculty of Computer Science, University of Vienna, Vienna, Austria
e-mail: Peter.Brezany@univie.ac.at; brezanyp@feec.vutbr.cz

M. Lenart
e-mail: Marek.Lenart@univie.ac.at

P. Brezany
SIX Research Center, Brno University of Technology, Brno, Czech Republic

O. Štěpánková · M. Uller
Katedra Kybernetiky FEL CVUT, Prague, Czech Republic
e-mail: step@labe.felk.cvut.cz

M. Uller
e-mail: uller@labe.felk.cvut.cz

M. Janatová
Joint Department of FBMI CVUT and CU, Prague, Czech Republic
e-mail: marketa.janatova@lf1.cuni.cz

A. Emrouznejad (ed.), *Big Data Optimization: Recent Developments and Challenges*, Studies in Big Data 18, DOI 10.1007/978-3-319-30265-2_13

Fig. 1 View of four paradigms composed of subfigures from [16, 17, 26, 34]

1 Introduction

Brain is the control center of our body. It is a part of a complex neurological system that includes the spinal cord and a vast network of nerves and neurons that control and implement the cognitive and motor functions we need for activities of daily living. Brain disorders occur when our brain is damaged or negatively influenced by injury, disease, or health conditions. Various professions may be involved in typically complex processes of medical care and rehabilitation of someone who suffers from impairment after a brain damage. Researchers continuously conduct studies investigating the impacts of many factors on the rehabilitation progress with the aim to improve healing processes. These efforts are incrementally dependent on the appropriate data recorded and advances in the utilization of such data resources. When considering these data-related issues, in the brain disorder research and rehabilitation development trajectory, three phases or paradigms can be observed and the fourth paradigm is specified as a vision for future[1] (Fig. 1):

- *1st Paradigm.* Information about the patients' condition was gained by means of observation, examination, based on individual experience and opinion of the therapist. The Edwin Smith Surgical Papyrus [51], written in the 17th century BC, contains the earliest recorded reference to the brain, from the neurosurgeon point of view. The corresponding information was typically manually recorded[2] and has been subjected to various research efforts (including generalization) resulting in development of different theories and hypothesis. Analogous developments can also be observed in other medical branches, e.g. epidemiology.[3]
- *2nd Paradigm.* The patient's condition was observed by means of (continuously improved) equipment-based methodologies, like EEG, CT, MRI, fMRI, etc.,[4] that have been used as a complement to the observational methods applied in the 1st Paradigm mentioned above and as effective tools for objective evaluation in the large clinical studies. The resulting (mostly analog) output or an image has been

[1]This categorization is analogous to [32] that, however, addresses the generic science development trajectory.

[2]The oldest known medical record was written in 2150 BC in Summeria.

[3]http://en.wikipedia.org/wiki/John_Snow_(physician), http://en.wikipedia.org/wiki/Typhoid_Mary.

[4]The diagnostic effect of X-rays, used for medical X-ray and computed tomography was discovered in 1895. Electrocardiograph was invented in 1903, electroencephalogram later in 1924.

stored in paper format and evaluated manually. This innovation brought great advance in quality of diagnostics and supported design of important abstract concepts and of new solution models resulting in novel therapy methods and theories. This approach helped to collect sufficient resources for development of evidence-based medicine during the 1990s, what was considered an important step forward in medical science supporting transition of scientific results into medical praxis.

- *3rd Paradigm.* In the last few decades, most of the devices characterizing the 2nd paradigm started to produce data in digital format what significantly simplified not only their archiving but first of all their further processing. Application of novel analytic techniques including data mining and related methods to (typically large) datasets[5] called attention to new domains of neuroscience research, namely brain simulations, development and testing of new medical devices, etc., and it has significantly enhanced our knowledge about the neuroscience domain. The results achieved so far through this paradigm are promising and it is believed that continuing and improving this approach by utilizing new methodologies based on artificial intelligence, semantic web, and BIG data and Cloud technologies will remove many existing gaps observed in the current state of the art and will result in improving research productivity as well as speed-up new discoveries and enable true individualization of treatment towards the needs of specific patients.
- *Towards the 4th Paradigm.* Recently, ambient-assisted living research started to provide easy to use means for on-line collection of complex information about mundane activities, e.g., of the selected inhabitant of a household. One of the ways towards individualization of care can be based on analysis of such streams of time-tagged data that are obtained through repeated measurements for an individual patient during the course of his/her treatment. The resulting data can reveal precious information about success or failure of applied therapy or about advance of the patient's disease. This information is crucial for making a decision on the course of further treatment that can be fine-tuned for the specific needs of the considered patient. This is fully in line with the philosophy of individualized and personalized health care. Such an approach has to be supported by novel software solutions that will adapt, extend, and optimize the techniques and technologies that are being developed for a lot of scientific disciplines (inclusive life-science domains) in data-intensive science [4] addressing three basic activities: capture, curation, and analysis of data. This vision is at the heart of our research and development effort as presented in this book chapter together with its preliminary results. Moreover, we believe that the full realization of this vision will also advance other medical and life-science application domains.

New techniques and technologies, like special sensor networks, active RFID-based equipments, etc. used in brain disorder rehabilitation will produce ever more data that can reveal hidden knowledge and help to discover and open new perspectives and chances for this science area. This new BIG data-based approach requires

---

[5] In the scientific data management research literature, "dataset" is more commonly used than "data set".

novel systematic solution for preservation and utilization of the treated data. Let us call this new approach, aligned with data-intensive science developments *Brain Disorder Rehabilitation Informatics (BDRI)*. It introduces new methods for organization and optimization of the entire data life cycle in the process of performing scientific studies, specifically, aiming at further enhancing reuse and dissemination of studies resulting from brain disorder detection and rehabilitation research. Besides optimization of the creation and management of scientific studies, BRDI also addresses many aspects of productive and optimized decision making in all the steps of the dynamic trajectories of rehabilitation processes associated with individual patients. This functionality is not only provided in the context of one research center, but, thanks to the support of Cloud Computing, it supports safe reuse of relevant data among multiple research centers that can be geographically distributed. This approach effectively promotes a large-scale, national and international, collaboration of actors involved in the rehabilitation and research of brain disorders.

This book chapter introduces the research and development work in progress and discusses the first results of our BDRI research project conducted by teams from the Czech Republic and Austria. The aim is to leverage state-of-the art digital capabilities for researchers, healthcare professionals, and patients in order to radically improve rehabilitation processes and make them more productive, accelerate scientific discoveries, and to enhance rehabilitation research altogether.

The kernel part of the chapter presents our novel approach to management and analysis of data produced by brain stimulation and balance training sensors that address therapy provided mainly for dementia, stroke, and cerebral palsy patients.

The part of the system, designed for diagnosis and therapy of balance disorders [7], consists of a portable force platform connected to a computer as an alternative cursor actuator (instead of the standard mouse). Special software is used to lead the patient through a series of specific balance tests producing valuable data from four transducers, located in each corner of the force platform. The resulting data provide unique and precise information related to immediate force and weight distribution of the patient, who (standing on the force platform) is instructed to move the item displayed on a computer screen by shifting his/her center of gravity. The collected data is targeted by analysis during various interactive therapeutic scenes [56]. Visual and auditory feedback enabling game-like form of training makes the therapy more effective and provides higher compliance and motivation to exercise. Difficulty or demands of a training session can be adjusted according to the current patient's state. Regular training has a beneficial effect especially on the balance, motor skills, spatial orientation, reaction time, memory, attention, confidence and mental well-being of the user.

Reminiscence therapy can be defined as "the use of life histories—written, oral, or both—to improve psychological well-being". The therapy helps to solve a number of problems associated with aging (gradual loss of short-term memory or discontinuity caused by the change of environment). One of favorite techniques reminiscence therapy utilizes is scrapbooking, i.e. making commemorative albums with photos, clippings and small decorations (scraps). We present eScrapBook [55], an online platform enabling easy creation and use of multimedia scrapbooks. The books may contain text, images, audio and video clips that are viewable either online or exportable as an archive to be used independently of the authoring application. The primary target group consists of seniors with an onset of Alzheimer disease or dementia and their family members and caretakers (who prepare the scrapbooks for their relatives/clients); however, the application is fairly generic so it could be used to create books for various audiences. All actvities accomplished by the user when working with a eScrapBook are automatically traced, the produced data is integrated with other available repositories and finally mined; key methods involve association analysis and time sequence analysis.

In summary, this chapter makes the following original contributions: (a) It outlines a vision and realization strategy for the future scientific research and patient treatment in the domain called Brain Disorder Rehabilitation Informatics (BRDI); (b) In details it describes the BRDI dataspace, a key realization paradigm of this vision; (c) It designs a service oriented architecture for implementation of this dataspace and its access services-the implementation is based on the current state-of-the-art of the BIG data and Cloud technologies; and (d) It discusses a novel visualization approach for selected data mining models extracted from the BIG and complex data involving a large number of attributes measured by the application considered.

The rest of the chapter is structured as follows. Section 2 introduces the scientific dataspace model, a novel scientific data management abstraction we developed [19] to support BDRI and other modern data-intensive e-Science applications during the whole scientific research life cycle. In its basic model, a scientific dataspace consists of a set of participants—datasets involved—and a set of semantically rich relationships between nodes or participants—the concrete specification of these relationships is based on an appropriate ontology. The BDRI dataspace (BDS) stores in a secure way patient's demographic, diagnostic and therapy data and the data flowing from a sensor network involving physical sensors, like body sensors, room sensors, and balance training sensors and special software sensors monitoring activities of the patient, like brain stimulation sensors (e.g. at her/his work with the eScrapBook, interactive therapy games, etc.). In response to the needs of conducted scientific studies, selected data resources are combined, integrated, and preprocessed by appropriate services to provide a high-quality input to data mining and other analytical tasks. Besides the data discussed above (called primary data), the BDS includes derived data (products of data mining and other analysis processes) and background data (information of applied processing workflows, provenance data, supporting knowledge base, etc.). BDS provides advanced searching operations. Searching the BDS is much more complex compared to a database search, which is typically done with one-shot queries. That is why, BDS search can be rather described as an iterative interactive process where, e.g., BDRI researchers first submit a keyword-based query, then retrieve a ranked list of included studies matching the keyword query and based on further selections made by the user, they may explore selected studies in more detail with all related datasets and semantic information connected to the study. The provenance support enables to reproduce the outcomes of treatment attempts and conclusions based on rehabilitation process data. The inclusion of the scientific dataspace paradigm in our framework is the key contribution to the optimization of the big BRDI data management. Section 3 briefly enumcrates use-cases we are currently focusing on, illustrates the associated dataflows, and discusses their models and appropriate analytical services considered. Section 4 introduces the BDRI-Cloud software architecture that is based on the CloudMiner concepts published in [28]. BDS is a central component of this architecture. BDRI-Cloud provides interfaces to the following kernel operations: (a) publishing entire brain disorder rehabilitation research studies with all relevant data and (b) submit keyword-based queries in order to search for studies conducted at a participating research center. There are several real usage scenarios, each with a brief motivation on how scientists can benefit from the information infrastructure provided by the BDRI-Cloud. One of them is automatic reproducibility of brain disorder rehabilitation research studies. In a dynamic research environment with scientists continuously entering and leaving research groups, it will hardly be possible to retrieve all relevant data of a specific study once the responsible scientist who conducted the study has left the group. In fact, all information about the study that is left back at the research group is stored either within scientific publications, technical reports or other kinds of documentations written by the corresponding researcher. The information represented in such documents however does not allow to reproduce the study. Conversely, if the scientists have published the study using the services provided by BDRI-Cloud, all treated data become accessible in the underlying space of data together with all relevant datasets such as the input dataset, the analytical methods applied and scientific publications related to the study. In addition, semantic information is available making the study better searchable and retrievable. Our preliminary results achieved in data exploration and visualization activities are presented in Sect. 5. Finally, we briefly conclude in Sect. 6.

2 Scientific Dataspace Model

In the past ten years. the data engineering and application specialists recognized the importance of raising the abstraction level at which data is managed in order to provide a system controlling different data sources, each with its own data model [24, 25]. The goal has been to manage a dataspace, rather than a single data resource (file, database, etc.). The initial ideas on managing dataspaces have started to evoke interests of the data management community, however, the initial effort was mainly related to the database research and application mainstream. Furthermore, most of the approaches towards realizing a dataspace system focussed on personal information management. In our research [18], we have addressed the dataspace paradigm in the context of advanced scientific data management. In the following, we first summarize our previous achievements, as a background information. Then we move to the BRDI specific issues.

In order to elaborate how dataspace concepts can support e-Science, we have investigated what happens with data, or better what should ideally happen to them in e-Science applications. Our effort focused firstly on analysis of modern

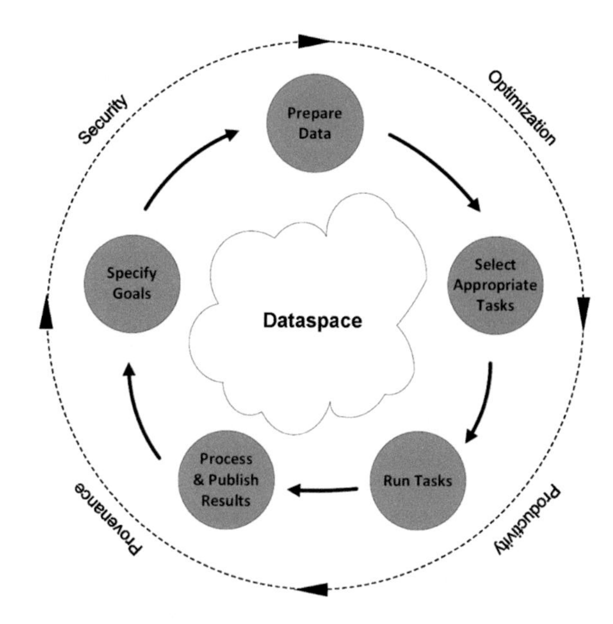

Fig. 2 Life-cycle model for modern data-intensive e-Science

scientific research processes and then, in details, on the research activities in our projects NIGM [20], ABA [21] and, later, especially, on the BDIR needs. The result of this investigation is an iterative and hierarchical model with five main activities that are supported by four continuously running activities; the model is represented in Fig. 2 and we define it in the following way[6]:

The e-Science life cycle—a domain independent ontology-based iterative model, tracing semantics about procedures in e-Science applications. Iterations of the model—so called e-Science life cycles—organized as instances of the e-Science life cycle ontology, are feeding a dataspace, allowing the dataspace to evolve and grow into a valuable, intelligent, and semantically rich space of scientific data [22].

First, we provide an overview of these activities that is followed by their more detailed explanation/discussion in Sect. 2.1.

At the beginning of the life cycle targeted goals (e.g., investigation of specific correlations, approvement or disapprovement of a hypothesis) are specified, followed by a data preparation step including pre-processing and integration tasks; here, the Hadoop framework that is optimized towards the BIG data processing can be used. Then appropriate data analysis tasks provided e.g., by the Weka or R-Systems are selected and applied to the prepared datasets of the previous step. Finally, achieved results are processed and published, which might provoke further experimentation and, consequentially, specification of new goals within the next iteration of the life cycle that can be optimized by the lessons learned within the former cycles. The outcome of this is a space of primary and derived data with semantically rich

[6]We are aware that with this e-Science life cycle definition we cannot stop the development of scientific research methodology and, therefore, it is assumed that it will be actualized in the future.

relationships among each other providing (a) easy determining of what data exists and where it resides, (b) searching the dataspacc for answers to specific questions, (c) discovering interesting new datasets and patterns, and (d) assisted and automated publishing of primary and derived data.

In addition to the above essential activities, there are four other activities that are related to the whole e-Science life cycle. The first one is *Provenance* [38], which tracks each activity performed during the life cycle to ensure the reproducibility of the processes. The second one is *Optimization*, which is applied on each performing task in order to ensure optimal utilization of available resources. The third one is *Security* that is extremely important because sensitive data, e.g. personal related data, could be involved in the addressed domain. Finally, the fourth one is *Productivity*. Productivity issues are an important research topic in the context of the development of modern scientific analysis infrastructures [43]. These four continuously running activies are not isolated from each other. For instance, the provenance information can be used by the optimization and security activities; further, it may improve productivity and quality of scientific research.

Each activity in the life cycle shown in Fig. 2 includes a number of tasks that again can contain a couple of subtasks. For instance, the activity Prepare Data covers, on a lower level of abstraction, a data integration task gathering data from multiple heterogeneous data resources that are participating within an e-Infrastructure (e.g., Grid, Cloud, etc.). This task consists of several steps that are organized into a workflow, which again is represented at different levels of abstraction—from a graphical high level abstraction representation down to a more detailed specific workflow language representation, which is further used to enact the workflow.

2.1 e-Science Life Cycle Activities

Now, we explain the e-Science lifecycle model in more details. We focus on five main activities.

1. *Specify Goals*—Scientists specify their research goals for a concrete experiment, which is one iteration of the entire life cycle. This is the starting activity in the life cycle. A textual description of the objectives, user name, corresponding user group, research domain and other optional fields like a selection of and/or references to an ontology representing the concrete domain is organized by this activity.
2. *Prepare Data*—Once the objectives for this life cycle are either specified or selected from a published life cycle that was executed in the past, the life cycle goes on with the data preparation activity. Here, it is specified which data sources are used in this life cycle in order to produce the final input dataset, by the data integration process. For example, the resource URI, name, and a reference to the Hadoop [58] repository might be recorded in case these technologies are used. Even information about the applied preprocessing algorithms and their parameter

settings have to be part of the stored data. The final dataset as well as the input datasets are acting as participants in the dataspace and are referenced with an unique id. Additionally, the user specifies a short textual description and optionally some keywords of the produced dataset.

3. *Select Appropriate Tasks*—In this activity the data analysis tasks to be applied on the prepared dataset are selected. In e-Science applications it is mostly the case that various analytical tasks, for instance the widely used data mining techniques, are executed successively. The selected tasks, which are available as web and grid services, are organized into workflows. For each service, its name and optionally a reference to an ontology describing the service more precisely is captured. Also for the created workflow, its name, a short textual description, and a reference to the document specifying the workflow are recorded.
4. *Run Tasks*—In this activity the composed workflow will be started, monitored and executed. A report showing a brief summary of the executed services and their output is produced. The output of the analytical services used might be represented in Predictive Model Markup Language (PMML) [29], which is a standard for representing statistical and data mining models. PMML documents represent derived datasets, thus they are managed as participants of the scientific dataspace and considered as resources by this activity.
5. *Process and Publish Results*—This is the most important activity in order to allow the underlying dataspace to evolve and grow into a valuable, powerful, semantically rich space of scientific data. Based on the settings of the user, one, with the support of appropriate tools, publishes the results of the analysis tasks as well as all semantic information captured in the previous activities. Different publishing modes allow to restrict access to selected collaborations, user groups, or research domains.

Already during one life-cycle run, huge data may be produced in an advanced scientific application. They are stored as a *life cycle resource (LCR)*; it could range from some Megabytes up to several Gigabytes. These issues are discussed in the following subsection.

2.2 The Environment of Dataspaces in e-Science—a BIG Data Challenge

After introducing the e-Science life-cycle model in the previous subsection, we discusss the main components of an e-Science dataspace and show that its management is a big data challenge issue.

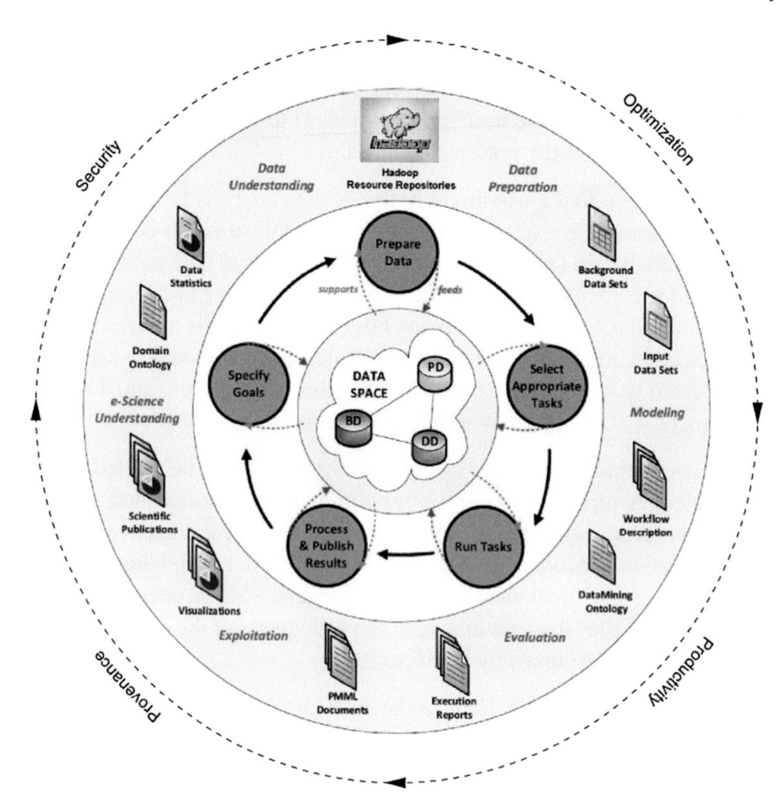

Fig. 3 Environment of a big scientific dataspace. Modification and extension of [18]

Figure 3 shows the environment of Dataspaces in e-Science.[7] In particular, there is a set of participants involved in one ore more activities of the e-Science life cycle. Each activity feeds the dataspace with new participants, as for example the activity Specify Goals adds new domain ontologies (stroke rehabilitation and dementia treatment ontologies) and a goal description in appropriate format, the activity Prepare Data adds new final input datasets as well as Hadoop resource repositories, and the activity Select Appropriate Tasks adds new workfow description documents, while the activity Run Tasks adds new PMML documents describing the data mining and statistics models produced, and finally the activity Process and Publish Results adds new documents visualizing the achieved outputs. All these participants belong to at least one or more e-Science life cycles, expressed as instances of the ontology describing its relationship and interconnection to a great extend.

[7] An additional data flow to the dataspace can originate from the Web considered as "the biggest database" and information resource. There already exist advanced Web data extraction tools, e.g. [47].

Each iteration of the life cycle will produce new instances and properties of the ontology. Based on the publishing mode, set by the scientist who accomplished the life cycle, the whole instance will automatically be published into the dataspace and thus become available to other users of a wider collaboration with respect to other research areas. We distinguish between four publication modes as listed in the following: (1) *free access*—the life cycle resource is publicly available, no access rights are defined; (2) *research domain*—the life cycle resource is restricted to members of the research domain the scientist who conducted the experiment belongs to; (3) *collaboration*—the life cycle resource is restricted to members of a collaboration defined among multiple research groups; and (4) *research group*—the life cycle resource is restricted to members of the research group the scientist who conducted the experiment belongs to.

Users will have access to sets of participants available in the scientific dataspace, depending on their assigned role. By this, the concept of managing sub-dataspaces is realized. A sub-dataspace contains a subset of participants and a subset of relationships of the overall dataspace. There can be sub-dataspaces setup for different domains, then for different research collaborations and even for single research groups. Scientific experiments that were published using the free access mode, will participate in the overall dataspace, thus its participants and the life cycle instances are accessible for every one having access to the scientific dataspace. In order to access data of a specific life cycle iteration, that was published using the research group mode, it will be necessary to be member of that specific research group, as the data will be only in the corresponding sub-dataspace.

Once a new iteration has been accomplished using at least some activities from other life cycle instances, both the new life cycle document and the one containing activities that were re-used will get an additional relationship. We can conclude from this, that the dataspace is evolving with an increasing number of life cycles. Catalogs and repositories for ontology instances that manage these LCRs organized in the Resource Description Framework (RDF) [40] trees will provide search and browsing features. The development of an appropriate repository providing rich functions to insert, update and delete as well as to semantically search LCRs on a multi-institutional level is also considered in our approach.

2.3 *Relationships in the Scientific Dataspace*

Scientific dataspaces will be set up to serve a special subject, which is on one hand to semantically enrich the relationship of primary and derived data in e-Science applications and on the other hand to integrate e-Science understandings into iterations of the life cycle model allowing scientists to understand the objectives of applied e-Science life cycles. Figure 4 illustrates what is considered as dataspace participant and relationship, respectively by breaking a typical very generic scientific experiment into its major pieces, which we organize into three categories (1) primary data, (2) background data, and (3) derived data. A typical scientific experiment consists of

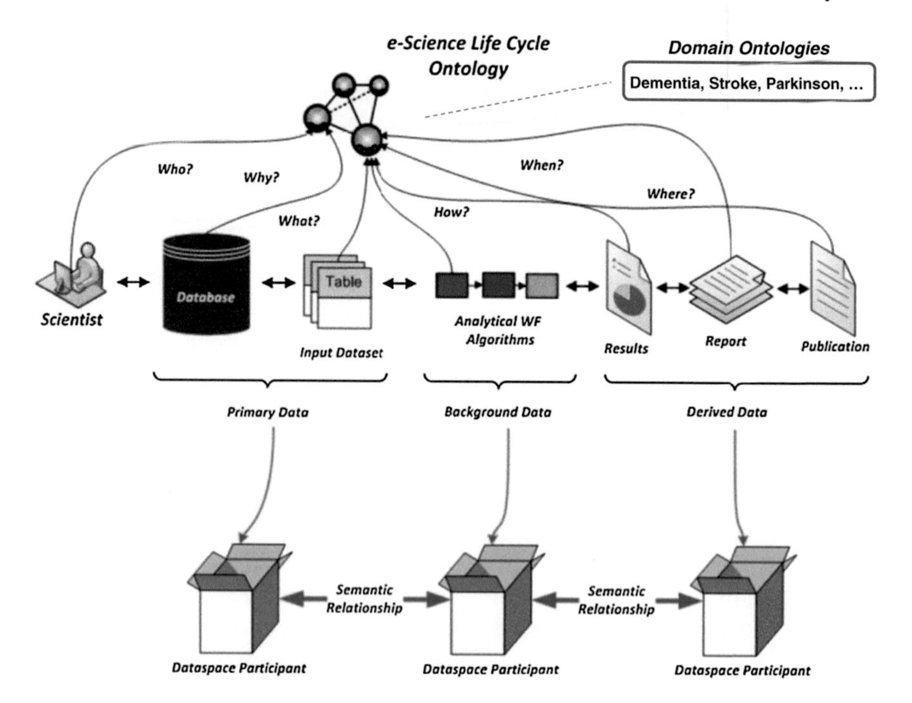

Fig. 4 Semantic relationship in e-Science applications organized be the e-Science life cycle ontology. Based on [18]

three types of dataspace participants: (1) an input dataset taken from a source database, datasets extracted from the Web (e.g., current dangerous places in the surroundings, weather condition, air quality), etc.; (2) a set of functions (analytical methods) used to analyze the input dataset (commonly organized into a scientific workflow); and (3) the derived results, which represent the outputs of the experiment i.e. plots and histograms, reports, or publications.

Those dataspace participants are stored in corresponding data repositories of the scientific dataspace. Their interconnection is semantically rich described by dataspace relationships. They are modeled in RDF as individuals and properties of the e-Science life cycle ontology and organized in multiple RDF trees within an RDF store supporting the SPARQL query language for RDF [48]. SPARQL contains capabilities for querying required and optional graph patterns along with their conjunctions and disjunctions. The results of SPARQL queries can be textual result sets or RDF graphs. Experiments described by the ontology are referred to as Life Cycle Resources (LCRs). A LCR in fact represents the semantic relationship among dataspace participants. In the following section we discuss how such scientific experiments can be described by the e-Science life cycle ontology.

In Fig. 4, besides the e-Science Life Cycle Ontology, a block of Domain Ontologies is introduced, in which several BRDI specific ontology examples are listed. They additionally support BRDI communities in exchanging knowledge by sharing data and results. More details on this issue are introduced in the following subsection.

2.4 The e-Science Life Cycle Ontology and Towards Conceptualization in the Brain Damage Restoration Domain

The e-Science life cycle ontology aims at providing a common language for sharing and exchanging scientific studies independent of any application domain. We use this ontology as a semantic model for the creation, organization, representation and maintenance of semantically rich relationships in Life Cycle Resources (LCRs) using the scientific dataspace model described in Sect. 2.1. The model involves essential concepts of the scientific dataspace paradigm. Thanks to its domain independent applicability it can easily be used in any e-Science application. These concepts are organized in the e-Science life cycle ontology. It provides the basis for presenting generic scientific studies as LCRs with well defined relationships among their participating datasets. On the other hand the e-Science life cycle ontology supports the scientific dataspace paradigm with primitives that can specify concrete relationships among primary, background, and derived data of these LCRs. The e-Science life cycle ontology can be seen as the heart of the underlying dataspace-based support platform. It is used to share common understanding of the structure of scientific studies among a research community. For example, suppose several different research centers conduct brain damage restoration analysis studies. If these research centers share and publish the same underlying ontology of concepts for conducting brain damage restoration studies, then software programs can extract and aggregate knowledge from these different research centers. The aggregated information can then be used to answer user queries or as input data to other applications (e.g. automation based analysis). Principally, ontologies are used for communication (between machines and/or humans), automated reasoning, and representation and re-use of knowledge [14]. To enable re-use of domain knowledge consolidated within scientific studies exposed as LCRs was one of the driving forces behind the development of the e-Science life cycle ontology. In this context domain knowledge is represented in semantic descriptions about scientific studies. In particular, activities of the e-Science life cycle such as the data preparation activity (providing descriptions about the input dataset used) or the select appropriate task activity (providing descriptions about the analytical methods applied to an input dataset) provide rich semantics about a conducted scientific study. With the help of the e-Science life cycle ontology we can describe scientific experiments according to the specification and implement a software program (e.g. a web service) that guides the scientist through the experiment in the way that

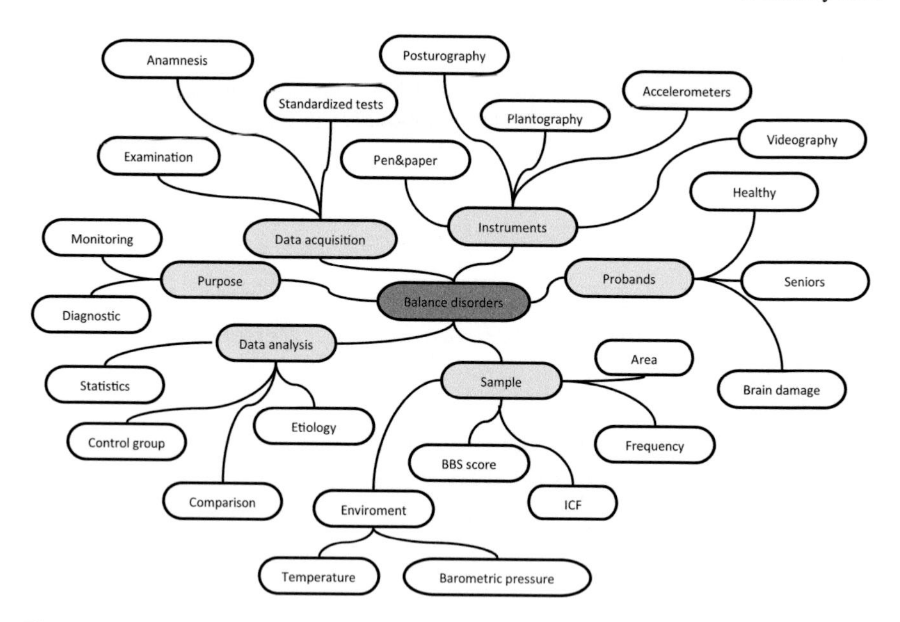

Fig. 5 Balance training research concept taxonomies

is conduction-independent of the e-Science application domain. It is further possible to analyze domain knowledge since a declarative specification of the process of conducting studies is available with the e-Science life cycle ontology.

However, the ontology itself is not the goal in itself. It is rather a definition of a set of concepts and their relations to be used by other software programs. In the research carried out in our project the ontology represents the concepts to model the entire life cycle in a scientific study. By this we provide with the e-Science life cycle ontology a framework for the management of semantically enriched relationships among datasets that participate in the data life cycle of the conduction of a scientific study. The software program that uses the ontology and its built knowledge base is represented by the software components of the scientific dataspace support platform.

In the previous subsection, we briefly mentioned the need of appropriate domain ontologies (see Fig. 4) dedicated to the BRDI research for productive knowledge sharing among research communities. The ontologies could be further used in conjunction with the e-Science life cycle ontology. A first step when developing such ontologies is the specification of appropriate concept taxonomies; they are knowledge structures organizing frequently used related concepts into a hierarchical index. Our initial specification of taxonomies involving the key concepts in the balance training research is depicted in Fig. 5. These concept taxonomies include Instruments, Probands, Sample, Data Analysis, Purpose, Data Acquisition. They represent a first vocabulary of terms (organized in a hierarchical order) for the balance disorder rehabilitation research domain. The resulting ontology will be presented and discussed in our subsequent publication.

2.4.1 Applied Methodology

As the first step in building the e-Science life cycle ontology we have selected a methodology supporting phases of the development process. Typically such a development process is organized in several phases. For the e-Science life cycle ontology development process we have selected the On-To-Knowledge methodology [52], as the most appropriate methodology because it provides the most accurate description of each phase through which an ontology passes during its lifetime.

In Fig. 6 we illustrate identified people who are involved in the e-Science life cycle ontology and show one usage scenario.

Typically, senior scientists will interact with the ontology in terms of submitting search and query requests (e.g. asking for stroke rehabilitation research studies from a particular person, organization, or research field), while Ph.D. and master students are continuously feeding the semantic repository with new breath research studies described according to the defined ontologies concepts. On the other hand, there is an ontology engineer, who is responsible for maintaining ontologies and for their evaluation in case changes were applied to the ontologies. A brain rehabilitation research expert provides domain knowledge in the form of concept taxonomies, defining a vocabulary of terms used in the brain rehabilitation research domain, as discussed in the previous subsection. The ontology engineer is also responsible for the

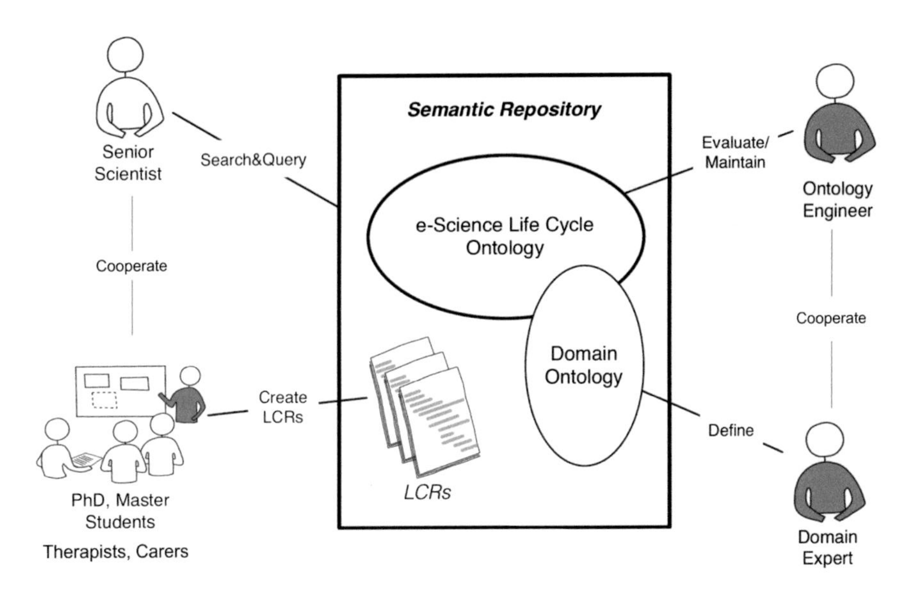

Fig. 6 Semantic relationship in e-Science applications organized by the e-Science life cycle ontology. Based on [21]

development of the brain rehabilitation research domain ontology[8] based on these defined concept taxonomies.

2.4.2 Searching and Querying Scientific Dataspaces

Based on our unified e-Science life cycle ontology, search and query services can be provided for all the participants of the scientific dataspace. Hence, it is possible to forward a keyword query to all participants, such an action has the aim to identify relevant datasets. However, each query submitted to the scientific dataspace will receive not only the matching data but also data of its follow-up e-Science activities. For instance it will be possible to receive information what mining task were applied on a discovered dataset, the concrete workflow, the workflow report, the results presented in PMML and its corresponding visualizations.

Using *SPARQL* query language for RDF and semantically rich described e-Science life cycles, consolidated within instances of the ontology, keeping relationships among each other, the dataspace is able to provide answers to rather diversed questions; let us specify wording of just a few typical examples (These queries have to be expressed in the SPARQL form and passed to a SPARQL Interpreter in the real application.).

A *I have found some interesting data, but I need to know exactly what corrections were applied before I can trust it.*
B *I have detected a model error and want to know which derived data products need to be recomputed.*
C *I want to apply a sequence pattern mining on balance-therapy data captured in the Middle European region in the last year. If the results already exist, I'll save hours of computation.*
D *What were the results of the examination of the patients, which didn't finish the long term therapeutic intervention and quit before the end of the study?*
E *If the patient is able to work with at least two different therapeutic games, would there be an evidence of a positive progress of patient's stability after the therapy in that study?*

3 Data Capture and Processing Model

In this part we introduce an application model that corresponds to the key use-cases of the considered application. Then the data capture and processing model and a high-level abstract view of the supporting software architecture are discussed.

[8]Publications [10, 49] deal with the development of an epilepsy and seizure ontology and a data mining ontology, respectively.

3.1 Use-Cases

The application use cases that we address can be briefly outlined in the following way.

- *Orientation support.* The aim is to localize persons through devices with integrated GPS and GPRS functions. However, this localization is not sufficient in some buildings due to signal loss. Here, Active Radio-Frequency Identification (aRFID) technology can be used in combination with GPS-tracking.
- *Preventing dangerous situations.* This can be achieved by combining the localization device with an alarm, triggered off at the border of a safe area; this is also based on GPS and aRFID technologies.
- *Talking key.* The aim is to inform the person with the aid of aRFID tags not to forget the keys when leaving the flat.
- *Finding things again.* The aim is to help find lost objects using aRFID technology and touch screens.
- *Brain stimulation.* This is our main focus in this project effort stage. We take two approaches:
  - Stimulation based on eScrapBooks. The aim is to create individual's memory books and reminders to be operated via touch screens. Individually prepared brain stimulation with positive memories can: prevent increase in dementia-process, slow down dementia-process, reactivate cogitation/thinking skills, increase ability to communicate, improve mental well-being, work against depressions, etc. [27].
  - Stimulation addressing balance disorders. The second part of the system is aimed for individual therapy of patients with balance and motor skill disorders. The use of visual feedback increases efficiency of the therapy. The evaluation of actual balance skills is made in the beginning and during the entire rehabilitation process. Data is analyzed in both time and frequency domain.

The treatment, data capturing and processing model that corresponds to these use-cases is depicted in Fig. 7. It includes an integrated outdoor and indoor treatment associated with production of potentially large volumes of data that with the appropriate linked data, like social networks could be a valuable source of knowledge. Discovery and further post-processing and usage of this knowledge is a challenging task pursued in our project. The following subsections are devoted to indoor and outdoor treatments, respectively. In the future other resources like MRI images, EEG investigation results, and appropriate services will be included into the BRDI-Cloud.

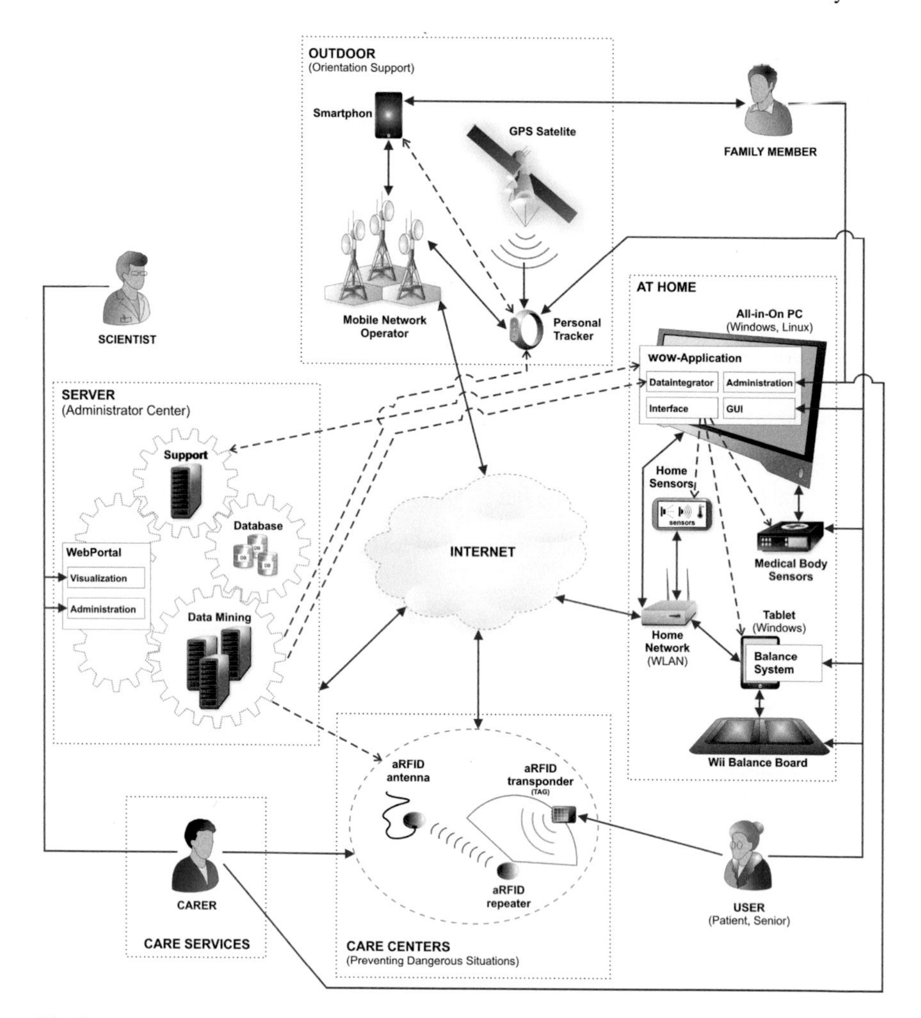

Fig. 7 Data capture and processing model associated with considered use cases

3.2 *Indoor Treatment*

This important treatment part is mainly based on utilization of sensor networks installed at the patients' homes and producing streams of data, which are analyzed on the fly. On demand some data can be stored persistently and then analyzed. In the example home instance depicted in Fig. 7, there are several categories of sensors:

- Touch screen. It produces a stream of data describing issued seniors'/patients' requirements;
- Medical body sensors. They continuously produce data about body temperature, body weight, blood pressure, heart beat rate, etc.

- aRFID tags. They are fixed to things, which are potential targets for searching (keys, bags, gloves, caps, etc.).
- Apartment environment status sensors. Door and window sensors, fridge door sensors, and gas sensors are good examples of this category.
- Software sensors monitoring behavior of the patient (e.g. during his/her work with an eScrapBook).
- Balance system sensors. The Wii balance board containing one transducer in each corner can be is used for diagnosis of balance disorders. It has been proven in [15] that this accessible and portable platform is a valid tool for assessing standing balance.

Figure 7 also provides a sketch of software architecture implementing indoor treatment. The kernel part involves a software module system called *WOW* that is aimed at administering and exploring various methods of brain stimulation. In this particular case, 'Brain Stimulation' motivates the test persons to use a touch screen PC to activate recall from their memory. An individually combined selection of video clips, songs, music and photos is accessed by touching the screen. Other reminders concern the test persons' daily routine. The WOW application is designed to suit these needs, to provide the users with simple and friendly user interface to access their uploaded media, browse YouTube videos, view current events, play games or browse eScrapbooks.

From the technological and architectural point of view, the WOW application combines HTML/SVG-based web GUI (best viewed by Google Chrome) with NodeJS server bundled together as a single package. The WOW application has two main parts: the administration GUI and the collection of user GUI pages. The application (framework) is modular and can be extended by developing additional plugins (containing single or more GUI pages and server logic) and addons (small embeddable applications or games).

3.3 Outdoor Treatment

The aim of using GPS for Alzheimer's patients is to give carers and families of those affected by Alzheimer's disease, as well as all the other dementia related conditions, a service that can, via SMS text message, notify them should their loved one leave their home. Through a custom website, it enables the carer to remotely manage a contour boundary that is specifically assigned to the patient as well as the telephone numbers of the carers. The technique makes liberal use of such as Google Maps.

The cost of keeping patients in nursing homes is considerable and is getting more substantial as the elderly population increases. On top of this there is often confusion in the patient's mind, generated simply through the process of moving to the nursing home. All of which is highly stressful for patient as well as their family and carers.

To tackle the problem, this research has considered the employment of a virtual safety border line (fence) from the point of view of care for a dementia patient. By using several different technologies the aim has been to enable carers to know when their patient or loved one has firstly left their house unassisted and secondly when/if they have traversed further than a geographically selected distance from their start point. Essentially the aim has been to throw down a virtual fence around the patient's known location and for notification to be given when the fence boundary is crossed.

The out-door treatment is based on the tracker XEXUN XT-107. The tracker that is carried by the patient is regularly determining its position from the GPS (Global Positioning System). The position data is sent through the GPRS (General Packet Radio Service) service and GSM (Global System for Mobile Communication) network to the Administration Centre server and Care Services. The whole communication and coordination is steered by the Care Services clients. The whole technical implementation of the system is described in deliverables of the project SPES [1].

In daycare centers, an active RFID alarm system is installed. For example, it is activated when an individual person with dementia leaves his/her personal safety area (building, garden, terrace, etc.) The respective person concerned is equipped with the active RFID tag/chip.

3.4 Data Model

All activities of the system actors are observed/registered and traced/recorded as events passed as data in the JSON format conforming to a specific JSON Schema. The event-based data model is discussed in the subsequent subsection.

Specifically, in the WOW application introduced in the previous section, for the purposes of evaluating user experience with the application and gathering feedback on various media, many events generated by the application may be logged into a database so the data can be analyzed later. The events can be categorized as follows:

- Administration events—events generated by admin (the user who manages content) of the application. Examples of events: new user created, application started, application exited, addon imported.
- User events—events associated with the user activity in the application. Examples: user logged in, user logged out, user page viewed.
- Addon events—general events generated by miniapplication/addon. Examples: addon started, addon closed.
- Game events—events generated by various games (a specific type of addons). Examples: game started, game ended, game paused/resumed, level up, score changed, settings changed.
- eScrapBook events—events generated by browsing eScrapBooks. Examples: book opened, book closed, page turned, video/audio started/stopped.

The events are implemented as class hierarchy (beginning with most abstract AbstractEvent, containing information about event type and a timestamp). The specific events can carry additional information relevant to their domain. For example, eScrapBook events like 'page turned' reveal and store information about content of the currently visible pages—how many images/videos are on which page, tags, word count, as well as the relevant time stamp. This may be useful for tracking which parts of the book are most interesting for the user or which parts/pages are only briefly skipped.

3.5 Event-Based Data

Event-based data are defined as data composed of sequences of ordered events, an event-sequence. Each event, or element, of an event-sequence has a start time and a duration and each begins when the previous one ends. The types of event present in such a sequence all belong to a set of predefined event types, an event alphabet. An event-based dataset, D, then consists of a set of event-sequences:

$$D = \{S_1, S_2, \dots, S_m\}$$

where m is the total number of event-sequences in the dataset; an empty sequence, denoted by ϵ, is also considered/allowed. Each sequence, S_i for $i = 1, 2, \dots, m$, is composed of an ordered list of events each of which has a start time, t, and a duration, dur. Considering the total set of event types E, each event is a triple (e, t, dur), where $e \in E$ is the event type, $t \in R$ is the start time in minutes and $dur \in R$ ($dur > 0$) the duration of the event in minutes. Each event-sequence is then described by a sequence of such events:

$$S = \; < (e_1, t_1, dur_1), (e_2, t_2, dur_2), \dots, (e_n, t_n, dur_n) >$$

where n is the total number of events in the sequence, the *length* of the sequence, and

$$e_i \in E \text{ for } i = 1, 2, \dots, n, \;\; \text{and} \;\; t_i + dur_i = t_{i+1} \text{ for } i = 1, 2, \dots, n-1$$

As a motivation let's take activity diary data addressed by Vrotsou [57]. In the case of activity diary data, each event-sequence in the dataset represents an individual's diary day consisting of a sequence of activities taken from a predefined activity coding scheme. The total duration of each diary corresponds to the predefined duration of the observation period and should usually be of the same length for all participants in the survey. Diaries collected through time-use surveys for everyday life studies usually include records per 24 h of a day (1440 min), so an additional restriction applies:

$$\sum_{i=1}^{n} dur_i = 1440.$$

Apart from activity diaries, event-based data is found in a wide range of application areas, e.g. streams of data produced by sensors, records about web site visits in the Internet, medical records, etc.

4 Towards the BDRI-Cloud

Our objective is to realize the data capture and processing model discussed in the previous section by a framework that involves appropriate technologies and methodologies. The kernel component of this framework is an advanced infrastructure called the *BRDI-Cloud* that is introduced in this section. It is based on the idea of the ABA-Cloud infrastructure [21] that was proposed for the breath research application we followed in our previous project.

4.1 Motivation and System Usage

BRDI research and rehabilitation processes generate huge amount of information, as discussed in Sect. 2.2 and illustrated in Fig. 3. The problem is how to extract knowledge from all this information. Data mining provides means for at least a partial solution to this problem. However, it would be much too expensive to all areas of human activity to develop their own data mining solutions, develop software and deploy it on their private infrastructure. This section presents the BRDI-Cloud that offers a cloud of computing, data mining, provenance and dataspace management services (Software as a Service approach) running on a cloud service provider infrastructure.

BRDI-Cloud provides interfaces to (a) publish entire brain research studies with all relevant data and to (b) submit keyword-based queries in order to search for studies conducted at a participating research center. Publishing a study can be done from within a Problem Solving Environment (PSE) such as Matlab [46], R [3], and Octave [2]; e.g., Matlab provides powerful support for publishing Matlab results to HTML, PowerPoint or XML.

Searching the underlying space of data that contains all the published breath research studies can be done either again from within a PSE or via specifically developed interfaces integrated into the BRDI-portal. The BRDI-portal provides special tools allowing us to search and browse for studies in the BRDI-Cloud. We develop the BRDI-Study Visualizer based on our previous work [54] as a handy tool for visualizing the most important semantic information about a single brain research study. Another novel tool is the BRDI-Study Browser [50], which aims at visually browsing studies available in the BRDI-Cloud.

4.2 System Design

The BRDI-Cloud design can be considered as a contribution to the realization of the data-intensive Sky Computing vision [37]. It is a system that includes multiple geographically distributed platforms, so called *BRDI-platforms*.

Each BRDI-platform involves a number of Web services that communicate with a semantic repository and a data repository. The semantic repository stores semantic descriptions about brain damage studies such as intended goals of the study, responsible persons, etc., while the data repository stores all its corresponding datasets (e.g. raw data, processed data, derived results, etc.). The semantic description is the backbone of the system enabling others to understand specific processes within a brain research study. They are represented by the e-Science Lifecycle Ontology, which formalizes the vocabularies of terms in the process of conducting scientific studies. The e-Science Lifecycle Ontology can be seen as the heart of the underlying BRDI-platform. It was discussed in Sect. 2.4.

BRDI-users are enabled to publish their scientific studies from within existing Problem Solving Environments (PSEs), thus do not have to switch to another tool in order to take full advantage of BRDI-Cloud. We basically distinguish among two kinds of BRDI Web-Services, (a) services to document complete BRDI-studies for the long run making the study repeatable for a period of at least 15 years, and (b) services to execute BRDI-studies on remote servers either within private or public clouds.[9] The possibility to execute BRDI-studies in the cloud can however be seen as an extension to the BRDI-platform, which might be interesting for complex computations. All services can be accessed from within existing PSEs. The integration of PSEs is realized by a specific BRDI-Cloud toolbox offering all needed functions to create, load, execute, search, update, and publish BRDI-Studies. From the research management point of view, each BRDI-platform represents a brain research group or lab, typically employing several persons including senior and post-doc researchers, master students, administrators, and technicians. These persons are acting as local users on their own platform, while they are guest-users in other platforms. Access rights can be defined by the research group leader on each platform. A Web-based portal (the BRDI-portal) represents a Web interface to the world including BRDI-platform users. The BRDI-portal also provides state-of-the-art social networking tools, which is the basis for a more intense cooperation among brain researchers from different BRDI-platforms. Figure 8 illustrates the main components of the BRDI-Cloud on a high abstraction level.

[9]Private cloud services operate solely for a single organization, typically managed and hosted internally or by a third-party, whereas public clouds services are offered by a service provider, they may be free or offered on a pay-per-usage model.

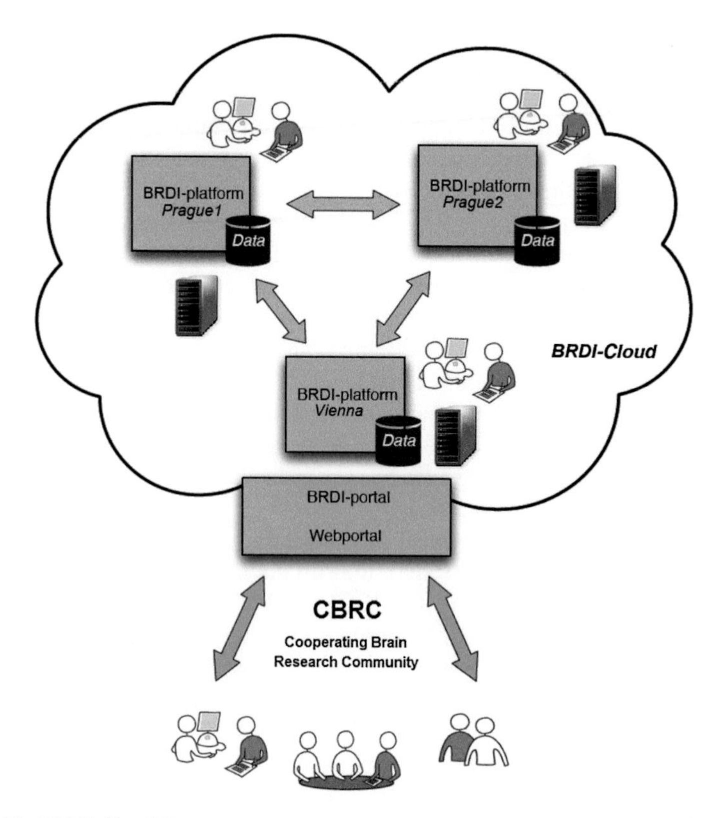

Fig. 8 The BRDI-Cloud illustrating a scenario with three collaborating brain damage research and rehabilitation centers (two in Prague and one in Vienna). At each research center an instance of the BRDI-platform and its underlying infrastructure is deployed. It provides services to local users who work at the corresponding center and preserves semantically enriched brain rehabilitation studies including all data (raw data, processed data analytical methods used to analyze the raw data, and derived results) that are gathered at the research center. This results in a distributed data and service environment for the Cooperating Brain Research Community (CBRC). Members of the community can (with appropriate access rights defined) get access to these resources through the BRDI Webportal

With the *BRDI-Portal*, a Web-based portal for the entire brain research community, we would like enable enhanced scientific discourse [5]. The portal integrates state-of-the-art social media tools, a forum, a wiki, as well as chat tools and user profiles for each registered member. By this, we aim at supporting the collaboration among the nationally and internationally distributed communities of brain researchers.

4.3 Architecture—a Service-Level View

The basic architecture of the BRDI-Cloud follows the Service Oriented Architecture (SOA), and as such contains three basic components. First, services developed by service providers and published to a registry. Second, the registry that stores and manages all information about published services. The third component of the architecture is a client (end-user).

Concerning knowledge management, we consider two areas: knowledge for the cloud—using semantic technologies, we describe the availability of cloud services (this is the semantic registry issue) and knowledge on the cloud—this is the knowledge to be queried by the client.

Normally, a Software as a Service (SaaS) cloud offers hundreds of services offered by one single service provider [33]. However, we propose here that many research institutions, and even individual users could offer their services needed for the BRDI domain. Thus, it is possible that many service providers will contribute their products to the BRDI-Cloud. The registry contains information about published services. A cloud may have one registry to store that information or could use a number of registries for reliability and performance reasons. We also assume that with the increase of amount of data being collected around the world and different areas of applications of those data, there could be more than one cloud each supported by its registry. These two cases lead to a solution of multi-registry infrastructure. Usually, a registry contains basic information about service location and service provider. We propose to contain more information about each individual published service. The additional information contains semantics of a mining service. This implies a need for the use of the concept of a broker offering intelligence, which is not provided by simple registries. The user, the third basic component of the SOA, should be able to learn about services offered by the cloud. We propose that the BRDI-Cloud offers through a portal a possibility to carry out operations on data and to learn about all services offered by service provider. The user accesses a registry to learn about computing and data mining services and select a service that could solve her problem. The management of computation using the selected service on specified data is being carried out directly by the end-user. The registry, made more intelligent, also could carry out this task. The BRDI-Cloud architecture offers both options.

Figure 9 illustrates the components of the BRDI-Cloud. The SOA-based architecture involves:

- ComputeCloud—it includes all published services involved in analytical tasks, like data mining, statistics computation, and visualization. These services can encapsulate established libraries, like those provided by Matlab, or specific codes written by providers.
- DataspaceCloud—it contains collected data and other dataspace components stored in data and semantic repositories. Cloud technologies enable realization of elastic large-scale dataspaces.

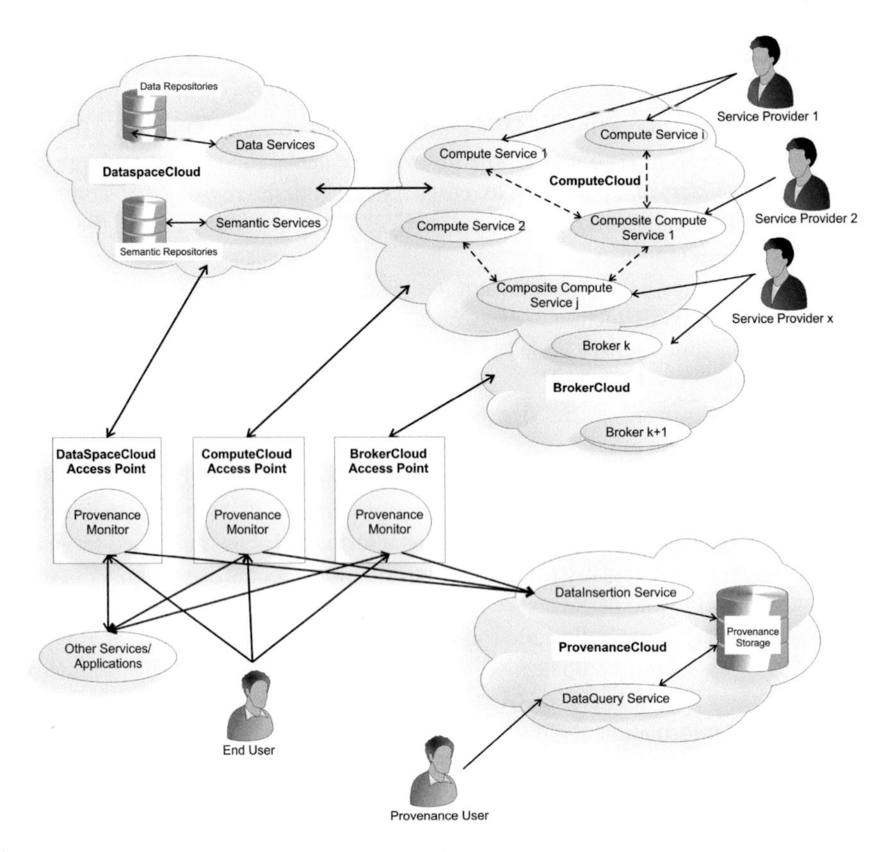

Fig. 9 Overview of the BRDI-Cloud service architecture

- ProvenanceCloud—it inserts provenance data captured by Provenance Monitors into the ProvenanceStorage and provides services for extracting required data from this storage.
- BrokerCloud—Service Providers publish services to it e.g., by the means of the OWL-S language [45].
- Access Points—they allow users to firstly access the BrokerCloud to discover and access needed services and then to steer other component clouds. An appropriate portal is supporting this functionality.

5 Data Analysis and Visualization Services

In the BRDI-Cloud, we plan to apply data analysis to different data repositories, in the first phase, to relational databases, flat files, and data streams. Dataflows involving these resources can be observed in Fig. 7. Data stream is relatively a new kind of

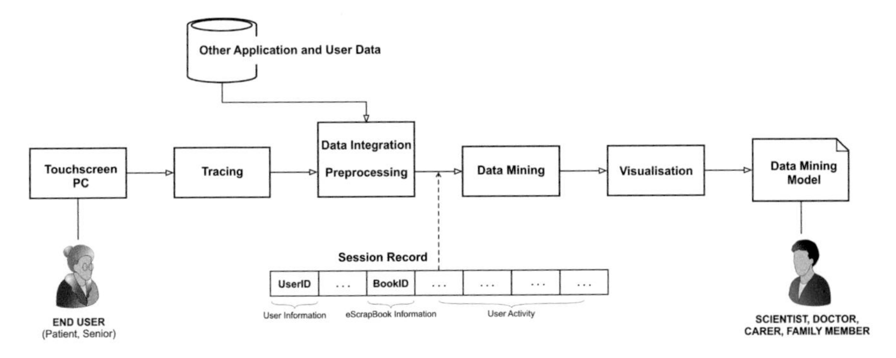

Fig. 10 Abstraction of the sequential pattern and association rule mining workflow

data. In our application, data captured by sensors is getting transformed to continuous data streams. Mining data streams helps to recognize activities, detect anomalies in the life of the patients, and enable automatic recognition of emergencies.

The services of the ComputeCloud (Fig. 9) are being developed by porting and adapting libraries and services developed in our previous projects and published in [9–13, 23, 31, 35, 39, 42, 44, 53, 60].

Our recent research has focused on services for mining sequence patterns and association rules in the data managed by the DataspaceCloud [6, 41]. A sketch of the applied workflow is depicted in Fig. 10. Our approach is based on the Weka data mining system [59], however enriched by an advanced visualization adapted to the needs of the end-users. The data captured from various sources are integrated by converting them to a common representation and by linking the sensor data with additional data—e.g. in case of eScrapBooks, the data describing which pages of which books were viewed are coupled with additional eScrapBook metadata and content stored on a separate server; in case of medical sensors, such as blood pressure, temperature, etc., those data are linked with patient profiles stored in a separate medical database.

5.1 Sequential Pattern Mining

The datasets of our application to be analyzed have an inherent sequential structure. Sequential nature means that the events occurring in such data are related to each other by relationships of the form before or after and together [36]. Sequential pattern mining is the mining of frequently occurring ordered events or subsequences as patterns [30]. An example of a sequential pattern expressed in a pseudo notation is:

Patients who primarily observe photos of their family and start to take the medicament M_code are likely to extend their mental horizont within a month significantly. [minimum_support = 10 %]

This means that the occurrence frequency of this pattern in the set of sequences is not less than 10 %. Another example comes from the stability training application:

Patients who undergo stability training focussed on wide range of movement have better results in functional reach test after 6 weeks of therapeutic intervention. [minimum_support = 20 %]

We have done the first data mining experiments with Weka. Very soon we realized that the native application domain independent Weka presentation is not suitable for application domain experts. Report [8] gives an overview of other considered solutions. Based on his work and motivated by the research results of Vrotsou [57], we developed a novel visualization approach that is briefly introduced by two figures and the accompanying text; further details can be found in [41].

Figures 11 and 12 show our approach providing the visualization of the end-user activity paths during their work with sScrapBooks on a higher, domain-oriented level. Right to the large presentation panel, there is control and information panel. In both figures, one can see that it is also possibile to gain information by a mouseover effect. Figure 11 represents a traditional summary activity path representation that shows the total amount of time spent on different activity categories during one eScrapBook session. In Fig. 12, the activity path is drawn in two dimensions (time and

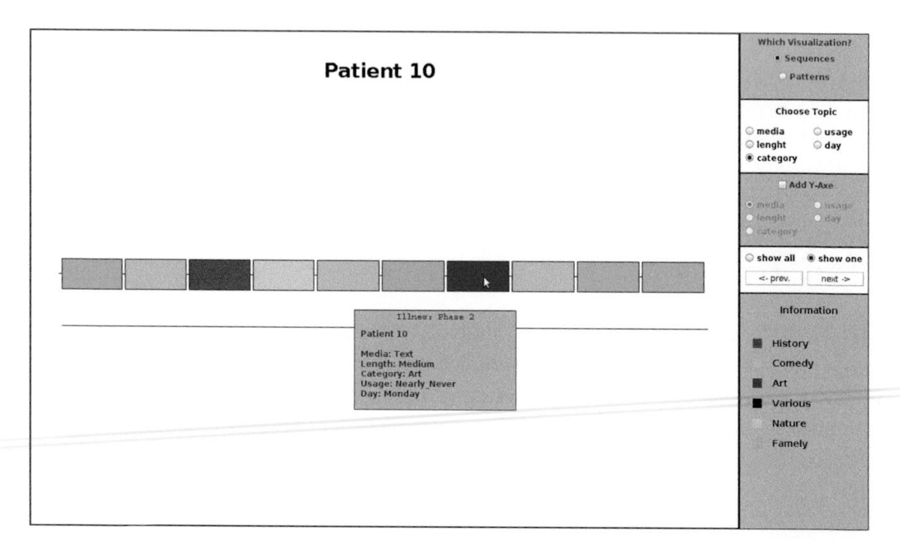

Fig. 11 Patient-focused sequence pattern (activity path) visualization

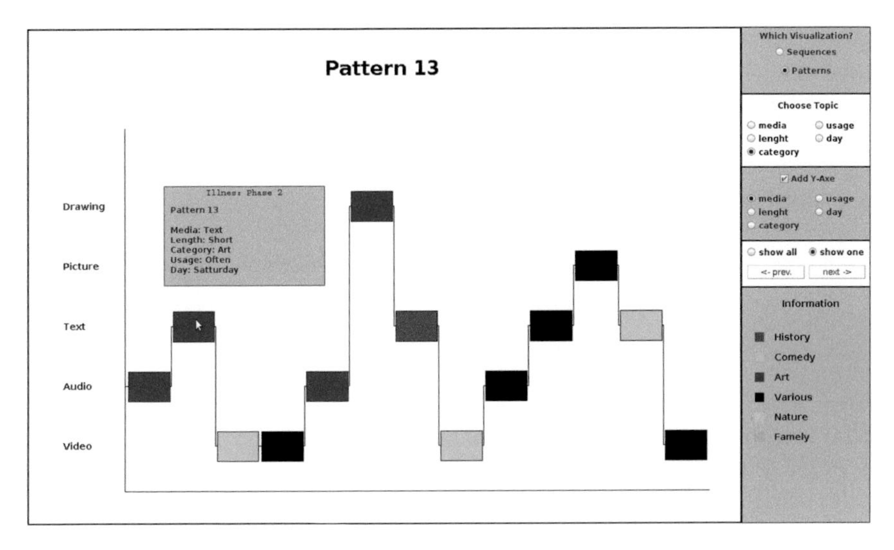

Fig. 12 Pattern-focused sequence pattern (activity path) extended by an axis labeled by considered activities

activity axes) as a trajectory showing the generalized 'movement' of a group of patients from one performed activity to the next one. The representation reveals further details concerning the type of activities performed by the group and shows how activities of a certain category are broken up into more specific descriptions. In this figure, one subsequence is shown for all patients who are in the disease phase 2. There are 13 events that are coincident for most of the patients. Among them, five show a partial coincident what is marked by a black color. The coincident strength (threshold) is specified as an input parameter of the analysis. Further details are explained in [41]; the document, together with the software prototype implementing the discussed functionality, can be downloaded from http://homepage.univie.ac.at/a1108558/.

5.2 *Association Rule Mining*

Association rule mining finds interesting association or correlation relationships among a large set of data items. The goal is to see if occurrence of certain items in a transaction can be used to deduce occurrence of other items or in other words to find associative relationships between items. These associations, or patterns, can be represented in the form of association rules.

The above challenges arise in the context of the data collected from scientific experiments or monitoring of physical systems such as telecommunications networks or from transactions at a supermarket. The problem was formulated originally in the context of the transaction data at supermarket. This market basket data as it is popularly known consists of transactions made by each customer. Each transaction contains items bought by the customer. The goal is to see if occurrence of certain items in a transaction can be used to deduce occurrence of other items or in other words to find associative relationships between items. These associations, or patterns, can be represented in the form of association rules. An example of an association rule expressed in a pseudo notation[10] is:

If patients received high education and have suffered from dementia less than 10 years, then they primarily view photos of the nature. [support = 35 %, confidence = 70 %]

The rule indicates that of the set of the observed patients, 35 % (support) have received high education, are dementia ill less than 10 years and primarily view photos of the nature. There is 70 % probability (confidence, or certainty) that a patient in this education and illness groups will primarily view photos of the nature.

Another example derived from the stability training application:

If a patient is less than 45 years old and more than 1 *year after brain damage, then he/she particularly chooses type of stability training combined with cognitive tasks. [support = 25 %, confidence = 60 %]*

Our visualization approach is illustrated in Fig. 13 that shows a coordinate system, whose x- and y-axes map the data of the log files that were captured by tracing the patients' activities during their work with eScrapBooks. More precisely, the x-axis denotes the antecedent and the y-axis is the consequent of the rule (Remark: Association rule's structure is antecedent → consequent.). We see twenty-two association rules, which have different red shade colors and a coloured scale aside the coordinate system. The scale gives more information about the value of the rectangles and shows the user of the visualization program the importance of each of them (the darker the more important). As shown above, this program offers one more additional function: if the user moves a mouse over a rule, he/she will get more information about the exact value. In this example '0.7' is the support of the association rule, but through a change of one parameter, the user can switch between support and confidence views. By clicking on the rectangle, the user gets a more detailed information about the whole rule. Example: First the user invokes the visualization program, then he/she chooses either the support or the confidence view. Through mouse over he/she gets additional information about the values. In the figure above the user moves the mouse over the following rule: AGE → BEHAVIOUR and sees that the support of this rule is 0.7, which means 70 %. But now he/she doesn't know the details of the rule. In [6], it is illustrated by several snapshots how to get access to such details; here we only give a brief description of the applicable activities. If the

[10]In a typical, almost standard notation, → is used instead 'then'.

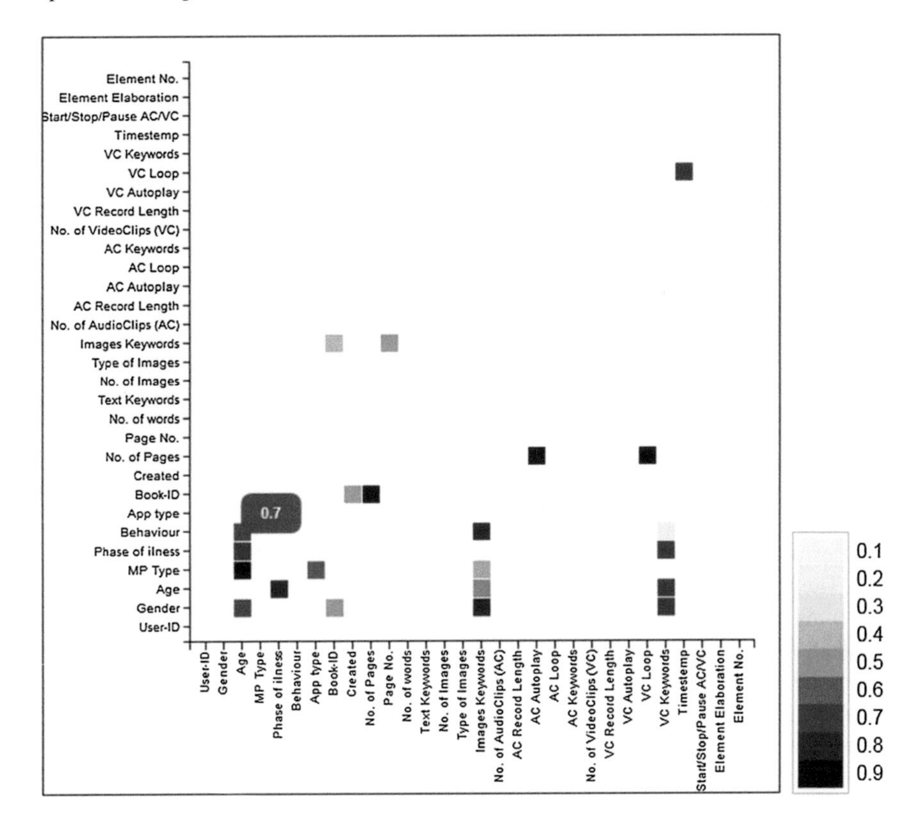

Fig. 13 Association rule visualization—base panel

user clicks on the rectangle mentioned above, a small window pops up and presens the following rule in a graphical form:

Elder senior (age greater than 87) → Behaviour (friendly, forgetful, slow) [support = 0.3, confidence = 0.7]

Thirty percent of the records obtained by the integration of eScrapbook session log files and other datasets bring the information that elder seniors have a specified behaviour—they are friendly, forgetful and slow. But from the record contents, it can be computed that seventy percents of data about elder seniors support the hypothesis that such a person is friendly, forgetful and slow.

6 Conclusions

In this chapter, we have introduced Brain Disorder Rehabilitation Informatics (BRDI) as an emerging field that uses informatics, especially BIG Data and Cloud technologies to collect and analyze data captured during rehabilitation processes. To fully reach the BRDI goals, it is important to provide optimized BIG data management for BRDI during the whole research life cycles and dealing with the knowledge extracted from the data. This is realized by means of the BRDI-Cloud, a novel research infrastructure for advanced management, reproducibility and execution of brain research studies. The infrastructure is organized into multiple BRDI-platforms. The kernel part of each BRDI platform enabling optimized BRDI data management is the BRDI dataspace realizing a novel scientific data management abstraction that we have developed. Each BRDI-platform is deployed in a single brain research and/or rehabilitation group (or organization) with its own instruments, data and other related resources. The designed infrastructure makes it possible to share selected resources that might include even entire research studies. In this way it enables all types of collaborative research studies, coordination of rehabilitation approaches, and exchange of experience.

It is no doubt that the BRDI-Cloud offers novel means than can significantly influence and change methodology of collaboration among distributed research groups and/or communities. The BRDI-Cloud can help to ensure a very important mission by offering natural and easy to be used tools for close interdisciplinary cooperation. The BRDI-Cloud makes it possible to combine services of data management and computational processing, two distinctive kinds of services necessary for full automation of workflows in most scientific studies that can be consequently shared and easily reproduced. Data management services are based on semantic web technologies, which enable us to collect, organize and represent semantic descriptions of conducted brain research studies. These descriptions, maintained within multiple semantic repositories on each BRDI-platform, can evolve into a large and distributed knowledge base—that has the potential to become a part of a large knowledge space envisioned e.g. in [61]—for the entire brain research community. Computing services provide the background for performing data-intensive brain research studies.

Design and implementation of our first BRDI infrastructure prototype builds on results we have gained in some former projects—in the BRDI-Cloud there are applied some services and libraries resulting from these projects. This chapter presents our approach to visualization of selected data mining models. A small core of early adopters is currently conducting their brain research studies and performs brain disorder rehabilitation according to methodology relying on the proposed model. Based on the data captured in these activities and feedback that we are receiving from its users, we will further improve the first prototype.

Acknowledgments The work described in this chapter is being carried out as part of four projects, namely the bilateral Austrian-Czech S&T cooperation project "Optimizing Large-Scale Dataflows" granted by the OeAD-GmbH/BMWFW, the project "SPES: Support Patients through E-service Solutions" supported by the CENTRAL EUROPE 3CE286P2 programme, the Czech National Sustainability Program grant LO1401 provided by the Czech and Austrian Ministries of Research and Education, and the project SGS16/231/OHK3/3T/13 provided by CVUT in Prague. We also express our deep gratitude to Dr. Ibrahim Elsayed; our research projects presented in this book chapter expands on his pioneering work on scientific dataspace and scientific research life cycle modelling, and his vision for new applications of these novel paradigms. Unfortunately, Dr. Ibrahim Elsayed suddenly passed away before the realization of these ideas was possible. This book chapter is dedicated to his memory.

References

1. EU-Project SPES. http://www.spes-project.eu/ (2014). Accessed Aug 2014
2. GNU Octave. http://www.gnu.org/software/octave/ (2012). Accessed Aug 2014
3. The R Project for Statistical Computing. http://www.r-project.org/ (2012). Accessed Aug 2014
4. Atkinson, M., Brezany, P., et al.: Data Bonanza—Improving Knowledge Discovery in BIG Data. Wiley (2013)
5. Bazerman, C.: Reading science: critical and functional perspectives on discourses of science, chapter 2. Emerging Perspectives on the Many Dimensions of Scientific Discourse, pp. 15–28. Routledge (1998)
6. Beneder, S.: Brain Stimulation of Dementia Patients—Automatic Tracing and Analysis of Their Activities. B.S. Thesis, Faculty of Computer Science, University of Vienna, 8 (2014)
7. Bohuncak, A., Janatova, M., Ticha, M., Svestkova, O., Hana, K.: Development of interactive rehabilitation devices. In: Smart Homes, vol. 2012, pp. 29–31 (2012)
8. Brezany, P., Ivanov, R.: Advanced Visualization of Data Mining and OLAP Results. Technical report, Aug 2005
9. Brezany, P., Janciak, I., Han, Y.: Cloud-enabled scalable decision tree construction. In: Proceedings of the International Conference on Semantic, Knowledge and Grid (2009)
10. Brezany, P., Janciak, I., Tjoa, A.M.: Chapter ontology-based construction of grid data mining workflows. Data Mining with Ontologies: Implementations, Findings, and Frameworks, pp. 182–210. IGI Global (2007)
11. Brezany, P., Janciak, I., Tjoa, A.M.: GridMiner: a fundamental infrastructure for building intelligent grid systems. In: Web Intelligence, pp. 150–156 (2005)
12. Brezany, P., Kloner, C., Tjoa, A.M.: Development of a grid service for scalable decision tree construction from grid databases. In: PPAM, pp. 616–624 (2005)
13. Brezany, P., Zhang, Y., Janciak, I., Chen, P., Ye, S.: An elastic OLAP cloud platform. In: DASC, pp. 356–363 (2011)
14. Cimiano, P., Hotho, A., Stumme, G., Tane, J.: Conceptual knowledge processing with formal concept analysis and ontologies. In: Eklund, P. (ed.) Concept Lattices. Lecture Notes in Computer Science, vol. 2961, pp. 189–207. Springer, Berlin (2004)
15. Clark, R.A., Bryant, A.L., Pua, Y., McCrory, P., Bennell, K., Hunt, M.: Validity and reliability of the Nintendo Wii Balance Board for assessment of standing balance. PubMed Gait Posture **2010**(31), 307–310 (2010)
16. Crystalinks. Metaphysics and Science Website. http://www.crystalinks.com/smithpapyrus700.jpg (2015). Accessed June 2015
17. Big Data-Careers. http://www.bigdata-careers.com/wp-content/uploads/2014/05/Big-Data-1.jpg?d353a9 (2015). Accessed June 2015
18. Elsayed, I.: Dataspace Support Platform for e-Science. PhD thesis, Faculty of Computer Science, University of Vienna, 2011. Supervised by P. Brezany, Revised version published by Südwestdeutscher Verlag für Hochschulschriften (https://www.svh-verlag.de), ISBN: 978-3838131573 (2013)
19. Elsayed, I., Brezany, P.: Dataspace support platform for e-science. Comput. Sci. **13**(1), 49–61 (2012)
20. Elsayed, I., Han, J., Liu, T., Whrer, A., Khan, F.A., Brezany, P.: Grid-enabled non-invasive blood glucose measurement. In: Bubak, M., van Albada, G., Dongarra, J., Sloot, P.M.A. (eds) Computational Science ICCS 2008, volume 5101 of Lecture Notes in Computer Science, pp. 76–85. Springer, Berlin (2008)
21. Elsayed, I., Ludescher, T., King, J., Ager, C., Trosin, M., Senocak, U., Brezany, P., Feilhauer, T., Amann, A.: ABA-Cloud: support for collaborative breath research. J. Breath Res. **7**(2), 026007–026007 (2013)
22. Elsayed, I., Muslimovic, A., Brezany, P.: Intelligent dataspaces for e-Science. In: Proceedings of the 7th WSEAS International Conference on Computational Intelligence, Man-machine Systems and Cybernetics, CIMMACS'08, pp. 94–100, Stevens Point, Wisconsin, USA (2008). World Scientific and Engineering Academy and Society (WSEAS)

23. Fiser, B., Onan, U., Elsayed, I., Brezany, P., Tjoa, A.M.: On-line analytical processing on large databases managed by computational grids. In: DEXA Workshops, pp. 556–560 (2004)
24. Franklin, M., Halevy, A., Maier, D.: From databases to dataspaces: a new abstraction for information management. SIGMOD Rec. **34**(4), 27–33 (2005)
25. Franklin, M., Halevy, A., Maier, D.: Principles of dataspace systems. In: PODS'06: Proceedings of the Twenty-fifth ACM SIGMOD-SIGACT-SIGART Symposium on Principles of Database Systems, pp. 1–9. ACM, New York, NY, USA (2006)
26. Gesundheitswissen. Gallery. http://www.fid-gesundheitswissen.de/bilder-responsive/gallery/768-Milz-milz-Fotolia-6856531-c-beerkoff.jpg (2015). Accessed June 2015
27. Gitlin, L.N.: Dementia (Improving Quality of Life in Individuals with Dementia: The Role of Nonpharmacologic Approaches in Rehabilitation). International Encyclopedia of Rehabilitation. http://cirrie.buffalo.edu/encyclopedia/en/article/28/ (2014). Accessed Aug 2014
28. Goscinski, A., Janciak, I., Han, Y., Brezany, P.: The cloudminer—moving data mining into computational cloud. In: Fiore, S., Aloisi, G. (eds.) Grid and Cloud Database Management, pp. 193–214. Springer, Berlin (2011)
29. Data Mining Group. The Predictive Model Markup Language (PMML). http://www.dmg.org/v3-2/ (2014). Accessed Aug 2014
30. Han, J., Kamber, M.: Data Mining: Concepts and Techniques. Morgan Kaufmann (2006)
31. Han, Y., Brezany, P., Goscinski, A.: Stream Management within the CloudMiner. In: ICA3PP (1), pp. 206–217 (2011)
32. Hey, T., Tansley, S., Tolle, K.M. (eds.) The Fourth Paradigm: Data-Intensive Scientific Discovery. Microsoft Research (2009)
33. Hoch, F., Kerr, M., Griffith, A.: Software as a Service: Strategic Backgrounder. http://www.siia.net/estore/ssb-01.pdf (2000). Accessed June 2015
34. Abirami Hospital. Facilities. http://www.abiramihospital.com/uploads/facilities/84977/t3_20120102005138.jpg (2015). Accessed June 2015
35. Janciak, I., Lenart, M., Brezany, P., Nováková, L., Habala, O.: Visualization of the mining models on a data mining and integration platform. In: MIPRO, pp. 215–220 (2011)
36. Joshi, M., Karypis, G., Kumar, V.: A Universal Formulation of Sequential Patterns. Technical report (1999)
37. Keahey, K., Tsugawa, M.O., Matsunaga, A.M., Fortes, J.A.B.: Sky computing. IEEE Internet Comput. **13**(5), 43–51 (2009)
38. Khan, F.A., Brezany, P.: Grid and Cloud Database Management, chapter Provenance Support for Data-Intensive Scientific Workflows, pp. 215–234. Springer, June 2011
39. Khan, F.A., Brezany, P.: Provenance support for data-intensive scientific workflows. In: Grid and Cloud Database Management, pp. 215–234 (2011)
40. Klyne, G., Carroll, J.J.: Resource Description Framework (RDF): Concepts and Abstract Syntax. World Wide Web Consortium, Recommendation REC-rdf-concepts-20040210, Feb 2004
41. Kühnel, J.: Mining Sequence Patterns from Data Collected by Brain Damage Rehabilitation. B.S. Thesis, Faculty of Computer Science, University of Vienna, Sept 2014
42. Liu, M.: Learning Decision Trees from Data Streams. B.S. Thesis, Faculty of Computer Science, University of Vienna, Oct 2010
43. Ludescher, T.: Towards High-Productivity Infrastructures for Time-Intensive Scientific Analysis. Ph.D. thesis, Faculty of Computer Science, University of Vienna (2013). Supervised by P. Brezany
44. Ludescher, T., Feilhauer, T., Amann, A., Brezany, P.:. Towards a high productivity automatic analysis framework for classification: an initial study. In: ICDM, pp. 25–39 (2013)
45. Martin, D. et al.: Bringing semantics to web services: the OWL-S approach. In: Proceedings of the First International Workshop on Semantic Web Services and Web Process Composition. San Diego, California (2004)
46. Matlab.: MATLAB—The Language of Technical Computing
47. Top Data Extraction Software Products
48. Prud'hommeaux, E., Seaborne, A.: SPARQL Query Language for RDF. http://www.w3.org/TR/rdf-sparql-query/ (2008). Accessed Jan 2008

49. Sahoo, S.S., Lhatoo, S.D., Gupta, D.K., Cui, L., Zhao, M., Jayapandian, C.P., Bozorgi, A., Zhang, G.-Q.: Epilepsy and seizure ontology: towards an epilepsy informatics infrastructure for clinical research and patient care. JAMIA, pp. 82–89 (2014)
50. Senocak. Design, Implementation and Evaluation of the e-Science Life-Cycle Browser. B.S. Thesis, Faculty of Computer Science, University of Vienna (2013)
51. Smith, E.: Surgical Papyrus. http://en.wikipedia.org/wiki/Edwin_Smith_Surgical_Papyrus (2014). Accessed Sept 2014
52. Sure, Y., et al.: On-To-Knowledge: Semantic Web-Enabled Knowledge Management, pp. 277–300. Springer, Berlin (2003)
53. Tian, Y.: Association Rules Mining in Data Stream. B.S. Thesis, Faculty of Computer Science, University of Vienna, June 2011
54. Trosin, M.: Design, Implementation and Evaluation of the e-Science Life-Cycle Visualizer. B.S. Thesis, Faculty of Computer Science, University of Vienna (2013)
55. Uller, M., Lenart, M., Stepankova, O.: eScrapBook: simple scrapbooking for seniors. In: Proceedings of the 1st Conference on Mobile and Information Technologies in Medicine, Prague, Czech Republic (2013)
56. Vogelova, M.: Evaluation of the Stabilometric Investigation in the Context of the Training of the Patients with Brain Damage. B.S. Thesis, Charles University Prague, Nov 2011
57. Vrotsou, K.: Everyday mining: exploring sequences in event-based data. Ph.D. thesis, Linköping University, Sweden (2010). Linköping Studies in Science and Technology. Dissertations No. 1331
58. White, T.: Hadoop: The Definitive Guide. 1st edn. O'Reilly Media Inc (2009)
59. Witten, I.H., Frank, E.: Data Mining: Practical Machine Learning Tools and Techniques, 2nd edn. Morgan Kaufmann, San Francisco (2005)
60. Ye, S., Chen, P., Janciak, I., Brezany, P.: Accessing and steering the elastic OLAP Cloud. In: MIPRO, pp. 322–327 (2012)
61. Zhuge, H.: Cyber-Physical society—The science and engineering for future society. Future Generation Comp. Syst. **32**, 180–186 (2014)

Author Biographies

Prof. Peter Brezany serves as a Professor of Computer Science at the University of Vienna, Austria and the Technical University of Brno, Czech Republic. He completed his doctorate in computer science at the Slovak Technical University Bratislava. His focus was first on the programming language and compiler design, automatic parallelization of sequential programs and later on high-performance input/output, parallel and distributed data analytics, and other aspects of data-intensive computing. Since 2001 he has led national and international projects developing Grid- and Cloud-based infrastructures for high-productivity data analytics. Dr. Brezany has published one book, several book chapters and over 120 papers. His recent book involvement addressing advanced e-Science data analytics appeared in Wiley in 2013.

Prof. Dr. Olga Štěpánková studied theoretical cybernetics at the Faculty of Mathematics and Physics of Charles University and in 1981 she gained her postgradual degree in Mathematical Logic. Currently, she is responsible for the master study program Biomedical engineering and informatics at the Faculty of Electrical Engineering and a head of the department Biomedical and assistive technologies (BEAT) in the Czech Institute of Informatics, Robotics and Cybernetics (CIIRC) of the Czech Technical University in Prague. Her current research covers data visualization and data mining, tele-medical systems design, design of various interdisciplinary ICT applications including that of novel assistive tools. She has reliable background in theoretical foundations of artificial intelligence, in mathematics and mathematical logic. She is author or co-author of more than 100 conference and journal papers, co-author or co-editor of 10 books.

Dr. Markéta Janatová graduated from the 1st Faculty of Medicine at the Charles University in Prague in 2008. Since 2010 she has worked at the Department of Rehabilitation Medicine of the First Faculty of Medicine of Charles University and General Teaching Hospital in Prague as a medical doctor. Since 2011 she is a member of interdisciplinary team at Joint Department of Biomedical Engineering, Czech Technical University with a specialization in development of innovative technical devices for rehabilitation. Since 2013 she works at Spin-off company and research results commercialization center at the 1st Faculty of Medicine, Charles University in Prague.

Miroslav Uller works as a researcher at the Department of Cybernetics, Faculty of Electrical Engineering of the Czech Technical University in Prague (CTU). He is a member of the Nature Inspired Technologies Group at CTU. He participated in the European projects OLDES (Older people's e-services at home) and SPES (Support Patients through E-service Solutions) and co-authored several publications resenting the results of this effort. His main interests are user interface design, data visualization in the big data science context, virtual reality, functional programming and domain specific languages.

Marek Lenart is an MSc candidate at the University of Vienna, at the Faculty of Computer Science. He was involved in the European projects ADMIRE (Advanced Data Mining and Integration Research for Europe) and SPES (Support Patients through E-service Solutions) and co-authored several publications addressing the results of these projects. His main focus was on advanced visualization of models produced by mining big scientific and business data. Now he works on ambitious industrial projects that are focused on simulation, analysis and optimization of data related to car production. The key objective of this effort is to minimize carbon dioxide emission in the context of the EU and international regulations.

Big Data Optimization in Maritime Logistics

Berit Dangaard Brouer, Christian Vad Karsten and David Pisinger

Abstract Seaborne trade constitutes nearly 80 % of the world trade by volume and is linked into almost every international supply chain. Efficient and competitive logistic solutions obtained through advanced planning will not only benefit the shipping companies, but will trickle down the supply chain to producers and consumers alike. Large scale maritime problems are found particularly within liner shipping due to the vast size of the network that global carriers operate. This chapter will introduce a selection of large scale planning problems within the liner shipping industry. We will focus on the solution techniques applied and show how strategic, tactical and operational problems can be addressed. We will discuss how large scale optimization methods can utilize special problem structures such as separable/independent subproblems and give examples of advanced heuristics using divide-and-conquer paradigms, decomposition and mathematical programming within a large scale search framework. We conclude the chapter by discussing future challenges of large scale optimization within maritime shipping and the integration of predictive big data analysis combined with prescriptive optimization techniques.

Keywords Large-scale optimization · Decision support tools · Prescriptive analytics · Maritime logistics

1 Introduction

Modern container vessels can handle up to 20,000 twenty-foot equivalent units (TEU) as seen on Fig. 1. The leading companies may operate a fleet of more than 500 vessels and transport more than 10,000,000 full containers annually that need

B.D. Brouer · C.V. Karsten · D. Pisinger (✉)
Department of Management Engineering, Technical University of Denmark, Produktionstorvet, Building 426, 2800 Kgs Lyngby, Denmark
e-mail: dapi@dtu.dk

B.D. Brouer
e-mail: berit@brouer.com

C.V. Karsten
e-mail: cvadkarsten@gmail.com

A. Emrouznejad (ed.), *Big Data Optimization: Recent Developments and Challenges*, Studies in Big Data 18, DOI 10.1007/978-3-319-30265-2_14

Fig. 1 Seaborne trade constitutes nearly 80 % of the world trade by volume, and calls for the solution of several large scale optimization problems involving big data. *Picture: Maersk Line*

to be scheduled through the network. There is a huge pressure to fill this capacity and utilize the efficiency benefits of the larger vessels but at the same time markets are volatile leading to ever changing conditions. Operating a liner shipping network is truly a big-data problem, demanding advanced decisions based on state-of-the art solution techniques. The digital footprint from all levels in the supply chain provides opportunities to use data that drive a new generation of faster, safer, cleaner, and more agile means of transportation. Efficient and competitive logistic solutions obtained through advanced planning will not only benefit the shipping companies, but will trickle down the supply chain to producers and consumers.

Maritime logistics companies encounter large scale planning problems at both the strategic, tactical, and operational level. These problems are usually treated separately due to complexity and practical considerations, but as will be seen in this chapter the decisions are not always independent and should not be treated as such. Large scale maritime problems are found both within transportation of bulk cargo, liquefied gasses and particularly within liner shipping due to the vast size of the network that global carriers operate. In 2014 the busiest container terminal in the world, Port of Shanghai, had a throughput of more than 35,000,000 TEU according to Seatrade Global, which is also approximately the estimated number of containers in circulation globally. This chapter will focus on the planning problems faced by a global carrier operating a network of container vessels and show how decision support tools based on mathematical optimization techniques can guide the process of adapting a network to the current market.

At the strategic level carriers determine their fleet size and mix along with which markets to serve thus deciding the layout of their network. The network spanning the globe serving tens of thousands of customers leads to a gazillion possible configurations for operating a particular network. At the tactical level schedules for the individual services and the corresponding fleet deployment is determined, while the routing of containers through the physical transportation network, stowage of containers on the vessels, berthing of the vessels in ports, and disruption management due to e.g. bad weather or port delays is handled at the operational level. In general these problems can be treated separately, but as the layout of the network will affect e.g. the routing of the containers the problems are far from independent.

Operational data can lead to better predictions of what will happen in the future and carriers are constantly receiving sensor data from vessels that can help predict e.g. disruptions or required maintenance and similarly, data received from terminals can be used to predict delays and help vessels adjust sailing speed to save fuel. But given a predicted future scenario it may still not be obvious what the best actions are neither at the strategic, tactical or operational level. A large shipping company may be capable of producing good estimates of future demand and oil price fluctuations, or predicting possible disruptions. Under certain circumstances these predictions may require simple independent actions to adjust the network, but it is more likely that the actions will be dependent on other factors in the network. In that case difficult and complex here-and-now decisions must be made to adjust the transportation network optimally to the new situation. When there is a large number of decisions to be made and when the decisions influence each other prescriptive models based on optimization can help make the best choice. Predictive and prescriptive methods combined can serve as decision support tools and help select the best strategy, where the predictions made by machine learning algorithms, can be fed into large scale optimization algorithms to guide the decision process faced by carriers.

Most data in liner shipping are associated with some degree of uncertainty. First of all, demands are fluctuating over the year, and even if customers have booked a time slot for their containers these data are affected by significant uncertainty. In liner shipping no fees are paid if the customer is not delivering the booked number of containers, so customers may at any time choose to use another shipping company, or to postpone the delivery. This stimulates overbooking which adds uncertainty to the models. Port availabilities are also highly uncertain. If a vessel sticks to the normal time table, it can generally be assumed that the time slot is available, but if a vessel is delayed or the company wants to change the route, all port calls must be negotiated with the port authorities. This substantially complicates planning, and makes it necessary to use a trial and force method to find a good solution.

There are several different approaches for solving large scale optimization problems. If a problem exhibit a special separable structure it can be decomposed and solved more efficiently by using either column generation if the complication involves the number of variables or row generation if the number of constraints is too large [5, 8, 18, 20], by dynamic programming [17], or constraint programming [36]. For less structured or extremely large problems it can be advantageous to use (meta)-heuristics to obtain solutions quickly, but often of unknown quality [15, 22]. Finally it is frequently possible, with a good modeling of a problem, to rely solely on Linear Programming, LP, or Mixed Integer Programming, MIP, solvers, see e.g. [42] for a discussion of modeling techniques and the trade-off between stronger versus smaller models. Algorithmic and hardware improvements have over the last three decades resulted in an estimated speed-up for commercial MIP solvers of a 200 billion factor [7], making it feasible not only to solve large linear models but also more advanced integer decision models of realistic size. In practice a combination of the different techniques is often seen and maritime logistics gives an illustrative case of the importance of all of these large scale optimization methods.

2 Liner Shipping Network Design

The Liner Shipping Network Design Problem, LSNDP, is a core planning problem facing carriers. Given an estimate of the demands to be transported and a set of possible ports to serve, a carrier wants to design routes for its fleet of vessels and select which demands of containers to satisfy. A route, or *service*, is a set of similarly sized vessels sailing on a non-simple cyclic itinerary of ports according to a fixed, usually weekly, schedule. Hence the round trip duration for a vessel is assumed to be a multiple of a week and to ensure weekly frequency in the serviced ports a sufficient number of vessels is assigned. If a round trip of the vessel takes e.g. 6 weeks, then 6 vessels are deployed on the same route. To make schedules more robust buffer time is included to account for delays. However, delays may still lead to local speed increases which increases the overall energy consumption. An example of a service can be seen in Fig. 2 which shows the Oceania-Americas Service with a round trip time of 10 weeks. The weekly departures may in some cases simplify the mathematical formulation of the problem, since customer demands and vessel capacities follow a weekly cycle. *Trunk services* serve central main ports and can be both inter and intra regional whereas *feeder services* serve a distinct market and typically visit one single main port and several smaller ports. When the network has been determined the containers can be routed according to a fixed schedule with a predetermined trip duration. A given demand is loaded on to a service at its departure port, which may bring the demand directly to the destination port or the container can be unloaded at one or several intermediate ports for *transshipment* to another service before finally reaching its final destination. Therefore, the design of the set of services is complex, as they interact through transshipments and the majority of containers are transshipped at least once during transport. A carrier aims for a network with high utilization, a low number of transshipments, and competitive transit times. Services are divided into a head- and a back-haul direction. The head haul direction is the most cargo intensive and vessels are almost full. Hence, the head haul generates the majority of the revenue and due to customer demand for fast delivery the head haul operates at increased speeds with nearly no buffer time for delays. The back haul operates at slower speeds with additional buffer time assigned. A delay incurred on the head haul is often recovered during the back-haul.

In practice a carrier will never re-design a network from scratch as there are significant costs associated with the reconfiguration [40]. Rather, the planners or network design algorithms will take the existing network and suggest incremental changes to adjust the network to the current economic environment. Most network changes requires evaluation of the full cargo routing problem to evaluate the quality

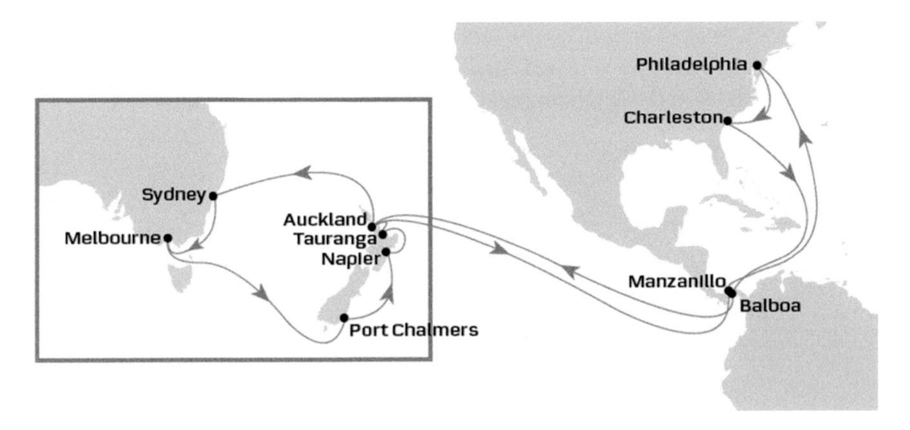

Fig. 2 The Oceania-Americas Service (OC1) from the 2014 Maersk Line Network. *Picture: Maersk Line*

of the network since regional changes can have unintended consequences in the entire network.

Routing of both vessels and containers are in most state-of-the-art methods considered simultaneously [1, 2, 11, 12, 35, 38], as these problems are completely interrelated. However, several of the before mentioned approaches exploit the fact that the problems are separable into two tiers and design algorithms utilizing this structure. The cargo routing reduces to a multicommodity flow problem, MCF, and serves as the lower tier where the revenue of the network is determined. The vessel routing problem reduces to a (more complex) problem of cycle generation and corresponds to the upper tier, where the cost of the network is determined. The following section gives insight to the container routing problem and its relation to the multicommodity flow problem.

2.1 Container Routing

We define $G = (N, A)$ to be a directed graph with nodes N and edges A. The node set N represents the geographical locations in the model i.e. ports and the arc set A connects the ports. The arcs are determined by the scheduled itineraries and the cargo capacity is determined by the assignment of vessels to the schedule. Let K be the set of commodities to transport, q_k be the amount of commodity $k \in K$ that is available for transport, and u_{ij} be the capacity of edge (i,j). We assume that each commodity has a single origin node, O_k, and a single destination node, D_k.

There are two commonly used formulations of the MCF based on either arc or path flow variables. The arc flow formulation can be stated as follows. For each node $i \in N$ and commodity $k \in K$ we define $q(i,k) = q^k$ if $i = O_k$, $q(i,k) = -q^k$ if $i = D_k$, and

$q(i,k) = 0$ otherwise. For each node $i \in N$ we define the set of edges with tail in node i as $\delta^+(i) = \{(j,j') \in A : j = i\}$ and head in node i as $\delta^-(i) = \{(j,j') \in A : j' = i\}$.

With this notation the MCF problem can be stated as the following LP

$$\min \sum_{(i,j)\in A} \sum_{k\in K} c_{ij}^k x_{ij}^k \tag{1}$$

$$\text{s.t.} \sum_{(j,j')\in\delta^+(i)} x_{jj'}^k - \sum_{(j,j')\in\delta^-(i)} x_{jj'}^k = q(i,k) \qquad i \in N, k \in K \tag{2}$$

$$\sum_{k\in K} x_{ij}^k \le u_{ij} \qquad (i,j) \in A \tag{3}$$

$$x_{ij}^k \ge 0 \qquad (i,j) \in A, k \in K \tag{4}$$

The objective function (1) minimizes the cost of the flow. The flow conservation constraint (2) ensures that commodities originates and terminates in the right nodes. The capacity constraint (3) ensures that the capacity of each edge is respected. The formulation has $|K||A|$ variables and $|A| + |K||N|$ constraints. The number of variables is hence polynomially bounded, but for large graphs like the ones seen in global liner shipping networks this formulation requires excessive computation time and may even be too large for standard LP-solvers (see e.g. [14]).

The *block-angular structure* of the constraint matrix in the arc-flow formulation can be exploited and by *Dantzig-Wolfe decomposition* it is possible to get a reformulation with a master problem considering paths for all commodities, and a subproblem defining the possible paths for each commodity $k \in K$. We note that in general any arc flow can be obtained as a convex combination of path flows. In the path-flow formulation each variable, f^p, in the model corresponds to a path, p, through the graph for a specific commodity. The variable states how many units of a specific commodity that is routed through the given path, the cost of each variable is given by the parameter c_p. Let P^k be the set of all feasible paths for commodity k, $P^k(a)$ be the set of paths for commodity k that uses edge a and $P(a) = \cup_{k\in K} P^k(a)$ is the set of all paths that use edge a. The model then becomes:

$$\min \sum_{k\in K} \sum_{p\in P^k} c_p f^p \tag{5}$$

$$\text{s.t.} \sum_{p\in P^k} f^p = q_k \qquad k \in K \tag{6}$$

$$\sum_{p\in P(a)} f^p \le u_{ij} \qquad (i,j) \in A \tag{7}$$

$$f^p \ge 0 \qquad k \in K, p \in P^k \tag{8}$$

The objective function (5) again minimizes the cost of the flow. Constraint (6) ensures that the demand of each commodity is met and constraint (7) ensures that the capacity limit of each edge is obeyed. The path-flow model has $|A| + |K|$ constraints, but the number of variables is, in general, growing exponentially with the size of the graph. However, using column generation the necessary variables can be generated dynamically and in practice the path-flow model can often be solved faster than the arc-flow model for large scale instances of the LSND problem [14].

Column generation operates with a reduced version of the LP (5)–(8), which is called the master problem. The master problem is defined by a reduced set of columns $Q^k \subseteq P^k$ for each commodity k such that a feasible solution to the LP (5)–(8) can be found using variables from $\cup_{k\in K} Q^k$. Solving this LP gives rise to dual variables π_k and λ_{ij} corresponding to constraint (6) and (7), respectively. For a variable $j \in \cup_{k\in K} P^k$ we let $\kappa(j)$ denote the commodity that a variable serves and let $p(j)$ represent the path corresponding to the variable j, represented as the set of edges traversed by the path. Then we can calculate the *reduced cost* $\bar{c}_j$ of each column $j \in \cup_{k\in K} P^k$ as follows

$$\bar{c}_j = \sum_{(i,j)\in p(j)} (c_{ij}^{\kappa(j)} - \lambda_{ij}) - \pi_{\kappa(j)}.$$

If we can find a variable $j \in \cup_{k\in K}(P^k \setminus Q^k)$ such that $\bar{c}_j < 0$ then this variable has the potential to improve the current LP solution and should be added to the master problem, which is resolved to give new dual values. If, on the other hand, we have that $\bar{c}_j \geq 0$ for all $j \in \cup_{k\in K}(P^k \setminus Q^k)$ then we know the master problem defined by Q^k provides the optimal solution to the complete problem (for more details see [24]). In order to find a variable with negative reduced cost or prove that no such variable exists we solve a sub-problem for each commodity. The sub-problem seeks the feasible path for commodity k with minimum reduced cost given the current dual values. Solving this problem amounts to solving a shortest path problem from source to destination of the commodity with edge costs given by $c_{ij} - \lambda_{ij}$ and subtracting π_k from this cost in order to get the reduced cost. As will be seen later we can extend the model to reject demands by including additional variables with an appropriate penalty. When solving the shortest path problem additional industry constraints such as number of transshipments, trade policies, or time limits on cargo trip duration can be included. Including such constraints will increase the complexity of the sub-problem as the resulting problem becomes a resource constrained shortest path problem. Karsten et al. [24] has made a tailored algorithm for a cargo routing problem considering lead times and show that it does not necessarily increase the solution time to include transit time constraints, mainly because the size of solution space is reduced. Additionally, Karsten et al. [24] give an overview of graph topologies accounting for transshipment operations when considering transit times.

To construct routes used in the upper tier of the network design problem we will go through a more recent approach in the next section which use an advanced mathematical programming based heuristic to solve the problem within a large scale search framework. In general, when a generic network has been designed it is transformed into a physical sailing network by determining a specific schedule, deploying vessels from the available fleet and deciding on the speed and actual flow of containers.

Some aspects of the tactical and operational decisions can of course be integrated in the network design process at the cost of computational tractability, but with the potential benefit of higher quality networks.

3 Mat-Heuristic for Liner Shipping Network Design

Mathematical programming models of the LSNDP are closely related to the capacitated fixed charge network design problem [23] in installing a discrete set of capacities for the set of commodities K. However, the capacity installed must reflect the routing of container vessels according to the specification of a service as defined in the beginning of this section. Therefore, it is also related to pick-up and delivery vehicle routing problems [41], however being significantly harder to solve as a consequence both of the non-simple cyclic routes, the multiple commodities and the vast size of real life networks. As a consequence optimal methods can only solve very insignificant instances of the LSNDP [2, 35] or provide lower bounds [33]. Several algorithms for solving larger instances of the LSNDP can be categorized as *matheuristics* combining mathematical programming with meta heuristics exploiting the two tier structure, where the variables of the upper tier describe a service and variables of the lower tier describe the container routing (for a reference model of the LSNDP see [10]). Agarwal and Ergun [1] apply a heuristic Benders' decomposition algorithm as well as a Branch and Bound algorithm for heuristicly generated routing variables, Alvarez [2] applies a tabu search scheme, where the routing variables are generated by a mathematical program based on the dual values of the lower tier MCF problem in each iteration. [10] use a heuristic column generation scheme, where the routing columns are generated by an integer program based on information from both tiers of the LSNDP along with a set of business rules. The integer program in [10] constructs a single, (possibly non-simple) cyclic route for a given service configuration of vessel class and speed. Route construction is based on the Miller-Tucker-Zemlin subtour elimination constraints known from the CVRP to enumerate the port calls in a non-decreasing sequence. This makes high quality routings for smaller instances of the LSNDP, but for large scale instances it becomes necessary to select a small cluster of related ports in order to efficiently solve the integer program used in the heuristic. A different matheuristic approach is seen in [11, 12], where the core component in a large scale neighborhood search is an integer program designed to capture the complex interaction of the cargo allocation between routes. The solution of the integer program provides a set of moves in the composition of port calls and fleet deployment. Meta-heuristics for the LSNDP are challenged by the difficulty of predicting the changes in the multicommodity flow problem for a given move in the solution space without reevaluating the MCF at the lower tier. The approach of [12] relies on estimation functions of changes in the flow and the fleet deployment related to inserting or removing a port call from a given service and network configuration. Flow changes and the resulting change in the revenue are estimated by solving a series of shortest path problems on the residual graph of

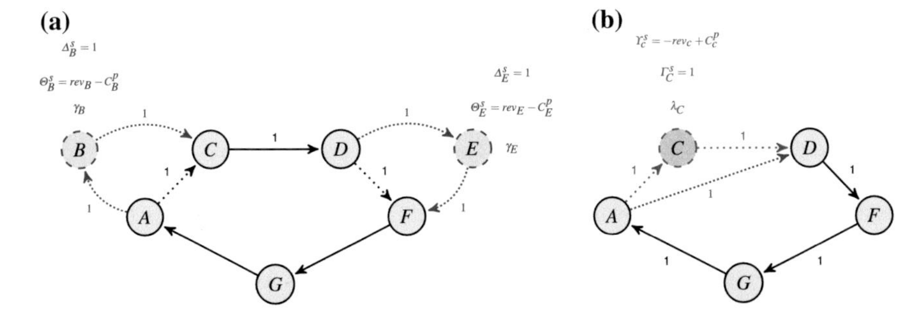

Fig. 3 Illustration of the estimation functions for insertion and removal of port calls. **a** *Blue* nodes are evaluated for insertion corresponding to variables γ_i for the set of ports in the neighborhood N^s of service s. **b** *Red* nodes are evaluated for removal corresponding to variables λ_i for the set of current port calls F^s on service s

the current network for relevant commodities to the insertion/removal of a port call along with an estimation of the change in the vessel related cost with the current fleet deployment.

Given a total estimated change in revenue of rev_i and port call cost of C_i^p Fig. 3a illustrate estimation functions for the change in revenue (Θ_i^s) and duration (Δ_i^s) increase for inserting port i into service s controlled by the binary variable γ_i. The duration controls the number of vessels needed to maintain a weekly frequency of service. Figure 3b illustrate the estimation functions for the change in revenue (Υ_i^s) and decrease in duration (Γ_i^s) for removing port i from service s controlled by the binary variable λ_i. Insertions/removals will affect the duration of the service in question and hence the needed fleet deployment modeled by the integer variable ω_s representing the change in the number of vessels deployed. The integer program (9)–(16) expresses the neighborhood of a single service, s.

$$\max \sum_{i \in N^s} \Theta_i \gamma_i + \sum_{i \in \mathbf{F^s}} \Upsilon_i \lambda_i - C_V^{e(s)} \omega_s \tag{9}$$

$$\text{s.t.} \quad T_s + \sum_{i \in N^s} \Delta_i^s \gamma_i - \sum_{i \in \mathbf{F^s}} \Gamma_i^s \lambda_i \leq 24 \cdot 7 \cdot (n_s^{e(s)} + \omega_s) \tag{10}$$

$$\omega_s \leq M_{e(s)} \tag{11}$$

$$\sum_{i \in N^s} \gamma_i \leq I_s \tag{12}$$

$$\sum_{i \in \mathbf{F^s}} \lambda_i \leq R_s \tag{13}$$

$$\sum_{j \in L_i} \lambda_j \leq |L_i|(1 - \gamma_i) \qquad i \in N^s \tag{14}$$

$$\sum_{j \in L_i} \lambda_j \leq |L_i|(1 - \lambda_i) \qquad i \in F^s \tag{15}$$

$$\lambda_i \in \{0, 1\}, i \in F^s, \qquad \gamma_i \in \{0, 1\}, i \in N^s, \qquad \omega_s \in \mathbb{Z}. \tag{16}$$

The objective function (9) accounts for the expected change in revenue of the considered insertions and removals along with the weekly vessel cost $C_V^{e(s)}$ of the vessel class $e(s)$ deployed to service s. Constraint (10) considers the expected change in the duration of the service, where T_s is the current duration and $n_s^{e(s)}$ is the number of vessels currently deployed to service s. The possible addition of vessels is bounded by the number of vessels available $M_{e(s)}$ of type e in constraint (11). A limit on the number of insertions/removals respectively are introduced in constraints (12)–(13) to reduce the error of the estimation functions for multiple insertions/removals. The estimation functions also depend on the existing port calls for unloading the commodities introduced by the insertions as well as the ports used for rerouting commodities when removing ports. This is handled by introducing a lockset L_i for each insertion/removal expressed in constraints (14)–(15). The integer program is solved iteratively for each service in the current network and the resulting set of moves are evaluated for acceptance in a simulated annealing framework. The procedure is an improvement heuristic [3] fine tuning a given network configuration. The algorithm in its entirety constructs an initial network using a simple greedy construction heuristic. The improvement heuristic is applied as a move operator for intensification of the constructed solution. To diversify the solution a perturbation step is performed at every tenth loop through the entire set of services. The perturbation step alters the service composition in the network by removing entire services with low utilization and introducing a set of new services based on the greedy construction heuristic for undeployed vessels. To evaluate the matheuristic the public benchmark suite, LINER-LIB, for liner shipping network design problems is used.

4 Computational Results Using LINER-LIB

LINER-LIB 2012 is a public benchmark suite for the LSNDP presented by [10]. The data instances of the benchmark suite are constructed from real-life data from the largest global liner-shipping company, Maersk Line, along with several industry and public stakeholders. LINER-LIB consists of seven benchmark instances available at http://www.linerlib.org/ (see [10] for details on the construction of the data instances). Each instance can be used in a low, medium, and high capacity case depending on the fleet of the instance. Table 1 presents some statistics on each instance ranging from smaller networks suitable for optimal methods to large scale instances spanning the globe. Currently published results are available for 6 of the 7 instances, leaving the *WorldLarge* instance unsolved.

LINER-LIB contains data on ports including port call cost, cargo handling cost and draft restrictions, distances between ports considering draft and canal traversal, vessel related data for capacity, cost, speed interval and bunker consumptions, and finally a commodity set with quantities, revenue, and maximal transit time. The commodity data reflects the current imbalance of world trade and the associated differentiated revenue. It is tailored for models of the LSNDP, but may provide useful data for related maritime transportation problems.

Table 1 The instances of the benchmark suite with indication of the number of ports ($|P|$), the number of origin-destination pairs ($|K|$), the number of vessel classes ($|E|$), the minimum (min v) and maximum number of vessels (max v)

Category	Instance and description	|P|	|K|	|E|	min v	max v
Single-hub instances	**Baltic** Baltic sea, Bremerhaven as hub	12	22	2	5	7
	WAF West Africa, Algeciras as hub	19	38	2	33	51
Multi-hub instance	**Mediterranean** Mediterranean, Algeciras, Tangier, Gioia Tauro as hubs	39	369	3	15	25
Trade-lane instances	**Pacific** (Asia-US West)	45	722	4	81	119
	AsiaEurope Europe, Middle East, and Far east regions	111	4000	6	140	212
World instance	**Small** 47 Main ports worldwide identified by Maersk Line	47	1764	6	209	317
	Large 197 ports worldwide identified by Maersk Line	197	9630	6	401	601

Computational results for LINER-LIB are presented in [10, 12, 33]. Brouer et al. [10] presented the first results for the benchmark suite using the reference model [10] with biweekly frequencies for the feeder vessel classes and weekly frequencies for remaining classes. The heuristic column generation algorithm is used to solve all instances but the Large world instance with promising results. [12] present computational results using the reference model with weekly frequencies for all vessel classes which has a more restricted solution space than [10]. As a consequence the solutions from [12] are feasible for the model used in [10], but not vice-versa. However, the computational results of [12] indicate that the matheuristic using an improvement heuristic based on integer programming scales well for large instances and holds the current best known results for the Pacific, World Small and AsiaEurope instances. [33] present a service flow model for the LSNDP using a commercial MIP solver presenting results for the two Baltic and WAF instances of LINER-LIB. For details on the results the reader is referred to the respective papers. LINER-LIB is currently used by researchers at a handful of different universities worldwide and may provide data for future results on models and algorithms for LSNDP.

5 Empty Container Repositioning

In extension of the network design process a liner shipping company must also consider revenue management at a more operational level. Requests for cargo can be rejected if it is not profitable to carry the containers, or if bottlenecks in the network make it infeasible. Moreover, empty containers tend to accumulate at importing regions due to a significant imbalance in world trade. Therefore, repositioning empty containers to exporting regions impose a large cost on liner shippers, and these costs need to be incorporated in the revenue model. Since larger shipping companies at any time have several millions of containers in circulation, these decisions are extremely complex and require advanced solution methods.

Alvarez [2] presented a study of large scale instances of the liner service network design problem. The cargo allocation problem is solved as a subproblem of the tabu search algorithm solving the network design problem. Meng and Wang [26] study a network design problem selecting among a set of candidate shipping lines while considering the container routing problem along with the repositioning of empty containers. The model is formulated as a minimum cost problem and as [21] the model handle loaded end empty containers simultaneously, however it does not allow load rejection and only seek to minimize the cost of transport. Song and Dong [39] consider a problem of joint cargo routing and empty container repositioning at the operational level accounting for the demurrage and inventory cost of empty containers. Like most other works on empty repositioning it is a cost minimizing problem where load rejection is not allowed.

Brouer et al. [14] present a revenue management model for strategic planning within a liner shipping company. A mathematical model is presented for maximizing the profit of cargo transportation while considering the possible cost of repositioning empty containers.

The booking decision of a liner shipper considering empty container repositioning can be described as a specialized multi-commodity flow problem with inter-balancing constraints to control the flow of empty containers.

Similarly to the pure cargo routing problem we can define a commodity as the tuple (O_k, D_k, q_k, r_k) representing a demand of q_k in number of containers from node O_k to node D_k with a sales price per unit of r_k. The unit cost of arc (i,j) for commodity k is denoted c^k_{ij}. The non-negative integer variable x^k_{ij} is the flow of commodity k on arc (i,j). The capacity of arc (i,j) is u_{ij}. To model the empty containers an empty super commodity k_e is introduced. The flow of the empty super commodity is defined for all $(i,j) \in A$ as the integer variables $x^{k_e}_{ij}$. The unit cost of arc (i,j) for commodity k_e is denoted $c^{k_e}_{ij}$. The empty super commodity has no flow conservation constraints and appear in the objective with a cost and in the bundled capacity and inter-balancing constraints. For convenience the commodity set is split into the loaded commodities and the empty super commodity: Let K_F be the set of loaded commodities. Let K_e be the set of the single empty super commodity. Finally, let $K = K_F \cup K_e$. The inter-balancing constraints also introduce a new set of variables representing leased containers at a node. The cost of leasing is modeled in the objective. Let c^i_l be the cost of leasing a container at port i, while l_i is the integer leasing variable at port i. Demand may be rejected, due to capacity constraints and unprofitability from empty repositioning cost. The slack variable γ_k represents the amount of rejected demand for commodity k.

5.1 Path Flow Formulation

In the following we introduce a path flow model which is an extension of model (5)–(8). Again, let p be a path connecting O_k and D_k and P_k be the set of all paths belonging to commodity k. The flow on path p is denoted by the variable f^p. The binary coefficient a^p_{ij} is one if and only if arc (i,j) is on the path p. Finally, $c^k_p = \sum_{(i,j)\in A} a^p_{ij} c^k_{ij}$ is the cost of path p for commodity k. The master problem is:

$$\max \quad \sum_{k\in K_F}\sum_{p\in P_k}(r_k - c^k_p) f^p - \sum_{(i,j)\in A} c^{k_e}_{ij} x^{k_e}_{ij} - \sum_{i\in N} c^i_l l^i \tag{17}$$

$$\text{s.t.} \quad \sum_{k\in K_F}\sum_{p\in P_k} a^p_{ij} f^p + x^{k_e}_{ij} \leq u_{ij} \qquad (i,j)\in A \tag{18}$$

$$\sum_{p\in P_k} f^p + \gamma_k = q_k \qquad k\in K_F \tag{19}$$

$$\sum_{k\in K_F}\sum_{p\in P_k}\sum_{j\in N}(a^p_{ij} - a^p_{ij}) f^p + x^{k_e}_{ij} - x^{k_e}_{ji} - l^i \leq 0 \qquad i\in N \tag{20}$$

$$f^p \in \mathbb{Z}_+,\ p\in P_k, \qquad \gamma_k \in \mathbb{Z}_+,\ k\in K_F \qquad x^{k_e}_{ij}\in \mathbb{Z}_+,\ (i,j)\in A, \quad l^i\in\mathbb{Z}_+,\ i\in N \tag{21}$$

where the x^k_{ij} variables can be replaced by $\sum_{p\in P_k} a^p_{ij} f^p$ for all $k \in K_F$. The convexity constraints for the individual subproblems (19) bound the flow between the (O_k, D_k) pair from above (a maximal flow of q_k is possible).

Paths are generated on the fly using delayed column generation. Brouer et al. [14] report computational results for eight instances based on real life shipping networks, showing that the delayed column generation algorithm for the path flow model clearly outperforms solving the arc flow model with the CPLEX barrier solver. In order to fairly compare the arc and path flow formulation a basic column generation algorithm is used for the path flow model versus a standard solver for the arc flow model. Instances with up to 234 ports and 293 vessels for 9 periods were solved in less than 35 min with the column generation algorithm. The largest instance solved for 12 periods contains 151 ports and 222 vessels and was solved in less than 75 min.

The algorithm solves instances with up to 16,000 commodities over a twelve month planning period within one hour. Integer solutions are found by simply rounding the LP solution. The model of Erera et al. [21] is solved to integer optimality using standard solvers as opposed to the rounded integer solution presented here. The problem sizes of [14] are significantly larger than those of [21] and the rounded integer solutions lead to a gap of at most 0.01 % from the LP upper bound of the path flow formulation, which is very acceptable, and far below the level of uncertainty in the data. The results of [21] confirm the economic rationale in simultaneously considering loaded and empty containers.

6 Container Vessel Stowage Plans

With vessels carrying up to 20,000 TEU, stowage of the containers on board is a non-trivial task demanding fast algorithms as the final load list is known very late. Stowage planning can be split into a master planning problem and a more detailed slot planning problem. The master planning problem should decide a proper mixture of containers, so that constraints on volume, weight, and reefer plugs are respected. The slot planning problem should assign containers to slots in the vessel so that the loading and unloading time in ports can be minimized. The vessel must be seaworthy, meaning that stability and stress constraints must be respected.

Figure 4 illustrates the arrangement of bays in a container vessel. Containers are loaded bottom-up in each bay up to a given stacking height limited by the line of sight and other factors. Some containers are loaded below deck, while other containers are loaded above the hatch cover. The overall weight sum of containers may not exceed a given limit, and the weight need to be balanced. Moreover, torsions should be limited, making it illegal to e.g. only load containers at the same front and end of the vessel. Refrigerated containers (reefers) need to be attached to an electric plug. Only a limited number of plugs are available, and these plugs are at specific positions.

A good stowage plan should make sure that it is not necessary to rearrange containers at each port call. All containers for the given port should be directly accessible when arriving to the port, and there should be sufficient free capacity for loading new containers. If several cranes are available in a port, it is necessary to ensure that all cranes can operate at the same time without blocking for each other.

Pacino [28] presents a MIP model for the master problem. The model is based on Pacino et al. [29, 30]. The model considers both 20' and 40' containers, assuming that two 20' containers can fit in the slot of a 40' container provided that the middle is properly supported. Four types of containers are considered: light, heavy, light reefer, and heavy reefer. Decision variables are introduced for each slot, indicating how many of each container type will be loaded in the slot.

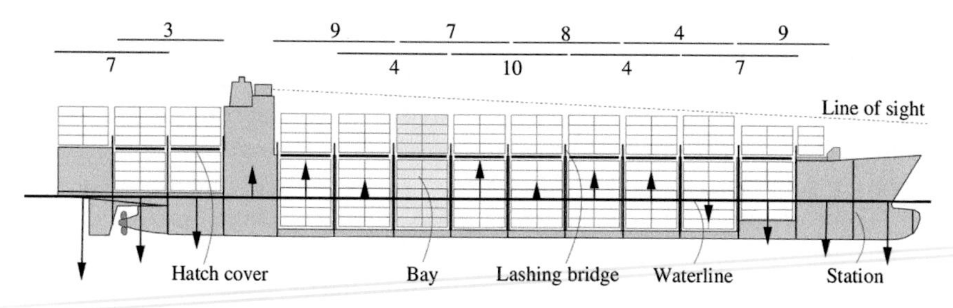

Fig. 4 The arrangement of bays in a small container vessel, and stacking heights. The *arrows* indicate forces. *Picture:* Pacino [28]

The MIP model has a large number of constraints: First of all, a load list and cargo estimates are used to calculate the number of containers of each type that needs to be stowed. Moreover, every slot has a capacity of dry containers and reefers. An overall weight limit given by the capacity of the vessel is also imposed. When calculating the weight limit, average values for light and heavy containers are used to ease the calculations.

Trim, draft, buoyancy and stability is calculated as a function of displacement and center of gravity of the vessel.

Finally, a number of penalties associated with a given loading are calculated. These include hatch-overstowage, overstowage in slots, time needed for loading, and excess of reefer containers. The objective of the model minimizes a weighted sum of the penalties.

Pacino [28] show that the master planning problem is NP-hard. Computational results are reported for instances with vessel capacity up to around 10,000 TEU, visiting up to 12 ports involving more than 25,000 lifts (crane moves of a container). Several of these instances can be solved within 5 min up to a 5 % gap, using a MIP-solver.

6.1 *Mathematical Model*

In the slot planning phase, the master plan is refined by assigning the containers to specific slots on board the vessel [31]. This problem involves handling of a number of stacking rules, as well as constraints on stack heights and stack weight. Since several of the containers are already stowed on board the vessel the objective is to arrange containers with the same destination port in the same stack, free as many stacks as possible, minimize overstowage, and minimize the number of non-reefer containers assigned to reefer slots. Due to the large number of logical constraints in this problem [19] proposed a logical model using the following notation. $\mathscr{S}$ is the set of stacks, $\mathscr{T}_s$ is the set of tiers for stack s, $\mathscr{P}$ represents the aft ($p = 1$) and fore ($p = 2$) of a cell, $\mathscr{C}$ is the set of containers to stow in the location and $\mathscr{C}^P \subset \mathscr{C}$ is the subset of containers in the release, i.e. the set of containers that are already on-board the vessel. $x_{stp} \in \mathscr{C} \cup \{\perp\}$ is a decision variable indicating the location of a container $c \in C$ or the empty assignment $\perp$. A^{40}_{stp} is a binary variable indicating if the cell in stack s, tier t, and position p can hold a 40' foot container and similarly A^{20}_{stp} is one if a slot can hold a 20' container. A^{R}_{stp} is a binary indicator for the position of reefer plugs. W_s and H_s is the maximum weight and height of stack s. The attribute functions use $w(c)$ and $h(c)$ for the weight and height of a container. $r(c)$ is true iff the container is a reefer, $\perp(c)$ is true iff $c = \perp$, $f(c)$ is true iff the container is 40', and $t(c)$ is true iff it is a 20' container. Then the logical model is:

$$|\{x_{stp} = c | s \in \mathscr{S}, t \in \mathscr{T}_s, p \in \mathscr{P}\}| = 1 \quad c \in \mathscr{C} \tag{22}$$

$$x_{s_c t_c p_c} = c \quad c \in \mathscr{C}^P \tag{23}$$

$$\neg f(x_{st1}) \wedge (f(x_{st2}) \implies \perp (x_{st1})) \quad s \in \mathscr{S}, t \in \mathscr{T}_s \tag{24}$$

$$t(x_{stp}) \implies A^{20}_{stp} \quad s \in \mathscr{S}, t \in \mathscr{T}_s, p \in \mathscr{P} \tag{25}$$

$$f(x_{st1}) \implies A^{40}_{st} \quad s \in \mathscr{S}, t \in \mathscr{T}_s \tag{26}$$

$$\sum_{t \in \mathscr{T}_s} (w(x_{st1}) + w(x_{st2})) \leq W_s \quad s \in \mathscr{S} \tag{27}$$

$$\sum_{t \in \mathscr{T}_s} \max(h(x_{st1}), h(x_{st2})) \leq H_s \quad s \in \mathscr{S} \tag{28}$$

$$\neg \perp (x_{stp}) \implies (t(x_{s(t-1)1}) \wedge t(x_{s(t-1)2})) \vee f(x_{s(t-1)1}) \quad s \in \mathscr{S}, t \in \mathscr{T}_s \setminus \{1\}, p \in \mathscr{P} \tag{29}$$

$$f(x_{st1}) \implies \perp t(x_{s(t+1)p}) \quad s \in \mathscr{S}, t \in \mathscr{T}_s \setminus \{N^T_s\}, p \in \mathscr{P} \tag{30}$$

$$r(x_{stp}) \wedge t(x_{stp}) \implies A^R_{stp} \quad s \in \mathscr{S}, t \in \mathscr{T}_s, p \in \mathscr{P} \tag{31}$$

$$r(x_{st1}) \wedge f(x_{st1}) \implies A^R_{st1} \vee A^R_{st2} \quad s \in \mathscr{S}, t \in \mathscr{T}_s \tag{32}$$

Constraints (22)–(23) ensure that each container is assigned to exactly one slot. Constraint (24) ensures that a 40' container occupies both the aft and fore position of a cell. The assignments need to respect cell capacity (25)–(26), stack height and stack weight limits (27)–(28). Two 20' containers can be stowed in a 40' slot, if properly supported from below (29). This means that 40' container can be stacked on top of two 20' containers, but not the other way around (30). Reefer containers need to be assigned to slots with a power plug (31)–(32).

In order to minimize the objective function [19] propose to use Constraint-Based Local Search. The framework combines local search algorithms with constraint programming. The constraint satisfaction part of the problem is transformed to an optimization problem where the objective is to minimize constraint violation. A hill-climbing method is used to optimize the slot planning. The neighborhood in the search consists of swapping containers between a pair of cells.

Pacino [28] report computational results for 133 real-life instances, showing that the local search algorithm actually finds the optimal solution in 86 % of the cases. The running times are below 1 second.

7 Bunker Purchasing

In a liner shipping network bunker fuel constitutes a very large part of the variable operating cost for the vessels. Also, the inventory holding costs of the bunker on board may constitute a significant expense to the liner shipping company.

Bunker prices are fluctuating and generally correlated with the crude oil price, but there are significant price differences between ports. This creates the need for frequent (daily) re-optimization of the bunker plan for a vessel, to ensure the lowest bunker costs.

Bunker can be purchased on the spot market when arriving to a port, but normally it is purchased some weeks ahead of arrival. Long-term contracts between a liner shipping company and a port can result in reduced bunkering costs by committing the company to purchase a given amount of bunker. Bunkering contracts may cover several vessels sailing on different services, making the planning quite complex.

The bunker purchasing problem is to satisfy the vessels consumption by purchasing bunkers at the minimum overall cost, while considering reserve requirements, and other operational constraints. Bunker purchasing problems involve big data. Real-life instances may involve more than 500 vessels, 40,000 port calls, and 750 contracts.

For a vessel sailing on a given port to port voyage at a given speed, the bunker consumption can be fairly accurately predicted. This gives an advantage in bunker purchasing, when a vessel has a stable schedule known for some months ahead. The regularity in the vessel schedules in liner shipping allows for detailed planning of a single vessel.

Besbes and Savin [9] consider different re-fueling policies for liner vessels and present some interesting considerations on the modeling of stochastic bunker prices using Markov processes. This is used to show that the bunkering problem in liner shipping can be seen as a stochastic capacitated inventory management problem. Capacity is the only considered operational constraint. More recently [43] examined re-fueling under a worst-case bunker consumption scenario.

The work of [34] considers multiple tanks in the vessel and stochasticity of both prices and consumption, as well as a range of operational constraints. [44] does not consider stochastic elements nor tanks, but has vessel speed as a variable of the model. The work of [25] minimizes bunker costs as well as startup costs and inventory costs for a single liner shipping vessel. This is done by choosing bunker ports and bunker volumes but also having vessel round trip speed (and thus the number of vessels on the service) as a variable of the model.

In [37] a model is developed which considers the uncertainty of bunker prices and bunker consumption, modeling their uncertainty by markov processes in a scenario tree. The work can be seen as an extension of [44], as it considers vessel speed as a variable within the same time window bounds. Capacity and fixed bunkering costs is considered, as is the holding / tied capital cost of the bunkers.

The studies described above do not consider bunker contracts, and all model the bunker purchasing for a single vessel.

7.1 Bunker Purchasing with Contracts

Plum et al. [32] presented a decomposition algorithm for the Bunker Purchasing with Contracts Problem, BPCP, and showed that the model is able to solve even very large real-life instances. The model is based on writing up all bunkering patterns, and hence may be of exponential size. Let I be the set of ports visited on an itinerary, B be the set of bunker types, and V be the set of vessels. A contract $c \in C$ has a minimal $\underline{q}_c$ and maximal $\overline{q}_c$ quantity that needs to be purchased. A contract c will give rise to a number of purchase options $m \in M$, i.e. discrete events where a specific vessel v calls a port within the time interval of a contract c, allowing it to purchase bunker at the specific price p_m. Each time a purchase is done at port i a startup cost sc_i is paid.

Let R_v be the set of all feasible bunkering patterns for a vessel v. A bunkering pattern is feasible if a sufficient amount of bunker is available for each itinerary, including reserves. Bunker is available in various grades, and it is allowed to substitute a lower grade with a higher grade. In some areas, only low-sulphur bunker may be used, and this needs to be respected by the bunkering plan. Moreover initial and terminal criteria for bunker volumes must be met. Finding a legal bunkering pattern can be formulated as a MIP model [32] and solved by commercial solvers. Each pattern $r \in R_v$ is denoted as a set of bunkerings.

Let $u_r = \sum_{m \in M}(p_m l_m) + \sum_{i \in I} \sum_{v \in V} \sum_{b \in B}(\delta_{i,b} sc_i)$ be the cost for pattern $r \in R_v$. In this expression, l_m is the purchase of bunker for each purchase option m. and p_m is the price of option m. The binary variable $\delta_{i,b}$ is set to one iff a purchase of bunker type b is made at port call i. Let λ_r be a binary variable, set to 1 iff the bunkering pattern r is used. Let $o_{r,c}$ be the quantity purchased of contract c by pattern r. The BPCP can then be formulated as

$$\min \quad \sum_{v \in V} \sum_{r \in R_v} \lambda_r u_r + \sum_{c \in C} (\underline{s}_c \underline{w} + \overline{s}_c \overline{w}) \tag{33}$$

$$\text{s.t.} \quad \underline{q}_c - \underline{s}_c \leq \sum_{v \in V} \sum_{r \in R_v} \lambda_r o_{r,c} \leq \overline{q}_c + \overline{s}_c \qquad c \in C \tag{34}$$

$$\sum_{r \in R_v} \lambda_r = 1 \qquad v \in V \tag{35}$$

$$\lambda_r \in \{0, 1\} \qquad r \in R_v \tag{36}$$

The objective minimizes the costs of purchased bunker, startup costs and slack costs. The parameters $\underline{w}$ and $\overline{w}$ denote a penalty for violating the minimal $\underline{q}_c$ and maximal $\overline{q}_c$ quantity imposed by contract c. Constraints (34) ensures that all contracts are fulfilled. Convexity constraints (35) ensure that exactly one bunker pattern is chosen for each vessel.

Due to the large number of columns in the model [32] proposed to solve the LP relaxed model by Column Generation. Using the generated columns from the LP-solution, the resulting problem is solved to integer optimality using a MIP solver, leading to a heuristic solution for the original problem.

Initially all dual variables are set to zero, a subproblem is constructed for each vessel and solved as a MIP problem. The first master problem is then constructed with one solution for each vessel as columns. This master is solved and the first values are found. The subproblems are resolved for all vessels (only the objective coefficients for the contracts needs updating) and new columns are generated for the master. This continues until no negative reduced cost columns can be generated, and the LP optimal solution is achieved.

The subproblems do not need to be solved to optimality since any column with negative reduced cost will ensure progress of the algorithm. Therefore the solver is allowed to return solutions to the subproblem having a considerable optimality gaps. As the algorithm progresses, the allowable subproblem gap is reduced.

A simple form of dual stabilization has been used in the implementation by [32] to speed up convergence. The Box-step method imposes a box around the dual variables, which are limited from changing more than π_{max} per iteration. This has been motivated by the dual variables only taking on values $\{-\underline{w}, \overline{w}, 0\}$ in the first iteration, these then stabilize at smaller numerical values in subsequent iterations.

The model is able to solve even very large real-life instances involving more than 500 vessels, 40,000 port calls, and 750 contracts. First, column generation is used to solve the linearized model, and then a MIP solver is used to find an integer solution only using the generated columns. This results in a small gap in the optimal solution compared to if all columns were known. However, computational results show that the gap is never more than around 0.5 % even for the largest instances. In practice the resulting gap of the algorithm, can be much smaller since the found solutions are benchmarked against a lower bound and not against the optimal solution.

An interesting side product of the model is the dual variables $\underline{\pi}_c$ and $\overline{\pi}_c$ for the upper and lower contract constraints (34). These values can be used to evaluate the gain of a given contract, which may be valuable information when (re)negotiating contracts.

Since bunker prices are stochastic of nature, future research should be focused on modeling the price fluctuation. However, the models tend to become quite complex and difficult to solve as observed by [34], while only adding small extra improvements to the results. So a trade-off must be done between model complexity and gain in bunker costs. The work of [37] shows some promising developments in this important direction.

Also, instruments from finance (bunker future or forward contracts, fixed price bunker fuel swaps) could be used to control risk in bunker purchasing, and to increase the margins on oil trade. Bunker purchasing for liner ships constitutes such a big market that it deserves a professional trading approach.

8 The Vessel Schedule Recovery Problem

It is estimated that approximately 70–80 % of vessel round trips experience delays in at least one port. The common causes are bad weather, strikes in ports, congestions in passageways and ports, and mechanical failures.

Currently when a disruption occur, the operator at the shipping companies manually decides what action to take. For a single delayed vessel a simple approach could be to speed up. However, the consumption of bunker fuel is close to a cubic function of speed and vessels' speeds are limited between a lower and upper limit. So even though an expensive speed increase strategy is chosen, a vessel can arrive late for connections, propagating delays to other parts of the network. Having more than 10,000 containers on board a large vessel, calculating the overall consequences of re-routing/delaying these containers demands algorithms for big data. Disruption management is well studied within the airline industry (see [4] or [16] for a review) and the network design of airlines resemble liner shipping networks inspiring the few works on disruption management found for liner shipping. Mulder et al. [27] presents a markov decision model to determine the optimal recovery policy. The core idea is to reallocate buffer time within a schedule in order to recover from disruptions. Brouer et al. [13] present the Vessel Schedule Recovery Problem (VSRP) handling a disruption in a liner shipping network by omitting port calls, swapping port calls or speeding up vessels in a predefined disruption scenario. The model and method will be presented in the following section.

8.1 Definitions

A given disruption scenario can mathematically be described by a set of vessels V, a set of ports P, and a time horizon consisting of discrete time slots $t \in T$. The time slots are discretized on port basis as terminal crews handling the cargo operate in shifts, which are paid for in full, even if arriving in the middle of a shift. Hence we only allow vessels arriving at the beginning of shifts. Reducing the graph to timeslots based on these shifts, also has the advantage of reducing the graph size, although this is a minor simplification of the problem. For each vessel $v \in V$, the current location and a planned schedule consisting of an ordered set of port calls $H_v \subseteq P$ are known within the recovery horizon, a port call A can precede a port call B, $A < B$ in H_v. A set of possible sailings, i.e. directed edges, L_h are said to *cover* a port call $h \in H_v$. Each L_h represent a sailing with a different speed.

The recovery horizon, T, is an input to the model given by the user, based on the disruption in question. Inter continental services will often recover by speeding during ocean crossing, making the arrival at first port after an ocean crossing a good horizon, severe disruptions might require two ocean crossings. Feeders recovering at arrival to their hub port call would save many missed transshipments giving an obvious horizon. In combination with a limited geographical dimension this ensures that the disruption does not spread to the entire network.

The disruption scenario includes a set of container groups C with planned transportation scenarios on the schedules of V. A feasible solution to an instance of the VSRP is to find a sailing for each $v \in V$ starting at the current position of v and ending on the planned schedule no later than the time of the recovery horizon. The solution must respect the minimum and maximum speed of the vessel and the constraints defined regarding ports allowed for omission or port call swaps. The optimal solution is the feasible solution of minimum cost, when considering the cost of sailing in terms of bunker and port fees along with a strategic penalty on container groups not delivered "on-time" or misconnecting altogether.

8.2 Mathematical Model

Brouer et al. [13] use a time space graph as the underlying network, but reformulate the model to address the set of available recovery techniques, which are applicable to the VSRP.

The binary variables x_e for each edge $e \in E_s$ are set to 1 iff the edge is sailed in the solution. Binary variables z_h for each port call $h \in H_v, v \in V$ are set to 1 iff call h is omitted. For each container group c we define binary variables $o_c \in \{0, 1\}$ to indicate whether the container group is delayed or not and y_c to account for container groups misconnecting. The parameter $O_e^c \in \{0, 1\}$ is 1 iff container group $c \in C$ is delayed when arriving by edge $e \in L_{T_c}$. $B_c \in H_v$ is defined as the origin port for a container group $c \in C$ and the port call where vessel v picks up the container group. Similarly, we define $T_c \in H_w$ as the destination port for container group $c \in C$ and the port call where vessel w delivers the container group. Intermediate planned transshipment points for each container group $c \in C$ are defined by the ordered set $I_c = (I_c^1, \dots, I_c^m)$. Here $I_c^i = (h_v^i, h_w^i) \in (H_v, H_w)$ is a pair of calls for different vessels $(v, w \in V | v \neq w)$ constituting a transshipment. Each container group c has m^c transshipments. M_c^e is the set of all non-connecting edges of $e \in L_h$ that result in miss-connection of container group $c \in C$. $M_c \in \mathbb{Z}_+$ is an upper bound on the number of transshipments for container group $c \in C$.

Let the demand of vessels v in a node n be given by $S_v^n = -1$ if $n = n_s^v$, $S_v^n = 1$ if $n = n_t^v$, while $S_v^n = 0$ for all other nodes. Then we get the following model:

$$\min \quad \sum_{v \in V} \sum_{h \in H_v} \sum_{e \in L_h} c_e^v x_e + \sum_{c \in C} \left(c_c^m y_c + c_c^d o_c \right) \tag{37}$$

$$\text{s.t.} \quad \sum_{e \in L_h} x_e + z_h = 1 \qquad v \in V, h \in H_v \tag{38}$$

$$\sum_{e \in n^-} x_e - \sum_{e \in n^+} x_e = S_v^n \qquad v \in V, n \in N_v \tag{39}$$

$$y_c \leq o_c \qquad c \in C \tag{40}$$

$$\sum_{e \in L_{T_c}} O_e^c x_e \leq o_c \qquad c \in C \tag{41}$$

$$z_h \leq y_c \qquad c \in C, h \in B_c \cup I_c \cup T_c \tag{42}$$

$$x_e + \sum_{\lambda \in M_c^e} x_\lambda \leq 1 + y_c \qquad c \in C, e \in \{L_h | h \in B_c \cup I_c \cup T_c\} \tag{43}$$

$$x_e \in \{0, 1\}, e \in E_s \qquad y_c, o_c \in \mathbb{R}_+, c \in C \qquad z_h \in \mathbb{R}_+, v \in V, h \in H_v \tag{44}$$

The objective function (37) minimizes the cost of operating vessels at the given speeds, the port calls performed along with the penalties incurred from delaying or misconnecting cargo.

Constraints (38) are set-partitioning constraints ensuring that each scheduled port call for each vessel is either called by some sailing or omitted. The next constraints (39) are flow-conservation constraints. Combined with the binary domain of variables x_e and z_h they define feasible vessel flows through the time-space network. A misconnection is by definition also a delay of a container group and hence the misconnection penalty is added to the delay penalty, as formulated in (40). Constraints (41) ensure that o_c takes the value 1 iff container group c is delayed when arriving via the sailing represented by edge $e \in E_s$. Constraints (42) ensure that if a port call is omitted, which had a planned (un)load of container group $c \in C$, the container group is misconnected. Constraints (43) are coherence constraints ensuring the detection of container groups' miss-connections due to late arrivals in transshipment ports. On the left-hand side the decision variable corresponding to a given sailing, x_e, is added to the sum of all decision variables corresponding to having onward sailing resulting in miss-connections, $\lambda \in M_c^e$.

In [13] the model has been tested on a number of real-life cases, including a delayed vessel, a port closure, a berth prioritization, and expected congestion. An analysis of the four real life cases, show that a disruption allowing to omit a port call or swap port calls may ensure timely delivery of cargo without having to increase speed and hence, a decision support tool based on the VSRP may aid in decreasing the number of delays in a liner shipping network, while maintaining a slow steaming policy. To operationalize this the rerouting of the actual flow and adjustment of the actual schedule must be incorporated in a real time system to enable here-and-now decisions. This is especially challenging for larger disruption scenarios than the ones described as the size of the problem grows exponentially.

9 Conclusion and Future Challenges

Maritime logistics companies operate in an environment which requires them to become more and more analytical. In general there are several insights to be gained from the data companies has available. Especially when companies start to use the forward looking analytical techniques rather than only using data for backward looking analysis (descriptive and diagnostic models) companies can unlock significant value from the collected data as shown in this chapter. Forward looking techniques (predictive models) can provide input for the decision making process where the best possible action is sought (prescriptive models). A pressing challenge in big data analysis today lies in the integration of predictive and prescriptive methods which combined can serve as valuable decision support tools. This chapter introduced a selection of large scale planning problems within maritime logistics with a primary focus on challenges found in the liner shipping industry. Focus has been on addressing strategic, tactical and operational problems by modern large scale optimization methods. However optimization within maritime logistics is complicated by the uncertainty and difficult accessibility of data. Most demands are only estimates, and for historic reasons even contracted cargo can be unreliable since there are no penalties associated with no-show cargo. To limit these uncertainties predictive machine learning techniques is an important tool. In particular, seasonal variations and similar trends can be predicted quite well and decision support systems should take such uncertainties into account. This can be done either by developing models where it is possible to re-optimize the problem quickly in order to meet new goals and use them interactively for decision support and for evaluating what-if scenarios suggested by a planner as there are still many decisions that will not be data-driven. Quantitative data can not always predict the future well in situations of e.g. one-time events and generally extrapolation is hard. But in situations where we operate in an environment where data can be interpolated mathematical models may serve as great decision support tools by integrating the predictive models directly in the prescriptive model. With the large volume of data generated by carriers, increased quality of forecasts, and algorithmic improvements it may also be beneficial and even tractable to include the uncertainties directly in the decision models. A relatively new way of handling data uncertainty is by introducing uncertainty sets in the definition of the data used for solving large-scale LP's. The standard LP found as a subproblem in many of the described problems can generically be stated as $\min_x\{c^T x : Ax \leq b\}$, where A, b, and c contain the data of the problem at hand. As described previously in this chapter most of the data is associated with uncertainties but in *Robust Optimization* this can be handled by replacing the original LP with an uncertain LP $\{\min_x\{c^T x : Ax \leq b\} : (A, b, c) \in \mathcal{U}\}$. The best robust solution to the problem can be found by solving the Robust Counterpart of the problem, which is an semi-infinite LP $\min_{x,t}\{t : c^T x \leq t, Ax \leq b \forall (A, b, c) \in \mathcal{U}\}$. Clearly this LP is larger than the original LP, but with good estimates of the uncertainty sets the size can be manageable, further details can be found in [6]. As the accuracy of predictive models increase it will be possible to come up with good estimates for the uncertainty sets and thereby actually making it feasible to solve robust versions of the planning problems. In the MIP case the problems usually become much harder and often intractable with a few exceptions. An alternative approach to Robust Optimization is to handle the uncertainties via probability distributions on the data and use *Stochastic Programming* and solve the chance constrained program $\min_{x,t}\{t : Prob_{(A,b,c)\sim P}\{c^T x \leq t, Ax \leq b\} \geq 1 - \epsilon\}$ or a two-stage stochastic program based on a set of scenarios. Again, machine learning algorithms can provide good estimates of the actual underlying distributions or expected scenarios and it may be possible to obtain results that are less conservative than the worst-case results provided by Robust Optimization, but the process can be more computationally extensive.

References

1. Agarwal, R., Ergun, Ö.: Ship scheduling and network design for cargo routing in liner shipping. Transp. Sci. **42**, 175–196 (2008)
2. Alvarez, J.F.: Joint routing and deployment of a fleet of container vessels. Marit. Econ. Logist. **11**, 186–208 (2009)
3. Archetti, C., Speranza, M.G.: A survey on matheuristics for routing problems. EURO J. Comput. Optim. **2**, 223–246 (2014)
4. Ball, M., Barnhart, C., Nemhauser, G., Odoni, A.: Chapter 1: Air Transportation—Irregular Operations and Control. In: Handbooks in Operations Research and Management Science: Transportation, vol. 14. Elsevier (2007)
5. Barnhart, C., Johnson, E.L., Nemhauser, G.L., Savelsbergh, M.W., Vance, P.H.: Branch-and-price: column generation for solving huge integer programs. Oper. Res. **46**(3), 316–329 (1998)
6. Ben-Tal, A., Nemirovski, A.: Robust optimization-methodology and applications. Math. Program. **92**(3), 453–480 (2002)
7. Bertsimas, D., King, A., Mazumder, R.: Best subset selection via a modern optimization lens. Submitted to Annals of Statistics (2014)
8. Bertsimas, D., Tsitsiklis, J.N.: Introduction to Linear Optimization, vol. 6. Athena Scientific, Belmont (1997)
9. Besbes, O., Savin, S.: Going bunkers: the joint route selection and refueling problem. Manuf. Serv. Oper. Manag. **11**, 694–711 (2009)
10. Brouer, B., Alvarez, J., Plum, C., Pisinger, D., Sigurd, M.: A base integer programming model and benchmark suite for liner-shipping network design. Transp. Sci. **48**, 281–312 (2014a)
11. Berit D. Brouer, J. Fernando Alvarez, Christian E. M. Plum, David Pisinger, Mikkel M. Sigurd 'A Base Integer Programming Model and Benchmark Suite for Liner-Shipping Network Design' Transportation Science, 48(2), pp. 281–312 (2014)
12. Brouer, B., Desaulniers, G., Pisinger, D.: A matheuristic for the liner shipping network design problem. Transp. Res. Part E Logist. Transp. Rev. **72**, 42–59 (2014b)
13. Brouer, B., Dirksen, J., Pisinger, D., Plum, C., Vaaben, B.: The vessel schedule recovery problem (VSRP) a MIP model for handling disruptions in liner shipping. Eur. J. Oper. Res. **224**, 362–374 (2013)
14. Brouer, B., Pisinger, D., Spoorendonk, S.: Liner shipping cargo allocation with repositioning of empty containers. INFOR: Inf. Syst. Oper. Res. **49**, 109–124 (2011)
15. Burke, E.K., Kendall, G.: Search Methodologies. Springer (2005)
16. Clausen, J., Larsen, A., Larsen, J., Rezanova, N.J.: Disruption management in the airline industry-concepts, models and methods. Comput. Oper. Res. **37**(5), 809–821 (2010)
17. Cormen, T.H., Leiserson, C.E., Rivest, R.L., Stein, C., et al.: Introduction to Algorithms, vol. 2. MIT Press, Cambridge (2001)
18. Costa, A.M.: A survey on benders decomposition applied to fixed-charge network design problems. Comput. Oper. Res. **32**(6), 1429–1450 (2005)
19. Delgado, A., Jensen, R.M., Janstrup, K., Rose, T.H., Andersen, K.H.: A constraint programming model for fast optimal stowage of container vessel bays. Eur. J. Oper. Res. **220**(1), 251–261 (2012)
20. Desrosiers, J., Lübbecke, M.E.: A Primer in Column Generation. Springer (2005)
21. Erera, A.L., Morales, J., Savelsbergh, M.: Global intermodal tank container management for the chemical industry. Transp. Res. Part E **41**(6), 551–566 (2005)

22. Gendreau, M., Potvin, J.-Y.: Handbook of Metaheuristics, vol. 2. Springer (2010)
23. Gendron, B., Crainic, T.G., Frangioni, A.: Multicommodity Capacitated Network Design. Springer (1999)
24. Karsten, C.V., Pisinger, D., Ropke, S., Brouer, B.D.: The time constrained multi-commodity network flow problem and its application to liner shipping network design. Transp. Res. Part E Logist. Transp. Rev. **76**, 122–138 (2015)
25. Kim, H.-J., Chang, Y.-T., Kim, K.-T., Kim, H.-J.: An epsilon-optimal algorithm considering greenhouse gas emissions for the management of a ships bunker fuel. Transp. Res. Part D Transp. Environ. **17**, 97–103 (2012)
26. Meng, Q., Wang, S.: Liner shipping service network design with empty container repositioning. Transp. Res. Part E Logist. Transp. Rev. **47**(5), 695–708 (2011)
27. Mulder, J., Dekker, R., Sharifyazdi, M.: Designing robust liner shipping schedules: optimizing recovery actions and buffer times. Report/Econometric Institute, Erasmus University Rotterdam (2012)
28. Pacino, D.: Fast generation of container vessel stowage plans. Ph.D. thesis, IT University of Copenhagen (2012)
29. Pacino, D., Delgado, A., Jensen, R., Bebbington, T.: Fast generation of near-optimal plans for eco-efficient stowage of large container vessels. Comput. Logist. **6971**, 286–301 (2011)
30. Pacino, D., Jensen, R.: Constraint-based local search for container stowage slot planning. Lect. Notes Eng. Comput. Sci. **2**, 1467–1472 (2012)
31. Pacino, D., Jensen, R.: Fast slot planning using constraint-based local search. IAENG Trans. Eng. Technol. **186**, 49–63 (2013)
32. Plum, C., Pisinger, D., Jensen, P.: Bunker purchasing in liner shipping. Handbook of Ocean Container Transport Logistics, vol. XVII, pp. 251–278. Springer (2015)
33. Plum, C., Pisinger, D., Sigurd, M.M.: A service flow model for the liner shipping network design problem. Eur. J. Oper. Res. **235**(2), 378–386 (2014)
34. Plum, C.E.M., Jensen, P.N.: Minimization of bunker costs. Master thesis, University of Copenhagen (2007)
35. Reinhardt, L.B., Pisinger, D.: A branch and cut algorithm for the container shipping network design problem. Flex. Serv. Manuf. J. **24**(3), 349–374 (2012)
36. Rossi, F., Van Beek, P., Walsh, T.: Handbook of Constraint Programming. Elsevier (2006)
37. Sheng, X., Lee, L., Chew, E.: Dynamic determination of vessel speed and selection of bunkering ports for liner shipping under stochastic environment. OR Spectr. **36**(2), 455–480 (2014)
38. Shintani, K., Imai, A., Nishimura, E., Papadimitriou, S.: The container shipping network design problem with empty container repositioning. Transp. Res. Part E Logist. Transp. Rev. **43**(1), 39–59 (2007)
39. Song, D.-P., Dong, J.X.: Cargo routing and empty container repositioning in multiple shipping service routes. Transp. Res. Part B Methodol. **46**(10), 1556–1575 (2012)
40. Tierney, K., Askelsdottir, B., Jensen, R.M., Pisinger, D.: Solving the liner shipping fleet repositioning problem with cargo flows. Transp. Sci. **49**(3), 652–674 (2015)
41. Toth, P., Vigo, D.: Vehicle Routing: Problems, Methods, and Applications. SIAM (2014)
42. Vielma, J.P.: Mixed integer linear programming formulation techniques. SIAM Rev. **57**(1), 3–57 (2015)
43. Wang, S., Meng, Q.: Robust bunker management for liner shipping networks. Eur. J. Oper. Res. **243**(3), 789–797 (2015)
44. Yao, Z., Ng, S.H., Lee, L.H.: A study on bunker fuel management for the shipping liner services. Comput. Oper. Res. **39**, 1160–1172 (2012)

Author Biographies

Berit Dangaard Brouer holds a Ph.D. in operations research from the Technical University of Denmark. Her research targets maritime logistics within the liner shipping industry. She has several publications in internationally recognized journals and her publications propose decision support tools for a range of planning problems within network design, cargo routing and disruption management.

Christian Vad Karsten is a Ph.D. student at the Technical University of Denmark where he is working on computational methods for large-scale optimization problems. His research is focused on advanced analytical methods and the application of these to data-driven problems to help make better decisions. More specifically he is working on how to introduce competitiveness in mathematical models addressing big data planning problems faced by container shipping companies. The project is part of a close cooperation between academia and industry aimed at developing effective models and methods to support planning and managerial decision-making. In addition to his mathematical training at the Technical University of Denmark, Christian has been a visiting student at Princeton University, UC Berkeley, and Massachusetts Institute of Technology.

David Pisinger is professor and head of research in Management Science, Technical University of Denmark (DTU). His principal focus of research is Maritime Optimization, Railway Optimization, and recently, Energy Models. In his research he tries to combine theoretical results with real-life applications. He has been principal investigator of the research projects ENERPLAN and GREENSHIP on maritime optimization in collaboration with Maersk Line. Professor Pisinger received the Ph.D. prize of the Danish Academy of Natural Sciences in 1995, and the award of teaching excellence from Faculty of Natural Sciences, University of Copenhagen in 2000. In 2013 he received Hedorf Fond prize for Transportation Research. He has been nominated as "Teacher of the year" at DTU several times, and several of his students have been awarded prizes for their master and Ph.D. theses.

Big Network Analytics Based on Nonconvex Optimization

Maoguo Gong, Qing Cai, Lijia Ma and Licheng Jiao

Abstract The scientific problems that Big Data faces may be network scientific problems. Network analytics contributes a great deal to networked Big Data processing. Many network issues can be modeled as nonconvex optimization problems and consequently they can be addressed by optimization techniques. In the pipeline of nonconvex optimization techniques, evolutionary computation gives an outlet to handle these problems efficiently. Because, network community discovery is a critical research agenda of network analytics, in this chapter we focus on the evolutionary computation based nonconvex optimization for network community discovery. The single and multiple objective optimization models for the community discovery problem are thoroughly investigated. Several experimental studies are shown to demonstrate the effectiveness of optimization based approach for big network community analytics.

Keywords Big data · Complex networks · Nonconvex optimization · Evolutionary computation · Multiobjective optimization

1 Introduction

Recent years have witnessed the growing enthusiasm for the concept of "Big Data" [86]. Big Data has been an active topic and has attracted great attention from every walk of life [18, 64, 89]. It should be noted that the scientific problems that Big Data faces may be that of network scientific problems, and complex network analytics should be an important cornerstone of data science [1, 71, 114, 125]. Network analytics undoubtedly can contribute a great deal to networked Big Data processing.

Network analytics contains many issues, to name a few, community structure discovery, network structural balance, network robustness, link prediction, network

M. Gong (✉) · Q. Cai · L. Ma · L. Jiao
Key Laboratory of Intelligent Perception and Image Understanding of Ministry of Education, International Research Center for Intelligent Perception and Computation, Xidian University, Shaanxi Province, Xi'an 710071, China
e-mail: gong@ieee.org

A. Emrouznejad (ed.), *Big Data Optimization: Recent Developments and Challenges*, Studies in Big Data 18, DOI 10.1007/978-3-319-30265-2_15

resource allocation, anomaly detection, network security, network recommendation, network propagation, and network ranking, etc. Most if not all of these issues can be modeled as nonconvex optimization problems and consequently they can be computed by optimization techniques. Because, those optimization models for network issues are nonconvex from mathematical view, thus, canonical mathematical optimization methods can hardly solve these problems. In the pipeline of optimization techniques, evolutionary computation gives an outlet to handle these nonconvex optimization problems efficiently.

Because network community discovery may be the cornerstone to the analytics of many other network issues, consequently this chapter focuses on the optimization based community structure discovery from networks. The rest of this chapter is organized as follows. Section 2 briefly talks about the issues that network analytics concerns and several eminent properties of networks. Section 3 discusses the basic definitions of optimization and evolutionary computation. Section 4 presents the related work of network community structure analytics, including the definition of a network community and the research progress of community discovery. Section 5 surveys the optimization models for network community discovery. The network data sets commonly used for community discovery benchmarking are listed in Sect. 6. Section 7 exhibits some experiments on network community discovery, and the conclusions are finally drawn in Sect. 8.

2 Network Issues, Properties and Notations

2.1 Issues Concerning Network Analytics

Network analytics is an essential research agenda of network and networked big data mining. Figure 1 shows the importance of network analytics to network and networked data mining. Network analysis not only may very likely result in the discovery of important patterns hidden beneath the networks, but also can potentially shed light on important properties that may control the growth of the networks. Network analytics involves many issues. To move forward, we show 12 critical issues that concern network analytics in Fig. 2.

Very often, to analyze a network issue one should consider the properties of the corresponding network. In the following, we are going to discuss several eminent properties of networks.

2.2 Eminent Properties of Network

Because structure always affects function, consequently, a substantial volume of work has been done to analyze the structural properties of complex networks

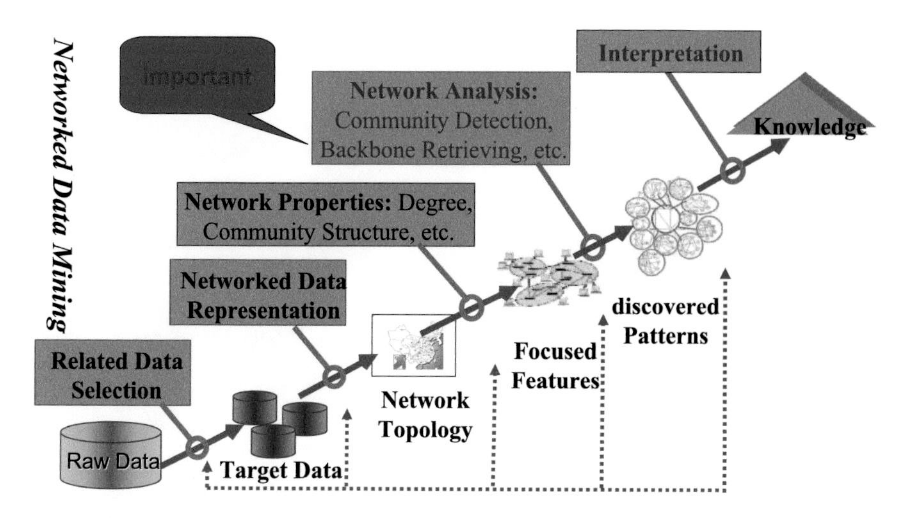

Fig. 1 Network analytics plays an important role in network and networked data mining. Reprinted figure with permission from Ref. [74]

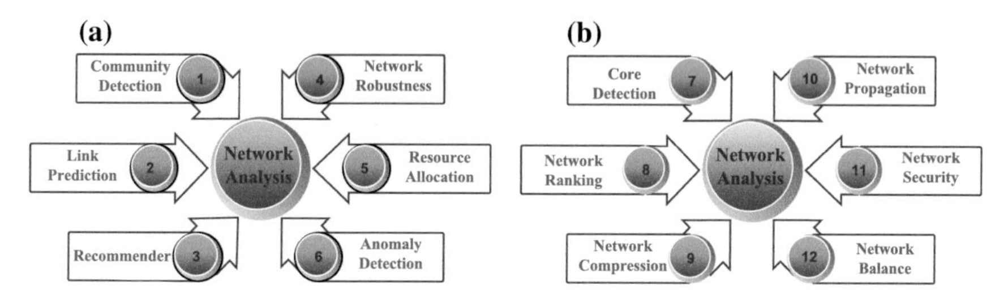

Fig. 2 Twelve critical issues that concern network analytics. **a** The first six issues and **b** the latter six issues

[16, 41, 94, 96, 97]. Networks have many notable properties, such as the small-world property [126], the scale-free property [14], the community structure property [45], etc.

The analysis of network properties is dispensable to network analytics. It is an essential part of network science. Figure 3 shows some representative properties of networks in the language of graph.

A scale-free network is a network whose degree distribution follows a power law, at least asymptotically. That is, the fraction $P(k)$ of nodes in the network having k connections to other nodes goes for large values of k as

$$P(k) \sim k^{-\gamma} \tag{1}$$

where γ is a parameter whose value is typically in the range $2 < \gamma < 3$, although occasionally it may lie outside these bounds.

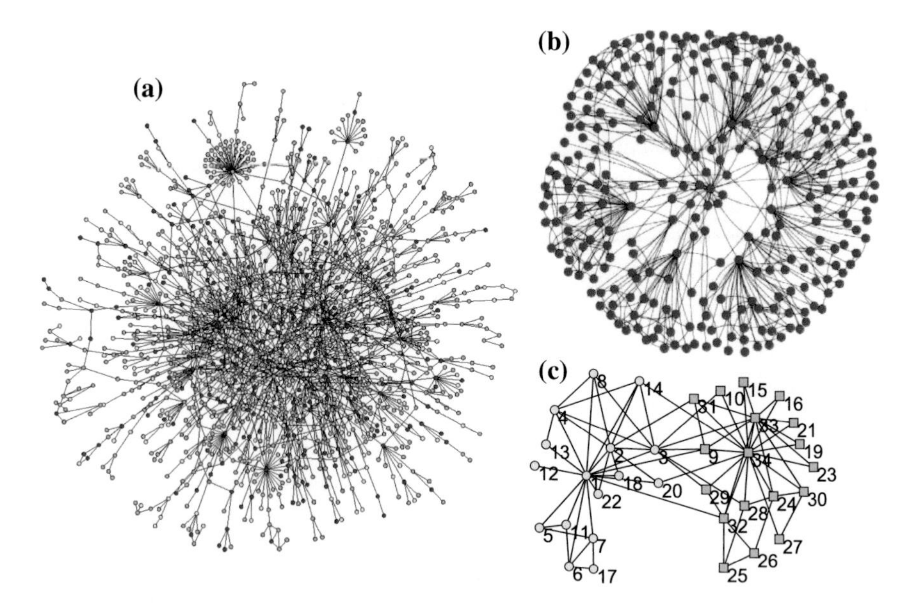

Fig. 3 **a** An example of a scale-free network. **b** An example of a small-world network. **c** An example of a network with two communities. Reprinted figure with permission from Ref. [74]

A small-world network is a type of mathematical graph in which most nodes are not neighbors of one another, but most nodes can be reached from every other by a small number of hops or steps.

A network with community structure means that the network can be separated into clusters with different sizes, and the similarities between nodes coming from the same cluster are large while from different clusters they are small.

2.3 Graph Based Network Notation

Data sets collected from many different realms can be represented in the form of interaction big networks in a very natural, concise and meaningful fashion. In order to better analyze a big network, one direct way is to represent a network with a graph denoted as $G = \{V, E\}$, where V representing the network objects is the aggregation of vertices, and E representing the relations between the objects is the aggregation of edges. Graph G can be denoted by an adjacency matrix $A_{n \times n}$ whose element a_{ij} is defined as:

$$\begin{cases} a_{ij} = \omega_{ij} \; if \; \exists L < i,j > \\ a_{ij} = 0 \quad if \; \nexists L < i,j > \end{cases} \tag{2}$$

where $L < i,j >$ represents the link between nodes i and j and ω_{ij} denotes the weight of $L < i,j >$.

In the field of social science, the networks that include both positive and negative edges are called signed social networks [37] or signed networks for short. In signed networks, the so called positive links (L^+) denote positive relationships such as friendship, common interests, and negative links (L^-) may denote negative relationships such as hostility, different interests, and so forth. A signed graph is normally denoted as $G = \{V, PE, NE\}$, where PE and NE represent the aggregations of positive and negative edges, respectively, and the element a_{ij} of the corresponding adjacency matrix $A_{n\times n}$ is defined as:

$$\begin{cases} a_{ij} = \omega_{ij} & if\ \exists L^+ < i,j > \\ a_{ij} = -\omega_{ij} & if\ \exists L^- < i,j > \\ a_{ij} = 0 & if\ \nexists L < i,j > \end{cases} \tag{3}$$

Matrix A is symmetric with the diagonal elements 0, but, if the corresponding network is directed, like the e-mail network, A is asymmetric.

3 Introduction to Nonconvex Optimization and Evolutionary Computation

3.1 *What is Optimization*

Optimization has long been an active research topic. Mathematically, a single objective optimization problem (assuming minimization) can be expressed as:

$$\begin{aligned} &\min\ f(\boldsymbol{x}),\ \boldsymbol{x} = [x_1, x_2, ..., x_d] \in \Phi \\ &s.t.\ g_i(\boldsymbol{x}) \leq 0,\ i = 1, ..., m \end{aligned} \tag{4}$$

where $\boldsymbol{x}$ is called the decision vector, d is the number of parameters to be optimized, Φ is the feasible region in decision space, and $g_i(\boldsymbol{x})$ is the constraint function.

Given that Φ is a convex set, $f(\boldsymbol{x})$ is said to be convex if $\forall \boldsymbol{x}_1, \boldsymbol{x}_2 \in \Phi, \forall \alpha \in [0, 1]$, and the following condition holds:

$$f\left(\alpha \boldsymbol{x}_1 + (1-\alpha)\boldsymbol{x}_2\right) \leq \alpha f(\boldsymbol{x}_1) + (1-\alpha) f(\boldsymbol{x}_2) \tag{5}$$

Particularly, $f(\boldsymbol{x})$ is strictly convex if $\forall \boldsymbol{x}_1 \neq \boldsymbol{x}_2 \in \Phi, \forall \alpha \in (0, 1)$, and the following condition holds:

$$f\left(\alpha \boldsymbol{x}_1 + (1-\alpha)\boldsymbol{x}_2\right) < \alpha f(\boldsymbol{x}_1) + (1-\alpha) f(\boldsymbol{x}_2) \tag{6}$$

If $f(\boldsymbol{x})$ and $g_i(\boldsymbol{x})$ are all convex, then we call Eq. 4 as a convex optimization problem. For a strictly convex optimization problem, there is at most one minimal solution which is also the global one. In real applications, the functions $f(\boldsymbol{x})$ and $g_i(\boldsymbol{x})$

may be nonconvex and there may exist many local and/or global minimum. In this respect, we call Eq. 4 as a nonconvex optimization problem. As a matter of fact, many real-world optimization problems arc nonconvex [56, 92].

In reality, many optimization problems involve multiple objectives, i.e., there are more than one $f(\boldsymbol{x})$ to be optimized. A multiobjective optimization problem can be mathematically formulated as:

$$\min \ F(\boldsymbol{x}) = (f_1(\boldsymbol{x}), f_2(\boldsymbol{x}), ..., f_k(\boldsymbol{x}))^T \tag{7}$$

The objectives in Eq. 7 often conflict with each other. Improvement of one objective may lead to deterioration of another. Thus, a single solution, which can optimize all objectives simultaneously, does not exist. For multi-objective optimization problems, the aim is to find good compromises (trade-offs) which are also called Pareto optimal solutions. The Pareto optimality concept was first proposed by Edgeworth and Pareto. To understand the concept, here are some related definitions.

- **Definition 1** (*Pareto Optimality*) A point $\boldsymbol{x}^* \in \Phi$ is Pareto optimal if for every $\boldsymbol{x} \in \Phi$ and $I = \{1, 2, ..., k\}$ either $\forall i \in I, f_i(\boldsymbol{x}) = f_i(\boldsymbol{x}^*)$ or, there is at least one $i \in I$ such that $f_i(\boldsymbol{x}) > f_i(\boldsymbol{x}^*)$.
- **Definition 2** (*Pareto Dominance*) Given two vectors $\boldsymbol{x}, \boldsymbol{y} \in \Phi$, where $\boldsymbol{x} = (x_1, x_2, ..., x_n)$ and $\boldsymbol{y} = (y_1, y_2, ..., y_n)$, we say that $\boldsymbol{x}$ dominates $\boldsymbol{y}$ (denoted as $\boldsymbol{x} \prec \boldsymbol{y}$), if $x_i \leqq y_i$ for $i = 1, 2, ..., n$, and $\boldsymbol{x} \neq \boldsymbol{y}$. $\boldsymbol{x}$ is nondominated with respect to Φ, if there does not exist another $\boldsymbol{x}' \in \Phi$ such that $F(\boldsymbol{x}') \prec F(\boldsymbol{x})$.
- **Definition 3** (*Pareto Optimal Set*) The set of all Pareto optimal solutions is called Pareto Optimal Set which is defined as:

$$PS = \{\boldsymbol{x} \in \Phi | \neg \exists \boldsymbol{x}^* \in \Phi,\ F(\boldsymbol{x}^*) \prec F(\boldsymbol{x})\} \tag{8}$$

- **Definition 4** (*Pareto Front*) The image of the Pareto set (PS) in the objective space is called the Pareto front (PF) which is defined as:

$$PF = \{F(\boldsymbol{x}) | \boldsymbol{x} \in PS\} \tag{9}$$

Figure 4 gives an example of the above mentioned definitions. Each dot except that labeled by C in the figure represents a nondominated solution to the optimization problem. The aim of a multiobjective optimization algorithm is to find the set of those nondominated solutions approximating the true PF.

3.2 How to Tackle Optimization Problems

In the field of optimization, evolutionary computation, a class of intelligent optimization techniques, has been proved to be an efficient tool for solving nonconvex

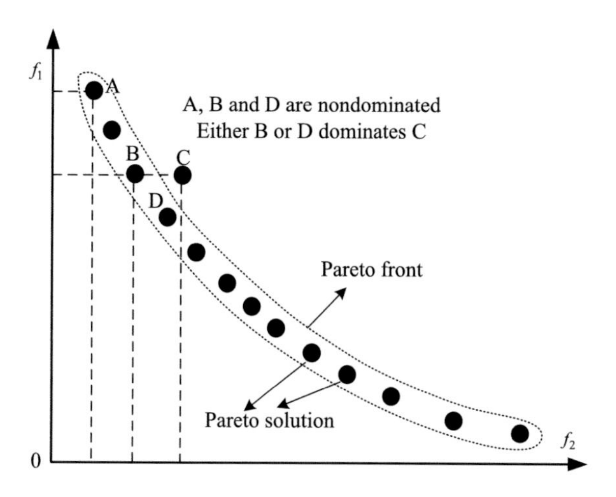

Fig. 4 Graphical illustration of Pareto optimal solution and Pareto front

optimization problems. In the last several decades, many evolutionary algorithms (EAs) originated from the evolution principles and behavior of living things, have sprung out and have found nationwide applications in the optimization domain [31, 34]. Most if not all of the EAs share the following commom properties:

1. They are population based stochastic searching methods. A population consists of a set of individuals, each individual represents a solution to the optimization problem. An evolutionary algorithm optimizes the problem by having a population of initialized solutions and then apply stochastic components to generate new solutions in the decision space.
2. They are recursively iterative methods. These methods iteratively search for optimal solutions in the search space. The search process will not stop until the maximum iteration number or a prescribed threshold is reached.

Algorithm 1 General framework of evolutionary algorithms.

Input: algorithm parameters, problem instance
Output: optimal solutions to the optimization problem

1. **Begin**
2. population initialization
3. store optimal solutions
4. **for** i=1 to max_iteration **do**
   (a) **for** each individual in the population, **do**
      i. generate a new individual through stochastic components
      ii. evaluate the fitness of the new individual
   (b) **end for**
   (c) update optimal solutions
5. **end for**
6. **End**

---

3. They have some inherent parameters, like the population size and the maximum iteration number, etc. These parameters are normally set empirically.

A general framework of EAs is shown in Algorithm 1. In the last few years, many efforts have been devoted to the application of EAs to the development of multiobjective optimization. A lot of multiobjective evolutionary algorithms (MOEAs) have been proposed, e.g., [11, 13, 30, 35, 53, 66, 131, 133, 135].

4 Community Structure Analytics

Community structure discovery is one of the cornerstones of network analytics. It can provide useful patterns and knowledge for further network analysis. This section is dedicated to summarizing the related works for community structure analytics.

4.1 Description of Community Discovery

Network community discovery plays an important role in the networked data mining field. Community discovery helps to discover latent patterns in networked data and it affects the ultimate knowledge presentation.

As illustrated above, a complex network can be expressed with a graph that is composed of nodes and edges. The task for network community discovery is to separate the whole network into small parts which are also called communities. There is no uniform definition for community in the literature, but in academic domain, a community, also called a cluster or a module, is normally regarded as a groups of vertices which probably share common properties and/or play similar roles within the graph. Figure 5 exhibits the community discovery problem under different network scenarios.

From Fig. 5 we can notice that community discovery under dynamic context is quite different from the others. In a dynamic network, the community structure is temporally changed. How to design algorithms to uncover time-varying communities is challenging.

4.2 Qualitative Community Definition

In order to formalize the qualitative community in unsigned network, Radicchi et al. in [107] gave a definition based on node degree. Given a network represented as $G = (V, E)$, where V is the set of nodes and E is the set of edges. Let k_i be the degree (the number of links that have connections with node i) of node i and A be the adjacency matrix of G. Given that $S \subset G$ is a subgraph, let $k_i^{in} = \sum_{i,j \in S} A_{ij}$ and

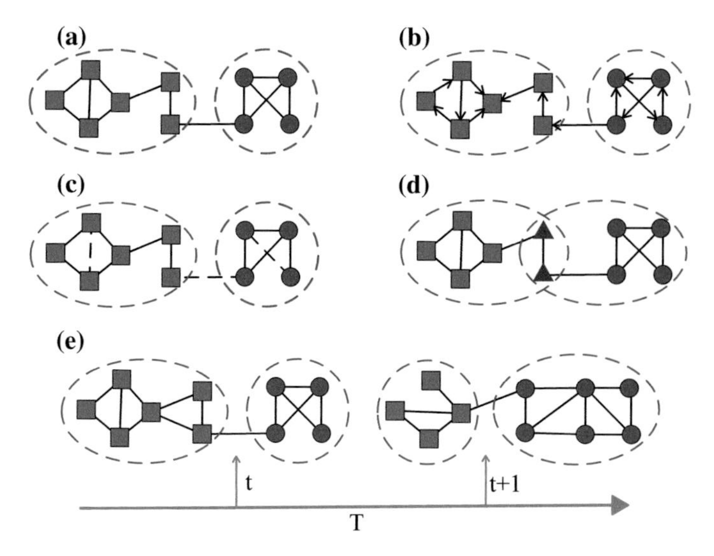

Fig. 5 Graphical illustration of community discovery. **a** Common model, **b** directed model, **c** signed model, **d** overlapping model and **e** dynamic model

$k_i^{out} = \sum_{i \in S, j \notin S} A_{ij}$ be the internal and external degree of node i, then S is a community in a strong sense if

$$\forall i \in S,\ k_i^{in} > k_i^{out} \tag{10}$$

S is a community in a weak sense if

$$\sum_{i \in S} k_i^{in} > \sum_{i \in S} k_i^{out} \tag{11}$$

The above community definition only fits for unsigned networks. In [48] the authors give a definition under signed context. Given a signed network modeled as $G = (V, PE, NE)$, where PE and NE are the set of positive and negative links, respectively. Given that $S \subset G$ is a subgraph, let $(k_i^+)^{in} = \sum_{j \in S, L_{ij} \in PE} A_{ij}$ and $(k_i^-)^{in} = \sum_{j \in S, L_{ij} \in NE} |A_{ij}|$ be the positive and negative internal degree of node i, respectively. Then S is a community in a strong sense if

$$\forall i \in S,\ (k_i^+)^{in} > (k_i^-)^{in} \tag{12}$$

Let $(k_i^-)^{out} = \sum_{j \notin S, L_{ij} \in NE} |A_{ij}|$ and $(k_i^+)^{out} = \sum_{j \notin S, L_{ij} \in PE} A_{ij}$ be the negative and positive external degree of node i, respectively. Then S is a community in a weak sense if

$$\begin{cases} \sum_{i \in S} (k_i^+)^{in} > \sum_{i \in S} (k_i^+)^{out} \\ \sum_{i \in S} (k_i^-)^{out} > \sum_{i \in S} (k_i^-)^{in} \end{cases} \tag{13}$$

Table 1 Representative non-optimization based methods for big network community discovery

Method	Ref.	Key technique	Network scale
CNM	[29]	Greedy optimization + sophisticated data structure	Medium
LPA	[12]	Mark each node with a label and then let them propagate	Very large
Infomod	[111]	Information compression, transmission, and decoding	Large
FEC	[130]	Random walk + cutoff function	Very large
BGLL	[15]	Fast hierarchical modularity optimization	Medium
Infomap	[112]	Clustering + information compression + random walks	Large

The above definitions only give the conditions that a community should satisfy, but they have not told how good on earth a community is. Therefore, there should have quantitative indexes that can measure the quality of a community. These indexes will be illustrated in Sect. 5.

4.3 *Existing Approaches for Community Discovery*

In the literature, a large amount of methods have been proposed to discover communities in big networks. Roughly, these methods can be divided into two categories: optimization based class and non-optimization based class.

For the non-optimization based avenues, in Table 1 we list several outstanding methods that can handle big networks. For more information about the existing community discovery methods developed before 2012, please refer to [41, 129].

As for the optimization based methods, most of them are nonconvex. The essence of them is to model the network community discovery task as different optimization problems and then design suitable nonconvex optimization methods such as EAs to deal with them. As what follows we will summarize the optimization models for community structure analytics.

5 Optimization Models for Community Structure Analytics

5.1 *Single Objective Optimization Model*

5.1.1 Modularity Based Model

The most popular evaluation criterion for community detection is the modularity (normally denoted as Q) proposed by Newman and Girvan in [98]. The modularity index can be given in the following form:

$$Q = \frac{1}{2m} \sum_{i,j}^{n} \left(A_{ij} - \frac{k_i \cdot k_j}{2m} \right) \delta(i,j) \tag{14}$$

where n and m are the number of nodes and edges of a network, respectively. $\delta(i,j) = 1$, if node i and j are in the same group, otherwise, 0. By assumption, higher values of Q indicate better partitions.

Q is very popular, a lof of bio-inspired metaheuristics have been utilized to optimize Q to find the community structure with biggest Q value [22, 43, 46, 60–63, 75–78, 82, 85, 88, 115, 119, 123, 124, 128]. However, Q has several drawbacks. First, to maximize Q is proved to be NP-hard [19]. Second, large Q value does not always make sense. Random networks with no community structures can also possess high Q values [59, 110]. Third, which is also the most important, Q has the resolution limitation [42], i.e., maximizing Q cannot discover communities whose sizes are smaller than a scale which depends on the total size of the network and on the degree of inter connectedness of the modules, even in the case scenario where modules are unambiguously defined.

To overcome these demerits, many researchers have devoted themselves to designing efficient operators for the optimization algorithms to enhance the exploration and exploitation; some scholars make efforts to design new evaluation criteria, such as extended modularity [10, 106, 110], multi-resolution index [80], and so forth. Because Q is originally designed for unsigned, unweighted, undirected, nonoverlapped and static networks, thus, many creative jobs have been done to extend Q to handle other types of networks.

Gómez et al. in [47] presented a reformulation of Q that allows the analysis of weighted, signed, and networks that have self-loops. The presented Q is formulized as:

$$Q_{sw} = \frac{1}{2(w^+ + w^-)} \sum_{i,j} \left[w_{ij} - \left(\frac{w_i^+ w_j^+}{2w^+} - \frac{w_i^- w_j^-}{2w^-} \right) \right] \delta(i,j) \tag{15}$$

where w_{ij} is the weight of the signed adjacency matrix, $w_i^+ (w_i^-)$ denotes the sum of all positive (negative) weights of node i. Based on the Q_{sw} metric, the authors in [23] suggested a discrete particle swarm optimization (DPSO) algorithm to detect communities from signed networks.

Q_{sw} can be easily changed to handle directed, weighted graphs [8, 72, 113], and the expression of directed and weighted Q reads:

$$Q_{dw} = \frac{1}{w} \sum_{i,j} \left(A_{ij} - \frac{w_i^{out} \cdot w_j^{in}}{w} \right) \delta(i,j) \tag{16}$$

where w_i^{out} (w_i^{in}) denotes the out-degrees (in-degrees) of node i. It can be noticed that the factor 2 is removed because the sum of the in-degrees (outdegrees), the number of non-vanishing elements of the asymmetric adjacency matrix, all equal w.

In the case when a node may belong to more than one community, Q has been modified to fit overlapping communities [99, 117, 134], and a general expression reads:

$$Q_{ov}(U_k) = \sum_{c=1}^{k} \left[\frac{A(\overline{V}_c, \overline{V}_c)}{A(V, V)} - \left(\frac{A(\overline{V}_c, V)}{A(V, V)} \right)^2 \right] \tag{17}$$

where $U_k = [u_1, u_2, \ldots, u_k]$ is a fuzzy parition of the nodes of the network into k clusters. $A(\overline{V}_c, \overline{V}_c) = \sum_{i \in \overline{V}_c} \sum_{j \in \overline{V}_c} ((u_{ic} + u_{jc})/2) w_{ij}$, where $\overline{V}_c$ is the set of vertices in community c, $A(\overline{V}_c, V) = A(\overline{V}_c, \overline{V}_c) + \sum_{i \in \overline{V}_c} \sum_{j \in V - \overline{V}_c} ((u_{ic} + (1 - u_{jc}))/2) w_{ij}$ and $A(V, V) = \sum_{i \in V} \sum_{j \in V} w_{ij}$. u_{ic} is the membership value that node i belongs to community c.

The existing overlapping community detection methods can be roughly divided into two categories, the node-based (directly cluster nodes) and the link-based (cluster links and then map link communities to node communities) ones, but the mainstream for single solution based overlapping community detection is to first utilize soft clustering technique such as fuzzy K-means to find a fuzzy partition of the nodes of a network into k clusters, and then apply a criterion to choose the best overlapping network partition [68, 70, 109, 134]. The key technique lies in the evaluation of an overlapped community. As long as an evaluation criterion is decided, bio-inspired metaheuristics can be easily utilized to solve this problem [24, 81, 84, 104]. For more information about the fitness evaluation for overlapping communities, please refer to [28, 129].

Other extended criteria such as the local modularity can be found in [90, 93], the triangle modularity in [9] and the bipartite modularity in [58].

5.1.2 Multi-resolution Model

To overcome the resolution limitation of modularity, many multi-resolution models have been developed. Pizzuti in [102] proposed a genetic algorithm for community detection. The highlight of the work is the suggested community score (CS) evaluation metric. Let $\mu_i = \frac{1}{|S|} k_i^{in}$ be the fraction of edges connecting node i to the other nodes in S and $M(S) = \frac{\sum_{i \in S} (\mu_i)^r}{|S|}$ be the power mean of S of order r. $|S|$ is the cardinality of S, i.e., the number of nodes in S. We further define $v_S = \frac{1}{2} \sum_i k_i^{in}$ be the volume of S, i.e., the number of edges connecting vertices inside S, then the score of S is defined as $score(S) = M(S) \times v_S$. Assume that G has a partition of k subgraphs, i.e., $\Omega = \{S_1, S_2, \ldots, S_k\}$, then CS can be written as:

$$CS = \sum_{i=1}^{k} score(S_i) \tag{18}$$

The *CS* metric takes one parameter r which is hard to tune. The author claims that higher values of the exponent r bias the *CS* towards matrices containing a low number of zeroes, i.e., higher values of r help in detecting communities.

Li et al. in [80] put forward the modularity density (D) index. D can break the resolution limitation brought by Q. For an unsigned network, let us define $L(S_a, S_b) = \sum_{i\in S_a, j\in S_b} A_{ij}$ and $L(S_a, \overline{S_a}) = \sum_{i\in S_a, j\in \overline{S_a}} A_{ij}$, where $\overline{S_a} = \Omega - S_a$. Then D is defined as:

$$D_\alpha = \sum_{i=1}^{k} \frac{2\alpha L(S_i, S_i) - 2(1-\alpha)L(S_i, \overline{S_i})}{|S_i|} \tag{19}$$

where $\alpha[0, 1]$ is a resolution control parameter. D_α can be viewed as a combination of the ratio association and the ratio cut [36]. Generally, optimize the ratio association algorithm often divides a network into small communities, while optimize the ratio cut often divides a network into large communities. By tuning the α value, we can use this general function to uncover more detailed and hierarchical organization of a complex network. Based on modularity density, many algorithms have emerged [21, 25, 27, 49, 51, 79].

5.2 *Multi-objective Optimization Model*

Many real-world optimization problems involve multiple objectives. From the statement of the community detection problem discussed earlier we can notice that, community detection can also be modeled as multiobjective optimization problems. Many multiobjective optimization based community detection methods have been developed in this respect. Each run of these methods can yield a set of community partitions for the decision maker to choose. The most important point for these methods should own to their abilities for breaking through the resolution limit of modularity. As stated earlier, components used in single objective optimization models, such as the individual representation, recombination, etc., serve multiobjective optimization models as well. This section primarily deals with the multiobjective community detection models.

5.2.1 General Model

As stated earlier, for an unsigned network, the links within a community should be dense while the links between communities should be sparse, as for a signed network, the inter and intra links should all be dense. On the basis of this property, many multiobjective community models are established.

Pizzuti in [103, 105] proposed a multiobjective genetic algorithm-based method called MOGA-Net. In this method, the author modeled the community detection task as a multiobjective optimization problem and then applied the fast elitist non-dominated sorting genetic algorithm (NSGA-II) [35] framework to solve it. The two objectives introduced are the *CS* and the *CF*. Thus, the proposed optimization model is:

$$\max \begin{Bmatrix} f_1 = CS \\ f_2 = -CF \end{Bmatrix} \tag{20}$$

CF (community fitness) is a criterion put forward by Lancichinetti in [68]. *CF* is formulated as:

$$CF = \sum_{S \in \Omega} \sum_{i \in S} \frac{k_i^{in}}{k_i} \tag{21}$$

From the formulation of *CF* and *CS* we may notice that, *CF* to some extent measures the link density within communities, while *CS* can be regarded as an index to measure the averaged degrees within communities.

An improved version of MOGA-Net can be found in [20]. To optimize the above model, other metaheuristics, such as the multi–objective enhanced firefly algorithm [6], hybrid evolutionary algorithm based on HSA (harmony search algorithm [44]) and CLS (chaotic local search) [4, 5, 7], non-dominated neighbor immune algorithm [52], have all find their niche in community detection.

In [54] the authors presented a multiobjective evolutionary algorithm based on decomposition (MOEA/D) based method. MOEA/D is proposed by Zhang and Li in [133]. The highlight of this work is the newly cranked out multiobjective community optimization model which optimizes two objectives termed as *NRA* (Negative Ratio Association) and *RC* (Ratio Cut). The optimization model is:

$$\min \begin{Bmatrix} NRA = -\sum_{i=1}^{k} \frac{L(S_i, S_i)}{|S_i|} \\ RC = \sum_{i=1}^{k} \frac{L(S_i, \overline{S_i})}{|S_i|} \end{Bmatrix} \tag{22}$$

It can be noticed that Eq. 22 is the decomposition of Eq. 19. *RC* measures the link density between two communities and *RA* calculates the link density within a community. To minimize *NRA* and *RC* we can ensure that the connections within a community is dense and the links between communities are sparse. A similar optimization model can be found in [50].

Other optimization models such as maximizing the combinations of *Q* and *CS* can be found in [2], and maximizing the two parts of the *Q* index, i.e., *Q* is decomposed into two objectives, can be found in [120]. A three objectives model can be found in [116]. Small surveys on the selection of objective functions in multiobjective community detection can be found in [121, 122].

5.2.2 Signed Model

Many social networks involve friendly and hostile relations between the objects that compose the networks. These networks are called signed networks. In [48] the authors put forward a novel discrete multiobjective PSO framework for community detection. To handle signed networks, the authors have suggested a signed optimization model which optimizes two objectives named as *SRA* (Signed Ratio Association) and *SRC* (Signed Ratio Cut). The optimization model reads:

$$\min \begin{Bmatrix} SRA = -\sum_{i=1}^{k} \frac{L^{+}(S_i, S_i) - L^{-}(S_i, S_i)}{|S_i|} \\ SRC = \sum_{i=1}^{k} \frac{L^{+}(S_i, \overline{S_i}) - L^{-}(S_i, \overline{S_i})}{|S_i|} \end{Bmatrix} \tag{23}$$

where $L^{+}(S_i, S_j) = \sum_{i \in S_i, j \in S_j} A_{ij}, (A_{ij} > 0)$ and $L^{-}(S_i, S_j) = \sum_{i \in S_i, j \in S_j} |A_{ij}|, (A_{ij} < 0)$. To minimize *SRA* and *SRC* we can make sure that the positive links within a community are dense while the negative links between communities are also dense, which is in accordance with the feature of signed community.

In [3] the authors put forward another signed optimization model which uses the NSGA-II framework to optimize it. The model reads:

$$\min \begin{Bmatrix} f_1 = -Q_{sw} \\ f_2 = frustration \end{Bmatrix} \tag{24}$$

where $frustration = \sum_{i,j}^{n} (A_{ij}^{+}(1 - \delta(i,j)) - A_{ij}^{-}\delta(i,j))$. The first objective Q_{sw} measures how good a signed community is and to minimize *frustration* we will ensure that the sum of the negative links within a community and the positive links between difference communities are minimum.

Recently, to detect communities from signed networks, the authors in [83] put forward a signed optimization model based on node similarity. The optimization model is as follows:

$$\max \begin{Bmatrix} f_{pos-in}(\Omega) = \frac{1}{k} \sum_{i=1}^{k} \frac{P_{in}^{S_i}}{P_{in}^{S_i} + P_{out}^{S_i}} \\ f_{neg-out}(\Omega) = \frac{1}{k} \sum_{i=1}^{k} \frac{N_{out}^{S_i}}{N_{in}^{S_i} + N_{out}^{S_i}} \end{Bmatrix} \tag{25}$$

where $P_{in}^{S_i}$ (or $P_{out}^{S_i}$) is the internal (or external) positive similarity of community S_i, and $N_{in}^{S_i}$ (or $N_{out}^{S_i}$) is the internal (or external) negative similarity of community S_i. See reference [83] for more information about the similarity of a community. To maximize f_{pos-in} we can ensure high positive similarities within communities, and to maximize $f_{neg-out}$ we can guarantee high negative similarities between different communities.

5.2.3 Overlapping Model

In real world, a node of a network may belong to more than one community, just like the friendship network. From the perspective of finding overlapping communities, intuitively, the nodes that connect multiple communities with similar strength are more likely to be overlapping nodes. For instance, if node i has both l links with community a and b, then we can regard i as an overlapping node. From the viewpoint of finding nonoverlapping or separated communities, the less the number of overlapping nodes, the more the separated communities.

Based on the above principle, the authors in [84] put forward a three objectives optimization model reads:

$$\max \begin{cases} f_1 = f_{\text{quality}}(\Omega) = \dfrac{CF}{k} \\ f_2 = f_{\text{separated}}(\Omega) = -\mid V_{overlap} \mid \\ f_3 = f_{\text{overlapping}}(\Omega) = \sum\limits_{i \in V_{overlap}} \min\limits_{s \in \Omega}\{\dfrac{k_i^s}{k_i}\} \end{cases} \tag{26}$$

where k_i^s denotes the number of edges connect node i and community s, $V_{overlap}$ is the set of the overlapping nodes. To maximize f_2 and f_3 one can get a tradeoff between nonoverlapping and overlapping communities.

5.2.4 Dynamical Model

In reality, networks may evolve with the time, the nodes and the links may disappear or new nodes may just come out, therefore, the community structures are also changing according to the time. However, traditional approaches mostly focuse on static networks for small groups. As the technologies move forward, in the presence of big data, how to design methods and tools for modeling and analyzing big dynamic networks is a challenging research topic in the years to come. To analyze the community structures of dynamical networks will help to predict the change tendency which may give support to the analysis of other network or networked scientific issues. Community detection in dynamic networks is challenging.

Dynamic community detection is normally based on a temporal smoothness framework which assumes that the variants of community division in a short time period are not desirable [39]. According to the temporal smoothness framework, the community detection in dynamic networks can be naturally modeled as a bi-objective optimization problem. The optimization of one objective is to reveal a community structure with high quality at this moment, and the optimization of the other objective is to uncover a community structure at the next moment which is highly similar with that at the previous time [26, 38–40, 55]. The commonly used dynamical optimization model can be written as:

$$\max \begin{cases} f_1 = CS \text{ } or \text{ } Q \text{ } or \text{ } D_\alpha \\ f_2 = NMI \end{cases} \tag{27}$$

NMI, Normalized Mutual Information [33], comes from the field of information theory. *NMI* can be regarded as a similarity index. For the community detection problem, given that A and B are two partitions of a network, respectively, C is a confusion matrix, C_{ij} equals to the number of nodes shared in common by community i in partition A and by community j in partition B. Then $NMI(A, B)$ is written as:

$$NMI = \frac{-2\sum_{i=1}^{C_A}\sum_{j=1}^{C_B} C_{ij}log(C_{ij}\cdot n/C_{i.}C_{.j})}{\sum_{i=1}^{C_A} C_{i.}log(C_{i.}/n) + \sum_{j=1}^{C_B} C_{.j}log(C_{.j}/n)} \tag{28}$$

where C_A (or C_B) is the number of clusters in partition A(or B), $C_{i.}$ (or $C_{.j}$) is the sum of elements of C in row i(or column j). $NMI(A, B) = 1$ means that A and B are identical and $NMI(A, B) = 0$ indicates that A and B are completely different.

The first objective in Eq. 27 is the snapshot cost which measures how well a community structure A is at time t and the second objective is the temporal cost which measures how similar the community structure B is at time $t + 1$ with the previous community structure A.

Another dynamical model which maximizes the Min-max cut and global silhouette index can be found in [65].

6 Network Data Sets

This section will list the network data sets commonly used in the literature for testing purpose. The data sets contain two types, artificial benchmark networks and real-world networks. Benchmark networks have controlled topologies. They are used to mimic real-world networks. Different real-world networks may have different properties. Hence, real-world networks are still needed for testing purpose.

6.1 Artificial Generated Benchmark Networks

6.1.1 GN Benchmark and Its Extended Version

Girvan and Newan (GN) in [45] put forward a benchmark network generator which is normally recognized as the GN benchmark. For a GN benchmark network, it was constructed with 128 vertices divided into four communities of 32 vertices each. Edges were placed between vertex pairs independently at random, with probability P_{in} for vertices belonging to the same community and P_{out} for vertices in different communities, with $P_{out} < P_{in}$. The probabilities were chosen so as to keep the average degree z of a vertex equal to 16.

An extended version of the GN model was introduced in [32]. The extended benchmark network also consists of 128 nodes divided into four communities of 32 nodes each. Every node has an average degree of 16 and shares a fraction γ of links with the rest in its community, and $1 - \gamma$ with the other nodes of the network. Here, γ is called the mixing parameter. When $\gamma < 0.5$, the neighbours of a vertex inside its community are more than the neighbors belonging to the rest groups.

6.1.2 LFR Benchmark

Standard benchmarks, like the GN benchmark or its extended version, do not account for important features in graph representations of real systems, like the fat-tailed distributions of node degree and community size, since on those benchmark networks, all vertices have approximately the same degree, moreover, all communities have exactly the same size by construction.

To overcome these drawbacks, a new class of benchmark graphs have been proposed by Lancichinetti, Fortunato, and Radicchi (LFR) in [69], in which the distributions of node degree and community size are both power laws with tunable exponents. They assume that the distributions of degree and community size are power laws, with exponents τ_1 and τ_2, respectively. Each vertex shares a fraction $1 - \mu$ of its edges with the other vertices of its community and a fraction μ with the vertices of the other communities; $0 \leq \mu \leq 1$ is the mixing parameter. The software to create the LFR benchmark graphs can be freely downloaded at http://santo.fortunato.googlepages.com/inthepress2. In our experiments, we generate 17 networks with the mixing parameter increasing from 0 to 0.8 with an interval of 0.05.

6.1.3 Signed LFR Benchmark

The LFR network generator is a reliable model for benchmarking. However, this model is originally designed for unsigned networks. In order to mimic signed social networks, The LFR model can be extended into signed version. Here we give a feasible way to do so.

A signed LFR model can be depicted by $SLFR(n, k_{avg}, k_{max}, \gamma, \beta, s_{min}, s_{max}, \mu, p-, p+)$, where n is the number of nodes; k_{avg} and k_{max} are the averaged and maximum degree of a node, respectively; γ and β are the exponents for the power law distribution of node degree and community size, respectively; s_{min} and s_{max} are the minimum and maximum community size, respectively. μ is a mixing parameter. Each node shares a fraction $1 - \mu$ of its links with the other nodes of its community and a fraction μ with the other nodes of the network. $p-$ is the fraction of negative edges within communities, and $p+$ is the fraction of positive edges between different communities.

6.2 Real-World Networks

Tables 2 and 3 list the parameters of 8 commonly tested unsigned and signed networks. In the Tables, m^+ and m^- denote the numbers of positive and negative edges, respectively. $\overline{k}$ is the averaged node degree.

6.3 Famous Websites

Apart from the above mentioned network data sets, many other network data sets are available on the Internet. In this part we list several famous websites as follows:

- http://www-personal.umich.edu/~mejn/ (Mark Newman Website)
- http://deim.urv.cat/~aarenas/data/welcome.htm (Alex Arenas Website)

Table 2 Eight commonly tested unsigned networks

Network	#Node	#Edge	#Clusters	$\overline{k}$	Ref.
Karate	34	78	2	4.588	[132]
Dolphin	62	159	2	5.129	[87]
Football	115	613	12	10.661	[45]
SFI	118	200	Unknown	3.390	[45]
E-mail	1133	5451	Unknown	9.622	[57]
Netscience	1589	2742	Unknown	3.451	[95]
Power grid	4941	6594	Unknown	2.669	[126]
PGP	10680	24340	Unknown	4.558	[17]

Table 3 Eight commonly tested signed networks

Network	#Node	#Edge	m^+	m^-	$\overline{k}$	Ref.
SPP	10	45	18	27	9.000	[67]
GGS	16	58	29	29	7.250	[108]
EGFR	329	779	515	264	4.736	[101]
Macrophage	678	1,425	947	478	4.204	[100]
Yeast	690	1,080	860	220	3.130	[91]
Ecoli	1,461	3,215	1,879	1,336	4.401	[118]
WikiElec	7,114	100,321	78,792	21,529	28.204	[73]
Slashdot	77,357	466,666	352,890	113,776	12.065	[73]

- http://snap.stanford.edu/index.html (Stanford Network Analysis Project. Diverse kinds of network data and graphical visualization softwares and tools and useful codes are available.)
- http://www.correlatesofwar.org/ (The Correlates of War Project. A large amount of signed networks mainly related to war are free to access.)
- http://www.gmw.rug.nl/~huisman/sna/software.html (A collection of softwares for social network analysis.)
- http://tuvalu.santafe.edu/~aaronc/hierarchy/ (Hierarchical Random Graphs)

7 Experimental Exhibition

In [22] we have suggested a greedy discrete particle swarm optimization algorithm (GDPSO) for big network community discovery. The GDPSO algorithm optimizes the modularity index. As what follows we will show its performance over several real-world networks.

Table 4 lists the averaged modularity values obtained by five methods over 30 independent runs on six networks. The GDPSO algorithm is an optimization based method. GDPSO is competitive to the rest four methods in terms of the modularity index.

On one hand, it is natural to model network community discovery as a multiobjective optimization problem. On the other hand, based on the preliminary shown in Sect. 3.1, we can get to know that a single run of a MOEA based community discovery method can output a set of solutions, as shown in Fig. 6.

As can be seen from Fig. 6 that each Pareto solution denotes a certain network community structure. However, each single run of the methods listed in Table 4 can only output one solution. There is no doubt that the MOEA based community discovery facilitates intelligent multi-criteria decision making. For more exhibitions about the MOEA based community discovery please refer to our recent work in [48].

It should be noted that based on the optimization models discussed in Sect. 5, one can design different single objective EAs or MOEAs to optimize those models. However, according to the NFL (No Free Lunch) theory [127], there is no one-for-all

Table 4 Averaged modularity values obtained by five methods over 30 independent runs

Network	GDPSO	CNM	BGLL	Infomap	LPA
Karate	0.4198	0.3800	0.4180	0.4020	0.3264
Dolphin	0.5280	0.4950	0.5188	0.5247	0.4964
Football	0.6041	0.5770	0.6046	0.6005	0.5848
E-mail	0.4783	0.4985	0.5412	0.5355	0.0070
Power grid	0.8368	0.9229	0.7756	0.8140	0.7476
PGP	0.8013	0.8481	0.9604	0.7777	0.7845

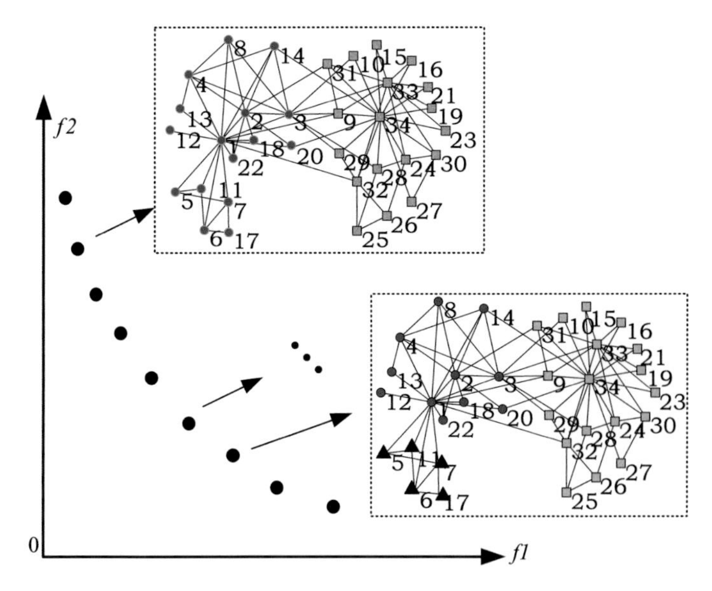

Fig. 6 An illustration of the Pareto front obtained by an MOEA for community discovery from the Karate network

method that can deal with all kinds of networks. For one thing, for different network issues, we can solve them well as long as we can establish a good optimization model that can well depict the nature of those problems. For another thing, we should make efforts to enhance the search abilities of the optimization algorithms. Meanwhile, different networks have different space-time properties. Consequently, we should take into account the special characters of the networks when designing algorithms to solve network issues.

8 Concluding Remarks

Network analysis is one of the theoretical underpinnings of big data. Network community discovery serves as the backbone of network analysis. The past decades have witnessed the prosperity of the research on community discovery. A large number of techniques have been cranked out to discover communities in the networks. Among the extant avenues for solving the network community discovery problem, many of them are nonconvex optimization based.

This chapter tries to investigate the network community discovery problem from the optimization view. Single objective and multiobjective optimization models for network community discovery problems are delineated. Experimental studies are also shown to demonstrate the promise of the optimization based idea for network analytics.

We expect that complex network analysis's scope will continue to expand and its applications to multiply. We are positive that methods and theories that work for community detection are helpful for other network issues. From both theoretical and technological perspectives, network community discovery technology will move beyond network analytics toward emphasizing network intelligence. We do hope that this chapter can benefit scholars who set foot in this field. Our future work will focus on more in-depth analysis of network issues. Such analysis is expected to shed light on how networks change the real world.

Acknowledgments This work was supported by the National Natural Science Foundation of China (Grant nos. 61273317, 61422209, and 61473215), the National Top Youth Talents Program of China, and the Specialized Research Fund for the Doctoral Program of Higher Education (Grant no. 20130203110011).

References

1. Aggarwal, C.C.: An introduction to social network data analytics. Springer (2011)
2. Agrawal, R.: Bi-objective community detection (bocd) in networks using genetic algorithm. In: Contemporary Computing, pp. 5–15 (2011)
3. Amelio, A., Pizzuti, C.: Community mining in signed networks: a multiobjective approach. In: Proceedings of the 2013 IEEE/ACM International Conference on Advances in Social Networks Analysis and Mining, pp. 95–99 (2013)
4. Amiri, B., Hossain, L., Crawford, J.: A hybrid evolutionary algorithm based on hsa and cls for multi-objective community detection in complex networks. In: 2012 IEEE/ACM International Conference on Advances in Social Networks Analysis and Mining (ASONAM), pp. 243–247 (2012a)
5. Amiri, B., Hossain, L., Crawford, J.W.: An efficient multiobjective evolutionary algorithm for community detection in social networks. In: 2011 IEEE Congress on Evolutionary Computation (CEC), pp. 2193–2199 (2011)
6. Amiri, B., Hossain, L., Crawford, J.W., Wigand, R.T.: Community detection in complex networks: multi-objective enhanced firefly algorithm. Knowl. Based Syst. **46**, 1–11 (2013)
7. Amiri, B., Hossain, L., Crowford, J.: A multiobjective hybrid evolutionary algorithm for clustering in social networks. In: Proceedings of the Fourteenth International Conference on Genetic and Evolutionary Computation Conference Companion. pp. 1445–1446 (2012b)
8. Arenas, A., Duch, J., Fernández, A., Gómez, S.: Size reduction of complex networks preserving modularity. New J. Phys. **9**(6), 176 (2007)
9. Arenas, A., Fernández, A., Fortunato, S., Gómez, S.: Motif-based communities in complex networks. J. Phys. A: Math. Theor. **41**(22), 224001 (2008a)
10. Arenas, A., Fernández, A., Gómez, S.: Analysis of the structure of complex networks at different resolution levels. New J.Phys. **10**(5), 053039 (2008b)
11. Bader, J., Zitzler, E.: Hype: an algorithm for fast hypervolume-based many-objective optimization. Evol. Comput. **19**(1), 45–76 (2011)
12. Bagrow, J.P., Bollt, E.M.: A local method for detecting communities. Phys. Rev. E **72**(4), 046108 (2005)
13. Bandyopadhyay, S., Saha, S., Maulik, U., Deb, K.: A simulated annealing-based multiobjective optimization algorithm: AMOSA. IEEE Trans. Evol. Comput. **12**(3), 269–283 (2008)
14. Barabási, A.-L., Albert, R.: Emergence of scaling in random networks. Science **286**(5439), 509–512 (1999)
15. Blondel, V.D., Guillaume, J.-L., Lambiotte, R., Lefebvre, E.: Fast unfolding of communities in large networks. J. Stat. Mech. Theory Exp. **2008**(10), P10008 (2008)

16. Boccaletti, S., Latora, V., Moreno, Y., Chavez, M., Hwang, D.-U.: Complex networks: structure and dynamics. Phys. Rep. **424**(4), 175–308 (2006)
17. Boguna, M., Pastor-Satorras, R., Díaz-Guilera, A., Arenas, A.: Models of social networks based on social distance attachment. Phys. Rev. E **70**(5), 056112 (2004)
18. Bollier, D., Firestone, C.M.: The promise and peril of big data. Aspen Institute, Washington, DC (2010). Communications and Society Program
19. Brandes, U., Delling, D., Gaertler, M., Görke, R., Hoefer, M., Nikoloski, Z., Wagner, D.: Maximizing Modularity is Hard (2006). arXiv preprint arXiv:physics/0608255
20. Butun, E., Kaya, M.: A multi-objective genetic algorithm for community discovery. In: 2013 IEEE 7th International Conference on Intelligent Data Acquisition and Advanced Computing Systems (IDAACS), vol. 1. pp. 287–292 (2013)
21. Cai, Q., Gong, M., Ma, L., Jiao, L.: A novel clonal selection algorithm for community detection in complex networks. Comput. Intell. 1–24 (2014a)
22. Cai, Q., Gong, M., Ma, L., Ruan, S., Yuan, F., Jiao, L.: Greedy discrete particle swarm optimization for large-scale social network clustering. Inf. Sci. **316**, 503–516 (2015)
23. Cai, Q., Gong, M., Shen, B., Ma, L., Jiao, L.: Discrete particle swarm optimization for identifying community structures in signed social networks. Neural Netw. **58**, 4–13 (2014b)
24. Cai, Y., Shi, C., Dong, Y., Ke, Q., Wu, B.: A novel genetic algorithm for overlapping community detection. In: Advanced Data Mining and Applications, pp. 97–108 (2011)
25. Chen, G., Guo, X.: A genetic algorithm based on modularity density for detecting community structure in complex networks. In: Proceedings of the 2010 International Conference on Computational Intelligence and Security, pp. 151–154 (2010)
26. Chen, G., Wang, Y., Wei, J.: A new multiobjective evolutionary algorithm for community detection in dynamic complex networks. Math. Probl. Eng. (2013)
27. Chen, G., Wang, Y., Yang, Y.: Community detection in complex networks using immune clone selection algorithm. Int. J. Digit. Content Technol. Appl. **5**, 182–189 (2011)
28. Chira, C., Gog, A.: Fitness evaluation for overlapping community detection in complex networks. In: IEEE Congress on Evolutionary Computation, pp. 2200–2206 (2011)
29. Clauset, A., Newman, M.E.J., Moore, C.: Finding community structure in very large networks. Phys. Rev. E **70**(6), 066111 (2004)
30. Coello, C., Pulido, G., Lechuga, M.: Handling multiple objectives with particle swarm optimization. IEEE Trans. Evol. Comput. **8**(3), 256–279 (2004)
31. Črepinšek, M., Liu, S.-H., Mernik, M.: Exploration and exploitation in evolutionary algorithms: a survey. ACM Comput. Surv. (CSUR) **45**(3), 35 (2013)
32. Danon, L., Díaz-Guilera, A., Arenas, A.: The effect of size heterogeneity on community identification in complex networks. J. Stat. Mech. Theory Exp. **2006**(11), P11010 (2006)
33. Danon, L., Díaz-Guilera, A., Duch, J., Arenas, A.: Comparing community structure identification. J. Stat. Mech. Theory Exp. **2005**(09), P09008 (2005)
34. Dasgupta, D., Michalewicz, Z.: Evolutionary algorithms in engineering applications. Springer Science & Business Media (2013)
35. Deb, K., Pratap, A., Agarwal, S., Meyarivan, T.: A fast and elitist multiobjective genetic algorithm: NSGA-II. IEEE Trans. Evol. Comput. **6**(2), 182–197 (2002)
36. Dhillon, I.S., Guan, Y., Kulis, B.: Kernel k-means: spectral clustering and normalized cuts. In: Proceedings of the 10th ACM SIGKDD international conference on Knowledge discovery and data mining, pp. 551–556 (2004)
37. Doreian, P., Mrvar, A.: A partitioning approach to structural balance. Soc. Netw. **18**(2), 149–168 (1996)
38. Folino, F., Pizzuti, C.: A multiobjective and evolutionary clustering method for dynamic networks. In: 2010 International Conference on Advances in Social Networks Analysis and Mining (ASONAM), pp. 256–263 (2010a)
39. Folino, F., Pizzuti, C.: Multiobjective evolutionary community detection for dynamic networks. In: Proceedings of the 12th annual conference on Genetic and evolutionary computation, pp. 535–536 (2010b)

40. Folino, F., Pizzuti, C.: An evolutionary multiobjective approach for community discovery in dynamic networks. IEEE Trans. Knowl. Data Eng. **26**(8), 1838–1852 (2014)
41. Fortunato, S.: Community Detect. Graphs Phys. Rep. **486**(3), 75–174 (2010)
42. Fortunato, S., Barthelemy, M.: Resolution limit in community detection. Proc. Natl. Acad. Sci. **104**(1), 36–41 (2007)
43. Gach, O., Hao, J.-K.: A memetic algorithm for community detection in complex networks. In: Parallel Probl. Solving Nat. PPSN XII, pp. 327–336 (2012)
44. Geem, Z.W., Kim, J.H., Loganathan, G.: A new heuristic optimization algorithm: harmony search. Simulation **76**(2), 60–68 (2001)
45. Girvan, M., Newman, M.E.J.: Community structure in social and biological networks. Proc. Natl. Acad. Sci. U.S.A **99**(12), 7821–7826 (2002)
46. Gog, A., Dumitrescu, D., Hirsbrunner, B.: Community detection in complex networks using collaborative evolutionary algorithms. In: Adv. Artif. Life. pp. 886–894 (2007)
47. Gómez, S., Jensen, P., Arenas, A.: Analysis of community structure in networks of correlated data. Phys. Rev. E **80**(1), 016114 (2009)
48. Gong, M., Cai, Q., Chen, X., Ma, L.: Complex network clustering by multiobjective discrete particle swarm optimization based on decomposition. IEEE Trans. Evol. Comput. **18**(1), 82–97 (2014)
49. Gong, M., Cai, Q., Li, Y., Ma, J.: An improved memetic algorithm for community detection in complex networks. In: Proceedings of 2012 IEEE Congress on Evolutionary Computation, pp. 1–8 (2012a)
50. Gong, M., Chen, X., Ma, L., Zhang, Q., Jiao, L.: Identification of multi-resolution network structures with multi-objective immune algorithm. Appl. Soft Comput. **13**(4), 1705–1717 (2013)
51. Gong, M., Fu, B., Jiao, L., Du, H.: Memetic algorithm for community detection in networks. Phys. Rev. E **84**(5), 056101 (2011a)
52. Gong, M., Hou, T., Fu, B., Jiao, L.: A non-dominated neighbor immune algorithm for community detection in networks. In: Proceedings of the 13th Annual Conference on Genetic and Evolutionary Computation, pp. 1627–1634 (2011b)
53. Gong, M., Jiao, L., Du, H., Bo, L.: Multiobjective immune algorithm with nondominated neighbor-based selection. Evol. Comput. **16**(2), 225–255 (2008)
54. Gong, M., Ma, L., Zhang, Q., Jiao, L.: Community detection in networks by using multiobjective evolutionary algorithm with decomposition. Phys. A **391**(15), 4050–4060 (2012b)
55. Gong, M., Zhang, L., Ma, J., Jiao, L.: Community detection in dynamic social networks based on multiobjective immune algorithm. J. Comput. Sci. Technol. **27**(3), 455–467 (2012c)
56. Grossmann, I.E.: Global Optimization in engineering design, vol. 9. Springer Science & Business Media (2013)
57. Guimerà, R., Danon, L., Díaz-Guilera, A.: Self-similar community structure in a network of human interactions. Phys. Rev. E **68**(6), 065103 (2003)
58. Guimerà, R., Sales-Pardo, M., Amaral, L.A.N.: Module identification in bipartite and directed networks. Phys. Rev. E **76**(3), 036102 (2007)
59. Guimerò, R., Sales-Pardo, M., Anaral, L.A.N.: Modularity from fluctuations in random graphs and complex networks. Phys. Rev. E **70**(2), 025101 (2004)
60. He, D., Wang, Z., Yang, B., Zhou, C.: Genetic algorithm with ensemble learning for detecting community structure in complex networks. In: Proceedings of the Fourth International Conference on Computer Sciences and Convergence Information Technology, pp. 702–707 (2009)
61. Huang, Q., White, T., Jia, G., Musolesi, M., Turan, N., Tang, K., He, S., Heath, J. K., Yao, X.: Community detection using cooperative co-evolutionary differential evolution. In: Parallel Problem Solving from Nature-PPSN XII, pp. 235–244 (2012)
62. Jia, G., Cai, Z., Musolesi, M., Wang, Y., Tennant, D. A., Weber, R. J., Heath, J. K., He, S.: Community detection in social and biological networks using differential evolution. In: Learning and Intelligent Optimization, pp. 71–85 (2012)

63. Jin, D., He, D., Liu, D., Baquero, C.: Genetic algorithm with local search for community mining in complex networks. In: 2010 22nd IEEE International Conference on Tools with Artificial Intelligence (ICTAI), vol. 1, pp. 105–112 (2010)
64. Kang, U., Tsourakakis, C. E., Appel, A. P., Faloutsos, C., Leskovec, J.: Radius plots for mining tera-byte scale graphs: Algorithms, patterns, and observations. In: SIAM International Conference on Data Mining (2010)
65. Kim, K., McKay, R.I., Moon, B.-R.: Multiobjective evolutionary algorithms for dynamic social network clustering. In: Proceedings of the 12th Annual Conference on Genetic and Evolutionary Computation, pp. 1179–1186 (2010)
66. Knowles, J.D., Corne, D.W.: Approxmating the nondominated front using the pareto archived evolution strategy. Evol. Comput. **8**(2), 149–172 (2000)
67. Kropivnik, S., Mrvar, A.: An analysis of the slovene parliamentary parties network. In: Ferligoj, A., Kramberger, A. (eds.) Developments in Statistics and Methodology, pp. 209–216 (1996)
68. Lancichinetti, A., Fortunato, S., Kertész, J.: Detecting the overlapping and hierarchical community structure in complex networks. New J. Phys. **11**(3), 033015 (2009)
69. Lancichinetti, A., Fortunato, S., Radicchi, F.: Benchmark graphs for testing community detection algorithms. Phys. Rev. E **78**(4), 046110 (2008)
70. Lázár, A., Ábel, D., Vicsek, T.: Modularity measure of networks with overlapping communities. Europhys. Lett. **90**(1), 18001 (2010)
71. Lazer, D., Pentland, A., Adamic, L., Aral, S., Barabasi, A.L., Brewer, D., Christakis, N., Contractor, N., Fowler, J., Gutmann, M., Jebara, T., King, G., Macy, M., Roy, D., Van Alstyne, M.: Social science. computational social science. Science **323**(5915), 721–723 (2009)
72. Leicht, E.A., Newman, M.E.: Community structure in directed networks. Phys. Rev. Lett. **100**(11), 118703 (2008)
73. Leskovec, J., Krevl, A.: SNAP Datasets: stanford large network dataset collection (2014). http://snap.stanford.edu/data
74. Li, D., Xiao, L., Han, Y., Chen, G., Liu, K.: Network thinking and network intelligence. In: Web Intelligence Meets Brain Informatics, pp. 36–58. Springer (2007)
75. Li, J., Song, Y.: Community detection in complex networks using extended compact genetic algorithm. Soft Comput. **17**(6), 925–937 (2013)
76. Li, S., Chen, Y., Du, H., Feldman, M.W.: A genetic algorithm with local search strategy for improved detection of community structure. Complexity **15**(4), 53–60 (2010)
77. Li, X., Gao, C.: A novel community detection algorithm based on clonal selection. J. Comput. Inf. Syst. **9**(5), 1899–1906 (2013)
78. Li, Y., Liu, G., Lao, S.-Y.: Complex network community detection algorithm based on genetic algorithm. In: The 19th International Conference on Industrial Engineering and Engineering Management, pp. 257–267 (2013)
79. Li, Y., Liu, J., Liu, C.: A comparative analysis of evolutionary and memetic algorithms for community detection from signed social networks. Soft Comput. **18**(2), 329–348 (2014)
80. Li, Z., Zhang, S., Wang, R.-S., Zhang, X.-S., Chen, L.N.: Quantitative function for community detection. Phys. Rev. E **77**(3), 036109 (2008)
81. Lin, C.-C., Liu, W.-Y., Deng, D.-J.: A genetic algorithm approach for detecting hierarchical and overlapping community structure in dynamic social networks. In: 2013 IEEE Wireless Communications and Networking Conference (WCNC), pp. 4469–4474 (2013)
82. Lipczak, M., Milios, E.: Agglomerative genetic algorithm for clustering in social networks. In: Proceedings of the 11th Annual Conference on Genetic and Evolutionary Computation, pp. 1243–1250 (2009)
83. Liu, C., Liu, J., Jiang, Z.: A multiobjective evolutionary algorithm based on similarity for community detection from signed social networks. IEEE Trans. Cybern. **44**(12), 2274–2287 (2014)
84. Liu, J., Zhong, W., Abbass, H.A., Green, D.G.: Separated and overlapping community detection in complex networks using multiobjective evolutionary algorithms. In: IEEE Congress on Evolutionary Computation, pp. 1–7 (2010)

85. Liu, X., Li, D., Wang, S., Tao, Z.: Effective algorithm for detecting community structure in complex networks based on ga and clustering. Comput. Sci. ICCS **2007**, 657–664 (2007)
86. Lohr, S.: The age of big data. New York Times 11 (2012)
87. Lusseau, D., Schneider, K., Boisseau, O.J., Haase, P., Slooten, E., Dawson, S.M.: The bottlenose dolphin community of doubtful sound features a large proportion of long-lasting associations. Behav. Ecol. Sociobiol. **54**(4), 396–405 (2003)
88. Ma, L., Gong, M., Liu, J., Cai, Q., Jiao, L.: Multi-level learning based memetic algorithm for community detection. Appl. Soft Comput. **19**, 121–133 (2014)
89. Manyika, J., Chui, M., Brown, B., Bughin, J., Dobbs, R., Roxburgh, C., Byers, A.H., Institute, M.G.: Big data: the next frontier for innovation, competition, and productivity (2011)
90. Massen, C.P., Doye, J.P.: Identifying communities within energy landscapes. Phys. Rev. E **71**(4), 046101 (2005)
91. Milo, R., Shen-Orr, S., Itzkovitz, S., Kashtan, N., Chklovskii, D., Alon, U.: Network motifs: simple building blocks of complex networks. Science **298**(5594), 824–827 (2002)
92. Mistakidis, E.S., Stavroulakis, G.E.: Nonconvex optimization in mechanics: algorithms, heuristics and engineering applications by the FEM, vol. 21. Springer Science & Business Media (2013)
93. Muff, S., Rao, F., Caflisch, A.: Local modularity measure for network clusterizations. Phys. Rev. E **72**(5), 056107 (2005)
94. Newman, M.E.J.: The structure and function of complex networks. SIAM Rev. **45**(2), 167–256 (2003)
95. Newman, M.E.J.: Finding community structure in networks using the eigenvectors of matrices. Phys. Rev. E **74**, 036104 (2006)
96. Newman, M.E.J.: Networks: An Introduction. Oxford University Press (2010)
97. Newman, M.E.J.: Complex Systems: A Survey (2011). arXiv preprint arXiv:1112.1440
98. Newman, M.E.J., Girvan, M.: Finding and evaluating community structure in networks. Phys. Rev. E **69**(2), 026113 (2004)
99. Nicosia, V., Mangioni, G., Carchiolo, V., Malgeri, M.: Extending the definition of modularity to directed graphs with overlapping communities. J. Stat. Mech. Theory Exp. **2009**(03), P03024 (2009)
100. Oda, K., Kimura, T., Matsuoka, Y., Funahashi, A., Muramatsu, M., Kitano, H.: Molecular interaction map of a macrophage. AfCS Res. Rep. **2**(14), 1–12 (2004)
101. Oda, K., Matsuoka, Y., Funahashi, A., Kitano, H.: A comprehensive pathway map of epidermal growth factor receptor signaling. Mol. Syst. Biol. **1**(1), 1–17 (2005)
102. Pizzuti, C.: GA-Net: a genetic algorithm for community detection in social networks. Parallel Probl. Solving Nat. (PPSN), **5199**, 1081–1090 (2008)
103. Pizzuti, C.: A multi-objective genetic algorithm for community detection in networks. In: 2009 ICTAI'09, 21st International Conference on Tools with Artificial Intelligence, pp. 379–386 (2009a)
104. Pizzuti, C.: Overlapped community detection in complex networks. GECCO **9**, 859–866 (2009b)
105. Pizzuti, C.: A multiobjective genetic algorithm to find communities in complex networks. IEEE Trans. Evol. Comput. **16**(3), 418–430 (2012)
106. Pons, P., Latapy, M.: Post-processing hierarchical community structures: quality improvements and multi-scale view. Theor. Comput. Sci. **412**(8–10), 892–900 (2011)
107. Radicchi, F., Castellano, C., Cecconi, F., Loreto, V., Parisi, D.: Defining and identifying communities in networks. Proc. Natl. Acad. Sci. U.S.A. **101**(9), 2658–2663 (2004)
108. Read, K.E.: Cultures of the central highlands, new guinea. Southwest. J. Anthropol. **10**(1), 1–43 (1954)
109. Rees, B.S., Gallagher, K.B.: Overlapping community detection using a community optimized graph swarm. Soc. Netw. Anal. Min. **2**(4), 405–417 (2012)
110. Reichardt, J., Bornholdt, S.: Statistical mechanics of community detection. Phys. Rev. E **74**(1), 016110 (2006)

111. Rosvall, M., Bergstrom, C.T.: An information-theoretic framework for resolving community structure in complex networks. Proc. Natl. Acad. Sci. U.S.A. **104**(18), 7327–7331 (2007)
112. Rosvall, M., Bergstrom, C.T.: Maps of random walks on complex networks reveal community structure. Proc. Natl. Acad. Sci. U.S.A. **105**(4), 1118–1123 (2008a)
113. Rosvall, M., Bergstrom, C.T.: Maps of random walks on complex networks reveal community structure. Proc. Natl. Acad. Sci. **105**(4), 1118–1123 (2008b)
114. Scott, J.: Social network analysis. Sage (2012)
115. Shang, R., Bai, J., Jiao, L., Jin, C.: Community detection based on modularity and an improved genetic algorithm. Phys. A Stat. Mech. Appl. **392**(5), 1215–1231 (2013)
116. Shelokar, P., Quirin, A., Cordón, Ó.: Three-objective subgraph mining using multiobjective evolutionary programming. J. Comput. Syst. Sci. **80**(1), 16–26 (2014)
117. Shen, H., Cheng, X., Cai, K., Hu, M.-B.: Detect overlapping and hierarchical community structure in networks. Phys. A **388**(8), 1706–1712 (2009)
118. Shen-Orr, S.S., Milo, R., Mangan, S., Alon, U.: Network motifs in the transcriptional regulation network of escherichia coli. Nat. Genet. **31**(1), 64–68 (2002)
119. Shi, C., Wang, Y., Wu, B., Zhong, C.: A new genetic algorithm for community detection. In: Complex Sciences, pp. 1298–1309 (2009)
120. Shi, C., Yan, Z., Cai, Y., Wu, B.: Multi-objective community detection in complex networks. Appl. Soft Comput. **12**(2), 850–859 (2012)
121. Shi, C., Yu, P. S., Cai, Y., Yan, Z., Wu, B.: On selection of objective functions in multiobjective community detection. In: Proceedings of the 20th ACM international conference on Information and knowledge management, pp. 2301–2304. ACM (2011)
122. Shi, C., Yu, P.S., Yan, Z., Huang, Y., Wang, B.: Comparison and selection of objective functions in multiobjective community detection. Comput. Intell. **30**(3), 562–582 (2014)
123. Shi, Z., Liu, Y., Liang, J.: PSO-based community detection in complex networks. In: Proceedings of the 2nd International Symposium on Knowledge Acquisition and Modeling, vol. 3. pp. 114–119 (2009)
124. Tasgin, M., Herdagdelen, A., Bingol, H.: Community detection in complex networks using genetic algorithms (2007). arXiv preprint arXiv:0711.0491
125. Wang, F.Y., Zeng, D., Carley, K.M., Mao, W.J.: Social computing: from social informatics to social intelligence. IEEE Intell. Syst. **22**(2), 79–83 (2007)
126. Watts, D.J., Strogatz, S.H.: Collective dynamics of 'small-world' networks. Nature **393**(6684), 440–442 (1998)
127. Wolpert, D.H., Macready, W.G.: No free lunch theorems for optimization. IEEE Trans. Evol. Comput. **1**(1), 67–82 (1997)
128. Xiaodong, D., Cunrui, W., Xiangdong, L., Yanping, L.: Web community detection model using particle swarm optimization. In: IEEE Congress on Evolutionary Computation, pp. 1074–1079 (2008)
129. Xie, J., Kelley, S., Szymanski, B.K.: Overlapping community detection in networks: the state-of-the-art and comparative study. ACM Comput. Surv. (CSUR) **45**(4), 43 (2013)
130. Yang, B., Cheung, W.K., Liu, J.M.: Community mining from signed social networks. IEEE Trans. Knowl. Data Eng. **19**(10), 1333–1348 (2007)
131. Yang, X.-S.: Multiobjective firefly algorithm for continuous optimization. Eng. Comput. **29**(2), 175–184 (2013)
132. Zachary, W.W.: An information flow model for confict and fission in small groups. J. Anthropol. Res. 452–473 (1977)
133. Zhang, Q., Li, H.: MOEA/D: a multiobjective evolutionary algorithm based on decomposition. IEEE Trans. Evol. Comput. **11**(6), 712–731 (2007)
134. Zhang, S., Wang, R.-S., Zhang, X.-S.: Identification of overlapping community structure in complex networks using fuzzy c-means clustering. Phys. A **374**(1), 483–490 (2007)
135. Zitzler, E., Laumanns, M., Thiele, L.: SPEA2: Improving the strength pareto evolutionary algorithm. In: Proceedings of Evolutionary Methods for Design Optimization and Control with Applications to Industrial Problems, pp. 95–100 (2002)

Author Biographies

Maoguo Gong received the B.S. degree and Ph.D. degree from Xidian University, Xi'an, China, in 2003 and 2009, respectively. Since 2006, he has been a Teacher with Xidian University. In 2008 and 2010, he was promoted as an Associate Professor and as a Full Professor, respectively, both with exceptive admission. His research interests are in the area of computational intelligence with applications to optimization, learning, data mining and image understanding. Dr. Gong has published over one hundred papers in journals and conferences, and holds over ten granted patents as the first inventor. He was the recipient of the prestigious National Program for Support of Top-notch Young Professionals (selected by the Central Organization Department of China), the Excellent Young Scientist Foundation (selected by the National Natural Science Foundation of China), the New Century Excellent Talent in University (selected by the Ministry of Education of China).

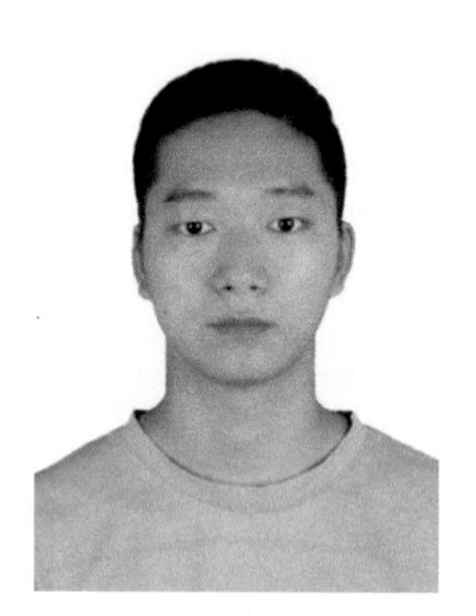

Qing Cai received the B.S. degree in electronic information engineering from Wuhan Textile University, Wuhan, China, in 2010, and the Ph.D. degree in pattern recognition and intelligent systems at the School of Electronic Engineering, Xidian University, Xi'an, China, in 2015. Currently, he is working as a research fellow in the Department of Computer Science, Hong Kong Baptist University, Kowloon Tong, Hong Kong. His current research interests are in the area of computational intelligence, complex network analytics and recommender systems.

Lijia Ma received the B.S. degree in communication engineering from Hunan Normal University, Hunan, China, in 2010. He is currently pursuing the Ph.D. degree in pattern recognition and intelligent systems at the School of Electronic Engineering, Xidian University, Xi'an, China. His research interests include evolutionary multiobjective optimization, data mining, and complex network analysis.

Licheng Jiao received the B.S. degree from Shanghai Jiaotong University, Shanghai, China, in 1982 and the M.S. and Ph.D. degrees from Xi'an Jiaotong University, Xi'an, China, in 1984 and 1990, respectively. Since 1992, he has been a Professor with the School of Electronic Engineering, Xidian University, Xi'an, where he is currently the Director of the Key Laboratory of Intelligent Perception and Image Understanding of the Ministry of Education of China. His research interests include image processing, natural computation, machine learning, and intelligent information processing.

Large-Scale and Big Optimization Based on Hadoop

Yi Cao and Dengfeng Sun

Abstract Integer Linear Programming (ILP) is among the most popular optimization techniques found in practical applications, however, it often faces computational issues in modeling real-world problems. Computation can easily outgrow the computing power of standalone computers as the size of problem increases. The modern distributed computing releases the computing power constraints by providing scalable computing resources to match application needs, which boosts large-scale optimization. This chapter presents a paradigm that leverages Hadoop, an open-source distributed computing framework, to solve a large-scale ILP problem that is abstracted from real-world air traffic flow management. The ILP involves millions of decision variables, which is intractable even with existing state-of-the-art optimization software package. Dual decomposition method is used to separate variables into a set of dual subproblems that are smaller ILPs with lower dimensions, the computation complexity is downsized. As a result, the subproblems are solvable with optimization tools. It is shown that the iterative update on Lagrangian multipliers in dual decomposition method can fit into the Hadoop's MapReduce programming model, which is designed to allocate computations to cluster for parallel processing and collect results from each node to report aggregate results. Thanks to the scalability of the distributed computing, parallelism can be improved by assigning more working nodes to the Hadoop cluster. As a result, the computational efficiency for solving the whole ILP problem is not impacted by the input size.

Keywords Integer linear programming · Distributed computing · Hadoop · MapReduce

Y. Cao (✉) · D. Sun
School of Aeronautics and Astronautics, Purdue University, West Lafayette
IN 47906-2045 USA
e-mail: cao20@purdue.edu

D. Sun
e-mail: dsun@purdue.edu

A. Emrouznejad (ed.), *Big Data Optimization: Recent Developments and Challenges*, Studies in Big Data 18, DOI 10.1007/978-3-319-30265-2_16

1 Introduction

Integer Linear Programing is a subclass of Linear Programming (LP) with integer constraints. A LP problem can be solved in polynomial-time but an ILP could be Non-deterministic Polynomial-time hard (NP-hard). It could be computationally demanding as the input size increases. The famous Traveling Salesman Problem (TSP) is a classic ILP problem that represents a wide range of optimization applications in real life, such as logistics, task scheduling, manufacturing. Many of these problems can be formulated in a similar way and solved by algorithms with exponential worst-case performance. That means even a small number of variables could lead to a considerable amount of computations. For many real-world problems with large number of variables, the computations are drastically beyond the capacity of general-purpose commodity computers. In the face of large-scale optimization problems, people often adopt tactic strategies, such as heuristics. But heuristics are quite problem-specific. It is hard to generalize the approach to benefit a broader set of ILP problems. Another popular strategy is to relax the integer constraints so that the LP relaxation can be solved in polynomial time. But such tactics must be compensated by well-designed rounding off strategies to avoid constraint violations. Another shortcoming due to relaxation is that optimality is sacrificed. Therefore, how to solve large-scale optimization posts a serious challenge.

Today, distributed computing has greatly changed the landscape of big data. Unlike traditional centralized computing that scales up hardware to escalate computing capacity, distributed computing scales out the hardware with cheaper hardware. This feature makes it prevalent in industry where cost-efficiency is a big concern. Hadoop is a paradigm of distributed computing. It is an open-source framework under Apache Software Foundation.[1] The philosophy of Hadoop is to ship the code to data stored in distributed file system and process the data locally and concurrently. Hadoop adopts MapReduce programming model that is devised for list processing. The output from a MapReduce job is a shorter list of values representing information extracted from a big input dataset. As the complexity of internal mechanism of job scheduling is transparent to developers, developers need not to master parallel computing before implementing the algorithms they want to parallelize. As a result, Hadoop has been leveraged in many computation intensive tasks since its birth, such as Discrete Fourier Transform used in signal and image processing [1], K-Means clustering used in cluster analysis [2], and logistic regression used in machine learning [3].

In optimization, researchers from Google have experimented distributed gradient optimization based on MapReduce framework [4]. The optimization serves logistic regression. Boyed et al. are among the first to explore solving large-scale optimization using MapReduce framework [5]. However, it was just a tentative discussion without proof of numeric results. But their effort has outlined the approach

[1]http://hadoop.apache.org.

that would address the massive computations arising from large-scale optimization. That is, if a large ILP can be decomposed into smaller subproblems, then the required memory and computing units to solve subproblems are downsized. Actually this idea is fully supported by dual decomposition method [6]. In this chapter, we will examine an ILP abstracted from air traffic flow management and show how to leverage the MapReduce programming model to achieve parallelism. In the following sections, we will first describe the ILP derived from an air traffic flow formulation and a dual decomposition method. Then the MapReduce programming model is introduced, followed by a discussion on fitting the decomposed ILP into MapReduce programming model. Finally, the computational efficiency is examined.

2 Air Traffic Flow Optimization

2.1 Problem Formulation

The National Airspace System (NAS) of United States is among the most complex aviation systems around the world. Minimizing delays while keeping safe operations has been the central topic in the air traffic management community. The entire airspace in the continental United States is vertically partitioned into three layers based on altitude ranges. In each layer, the airspace is partitioned into small control sectors, as illustrated in Fig. 1. A flight path can be represented as a sequence of links that connect origin and destination airports, with each link being an abstraction of passage through a sector. Flight paths are indexed by $k \in \{0, \ldots K\}$, and links are indexed by $i \in \{0, \ldots n^k\}$. The length of a link is equal to the amount of traversal time rounded to the nearest integer minutes, denoted as $T_i^k \in N$. Statistically, the observed traversal time is Gaussian without interruptions arising from stochastic factors, like extreme weather, temporary airport or sector closure, and many other emergencies [7]. We only focus on deterministic scheduling here, which is the backbone for stochastic models [8].

Consider traffic with a planning time horizon $t \in \{0, \ldots T\}$. The variable $x_i^k(t)$ represents the number of flights in link $\boldsymbol{i}$ on route $\boldsymbol{k}$. The dynamics of a path is a description of flows moving forward from upstream link to the downstream link at each time step t, and the flows are in compliance with the principle of flow conservation. The inflow rate from link $i-1$ to link $\boldsymbol{i}$ at time $\boldsymbol{t}$ is denoted by $q_i^k(t)$. The flow dynamics of a path is given by:

$$x_i^k(t+1) = x_i^k(t) - q_i^k(t) + q_{i-1}^k(t) \qquad \forall \mathrm{t} \in \{0, \ldots, \mathrm{T}\} \tag{1}$$

For the first link, the inflow is scheduled departures $q_0^k = f^k(t)$. To guarantee reasonable flow rate moving forward along a path, we construct the following constraints:

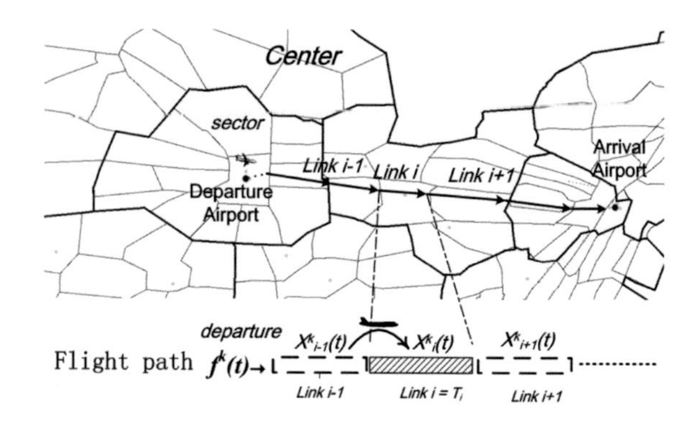

Fig. 1 Link transmission model for air traffic flows

$$\sum_{t=0}^{T} q_0^k(t) = \sum_{t=0}^{T} q_{n^k}^k(t) = \sum_{t=0}^{T} f^k(t) \tag{2}$$

$$\sum_{t=T_0^k+T_1^k+\ldots+T_i^k}^{T_*^k} q_i^k(t) \le \sum_{t=T_0^k+T_1^k+\ldots+T_{i-1}^k}^{T_*^k-T_i^k} q_{i-1}^k(t) \qquad \forall T_*^k > T_0^k + T_1^k + \cdots + T_i^k \tag{3}$$

$$q_i^k(0) = q_i^k, \quad x_i^k(0) = x_i^k \tag{4}$$

Equality (2) states that the total inflow into a path is equal to its total outflow. Inequality (3) dictates that every flight must dwell in link $\boldsymbol{i}$ for at least T_i^k minutes. Equalities (4) are initial states for each link, where q_i^k and x_i^k are constants. Besides the flow dynamics, the most important constraint in flow management is capacity. For each sector, the Federal Aviation Administration (FAA) establishes a so-called Monitor Alert Parameter (MAP) that would trigger notification to the air traffic controllers if the number of flights in a control volume exceeds the MAP value [9]. We denote the value for sector $\boldsymbol{s}$ as $MAP_s(t)$. It is a time varying parameters but could be derived through FAA's system, so it is an input. The capacity is formulated as:

$$\sum_{(i,k)\in s} x_i^k(t) \le MAP_s(t) \qquad \forall t \in \{0, \ldots, \mathrm{T}\} \tag{5}$$

Likewise, airports also have arrival capacities $C_{arr}(t)$ and departure capacities $C_{dep}(t)$, formulated as:

$$\sum_{(0,k)\in arr} x_0^k(t) \le C_{arr}(t) \qquad \sum_{(n^k,k)\in dep} x_{n^k}^k(t) \le C_{dep}(t) \qquad \forall t \in \{0, \ldots, \mathrm{T}\} \tag{6}$$

As mentioned above, the goal is to minimize delays. The objective is given by:

$$Objective \quad min\ d(x_i^k(t)) = \sum_t \sum_k \sum_i c_i^k x_i^k(t) \tag{7}$$

where c_i^k is a weight associated with link $\boldsymbol{i}$. This objective function can be interpreted as the total flight time of all flights in the planning time horizon. It is equivalent to minimizing total delays in the NAS. Equality (2) drives the flow moving forward, and Inequality (3), on the other hand, restricts the flow rate. A feasible solution is that flights are delayed in upstream sectors if downstream sectors are full. $x_i^k(t)$ tends to stay at current level for longer when delay happens. Flights continue to move on once the downstream sectors become available, then $x_i^k(t)$ begins to decrease. Objective function (7) ensures that $x_i^k(t)$ keeps at low level as much as possible, which is equivalent to minimizing delays.

By definition, the variables $x_i^k(t)$ and $q_i^j(t)$ are all integers as they are numbers of flights. Note that all constraints are linear, therefore we obtain an ILP. Also note that all the coefficients associated with the variables in constraint (1)–(6) are 1 or -1, hence the constraint matrix preserves total unimodularity. In addition, all the right hand side parameters, namely $f^k(t)$, $MAP_s(t)$, $C_{arr}(t)$ and, $C_{dep}(t)$, are integers, it can be proved that an ILP of this kind is equivalent to its relaxation so it can be solved with polynomial time algorithms like simplex [10]. But even so the problem is too big to solve as a whole. Objective (7) indicates that the size of vector $x_i^k(t)$ is proportional to the length of planning time horizon, the number of paths involved, and the number of links on each path. Strategic air traffic planning usually plan 2–3 h into the future, and roughly 2400 flight paths were identified during the peak hours in the NAS, and each path comprises 15 links on average [11]. As a result, there could be up to $120 \times 2400 \times 15 = 4.32$ million variables for $x_i^k(t)$ alone. In addition, the flow rate $q_i^k(t)$ is also considered variable and has the same population as $x_i^k(t)$. The total number of variables could be up to 9 million. The best know LP algorithm has a runtime complexity of $O(n^{3.5}L)$ [12], where $\boldsymbol{n}$ is the number of variables and $\boldsymbol{L}$ is the length of the data. The estimate computational complexity is so huge, thus the original problem is considered intractable.

2.2 Dual Decomposition Method

A feasible solution is to decompose the original problem to derive subproblems that are smaller in size, which can be done by employing Lagrangian multiplier [13]. Note that constraints (1)–(4) are arranged by path, and constraints (5) and (6) are the coupling constraints where paths are coupled together. We can introduce Lagrangian multipliers and obtain the following Lagrangian:

$$L(x_i^k(t),\lambda_i^k(t),\lambda_{arr}(t),\lambda_{dep}(t))=\sum_t\sum_k\sum_i c_i^k x_i^k(t)+\sum_t\sum_{(i,k)\in s}\lambda_i^k(t)(x_i^k(t)-MAP_s(t))$$
$$+\sum_t\sum_{(0,k)\in arr}\lambda_{arr}(t)(x_0^k(t)-C_{arr}(t))+\sum_t\sum_{(n^k,k)\in dep}\lambda_{dep}(t)(x_{n^k}^k(t)-C_{dep}(t)).$$

Reorganize the variables by path:

$$min\ L(x_i^k(t),\lambda_i^k(t),\lambda_{arr}(t),\lambda_{dep}(t))=\sum_t\left[\sum_k\sum_i c_i^k x_i^k(t)+\sum_{(i,k)\in s}\lambda_i^k(t)x_i^k(t)\right.$$
$$\left.+\sum_{(0,k)\in arr}\lambda_{arr}(t)x_0^k(t)+\sum_{(n^k,k)\in dep}\lambda_{dep}(t)x_{n^k}^k(t)\right]$$
$$-\sum_t\left[\sum_{(i,k)\in s}\lambda_i^k(t)MAP_s(t)+\sum_{(n^k,k)\in arr}\lambda_{arr}(t)C_{arr}(t)+\right.$$
$$\left.\sum_{(n^k,k)\in dep}\lambda_{dep}(t)C_{dep}(t)\right]$$
$$=\sum_k\left[\sum_t\sum_{t=1}^{n^k-1}(c_i^k+\lambda_i^k(t))x_i^k(t)+\sum_t(c_0^k+\lambda_{arr}(t))x_0^k(t)\right.$$
$$\left.+\sum_t(c_0^k+\lambda_{dep}(t))x_{n^k}^k(t)\right.$$
$$-\sum_t\left[\sum_{(i,k)\in s}\lambda_i^k(t)MAP_s(t)+\sum_{(n^k,k)\in arr}\lambda_{arr}(t)C_{arr}(t)+\right.$$
$$\left.\sum_{(n^k,k)\in dep}\lambda_{dep}(t)C_{dep}(t)\right]$$

If we denote:

$$\underset{\substack{\lambda_i^k(t)\geq 0,\\ \lambda_{arr}(t)\geq 0,\\ \lambda_{dep}(t)\geq 0}}{arg\ min}\quad d^k(x_i^k(t))=\sum_t\left[\sum_{i=1}^{n^k-1}(c_i^k+\lambda_i^k(t))x_i^k(t)+(c_0^k+\lambda_{arr}(t))x_0^k(t)+(c_0^k+\lambda_{dep}(t))x_{n^k}^k(t)\right],$$

and have the Lagrangian multipliers fixed for some non-negative values, this along with constraint (1)–(4) forms a set of independent subproblems that are related to each path only. The number of variables is $120\times15\times2=3600$ for each subproblem on average, which is within reach of normal optimization tool. Once subproblems are solved, the original problem is a function of Lagrangian multipliers only, which can be updated using subgradient method. The global optimum is approached by updating the subproblems and Lagrangian multipliers iteratively. The algorithm is summarized in Table 1.

Thanks to the decomposition method, we derive a parallel algorithm that enables solving subproblems in parallel. The subproblems can be solved using off-the-shelf

Table 1 Parallel algorithm for the traffic flow optimization

while $\left|\frac{L(x_i^k(t)^*, \lambda_i^k(t), \lambda_{arr}(t))^j - L(x_i^k(t)^*, \lambda_i^k(t), \lambda_{arr}(t), \lambda_{dep}(t))^{j-1}}{L(x_i^k(t)^*, \lambda_i^k(t), \lambda_{arr}(t), \lambda_{dep}(t))^{j-1}}\right| > \varepsilon$:

for path k:

solve the kth ILP:

$$d^k(x_i^k(t))^* = min \sum_t \left[\sum_{i=1}^{n^k-1} (c_i^k + \lambda_i^k(t)) x_i^k(t) + (c_0^k + \lambda_{arr}(t)) x_0^k(t) + (c_0^k + \lambda_{dep}(t)) x_{n^k}^k(t) \right]$$

s. t.

$$x_i^k(t+1) = x_i^k(t) - q_i^k(t) + x_{i-1}^k(\mathrm{t}) \quad \forall\, \mathrm{t} \in \{0, \rightleftharpoons \mathrm{T}\}$$

$$\sum_{t=0}^{T} q_0^k(t) = \sum_{t=0}^{T} q_{n^k}^k(t) = \sum_{t=0}^{T} f^k(t)$$

$$\sum_{t=T_0^k+T_1^k+\cdots+T_i^k}^{T_*^k} q_i^k(t) \le \sum_{t=T_0^k+T_1^k+\cdots+T_{i-1}^k}^{T_*^k - T_i^k} q_{i-1}^k(t) \quad \forall\, T_*^k > T_0^k + T_1^k + \cdots + T_i^k$$

$$q_i^k(0) = q_i^k, \quad x_i^k(0) = x_i^k$$

$$x_i^k(t) \ge 0, \quad q_i^k(t) \ge 0$$

Update master problem:

$$L(x_i^k(t)^*, \lambda_i^k(t), \lambda_{arr}(t), \lambda_{dep}(t)) = \sum_k d^k(x_i^k(t))^* - \sum_t \left[\sum_{(i,k)\in s} \lambda_i^k(t) MAP_s(t) + \sum_{(n^k,k)\in arr} \lambda_{arr}(t) C_{arr}(t) + \sum_{(n^k,k)\in dep} \lambda_{dep}(t) C_{dep}(t) \right]$$

$$\lambda_i^k(t) \quad := max\left\{0, \lambda_i^k(t) + \frac{1}{j+1}\Big(\sum_{(i,k)\in s} x_i^k(t) - MAP_s(t)\Big)\right\}$$

$$\lambda_{arr}(t) \quad := max\left\{0, \lambda_{arr} + \frac{1}{j+1}\Big(\sum_{(0,\mathrm{k})\in \mathrm{s}} x_0^k(t) - C_{arr}(t)\Big)\right\}$$

$$\lambda_{dep}(t) \quad := max\left\{0, \lambda_{dep}(t) + \frac{1}{j+1}\Big(\sum_{(n^i,k)\in s} x_{n^k}^k(t) - C_{dep}(t)\Big)\right\}$$

$\mathrm{j} = \mathrm{j} + 1$

optimization package, such as COIN-OR[2] or IBM ILOG CPLEX.[3] Efficient implementation of the parallel algorithms relies on parallel computing technology.

3 Hadoop MapReduce Programming Model

Hadoop has gained popularity in processing big data worldwide. Two most important components in Hadoop are Hadoop Distributed File System (HDFS) and Hadoop MapReduce, which are strongly correlated. The HDFS manages data

[2]http://www.coin-or.org.

[3]http://www-01.ibm.com/software/commerce/optimization/cplex-optimizer/.

storage and MapReduce governs data processing. Hadoop clusters adopt master/slave topology, as shown in Fig. 2. In its own terminology, the master is called NameNode and slave is called DataNodc. When files (data) are loaded into a Hadoop cluster, they are split into chunks with multiple replicas stored on different DataNodes. This redundancy provides fault tolerance for DataNode failures. The NameNode keeps track of these data splits and coordinates data transfer when necessary. Both NameNode and DataNode run daemon processes when the Hadoop system is up. The one running on NameNode is called Job tracker, and the ones running on DataNodes are called Task tracker. Job tracker is in charge of allocating jobs to Task tracker and gathers results, and Task tracker monitors the jobs on its local DataNode and sends results back. This infrastructure forms the basis for distributed computing.

Jobs are defined by MapReduce, which is a programming model written in Java [14]. MapReduce job conceptualizes data processing as list processing and is primarily fulfilled by two interfaces: *mapper* and *reducer* [15], as illustrated in Fig. 3. Input data are splits on DataNodes. Each split is read and parsed as a list of tuples in

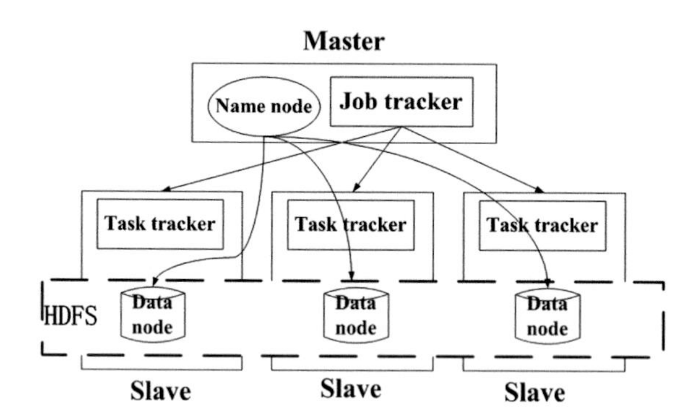

Fig. 2 Hadoop cluster topology

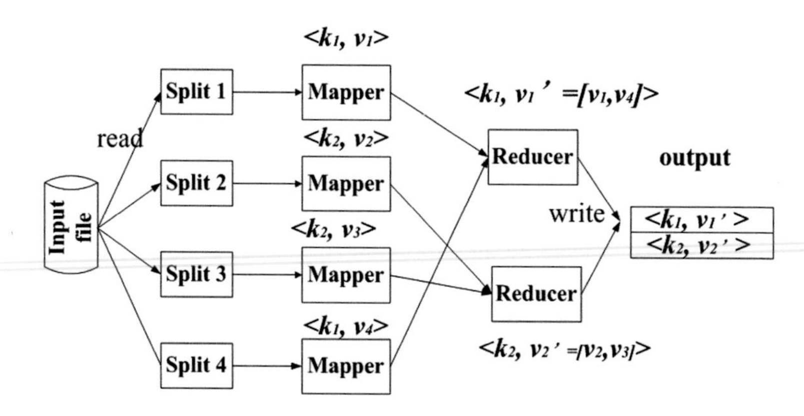

Fig. 3 MapReduce process

a format of <*key*, *value*> by the *mapper*. The *key* and *value* can be any data type. The list undergoes some filtering logic defined by users. This digestion process takes place on each DataNode in parallel. The intermediate results are sent to *reducer*, which aggregates the results and emits another list of <*key*′, *value*′> as the final results.

Essentially, the *map* step is a process where data are parsed and processed, and *reduce* step is a process where information of interest is extracted, aggregated and summarized. At a high level, the pseudo-code of *mapper* and *reducer* can be expressed as follows:

```
mapper (IntWritable k, TextWritable v){
      <k', v'> processing1(k, v);
      Emit(k', v');
}
reducer (Writable type k', Writable set [v1',v2', ...]){
      <k'',v''> processing2(k', [v1',v2', ...]);
      Emit(k'',v'');
}
```

The outputs from *mapper* are grouped by *key*, and each reducer receives a group of values per *key*. The processing in *mapper* and *reducer* is custom code by users. Thanks to MapReduce's standardized data structure for input and output, developers only need to design the processing logic applied to each <*key*, *value*> pair. The framework automatically handles sorting, data exchange and synchronization between nodes, which significantly simplifies the development of distributed computing applications.

The parallel algorithm described in the last section could benefit from this programming model. In implementation, the *mapper* is a wrapper of an optimizer. A *mapper* handles subproblems and returns sector counts contributed by each path it handled. A *reducer* aggregates sector counts and updates the master problem and Lagrangian multipliers. A <*key*, *value*> mapping of the algorithm is shown below.

mapper input key k	path ID k
mapper input value v	$\lambda_i^k(t), \lambda_{arr}(t), \lambda_{dep}(t), f^k(t)$
mapper output key k'	sector ID s
mapper output value v'	$x_i^k(t)$
reducer input key k'	sector ID s
reducer input value [$v1'$, $v2'$, …]	$\left[x_i^k(t) \middle| (i,k) \in s\right]$
reducer output key k"	sector ID s
reduce output value v"	$\lambda_i^k(t), \lambda_{arr}(t), \lambda_{dep}(t)$

Figure 4 shows the data flow in a MapReduce cycle. Because link information, such as length of links T_i^k, sector capacities $MAP_s(t)$ and airport capacities $C_{arr}(t)$ and $C_{dep}(t)$, are constants, these parameters are stored in a relational database and retrieved when subproblems are set up in each iteration.

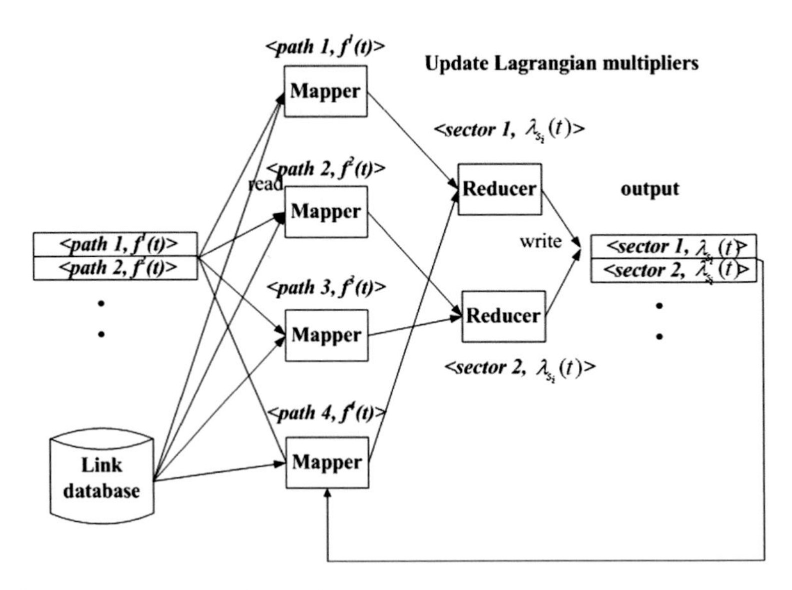

Fig. 4 Data flow in a MapReduce cycle

4 Numeric Results

The following analysis is based on real-world air traffic data extracted from Aircraft Situation Display to Industry (ASDI). The ASDI data are collected by Enhanced Traffic Management System (ETMS), which receives flight status from Air Traffic Control System Command Centers [16]. ASDI feed contains planned flight schedules and routes, 3-dimension position and associated timestamp of flights, and many other related flight information. The flight schedules and routes are of interest here, which can be used to parameterize $f^k(t)$ and T_i^k. We set the planning time horizon to be 2 h and choose high traffic period in which 2423 flight routes are involved in the optimization. This is a NAS-wide instance of air traffic flow optimization, which covers 1258 sectors and 3838 general public airports in the United States. For validation purpose, we purposely decrease the MAP values and airport capacities to 70 % of their normal operation values to stress the optimization model.

The optimization runs on a Hadoop cluster with six nodes. Each node is a DELL workstation configured with an Intel i7 CPU and 16 GB of RAM. Ubuntu 10.04 and Hadoop v0.20.2 are deployed on the cluster. We use IBM ILOG CPLEX as the optimizer. CPLEX provides a Java API library, hence the language is in line with Hadoop MapReduce framework. In addition, CPLEX supports multi-processing. With a parallel option license, multiple optimizers can be launched simultaneously.

Figure 5 shows the number of flights in a sector during the 2-h traffic optimization in Cleveland Air Route Traffic Control Center. The optimized traffic (ASDI data) does not exceed the reduced MAP during the planning time horizon so the capacity constraints are well respected. But as a trade-off, this results in delays.

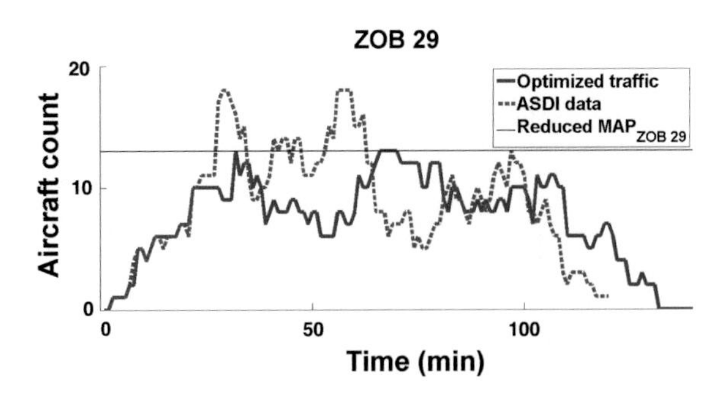

Fig. 5 Aircraft count in a sector (ZOB 29) in Cleveland Air Route Traffic Control Center

We can see that the blue line extends beyond the 2-h planning time horizon, which reflects the delayed traffic. Figure 6 shows the development of traffic during the 2-h planning time horizon. It compares the 'uncontrolled' traffic (ASDI data) and optimized traffic in high altitude sectors of the NAS. We use different coloring to represent sector loads. When the flight count in a sector is equal to the reduced MAP value, the sector load is said to be 1 and shown as red. It can be seen that the optimization effectively reduces occurrence of capacity violation.

To examine the efficiency of distributed platform, we first set up a baseline where the optimization runs on a standalone computer. In a monolithic implementation, the subproblems are solved sequentially, which takes about 135 min to finish the optimization. In the experiment, we use two cluster configurations to run the optimization, one with 3 DataNodes and the other with 6 DataNodes. The running time of the optimization is shown in Fig. 7. As it shows, the larger cluster gains higher efficiency.

The concurrency is mainly determined by the number of *mappers* in this specific algorithm. As a configurable parameter, the maximum number of *mappers* allowed on a node can be tuned by users. The running time decreases as more *mappers* are launched. But, the efficiency stops increasing when the number of mappers per node is more than eight. This is due to the fact that an Intel i7 CPU has four physical cores, with each being capable of handling two threads simultaneously. As a result, the maximum number of threads a node can run in parallel is eight. In theory, the Hadoop cluster can have at maximum of $6 \times 8 = 48$ threads running concurrently. When more *mappers* are launched, threads have to share computing resources on the same node. Concurrency hits the cap unless more DataNodes are added to the cluster. The trend of the curve also indicates that the speedup is not linear to the number of *mappers*. Only 16 times the speedup was achieved. There are two reasons to account for this. First, the parallel programming model has inherent overheads such as communication and synchronization between nodes. Second, unbalanced workloads cause idle time for some nodes. Figure 8 shows the

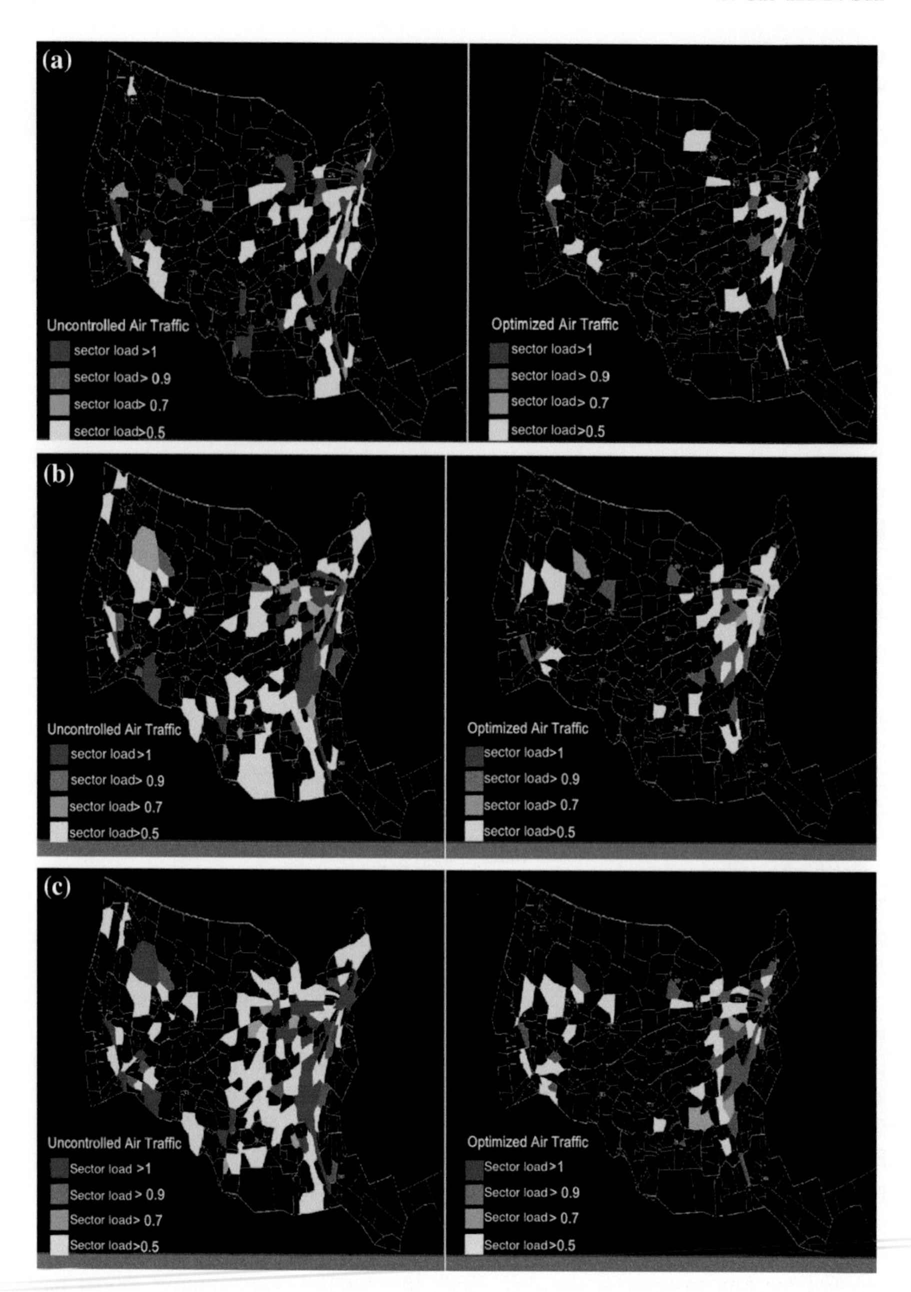

Fig. 6 Sector payload status during the planning time horizon. *Left* are ASDI data. *Right* optimized traffic. **a** 5:00 PM. **b** 6:00 PM. **c** 7:00 PM

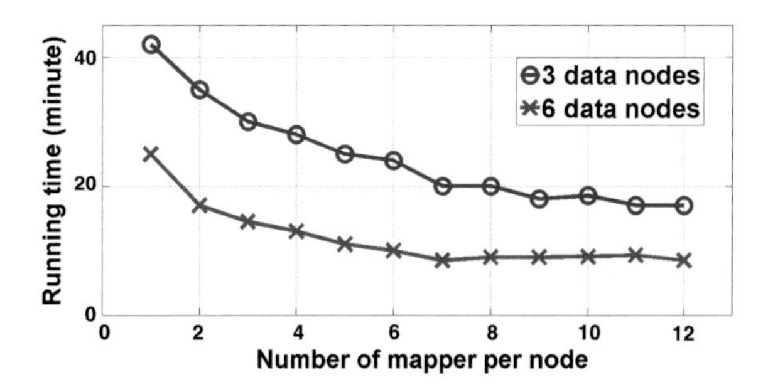

Fig. 7 Running time reduction

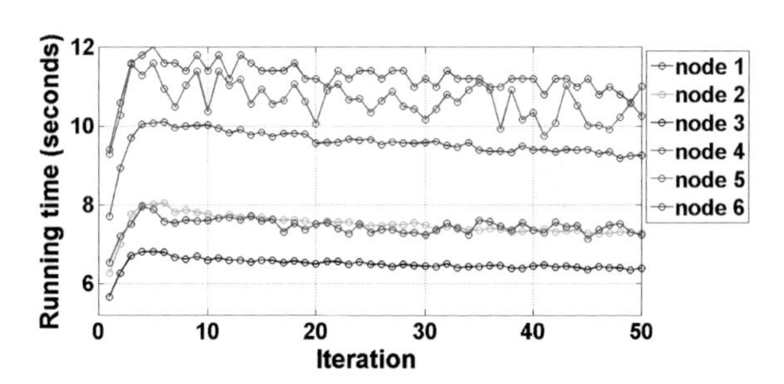

Fig. 8 Running time for each iterations

running time of solving subproblems on each node. Node 3 is about 4.5 s ahead of node 4 in each iteration. When setting up the optimization, we evenly distributed subproblems to each node. Although all the nodes have the same configuration, the complexity of subproblems may have big variations due to the number of constraints and variables.

5 Conclusion

In this chapter, we present a model that solves large-scale Integer Linear Programming problem by using a combination of parallel algorithm and distributed platform. We first seek arithmetical approach to decompose the original problem such that computational complexity is decreased. Then we leverage Hadoop MapReduce framework to achieve parallel computation. This approach is validated through tackling a large air traffic flow optimization problem. It is shown that the parallelism level is scalable.

Leveraging Hadoop MapReduce in parallel algorithm implementation is cost-efficient. On one hand, the hardware requirement is relatively low. The framework can be deployed on commodity computer clusters with easy configuration. On the other hand, programming effort is minimized since the framework automatically schedules parallel computation tasks across cluster nodes. With the flexible data model and well-designed interface of MapReduce, many parallel algorithms can be easily programmed and applied to real-world applications.

More recently, a Hadoop-derivative framework, Apache Spark, has emerged and is attracting more and more interests from the big data community. As it holds intermediate results in memory rather than storing them into disk, multi-pass MapReduce jobs can be stringed together with minimized Input/Output operations in the distributed file system. Therefore it achieves higher performance than MapReduce's disk-based paradigm. Applications under Apache Spark framework are reportedly 100 times faster. As a future direction, it is worthwhile to translate MapReduce-based optimizations into Apache Spark to further reduce the running time. The overall performance of parallel optimization should greatly benefit from the streaming functionality of Apache Spark.

References

1. Yan, Y., Huang, L.: Large-scale image processing research cloud. In: Cloud Computing, pp. 88–93 (2014)
2. Kang, Y., Park, Y.B.: The performance evaluation of k-means by two MapReduce frameworks, Hadoop vs. Twister. In: International Conference on Information Networking (ICOIN), pp. 405–406 (2015)
3. Chu, C., Kim, S.K., Lin, Y.A., Yu, Y., Bradski, G., Ng, A.Y., Olukotun, K.: Map-reduce for machine learning on multicore. Adv. Neural Inf. Process. Syst. **19**, 281–288 (2007)
4. Hall, K.B., Gilpin, S., Mann, G.: MapReduce/Bigtable for distributed optimization. In: NIPS LCCC Workshop (2010)
5. Boyd, S., Parikh, N., Chu, E., Peleato, B., Eckstein, J.: Distributed optimization and statistical learning via the alternating direction method of multipliers. In: Foundations and Trends® in Machine Learning, vol. 3, no. 1, pp. 1–122 (2011)
6. Palomar, D.P., Chiang, M.: A tutorial on decomposition methods for network utility maximization. IEEE J. Sel. Areas Commun. **24**(8), 1439–1451 (2006)
7. Cao, Y., Sun, D., Zhang, L.: Air traffic prediction based on the kernel density estimation. In: American Control Conference, Washington D.C., 17–19 June 2013
8. Bosson, C.S., Sun, D.: An aggregate air traffic forecasting model subject to stochastic inputs. In: AIAA Guidance, Navigation, and Control and Co-located Conferences, Boston, MA, 19–22 Aug 2013
9. U.S. Department of Transportation Federal Aviation Administration: Facility Operation and Administration, Washington, DC, Order JO 7210.3W, Feb. 2010
10. Wei, P., Cao, Y., Sun, D.: Total unimodularity and decomposition method for large-scale air traffic cell transmission model. Transp. Res. Part B **53**, 1–16 (2013)
11. Cao, Y., Sun, D.: A parallel computing framework for large-scale traffic flow optimization. IEEE Trans. Intell. Transp. Syst. **13**(14), 1855–1864 (2012)
12. Ye, Y.: An O(n^3L) potential reduction algorithm for linear programming. Math. Program. **50** (2), 239–258 (1991)
13. Bertsekas, D.P., Tsitsiklis, J.N.: Parallel and Distributed Computation: Numerical Methods. Athena Scientific (1997)
14. White, T.: Hadoop: The Definitive Guide. Yahoo! Press, Sebastapool, CA (2009)

15. Dean, J., Ghemawat, S.: MapReduce: simplified data processing on large cluster. Commun. ACM **51**(1) (2008)
16. National Transportation Center Volpe: Enhanced Traffic Management System (ETMS). Number Technical Report VNTSC-DTS56-TMS-002, United States Department of Transportation, Cambridge, MA (2005)

Author Biographies

Yi Cao received the B.S. degree in instrumentation science and engineering and the M.S. degree in navigation, guidance, and control from Shanghai Jiao Tong University, Shanghai, China, in 2006 and 2009, respectively. He is currently working toward the Ph.D. degree with the School of Aeronautics and Astronautics, Purdue University, West Lafayette, IN. His research interests include modeling, optimization, and simulation, with emphasis on air traffic flow management.

Dengfeng Sun is an Associate Professor in School of Aeronautics and Astronautics at Purdue University, West Lafayette, Indiana. Before joining Purdue, he was an Associate Scientist with University of California Santa Cruz at NASA Ames Research Center. He received a bachelor's degree in precision instruments and mechanology from Tsinghua University in China, a master's degree in industrial and systems engineering from the Ohio State University, and a Ph.D. degree in civil engineering from University of California at Berkeley. Dr. Sun's research interests include control and optimization for large-scale systems and their applications in aerospace systems.

Computational Approaches in Large-Scale Unconstrained Optimization

Saman Babaie-Kafaki

Abstract As a topic of great significance in nonlinear analysis and mathematical programming, unconstrained optimization is widely and increasingly used in engineering, economics, management, industry and other areas. Unconstrained optimization also arises in reformulation of the constrained optimization problems in which the constraints are replaced by some penalty terms in the objective function. In many big data applications, solving an unconstrained optimization problem with thousands or millions of variables is indispensable. In such situations, methods with the important feature of low memory requirement are helpful tools. Here, we study two families of methods for solving large-scale unconstrained optimization problems: conjugate gradient methods and limited-memory quasi-Newton methods, both of them are structured based on the line search. Convergence properties and numerical behaviors of the methods are discussed. Also, recent advances of the methods are reviewed. Thus, new helpful computational tools are supplied for engineers and mathematicians engaged in solving large-scale unconstrained optimization problems.

Keywords Unconstrained optimization · Large-scale optimization · Line search · Memoryless quasi-Newton method · Conjugate gradient method

1 Introduction

We consider the minimization of a smooth nonlinear function $f : \mathbb{R}^n \to \mathbb{R}$, that is,

$$\min_{x \in \mathbb{R}^n} f(x), \tag{1}$$

in the case where the number of variables n is large and analytic expressions for the function f and its gradient ∇f are available. Although the minimizer of f is a solution

S. Babaie-Kafaki (✉)
Department of Mathematics, Faculty of Mathematics, Statistics and Computer Science, Semnan University, P.O. Box: 35195-363, Semnan, Iran
e-mail: sbk@semnan.ac.ir

A. Emrouznejad (ed.), *Big Data Optimization: Recent Developments and Challenges*, Studies in Big Data 18, DOI 10.1007/978-3-319-30265-2_17

of the system $\nabla f(x) = 0$, solving this generally nonlinear and complicated system is not practical.

Among the most useful tools for solving large-scale cases of (1) there are the conjugate gradient methods and the limited-memory quasi-Newton methods, because the amount of memory storage required by the methods is low. In addition, the methods possess the attractive features of simple iterative formula and strong global convergence as well as applying the Hessian information. The methods can also be straightly employed in penalty function methods, a class of efficient methods for solving constrained optimization problems.

Generally, iterations of the above-mentioned methods are in the following form:

$$x_0 \in \mathbb{R}^n, \; x_{k+1} = x_k + s_k, \; s_k = \alpha_k d_k, \; k = 0, 1, \ldots, \tag{2}$$

where d_k is a search direction to be computed by a few inner products and α_k is a step length to be determined by a line search procedure. The search direction d_k should be a descent direction, i.e.,

$$g_k^T d_k < 0, \tag{3}$$

where $g_k = \nabla f(x_k)$, to ensure that the function f can be reduced along the search direction d_k. The most reduction is achieved when the exact (optimal) line search is used in which

$$\alpha_k = \arg\min_{\alpha \geq 0} f(x_k + \alpha d_k).$$

Hence, in the exact line search the step length α_k can be considered as a solution of the following equation:

$$\nabla f(x_k + \alpha d_k)^T d_k = 0. \tag{4}$$

Since the exact line search is not computationally tractable, inexact line search techniques have been developed [74, 86], most of them structured based on quadratic or cubic polynomial interpolations of the one-dimensional function $\varphi(\alpha) = f(x_k + \alpha d_k)$. Finding minimizers of the polynomial approximations of $\varphi(\alpha)$, inexact line search procedures try out a sequence of candidate values for the step length, stopping to accept one of these values when certain conditions are satisfied.

Among the stopping conditions for the inexact line search procedures, the so-called Wolfe conditions [91, 92] have attracted especial attention in convergence analyses and implementations of the unconstrained optimization algorithms, requiring that

$$f(x_k + \alpha_k d_k) - f(x_k) \leq \delta \alpha_k g_k^T d_k, \tag{5}$$

$$\nabla f(x_k + \alpha_k d_k)^T d_k \geq \sigma g_k^T d_k, \tag{6}$$

where $0 < \delta < \sigma < 1$. The first condition, called the Armijo condition, ensures adequate reduction of the objective function value while the second condition, called the curvature condition, ensures unacceptably of the short step lengths. However, a step length may fulfill the Wolfe conditions without being sufficiently close to a minimizer of $\varphi(\alpha)$. To overcome this problem, the strong Wolfe conditions have been proposed which consist of (5) and the following strengthened version of (6):

$$|\nabla f(x_k + \alpha_k d_k)^T d_k| \leq -\sigma g_k^T d_k. \tag{7}$$

Considering (4), if $\sigma \to 0$, then the step length which satisfies the strong Wolfe conditions (5) and (7) tends to the optimal step length.

In practical computations, the Wolfe condition (5) may never be satisfied due to the existence of numerical errors. This computational drawback of the Wolfe conditions was carefully analyzed in [59] on a one-dimensional quadratic function. Based on the insight gained by the numerical example of [59], one of the most accurate and efficient inexact line search algorithms has been proposed in [59, 60], using a quadratic interpolation scheme and the following approximate Wolfe conditions:

$$\sigma g_k^T d_k \leq \nabla f(x_k + \alpha_k d_k)^T d_k \leq (2\delta - 1) g_k^T d_k, \tag{8}$$

where $0 < \delta < \frac{1}{2}$ and $\delta \leq \sigma < 1$. The line search algorithm of [60] has been further improved in [42].

In what follows, at first we discuss several basic choices for the search direction d_k in (2) corresponding to the steepest descent method, Newton method, conjugate direction methods and quasi-Newton methods, together with their advantages and disadvantages as well as their relationships. Then, we focus on the conjugate gradient methods and the limited-memory quasi-Newton methods which are proper algorithms for large-scale unconstrained optimization problems. For all of these methods, the line search procedure of [60] can be applied efficiently. Also, a popular stopping criterion for the iterative method (2) is given by

$$||g_k|| < \varepsilon,$$

in which ε is a small positive constant and $||.||$ stands for the Euclidean norm.

2 Basic Unconstrained Optimization Algorithms

Here, we briefly study basic algorithms in the field of unconstrained optimization, all of them are iterative in the form of (2) with especial choices for the search direction d_k. A detailed discussion can be found in [86].

2.1 Steepest Descent Method

One of the simplest and most fundamental methods for solving the unconstrained optimization problem (1) is the steepest descent (or the gradient) method [39] in which the search direction is computed by

$$d_k = -g_k,$$

that is trivially a descent direction. Although the steepest descent method is globally convergent under a variety of inexact line search conditions, the method performs poorly, converges linearly and is badly affected by ill conditioning [1, 55].

2.2 Newton Method

Based on a quadratic interpolation of the objective function at the kth iteration, search direction of the Newton method can be computed by

$$d_k = -\nabla^2 f(x_k)^{-1} g_k,$$

where $\nabla^2 f$ is the Hessian matrix of the objective function f. If $\nabla^2 f(x_k)$ is a positive definite matrix, then the Newton search direction is a descent direction and in such situation, it can be effectively computed by solving the following linear system using Cholesky decomposition [88]:

$$\nabla^2 f(x_k) d_k = -g_k.$$

In the Newton method, the Hessian information is employed in addition to the gradient information. Also, if the starting point x_0 is adequately close to the optimal point x^*, then the sequence $\{x_k\}_{k \geq 0}$ generated by the Newton method converges to x^* with a quadratic rate. However, since in the Newton method it is necessary to compute and save the Hessian matrix $\nabla^2 f(x_k) \in \mathbb{R}^{n \times n}$, the method is not proper for large-scale problems. Moreover, far from the solution, the Hessian $\nabla^2 f(x_k)$ may not be a positive definite matrix and consequently, the Newton search direction may not be a descent direction. To overcome this problem, a variant of modified Newton methods have been proposed in the literature [86].

2.3 Conjugate Direction Methods

Consider the problem of minimizing a strictly convex quadratic function, i.e.,

$$\min_{x \in \mathbb{R}^n} q(x), \tag{9}$$

in which

$$q(x) = \frac{1}{2}x^T Ax - b^T x, \tag{10}$$

where the Hessian $A \in \mathbb{R}^{n\times n}$ is a symmetric positive definite matrix and $b \in \mathbb{R}^n$. To find the optimal solution x^*, the following system of linear equations can be solved:

$$\nabla q(x) = Ax - b = 0, \tag{11}$$

or equivalently,

$$Ax = b.$$

Although the problem can be solved by Cholesky decomposition, conjugate direction methods are a class of efficient algorithms for finding the minimizer of a strictly convex quadratic function in large-scale cases.

Definition 1 Let $A \in \mathbb{R}^{n\times n}$ be a symmetric positive definite matrix and $\{d_k\}_{k=1}^m, m \leq n$, be a set of nonzero vectors in $\mathbb{R}^n$. If

$$d_i^T Ad_j = 0, \ \forall i \neq j,$$

then the vectors $\{d_k\}_{k=1}^m$ are called A-conjugate, or simply called conjugate.

Exercise 1 *(i) Show that a set of conjugate vectors are linearly independent.*

(ii) Assume that a symmetric positive definite matrix $A \in \mathbb{R}^{n\times n}$ and a set of linearly independent vectors $\{d'_k\}_{k=1}^m \subseteq \mathbb{R}^n$ are available. Describe how a set of A-conjugate vectors $\{d_k\}_{k=1}^m$ can be constructed from $\{d'_k\}_{k=1}^m$.
(Hint: Use the Gram-Schmidt orthogonalization scheme [88].)

In each iteration of a conjugate direction method for solving (9), the function $q(x)$ given by (10) is minimized along the search direction d_k for which we have

$$d_k^T Ad_i = 0, \ \forall i < k.$$

Here, since the objective function is quadratic, the exact line search can be used. The following theorem shows that under the exact line search, the conjugate direction methods have quadratic termination property which means that the methods terminate in at most n steps when they are applied to a strictly convex quadratic function.

Theorem 1 *For a quadratic function with the positive definite Hessian A, the conjugate direction method terminates in at most n exact line searches. Also, each x_{k+1} is the minimizer in subspace $\mathcal{S}_k$ generated by x_0 and the directions $\{d_i\}_{i=0}^k$, that is, $\mathcal{S}_k = x_0 + span\{d_0, \ldots, d_k\}$.*

Exercise 2 *Prove Theorem 1.*
(Hint: By induction show that $\nabla q(x_{k+1}) \perp d_i$, $i = 0, 1, \ldots, k$.)

2.4 Quasi-Newton Methods

As known, quasi-Newton methods are of particular performance for solving unconstrained optimization problems since they do not require explicit expressions of the second derivatives and their convergence rate is often superlinear [86]. The methods are sometimes referred to variable metric methods.

In the quasi-Newton methods, the search direction is often calculated by

$$d_k = -H_k g_k, \tag{12}$$

in which $H_k \in \mathbb{R}^{n \times n}$ is an approximation of the inverse Hessian; more precisely, $H_k \approx \nabla^2 f(x_k)^{-1}$. The methods are characterized by the fact that H_k is effectively updated to achieve a new matrix H_{k+1} as an approximation of $\nabla^2 f(x_{k+1})^{-1}$, in the following general form:

$$H_{k+1} = H_k + \Delta H_k,$$

where ΔH_k is a correction matrix. The matrix H_{k+1} is imposed with the scope of satisfying a particular equation, namely secant (quasi-Newton) equation, which includes the second order information. The most popular equation is the standard secant equation, that is,

$$H_{k+1} y_k = s_k, \tag{13}$$

in which $y_k = g_{k+1} - g_k$. Note that the standard secant equation is obtained based on the mean-value theorem, or equivalently, the following approximation:

$$\nabla^2 f(x_{k+1}) s_k \approx y_k,$$

which holds exactly for the quadratic objective functions.

Among the well-known quasi-Newton update formulas there are the BFGS (Broyden-Fletcher-Goldfarb-Shanno) and DFP (Davidon-Fletcher-Powell) updates [86] given by

$$H_{k+1}^{BFGS} = H_k - \frac{s_k y_k^T H_k + H_k y_k s_k^T}{s_k^T y_k} + \left(1 + \frac{y_k^T H_k y_k}{s_k^T y_k}\right) \frac{s_k s_k^T}{s_k^T y_k}, \tag{14}$$

and

$$H_{k+1}^{DFP} = H_k + \frac{s_k s_k^T}{s_k^T y_k} - \frac{H_k y_k y_k^T H_k}{y_k^T H_k y_k},$$

in which the initial approximation H_0 can be considered as an arbitrary positive definite matrix. In a generalization scheme, the BFGS and DFP updates have been combined linearly and the Broyden class of quasi-Newton update formulas [86] has been proposed as follows:

$$\begin{aligned} H_{k+1}^{\phi} &= (1-\phi)H_{k+1}^{DFP} + \phi H_{k+1}^{BFGS} \\ &= H_k + \frac{s_k s_k^T}{s_k^T y_k} - \frac{H_k y_k y_k^T H_k}{y_k^T H_k y_k} + \phi v_k v_k^T, \end{aligned} \tag{15}$$

in which ϕ is a real parameter and

$$v_k = \sqrt{y_k^T H_k y_k}\left(\frac{s_k}{s_k^T y_k} - \frac{H_k y_k}{y_k^T H_k y_k}\right). \tag{16}$$

It can be seen that if H_k is a positive definite matrix and the line search ensures that $s_k^T y_k > 0$, then H_{k+1}^{ϕ} with $\phi \geq 0$ is also a positive definite matrix [86] and consequently, the search direction $d_{k+1} = -H_{k+1}^{\phi} g_{k+1}$ is a descent direction. Moreover, for a strictly convex quadratic objective function, search directions of a quasi-Newton method with the update formulas of the Broyden class are conjugate directions. So, in this situation the method possesses the quadratic termination property. Also, under convexity assumption on the objective function and when $\phi \in [0, 1]$, it has been shown that the method is globally and locally superlinearly convergent [86]. It is worth noting that among the quasi-Newton update formulas of the Broyden class, the BFGS update is superior with respect to the computational performance. A nice survey on the quasi-Newton methods has been provided in [93].

Similar to the quasi-Newton approximations $\{H_k\}_{k\geq 0}$ for the inverse Hessian satisfying (13), quasi-Newton approximations $\{B_k\}_{k\geq 0}$ for the Hessian can be proposed for which the following equivalent version of the standard secant equation (13) should be satisfied:

$$B_{k+1}s_k = y_k. \tag{17}$$

In such situation, considering (12), search directions of the quasi-Newton method can be computed by solving the following linear system:

$$B_k d_k = -g_k. \tag{18}$$

Exercise 3 *(i) Prove that if the search direction d_k is a descent direction and the line search fulfills the Wolfe conditions (5) and (6), then $s_k^T y_k > 0$.*
(ii) For the Broyden class of update formulas (15), prove that if H_k is a positive definite matrix, $s_k^T y_k > 0$ and $\phi \geq 0$, then H_{k+1}^{ϕ} is also a positive definite matrix.

Exercise 4 *(i) (Sherman-Morrison Theorem) Let $A \in \mathbb{R}^{n\times n}$ be a nonsingular matrix and $u, v \in \mathbb{R}^n$ be arbitrary vectors. Prove that if $1 + v^T A^{-1} u \neq 0$, then the rank-one update $A + uv^T$ of A is nonsingular, and*

$$(A + uv^T)^{-1} = A^{-1} - \frac{A^{-1}uv^T A^{-1}}{1 + v^T A^{-1} u}.$$

(ii) Compute ${H_{k+1}^{DFP}}^{-1}$ and find its relationship with H_{k+1}^{BFGS}.

2.4.1 Scaled Quasi-Newton Updates

In order to achieve an ideal distribution of the eigenvalues of quasi-Newton updates of the Broyden class, improving the condition number of successive approximations of the inverse Hessian and consequently, increasing the numerical stability in the iterative method (2), the scaled quasi-Newton updates have been developed [86]. In this context, replacing H_k by $\theta_k H_k$ in (15), where $\theta_k > 0$ is called the scaling parameter, the scaled Broyden class of quasi-Newton updates can be achieved as follows:

$$H_{k+1}^{\phi,\theta_k} = \left(H_k - \frac{H_k y_k y_k^T H_k}{y_k^T H_k y_k} + \phi v_k v_k^T \right) \theta_k + \frac{s_k s_k^T}{s_k^T y_k}, \tag{19}$$

where v_k is defined by (16). The most effective choices for θ_k in (19) have been proposed by Oren and Spedicato [75, 77],

$$\theta_k = \frac{s_k^T y_k}{y_k^T H_k y_k}, \tag{20}$$

and, Oren and Luenberger [75, 76],

$$\theta_k = \frac{s_k^T H_k^{-1} s_k}{s_k^T y_k}. \tag{21}$$

A scaled quasi-Newton update in the form of (19) with one of the parameters (20) or (21) is called a self-scaling quasi-Newton update.

Although the self-scaling quasi-Newton methods are numerically efficient, as an important defect the methods need to save the matrix $H_k \in \mathbb{R}^{n \times n}$ in each iteration, being improper for solving large-scale problems. Hence, in a simple modification in the sense of replacing H_k by the identity matrix in (19), self-scaling memoryless update formulas of the Broyden class have been proposed as follows:

$$\tilde{H}_{k+1}^{\phi,\theta_k} = \left(I - \frac{y_k y_k^T}{y_k^T y_k} + \phi \tilde{v}_k \tilde{v}_k^T \right) \theta_k + \frac{s_k s_k^T}{s_k^T y_k},$$

where

$$\tilde{v}_k = \sqrt{y_k^T y_k} \left(\frac{s_k}{s_k^T y_k} - \frac{y_k}{y_k^T y_k} \right).$$

Similarly, memoryless version of the scaling parameters (20) and (21) can be respectively written as:

$$\theta_k = \frac{s_k^T y_k}{||y_k||^2}, \tag{22}$$

and

$$\theta_k = \frac{||s_k||^2}{s_k^T y_k}. \tag{23}$$

The scaling parameter (23) can also be determined based on a two-point approximation of the standard secant equation (13) [35].

Exercise 5 *(i) Find all the eigenvalues of the scaled memoryless BFGS update formula with the parameters (22) or (23).*

(ii) Assume that ∇f is Lipschitz continuous on a nonempty open convex set $\mathcal{N}$; that is, there exists a positive constant L such that

$$||\nabla f(x) - \nabla f(y)|| \leq L||x - y||, \ \forall x, y \in \mathcal{N}. \tag{24}$$

Prove that if the objective function f is uniformly convex, then there exists a positive constant c such that for the sequence $\{x_k\}_{k\geq 0}$ generated by the scaled memoryless BFGS method with the parameter (23) we have

$$g_k^T d_k \leq -c||g_k||^2, \ \forall k \geq 0. \tag{25}$$

(Hint: Note that a differentiable function f is said to be uniformly (or strongly) convex on a nonempty open convex set $\mathcal{S}$ if and only if there exists a positive constant μ such that

$$(\nabla f(x) - \nabla f(y))^T (x - y) \geq \mu||x - y||^2, \ \forall x, y \in \mathcal{S} \text{ [86].)}$$

Definition 2 Inequality (25) is called the sufficient descent condition.

2.4.2 Modified Secant Equations

The standard secant equation (13), or its equivalent form (17), only uses the gradient information and ignores the function values. So, efforts have been made to modify the Eq. (17) such that more available information be employed and consequently, better approximations for the (inverse) Hessian be obtained (see [16] and the references therein).

Assume that the objective function f is smooth enough and let $f_k = f(x_k)$, $\forall k \geq 0$. From Taylor's theorem we get

$$f_k = f_{k+1} - s_k^T g_{k+1} + \frac{1}{2} s_k^T \nabla^2 f(x_{k+1}) s_k - \frac{1}{6} s_k^T (T_{k+1} s_k) s_k + O(||s_k||^4), \tag{26}$$

where

$$s_k^T (T_{k+1} s_k) s_k = \sum_{i,j,l=1}^{n} \frac{\partial^3 f(x_{k+1})}{\partial x^i \partial x^j \partial x^l} s_k^i s_k^j s_k^l. \tag{27}$$

So, after some algebraic manipulations it can be seen that

$$s_k^T \nabla^2 f(x_{k+1}) s_k = s_k^T y_k + 2(f_k - f_{k+1}) + s_k^T (g_k + g_{k+1}) + \frac{1}{3} s_k^T (T_{k+1} s_k) s_k + O(||s_k||^4).$$

Hence, the following approximation can be proposed:

$$s_k^T \nabla^2 f(x_{k+1}) s_k \approx s_k^T y_k + \vartheta_k,$$

where

$$\vartheta_k = 2(f_k - f_{k+1}) + s_k^T (g_k + g_{k+1}), \tag{28}$$

which leads to the following modified secant equation [89, 90]:

$$B_{k+1} s_k = z_k, \; z_k = y_k + \frac{\vartheta_k}{s_k^T u_k} u_k, \tag{29}$$

where $u_k \in \mathbb{R}^n$ is a vector parameter satisfying $s_k^T u_k \neq 0$ (see also [98, 99]).

Again, from Taylor's theorem we can write:

$$s_k^T g_k = s_k^T g_{k+1} - s_k^T \nabla^2 f(x_{k+1}) s_k + \frac{1}{2} s_k^T (T_{k+1} s_k) s_k + O(||s_k||^4). \tag{30}$$

Now, considering (26) and (30), by canceling the terms which include tensor we get

$$s_k^T \nabla^2 f(x_{k+1}) s_k = s_k^T y_k + 3\vartheta_k + O(||s_k||^4),$$

where ϑ_k is defined by (28). Hence, the following secant equation can be proposed [100]:

$$B_{k+1} s_k = w_k, \; w_k = y_k + \frac{3\vartheta_k}{s_k^T u_k} u_k, \tag{31}$$

where $u_k \in \mathbb{R}^n$ is a vector parameter satisfying $s_k^T u_k \neq 0$.

For a quadratic objective function f, we have $\vartheta_k = 0$, and consequently, the modified secant equations (29) and (31) reduce to the standard secant equation. For the vector parameter u_k, we can simply let $u_k = s_k$, or $u_k = y_k$ provided that the line search fulfills the Wolfe conditions. To guarantee positive definiteness of the successive quasi-Newton approximations for the (inverse) Hessian obtained based on the modified secant equations (29) and (31) we should respectively have $s_k^T z_k > 0$ and $s_k^T w_k > 0$ which may not be necessarily satisfied for general functions. To overcome this problem, in a simple modification we can replace ϑ_k in (29) and (31) by $\max\{\vartheta_k, 0\}$. The modified secant equations (29) and (31) are justified by the following theorem [89, 100, 104], demonstrating their accuracy in contrast to the standard secant equation (17).

Theorem 1 *If f is sufficiently smooth and $||s_k||$ is small enough, then the following estimating relations hold:*

$$s_k^T(\nabla^2 f(x_{k+1})s_k - y_k) = \frac{1}{2}s_k^T(T_{k+1}s_k)s_k + O(||s_k||^4),$$
$$s_k^T(\nabla^2 f(x_{k+1})s_k - z_k) = \frac{1}{3}s_k^T(T_{k+1}s_k)s_k + O(||s_k||^4),$$
$$s_k^T(\nabla^2 f(x_{k+1})s_k - w_k) = O(||s_k||^4),$$

where T_{k+1} is defined by (27).

Convexity assumption on the objective function plays an important role in convergence analysis of the quasi-Newton methods with secant equations (17), (29) and (31). However, in [64] a modified BFGS method has been proposed which is globally and locally superlinearly convergent for nonconvex objective functions (see also [58, 65]). The method has been designed based on the following modified secant equation:

$$B_{k+1}s_k = \bar{y}_k, \ \bar{y}_k = y_k + h_k||g_k||^r s_k, \tag{32}$$

where r is a positive constant and $h_k > 0$ is defined by

$$h_k = C + \max\{-\frac{s_k^T y_k}{||s_k||^2}, 0\}||g_k||^{-r},$$

with some positive constant C. As an interesting property, for the modified secant equation (32) we have $s_k^T\bar{y}_k > 0$, independent of the line search conditions and the objective function convexity, which guarantees heredity of positive definiteness for the related BFGS updates. Recently, scaled memoryless BFGS methods have been proposed based on the modified secant equations (29), (31) and (32) which possess the sufficient descent property (25) [17–19, 22, 28]. In addition to the modified secant equations (29), (31) and (32) which apply information of the current iteration, the multi-step secant equations have been developed by Ford et al. [51–53] based on the polynomial interpolation using available data from the m recent steps.

3 Conjugate Gradient Methods

Conjugate gradient methods comprise a class of algorithms which are between the steepest descent method and the Newton method. Utilizing the Hessian information implicitly, the methods deflect the steepest descent direction by adding to it a multiple of the direction used in the last step, that is,

$$d_{k+1} = -g_{k+1} + \beta_k d_k, \ k = 0, 1, \ldots, \tag{33}$$

with $d_0 = -g_0$, where β_k is a scalar called the conjugate gradient (update) parameter. Although the methods only require the first-order derivatives, they overcome the slow convergence of the steepest descent method. Also, the methods need not to save and compute the second-order derivatives which are needed in the Newton method. Hence, they are widely used to solve large-scale optimization problems.

Different conjugate gradient methods mainly correspond to different choices for the conjugate gradient parameter [61]. Although the conjugate gradient methods are equivalent in the linear case, that is, when f is a strictly convex quadratic function and the line search is exact, their behavior for general functions may be quite different [3, 45, 82]. It is worth noting that search directions of the linear conjugate gradient methods are conjugate directions. In what follows, we deal with several essential conjugate gradient methods.

3.1 The Hestenes-Stiefel Method

Conjugate gradient methods were originally developed in the 1950s by Hestenes and Stiefel [62] (HS) as an alternative to factorization methods for solving linear systems. Conjugate gradient parameter of the HS method is given by

$$\beta_k^{HS} = \frac{g_{k+1}^T y_k}{d_k^T y_k}.$$

From the mean-value theorem, there exists some $\xi \in (0, 1)$ such that

$$d_{k+1}^T y_k = d_{k+1}^T (g_{k+1} - g_k) = \alpha_k d_{k+1}^T \nabla^2 f(x_k + \xi \alpha_k d_k) d_k.$$

Hence, the condition

$$d_{k+1}^T y_k = 0, \tag{34}$$

can be considered as a conjugacy condition for the nonlinear objective functions since it shows that search directions d_k and d_{k+1} are conjugate directions. As an attractive feature, considering (33) it can be seen that search directions of the HS method satisfy the conjugacy condition (34), independent of the line search and the objective function convexity.

Perry [78] noted that the search direction d_{k+1} of the HS method can be written as:

$$d_{k+1} = -\left(I - \frac{s_k y_k^T}{s_k^T y_k} \right) g_{k+1}. \tag{35}$$

Then, he made a modification on the search direction (35) as follows:

$$d_{k+1} = -\underbrace{\left(I - \frac{s_k y_k^T}{s_k^T y_k} + \frac{s_k s_k^T}{s_k^T y_k}\right)}_{P_{k+1}} g_{k+1} = -P_{k+1} g_{k+1}.$$

Perry justified the addition of the correction term $\frac{s_k s_k^T}{s_k^T y_k}$ by noting that the matrix P_{k+1} satisfies the following equation:

$$y_k^T P_{k+1} = s_k^T,$$

which is similar, but not identical, to the standard secant equation (13). To improve Perry's approach, Shanno [84] modified the matrix P_{k+1} as follows:

$$P_{k+1}^S = I - \frac{s_k y_k^T + y_k s_k^T}{s_k^T y_k} + \left(1 + \frac{y_k^T y_k}{s_k^T y_k}\right) \frac{s_k s_k^T}{s_k^T y_k}.$$

Thus, the related conjugate gradient method is precisely the BFGS method in which the approximation of the inverse Hessian is restarted as the identity matrix at every step and so, no significant storage is used to develop a better approximation for the inverse Hessian. Hence, the HS method can be extended to the memoryless BFGS method. This idea was also discussed by Nazareth [71] and Buckley [38]. A nice survey concerning the relationship between conjugate gradient methods and the quasi-Newton methods has been provided in [72].

Although the HS method is numerically efficient, its search directions generally fail to satisfy the descent condition (3), even for strictly convex objective functions [40]. It is worth noting that when in an iteration of a conjugate gradient method the search direction does not satisfy the descent condition (3), i.e., when encountering with an uphill search direction, the steepest descent direction can be used. This popular scheme for the conjugate gradient methods is called the restart procedure. In another approach, Powell [82] suggested to restart the conjugate gradient method if the following inequality is violated:

$$g_k^T g_{k-1} \le \varsigma ||g_k||^2,$$

where ς is a small positive constant (see also [56]).

As another defect of the HS method that will be discussed in the next parts of this section, it can be stated that the method lacks global convergence in certain circumstances in the sense of cycling infinitely [82].

3.2 The Fletcher-Reeves Method

Since solving a linear system is equivalent to minimizing a quadratic function, in the 1960s Fletcher and Reeves [50] (FR) modified the HS method and developed a conjugate gradient method for unconstrained minimization with the following parameter:

$$\beta_k^{FR} = \frac{||g_{k+1}||^2}{||g_k||^2}.$$

Although search directions of the FR method generally are not descent directions, convergence analysis of the method has been appropriately developed. As a brief review, at first Zoutendijk [105] established a convergence result for the FR method under the exact line search. Then, Al-Baali [2] dealt with convergence of the FR method when the line search fulfills the strong Wolfe conditions (5) and (7), with $0 < \delta < \sigma < 1/2$. Liu et al. [68] extended the Al-Baali's result for $\sigma = 1/2$. A comprehensive study on the convergence of the FR method has been made by Gilbert and Nocedal [56]. Notwithstanding the strong convergence properties, numerical performance of the FR method is essentially affected by jamming [56, 81], i.e., generating many short steps without making significant progress to the solution because the search directions became nearly orthogonal to the gradient.

3.3 The Polak-Ribière-Polyak Method

One of the efficient conjugate gradient methods has been proposed by Polak et al. [79, 80] (PRP) where its parameter is computed by

$$\beta_k^{PRP} = \frac{g_{k+1}^T y_k}{||g_k||^2}.$$

It is important that when the iterations jam, the step s_k is small. So, the factor y_k in the numerator of β_k^{PRP} tends to zero and consequently, β_k^{PRP} becomes small. Therefore, the search direction d_{k+1} tends to the steepest descent direction and an automatic restart occurs. This favorable numerical feature of jamming prevention also occurs for the HS method.

In spite of numerical efficiency of the PRP method, the method lacks the descent property. Also, Powell [82] constructed a three-dimensional counter example with the exact line search, demonstrating the method can cycle infinitely without convergence to a solution. Nevertheless, based on the insight gained by his counter example, Powell [82] suggested the following truncation of β_k^{PRP}:

$$\beta_k^{PRP+} = \max\{\beta_k^{PRP}, 0\},$$

which yields a globally convergent conjugate gradient method [56], being also computationally efficient [3].

Since under the exact line search the PRP and the HS methods are equivalent, the cycling phenomenon may occur for the HS method. The following truncation of β_k^{HS} has been shown to lead to a globally convergent conjugate gradient method [43, 56]:

$$\beta_k^{HS+} = \max\{\beta_k^{HS}, 0\},$$

which is also more efficient than the HS method [3].

3.4 The Dai-Yuan Method

Another essential conjugate gradient method has been proposed by Dai and Yuan [47] (DY) with the following parameter:

$$\beta_k^{DY} = \frac{||g_{k+1}||^2}{d_k^T y_k}.$$

It is notable that under mild assumptions on the objective function, the DY method has been shown to be globally convergent under a variety of inexact line search conditions. Also, in addition to the generation of descent search directions when $d_k^T y_k > 0$, as guaranteed by the Wolfe conditions (5) and (6), the DY method has been proved to have a certain self-adjusting property, independent of the line search and the objective function convexity [41]. More exactly, if there exist positive constants γ_1 and γ_2 such that $\gamma_1 \leq ||g_k|| \leq \gamma_2$, for all $k \geq 0$, then, for any $p \in (0, 1)$, there exists a positive constant c such that the sufficient descent condition $g_i^T d_i \leq -c||g_i||^2$ holds for at least $\lfloor pk \rfloor$ indices $i \in [0, k]$, where $\lfloor j \rfloor$ denotes the largest integer less than or equal to j. However, similar to the FR method, in spite of strong theoretical properties the DY method has a poor computational performance due to the jamming phenomenon.

3.5 The Dai-Liao Method

In order to employ quasi-Newton aspects in the conjugacy condition (34), Dai and Liao [43] (DL) noted that considering $B_{k+1} \in \mathbb{R}^{n\times n}$ as an approximation of $\nabla^2 f(x_{k+1})$ given by a quasi-Newton method, from the standard secant equation (17) and the linear system (18) we can write

$$d_{k+1}^T y_k = d_{k+1}^T (B_{k+1} s_k) = -g_{k+1}^T s_k. \tag{36}$$

If the line search is exact, then $g_{k+1}^T s_k = 0$, and consequently (36) reduces to (34). However, in practice the algorithms normally adopt inexact line searches. Hence, the following extension of the conjugacy condition (34) has been proposed in [43]:

$$d_{k+1}^T y_k = -t g_{k+1}^T s_k, \tag{37}$$

where t is a nonnegative parameter. If $t = 0$ or the line search is exact, then (37) reduces to (34), and if $t = 1$, then (37) reduces to (36) which implicitly contains the effective standard secant equation (17). Also, for small values of t, the conjugacy condition (37) tends to the conjugacy condition (34). Thus, the conjugacy condition (37) can be regarded as a generalization of the conjugacy conditions (34) and (36).

Taking inner product of (33) with y_k and using (37), Dai and Liao [43] obtained the following formula for the conjugate gradient parameter:

$$\beta_k^{DL} = \frac{g_{k+1}^T y_k}{d_k^T y_k} - t\frac{g_{k+1}^T s_k}{d_k^T y_k}, \tag{38}$$

shown to be globally convergent for uniformly convex objective functions. Theoretical and numerical features of the DL method is very dependent on the parameter t for which there is no any optimal choice [15]. It is worth noting that if

$$t = 2\frac{||y_k||^2}{s_k^T y_k}, \tag{39}$$

then the conjugate gradient parameter proposed by Hager and Zhang [59] is achieved. Also, the choice

$$t = \frac{||y_k||^2}{s_k^T y_k}, \tag{40}$$

yields another conjugate gradient parameter suggested by Dai and Kou [42]. The choices (39) and (40) are effective since they guarantee the sufficient descent condition (25), independent of the line search and the objective function convexity, and lead to numerically efficient conjugate gradient methods [21, 42, 60]. Recently, Babaie-Kafaki and Ghanbari [25, 27, 32] dealt with other proper choices for the parameter t in the DL method.

Based on Powell's approach of nonnegative restriction of the conjugate gradient parameters [82], Dai and Liao proposed the following modified version of β_k^{DL}:

$$\beta_k^{DL+} = \beta_k^{HS+} - t\frac{g_{k+1}^T s_k}{d_k^T y_k},$$

and showed that the DL+ method is globally convergent for general objective functions [43]. In several other attempts to make modifications on the DL method, modified secant equations have been applied in the Dai-Liao approach. In this context, in order to employ the objective function values in addition to the gradient information, Yabe and Takano [94] used the modified secant equation (31). Also, Li et al. [66] used the modified secant equation (29). Babaie-Kafaki et al. [33] applied a revised form of the modified secant equation (31), and the modified secant equation proposed in [98]. Ford et al. [54] employed the multi-step quasi-Newton equations proposed by Ford and Moghrabi [51]. In another attempt to achieve global convergence without convexity assumption on the objective function, Zhou and Zhang [104] applied the modified secant equation (32).

Exercise 6 *For the DL method, assume that* $s_k^T y_k > 0$ *and* $t > 0$.

(i) Find the matrix Q_{k+1} *for which search directions of the DL method can be written as* $d_{k+1} = -Q_{k+1} g_{k+1}$. *The matrix* Q_{k+1} *is called the search direction matrix.*

(ii) Find all the eigenvalues of the matrix $A_{k+1} = \dfrac{Q_{k+1}^T + Q_{k+1}}{2}$.

(iii) Prove that if

$$t > \frac{1}{4}\left(\frac{||y_k||^2}{s_k^T y_k} - \frac{s_k^T y_k}{||s_k||^2}\right),$$

then search directions of the DL method satisfy the descent condition (3).

Exercise 7 *For the DL method, assume that* $s_k^T y_k > 0$ *and* $t > 0$.

(i) Prove that the search direction matrix Q_{k+1} *is nonsingular. Then, find the inverse of* Q_{k+1}.

(ii) Find $||Q_{k+1}||_F^2$ *and* $||Q_{k+1}^{-1}||_F^2$, *where* $||.||_F$ *stands for the Frobenius norm.*

(iii) Prove that if $n \to \infty$, *then* $t^* = \sqrt{\dfrac{||y_k||(s_k^T y_k)}{||s_k||^3}}$ *is the minimizer of* $\kappa_F(Q_{k+1}) = ||Q_{k+1}||_F ||Q_{k+1}^{-1}||_F$.

3.6 The CG-Descent Algorithm

In an attempt to make a modification of the HS method in order to achieve the sufficient descent property, Hager and Zhang [59] proposed the following conjugate gradient parameter:

$$\beta_k^N = \frac{1}{d_k^T y_k}\left(y_k - 2d_k \frac{||y_k||^2}{d_k^T y_k}\right)^T g_{k+1} = \beta_k^{HS} - 2\frac{||y_k||^2}{d_k^T y_k}\frac{d_k^T g_{k+1}}{d_k^T y_k},$$

which can be considered as an adaptive version of β_k^{DL} given by (38). The method has been shown to be globally convergent for uniformly convex objective functions. In order to achieve the global convergence for general functions, the following truncation of β_k^N has been proposed in [59]:

$$\bar{\beta}_k^N = \max\{\beta_k^N, \eta_k\},\ \eta_k = \frac{-1}{||d_k|| \min\{\eta, ||g_k||\}},$$

where η is a positive constant. A conjugate gradient method with the parameter $\bar{\beta}_k^N$ in which the line search fulfills the approximate Wolfe conditions given by (8) is called the CG-Descent algorithm [60]. Search directions of the CG-Descent algorithm satisfy the sufficient descent condition (25) with $c = \frac{7}{8}$. The CG-Descent algorithm is one of the most efficient and popular conjugate gradient methods, widely used by engineers and mathematicians engaged in solving large-scale unconstrained optimization problems.

Based on the Hager-Zhang approach [59], Yu et al. [96] proposed a modified form of β_k^{PRP} as follows:

$$\beta_k^{DPRP} = \beta_k^{PRP} - C\frac{||y_k||^2}{||g_k||^4} g_{k+1}^T d_k,$$

with a constant $C > \frac{1}{4}$, guaranteeing the sufficient descent condition (25) (see also [20]). Afterwards, several other descent extensions of the PRP method have been proposed in [26, 97], using the conjugate gradient parameter β_k^{DPRP}.

Exercise 8 *Prove that if* $d_k^T y_k > 0$, $\forall k \geq 0$, *then search directions of a conjugate gradient method with the following parameter:*

$$\beta_k^{\tau} = \beta_k^{HS} - \tau_k \frac{||y_k||^2 (g_{k+1}^T d_k)}{(d_k^T y_k)^2},$$

in which $\tau_k \geq \bar{\tau}$, *for some positive constant* $\bar{\tau} > \frac{1}{4}$, *satisfy the sufficient descent condition (25).*

Exercise 9 *Prove that search directions of the DPRP method with* $C > \frac{1}{4}$ *satisfy the sufficient descent condition (25).*

3.7 Hybrid Conjugate Gradient Methods

Essential conjugate gradient methods generally can be divided into two categories. In the first category, all the conjugate gradient parameters have the common numerator $g_{k+1}^T y_k$; such as the HS and PRP methods, and also, the conjugate gradient method proposed by Liu and Storey [69] (LS) with the following parameter:

$$\beta_k^{LS} = -\frac{g_{k+1}^T y_k}{d_k^T g_k}.$$

In the second category, all the conjugate gradient parameters have the common numerator $||g_{k+1}||^2$; such as the FR and DY methods, and also, the conjugate descent

(CD) method proposed by Fletcher [49] with the following conjugate gradient parameter:

$$\beta_k^{CD} = -\frac{||g_{k+1}||^2}{d_k^T g_k}.$$

There are some advantages and disadvantages for the conjugate gradient methods in each category. As mentioned before, generally the methods of the first category are numerically efficient because of an automatic restart feature which avoids jamming while the methods of the second category are theoretically strong in the sense of (often) generating descent search directions and being globally convergent under a variety of line search conditions and some mild assumptions. To attain good computational performance and to maintain the attractive feature of strong global convergence, researchers paid especial attention to hybridize the conjugate gradient parameters of the two categories. Hybrid conjugate gradient methods are essentially designed based on an adoptive switch from a conjugate gradient parameter in the second category to one in the first category when the iterations jam. Well-known hybrid conjugate gradient parameters can be listed as follows:

- $\beta_k^{HuS} = \max\{0, \min\{\beta_k^{PRP}, \beta_k^{FR}\}\}$, proposed by Hu and Storey [63];
- $\beta_k^{TaS} = \begin{cases} \beta_k^{PRP}, & 0 \leq \beta_k^{PRP} \leq \beta_k^{FR}, \\ \beta_k^{FR}, & \text{otherwise}, \end{cases}$ which has been proposed by Touati-Ahmed and Storey [87];
- $\beta_k^{GN} = \max\{-\beta_k^{FR}, \min\{\beta_k^{PRP}, \beta_k^{FR}\}\}$, proposed by Gilbert and Nocedal [56];
- $\beta_k^{hDYz} = \max\{0, \min\{\beta_k^{HS}, \beta_k^{DY}\}\}$, proposed by Dai and Yuan [48];
- $\beta_k^{hDY} = \max\{-\frac{1-\sigma}{1+\sigma}\beta_k^{DY}, \min\{\beta_k^{HS}, \beta_k^{DY}\}\}$, with the positive constant σ used in the Wolfe condition (6) [48];
- $\beta_k^{LS-CD} = \max\{0, \min\{\beta_k^{LS}, \beta_k^{CD}\}\}$, proposed by Andrei [7] (see also [95]).

In all of the above hybridization schemes, discrete combinations of the conjugate gradient parameters of the two categories have been considered. Recently, Andrei [8, 9, 11, 12] dealt with convex combinations of the conjugate gradient parameters of the two categories which are continuous hybridizations. More exactly, in [8] the following hybrid conjugate gradient method has been proposed:

$$\beta_k^C = (1 - \mu_k)\beta_k^{HS} + \mu_k \beta_k^{DY},$$

in which $\mu_k \in [0, 1]$ is called the hybridization parameter. As known, if the point x_{k+1} is close enough to a local minimizer x^*, then a good direction to follow is the Newton direction, that is, $d_{k+1} = -\nabla^2 f(x_{k+1})^{-1} g_{k+1}$. So, considering search directions of the hybrid conjugate gradient method with the parameter β_k^C we can write:

$$-\nabla^2 f(x_{k+1})^{-1} g_{k+1} = -g_{k+1} + (1 - \mu_k)\frac{g_{k+1}^T y_k}{s_k^T y_k} s_k + \mu_k \frac{g_{k+1}^T g_{k+1}}{s_k^T y_k} s_k.$$

After some algebraic manipulations we get

$$\mu_k = \frac{s_k^T \nabla^2 f(x_{k+1}) g_{k+1} - s_k^T g_{k+1} - \dfrac{g_{k+1}^T y_k}{s_k^T y_k} s_k^T \nabla^2 f(x_{k+1}) s_k}{\dfrac{g_{k+1}^T g_k}{s_k^T y_k} s_k^T \nabla^2 f(x_{k+1}) s_k}.$$

Due to the essential property of low memory requirement for the conjugate gradient methods, Andrei applied the secant equations in order to avoid exact computation of $\nabla^2 f(x_{k+1}) s_k$ [8, 9, 12] (see also [24, 34]). In a different approach, recently Babaie-Kafaki and Ghanbari [30, 31] proposed two other continuous hybrid conjugate gradient methods in which the hybridization parameter is computed in a way to make the search directions of the hybrid method as closer as possible to the search directions of the descent three-term conjugate gradient methods proposed by Zhang et al. [102, 103].

3.8 Spectral Conjugate Gradient Methods

In the stream of overcoming drawbacks of the steepest descent method, Barzilai and Borwein [35] developed the two-point stepsize gradient algorithms in which the search directions are computed by

$$d_0 = -g_0,\ d_{k+1} = -\theta_k g_{k+1},\ k = 0, 1, \ldots,$$

where the positive parameter θ_k, called the scaling parameter, is computed by solving the following least-squares problem:

$$\min_{\theta \geq 0} ||\frac{1}{\theta} s_k - y_k||, \tag{41}$$

being a two-point approximation of the standard secant equation (17). After some algebraic manipulations, it can be seen that the solution of (41) is exactly the scaling parameter θ_k given by (23), used in the scaled memoryless quasi-Newton methods. Convergence of the two-point stepsize gradient algorithms has been studied in [44]. Using a nonmonotone line search procedure [57], Raydan [83] showed that the two-point stepsize gradient algorithms can be regarded as an efficient approach for solving large-scale unconstrained optimization problems. In [23, 46] the modified secant equations (29), (31) and (32) have been employed in the two-point stepsize gradient algorithms.

Combining search directions of the conjugate gradient methods and the two-point stepsize gradient algorithms, the spectral conjugate gradient methods [37] have been proposed in which the search directions are given by

$$d_{k+1} = -\theta_k g_{k+1} + \beta_k d_k,\ k = 0, 1, \ldots,$$

with $d_0 = -g_0$ and the scaling parameter θ_k often computed by (23) (see also [4–6, 10, 13]).

3.9 Three-Term Conjugate Gradient Methods

Although the concept of three-term conjugate gradient methods has been originally developed in 1970s [36, 70], recently researchers dealt with them in order to achieve the sufficient descent property. As known, some of the conjugate gradient methods such as HS, FR and PRP generally can not guarantee the descent condition (3). To overcome this problem, three-term versions of the HS, FR and PRP methods have been proposed respectively with the following search directions [101–103]:

$$d_{k+1} = -g_{k+1} + \beta_k^{HS} d_k - \frac{g_{k+1}^T d_k}{d_k^T y_k} y_k,$$

$$d_{k+1} = -g_{k+1} + \beta_k^{FR} d_k - \frac{g_{k+1}^T d_k}{||g_k||^2} g_{k+1},$$

$$d_{k+1} = -g_{k+1} + \beta_k^{PRP} d_k - \frac{g_{k+1}^T d_k}{||g_k||^2} y_k,$$

for all $k \geq 0$, with $d_0 = -g_0$. When the line search is exact, the above three-term conjugate gradient methods respectively reduce to the HS, FR and PRP methods. Also, for all of these methods we have the sufficient descent condition $d_k^T g_k = -||g_k||^2$, $\forall k \geq 0$, independent of the line search and the objective function convexity. A nice review of different three-term conjugate gradient methods has been presented in [85] (see also [14, 29]).

4 Limited-Memory Quasi-Newton Methods

As known, since quasi-Newton methods save an $n \times n$ matrix as an approximation of the inverse Hessian, they are not useful for solving large-scale unconstrained optimization problems. However, limited-memory quasi-Newton methods maintain a compact approximation of the inverse Hessian, saving only a few vectors of length n available from a certain number of the most recent iterations, and so, being useful in large-scale cases [74]. Convergence properties of the methods are often acceptable [13, 67, 73]. Although various limited-memory quasi-Newton methods have been proposed in the literature, here we deal with the limited-memory BFGS method, briefly called the L-BFGS method.

Note that the BFGS updating formula (14) can be written as:

$$H_{k+1}^{BFGS} = V_k^T H_k V_k + \rho_k s_k s_k^T, \tag{42}$$

where

$$\rho_k = \frac{1}{s_k^T y_k}, \text{ and } V_k = I - \rho_k y_k s_k^T.$$

In the limited-memory approach, a modified version of H_{k+1} is implicitly stored, saving a set of vector pairs $\{s_i, y_i\}$ available from the $m > 1$ recent iterations. More precisely, by repeated application of the formula (42), we get

$$\begin{aligned} H_{k+1}^{L-BFGS} &= (V_k^T \cdots V_{k-m+1}^T) H_{k-m+1} (V_{k-m+1} \cdots V_k) \\ &+ \rho_{k-m+1} (V_k^T \cdots V_{k-m+2}^T) s_{k-m+1} s_{k-m+1}^T (V_{k-m+2} \cdots V_k) \\ &+ \cdots \\ &+ \rho_k s_k s_k^T, \end{aligned}$$

in which in order to use a low memory storage, H_{k-m+1} is computed by

$$H_{k-m+1} = \theta_k I,$$

where θ_k is often calculated by (22), proved to be practically effective [67]. Also, the search direction $d_{k+1} = -H_{k+1}^{L-BFGS} g_{k+1}$ can be effectively computed by the following recursive procedure [74].

Algorithm 1 (*Computing search directions of the L-BFGS method*)

$q = g_{k+1}$;
for $i = k, k-1, \ldots, k-m+1$

$\gamma_i \leftarrow \rho_i s_i^T q$;
$q \leftarrow q - \gamma_i y_i$;

end
$r \leftarrow \theta_k q$;
for $i = k-m+1, k-m+2, \ldots, k$

$\xi \leftarrow \rho_i y_i^T r$;
$r \leftarrow r + s_i(\gamma_i - \xi)$;

end
$d_{k+1} = -r$.

Remark 1 Practical experiences have shown that the values of m between 3 and 20 often produce satisfactory numerical results [74].

5 Conclusions

Recent line search-based approaches in large-scale unconstrained optimization have been studied. Especially, the conjugate gradient methods and the memoryless quasi-Newton methods have been focused on. At first, after introducing the essential unconstrained optimization algorithms, merits and demerits of the classical conjugate gradient methods have been reviewed. Then, their descent extensions, their hybridizations based on the secant equations, and their three-term versions with sufficient descent property have been discussed. Finally, a limited-memory quasi-Newton method has been presented. So, recent efficient tools for big data applications have been provided.

Acknowledgments This research was supported by Research Council of Semnan University. The author is grateful to Professor Ali Emrouznejad for his valuable suggestions helped to improve the presentation.

References

1. Akaike, H.: On a successive transformation of probability distribution and its application to the analysis of the optimum gradient method. Ann. Inst. Statist. Math. Tokyo **11**(1), 1–16 (1959)
2. Al-Baali, M.: Descent property and global convergence of the Fletcher-Reeves method with inexact line search. IMA J. Numer. Anal. **5**(1), 121–124 (1985)
3. Andrei, N.: Numerical comparison of conjugate gradient algorithms for unconstrained optimization. Stud. Inform. Control **16**(4), 333–352 (2007)
4. Andrei, N.: A scaled BFGS preconditioned conjugate gradient algorithm for unconstrained optimization. Appl. Math. Lett. **20**(6), 645–650 (2007)
5. Andrei, N.: Scaled conjugate gradient algorithms for unconstrained optimization. Comput. Optim. Appl. **38**(3), 401–416 (2007)
6. Andrei, N.: Scaled memoryless BFGS preconditioned conjugate gradient algorithm for unconstrained optimization. Optim. Methods Softw. **22**(4), 561–571 (2007)
7. Andrei, N.: 40 conjugate gradient algorithms for unconstrained optimization—a survey on their definition. ICI Technical Report No. 13/08 (2008)
8. Andrei, N.: Another hybrid conjugate gradient algorithm for unconstrained optimization. Numer. Algorithms **47**(2), 143–156 (2008)
9. Andrei, N.: A hybrid conjugate gradient algorithm for unconstrained optimization as a convex combination of Hestenes-Stiefel and Dai-Yuan. Stud. Inform. Control **17**(1), 55–70 (2008)
10. Andrei, N.: A scaled nonlinear conjugate gradient algorithm for unconstrained optimization. Optimization **57**(4), 549–570 (2008)
11. Andrei, N.: Hybrid conjugate gradient algorithm for unconstrained optimization. J. Optim. Theory Appl. **141**(2), 249–264 (2009)
12. Andrei, N.: Accelerated hybrid conjugate gradient algorithm with modified secant condition for unconstrained optimization. Numer. Algorithms **54**(1), 23–46 (2010)
13. Andrei, N.: Accelerated scaled memoryless BFGS preconditioned conjugate gradient algorithm for unconstrained optimization. Eur. J. Oper. Res. **204**(3), 410–420 (2010)
14. Andrei, N.: A modified Polak-Ribière-Polyak conjugate gradient algorithm for unconstrained optimization. Optimization **60**(12), 1457–1471 (2011)
15. Andrei, N.: Open problems in conjugate gradient algorithms for unconstrained optimization. B. Malays. Math. Sci. So. **34**(2), 319–330 (2011)

16. Babaie-Kafaki, S.: A modified BFGS algorithm based on a hybrid secant equation. Sci. China Math. **54**(9), 2019–2036 (2011)
17. Babaie-Kafaki, S.: A note on the global convergence theorem of the scaled conjugate gradient algorithms proposed by Andrei. Comput. Optim. Appl. **52**(2), 409–414 (2012)
18. Babaie-Kafaki, S.: A modified scaled memoryless BFGS preconditioned conjugate gradient method for unconstrained optimization. 4OR **11**(4):361–374 (2013)
19. Babaie-Kafaki, S.: A new proof for the sufficient descent condition of Andrei's scaled conjugate gradient algorithms. Pac. J. Optim. **9**(1), 23–28 (2013)
20. Babaie-Kafaki, S.: An eigenvalue study on the sufficient descent property of a modified Polak-Ribière-Polyak conjugate gradient method. Bull. Iranian Math. Soc. **40**(1), 235–242 (2014)
21. S. Babaie-Kafaki. On the sufficient descent condition of the Hager-Zhang conjugate gradient methods. 4OR **12**(3):285–292 (2014)
22. Babaie-Kafaki, S.: Two modified scaled nonlinear conjugate gradient methods. J. Comput. Appl. Math. **261**(5), 172–182 (2014)
23. Babaie-Kafaki, S., Fatemi, M.: A modified two-point stepsize gradient algorithm for unconstrained minimization. Optim. Methods Softw. **28**(5), 1040–1050 (2013)
24. Babaie-Kafaki, S., Fatemi, M., Mahdavi-Amiri, N.: Two effective hybrid conjugate gradient algorithms based on modified BFGS updates. Numer. Algorithms **58**(3), 315–331 (2011)
25. Babaie-Kafaki, S., Ghanbari, R.: The Dai-Liao nonlinear conjugate gradient method with optimal parameter choices. Eur. J. Oper. Res. **234**(3), 625–630 (2014)
26. Babaie-Kafaki, S., Ghanbari, R.: A descent extension of the Polak-Ribière-Polyak conjugate gradient method. Comput. Math. Appl. **68**(12), 2005–2011 (2014)
27. Babaie-Kafaki, S., Ghanbari, R.: A descent family of Dai-Liao conjugate gradient methods. Optim. Methods Softw. **29**(3), 583–591 (2014)
28. Babaie-Kafaki, S., Ghanbari, R.: A modified scaled conjugate gradient method with global convergence for nonconvex functions. Bull. Belg. Math. Soc. Simon Stevin **21**(3), 465–477 (2014)
29. Babaie-Kafaki, S., Ghanbari, R.: Two modified three-term conjugate gradient methods with sufficient descent property. Optim. Lett. **8**(8), 2285–2297 (2014)
30. Babaie-Kafaki, S., Ghanbari, R.: A hybridization of the Hestenes-Stiefel and Dai-Yuan conjugate gradient methods based on a least-squares approach. Optim. Methods Softw. **30**(4), 673–681 (2015)
31. Babaie-Kafaki, S., Ghanbari, R.: A hybridization of the Polak-Ribière-Polyak and Fletcher-Reeves conjugate gradient methods. Numer. Algorithms **68**(3), 481–495 (2015)
32. Babaie-Kafaki, S., Ghanbari, R.: Two optimal Dai-Liao conjugate gradient methods. Optimization **64**(11), 2277–2287 (2015)
33. Babaie-Kafaki, S., Ghanbari, R., Mahdavi-Amiri, N.: Two new conjugate gradient methods based on modified secant equations. J. Comput. Appl. Math. **234**(5), 1374–1386 (2010)
34. Babaie-Kafaki, S., Mahdavi-Amiri, N.: Two modified hybrid conjugate gradient methods based on a hybrid secant equation. Math. Model. Anal. **18**(1), 32–52 (2013)
35. Barzilai, J., Borwein, J.M.: Two-point stepsize gradient methods. IMA J. Numer. Anal. **8**(1), 141–148 (1988)
36. Beale, E.M.L.: A derivation of conjugate gradients. In: Lootsma, F.A. (ed.) Numerical Methods for Nonlinear Optimization, pp. 39–43. Academic Press, NewYork (1972)
37. Birgin, E., Martínez, J.M.: A spectral conjugate gradient method for unconstrained optimization. Appl. Math. Optim. **43**(2), 117–128 (2001)
38. Buckley, A.G.: Extending the relationship between the conjugate gradient and BFGS algorithms. Math. Program. **15**(1), 343–348 (1978)
39. Cauchy, A.: Méthodes générales pour la résolution des systèmes déquations simultanées. C. R. Acad. Sci. Par. **25**(1), 536–538 (1847)
40. Dai, Y.H.: Analyses of conjugate gradient methods. Ph.D. Thesis, Mathematics and Scientific/Engineering Computing, Chinese Academy of Sciences (1997)
41. Dai, Y.H.: New properties of a nonlinear conjugate gradient method. Numer. Math. **89**(1), 83–98 (2001)

42. Dai, Y.H., Kou, C.X.: A nonlinear conjugate gradient algorithm with an optimal property and an improved Wolfe line search. SIAM J. Optim. **23**(1), 296–320 (2013)
43. Dai, Y.H., Liao, L.Z.: New conjugacy conditions and related nonlinear conjugate gradient methods. Appl. Math. Optim. **43**(1), 87–101 (2001)
44. Dai, Y.H., Liao, L.Z.: R-linear convergence of the Barzilai and Borwein gradient method. IMA J. Numer. Anal. **22**(1), 1–10 (2002)
45. Dai, Y.H., Ni, Q.: Testing different conjugate gradient methods for large-scale unconstrained optimization. J. Comput. Math. **22**(3), 311–320 (2003)
46. Dai, Y.H., Yuan, J., Yuan, Y.X.: Modified two-point stepsize gradient methods for unconstrained optimization. Comput. Optim. Appl. **22**(1), 103–109 (2002)
47. Dai, Y.H., Yuan, Y.X.: A nonlinear conjugate gradient method with a strong global convergence property. SIAM J. Optim. **10**(1), 177–182 (1999)
48. Dai, Y.H., Yuan, Y.X.: An efficient hybrid conjugate gradient method for unconstrained optimization. Ann. Oper. Res. **103**(1–4), 33–47 (2001)
49. Fletcher, R.: Practical Methods of Optimization. Wiley, New York (1987)
50. Fletcher, R., Reeves, C.M.: Function minimization by conjugate gradients. Comput. J. **7**(2), 149–154 (1964)
51. Ford, J.A., Moghrabi, I.A.: Multi-step quasi-Newton methods for optimization. J. Comput. Appl. Math. **50**(1–3), 305–323 (1994)
52. Ford, J.A., Moghrabi, I.A.: Minimum curvature multistep quasi-Newton methods. Comput. Math. Appl. **31**(4–5), 179–186 (1996)
53. Ford, J.A., Moghrabi, I.A.: Using function-values in multi-step quasi-Newton methods. J. Comput. Appl. Math. **66**(1–2), 201–211 (1996)
54. Ford, J.A., Narushima, Y., Yabe, H.: Multi-step nonlinear conjugate gradient methods for unconstrained minimization. Comput. Optim. Appl. **40**(2), 191–216 (2008)
55. Forsythe, G.E.: On the asymptotic directions of the s-dimensional optimum gradient method. Numer. Math. **11**(1), 57–76 (1968)
56. Gilbert, J.C., Nocedal, J.: Global convergence properties of conjugate gradient methods for optimization. SIAM J. Optim. **2**(1), 21–42 (1992)
57. Grippo, L., Lampariello, F., Lucidi, S.: A nonmonotone line search technique for Newton's method. SIAM J. Numer. Anal. **23**(4), 707–716 (1986)
58. Guo, Q., Liu, J.G., Wang, D.H.: A modified BFGS method and its superlinear convergence in nonconvex minimization with general line search rule. J. Appl. Math. Comput. **28**(1–2), 435–446 (2008)
59. Hager, W.W., Zhang, H.: A new conjugate gradient method with guaranteed descent and an efficient line search. SIAM J. Optim. **16**(1), 170–192 (2005)
60. Hager, W.W., Zhang, H.: Algorithm 851: CG-Descent, a conjugate gradient method with guaranteed descent. ACM Trans. Math. Softw. **32**(1), 113–137 (2006)
61. Hager, W.W., Zhang, H.: A survey of nonlinear conjugate gradient methods. Pac. J. Optim. **2**(1), 35–58 (2006)
62. Hestenes, M.R., Stiefel, E.: Methods of conjugate gradients for solving linear systems. J. Research Nat. Bur. Standards **49**(6), 409–436 (1952)
63. Hu, Y.F., Storey, C.: Global convergence result for conjugate gradient methods. J. Optim. Theory Appl. **71**(2), 399–405 (1991)
64. Li, D.H., Fukushima, M.: A modified BFGS method and its global convergence in nonconvex minimization. J. Comput. Appl. Math. **129**(1–2), 15–35 (2001)
65. Li, D.H., Fukushima, M.: On the global convergence of the BFGS method for nonconvex unconstrained optimization problems. SIAM J. Optim. **11**(4), 1054–1064 (2001)
66. Li, G., Tang, C., Wei, Z.: New conjugacy condition and related new conjugate gradient methods for unconstrained optimization. J. Comput. Appl. Math. **202**(2), 523–539 (2007)
67. Liu, D.C., Nocedal, J.: On the limited memory BFGS method for large-scale optimization. Math. Program. **45**(3, Ser. B), 503–528 (1989)
68. Liu, G.H., Han, J.Y., Yin, H.X.: Global convergence of the Fletcher-Reeves algorithm with an inexact line search. Appl. Math. J. Chin. Univ. Ser. B **10**(1), 75–82 (1995)

69. Liu, Y., Storey, C.: Efficient generalized conjugate gradient algorithms. I. Theory. J. Optim. Theory Appl. **69**(1), 129–137 (1991)
70. Nazareth, J.L.: A conjugate direction algorithm without line searches. J. Optim. Theory Appl. **23**(3), 373–387 (1977)
71. Nazareth, J.L.: A relationship between the BFGS and conjugate gradient algorithms and its implications for the new algorithms. SIAM J. Numer. Anal. **16**(5), 794–800 (1979)
72. Nazareth, J.L.: Conjugate gradient methods less dependent on conjugacy. SIAM Rev. **28**(4), 501–511 (1986)
73. Nocedal, J.: Updating quasi-Newton matrices with limited storage. Math. Comput. **35**(151), 773–782 (1980)
74. Nocedal, J., Wright, S.J.: Numerical Optimization. Springer, New York (2006)
75. Oren, S.S.: Self-scaling variable metric (SSVM) algorithms. II. Implementation and experiments. Manage. Sci. **20**(5), 863–874 (1974)
76. S.S. Oren and D.G. Luenberger. Self-scaling variable metric (SSVM) algorithms. I. Criteria and sufficient conditions for scaling a class of algorithms. Manage. Sci. **20**(5), 845–862 (1973/1974)
77. Oren, S.S., Spedicato, E.: Optimal conditioning of self-scaling variable metric algorithms. Math. Program. **10**(1), 70–90 (1976)
78. Perry, A.: A modified conjugate gradient algorithm. Oper. Res. **26**(6), 1073–1078 (1976)
79. Polak, E., Ribière, G.: Note sur la convergence de méthodes de directions conjuguées. Rev. Française Informat. Recherche Opérationnelle **3**(16), 35–43 (1969)
80. Polyak, B.T.: The conjugate gradient method in extreme problems. USSR Comput. Math. Math. Phys. **9**(4), 94–112 (1969)
81. Powell, M.J.D.: Restart procedures for the conjugate gradient method. Math. Program. **12**(2), 241–254 (1977)
82. Powell, M.J.D.: Nonconvex minimization calculations and the conjugate gradient method. In: Griffiths, D.F. (ed.) Numerical Analysis (Dundee, 1983), Lecture Notes in Mathematics, vol. 1066, pp. 122–141. Springer, Berlin (1984)
83. Raydan, M.: The Barzilai and Borwein gradient method for the large-scale unconstrained minimization problem. SIAM J. Optim. **7**(1), 26–33 (1997)
84. Shanno, D.F.: Conjugate gradient methods with inexact searches. Math. Oper. Res. **3**(3), 244–256 (1978)
85. Sugiki, K., Narushima, Y., Yabe, H.: Globally convergent three-term conjugate gradient methods that use secant conditions and generate descent search directions for unconstrained optimization. J. Optim. Theory Appl. **153**(3), 733–757 (2012)
86. Sun, W., Yuan, Y.X.: Optimization Theory and Methods: Nonlinear Programming. Springer, New York (2006)
87. Touati-Ahmed, D., Storey, C.: Efficient hybrid conjugate gradient techniques. J. Optim. Theory Appl. **64**(2), 379–397 (1990)
88. Watkins, D.S.: Fundamentals of Matrix Computations. Wiley, New York (2002)
89. Wei, Z., Li, G., Qi, L.: New quasi-Newton methods for unconstrained optimization problems. Appl. Math. Comput. **175**(2), 1156–1188 (2006)
90. Wei, Z., Yu, G., Yuan, G., Lian, Z.: The superlinear convergence of a modified BFGS-type method for unconstrained optimization. Comput. Optim. Appl. **29**(3), 315–332 (2004)
91. Wolfe, P.: Convergence conditions for ascent methods. SIAM Rev. **11**(2), 226–235 (1969)
92. Wolfe, P.: Convergence conditions for ascent methods. II. Some corrections. SIAM Rev. **13**(2), 185–188 (1971)
93. Xu, C., Zhang, J.Z.: A survey of quasi-Newton equations and quasi-Newton methods for optimization. Ann. Oper. Res. **103**(1–4), 213–234 (2001)
94. Yabe, H., Takano, M.: Global convergence properties of nonlinear conjugate gradient methods with modified secant condition. Comput. Optim. Appl. **28**(2), 203–225 (2004)
95. Yang, X., Luo, Z., Dai, X.: A global convergence of LS-CD hybrid conjugate gradient method. Adv. Numer. Anal. Article ID 517452 (2013)

96. Yu, G., Guan, L., Li, G.: Global convergence of modified Polak-Ribière-Polyak conjugate gradient methods with sufficient descent property. J. Ind. Manage. Optim. **4**(3), 565–579 (2008)
97. Yuan, G.L.: Modified nonlinear conjugate gradient methods with sufficient descent property for large-scale optimization problems. Optim. Lett. **3**(1), 11–21 (2009)
98. Yuan, Y.X.: A modified BFGS algorithm for unconstrained optimization. IMA J. Numer. Anal. **11**(3), 325–332 (1991)
99. Yuan, Y.X., Byrd, R.H.: Non-quasi-Newton updates for unconstrained optimization. J. Comput. Math. **13**(2), 95–107 (1995)
100. Zhang, J.Z., Deng, N.Y., Chen, L.H.: New quasi-Newton equation and related methods for unconstrained optimization. J. Optim. Theory Appl. **102**(1), 147–167 (1999)
101. Zhang, L., Zhou, W., Li, D.H.: A descent modified Polak-Ribière-Polyak conjugate gradient method and its global convergence. IMA J. Numer. Anal. **26**(4), 629–640 (2006)
102. Zhang, L., Zhou, W., Li, D.H.: Global convergence of a modified Fletcher-Reeves conjugate gradient method with Armijo-type line search. Numer. Math. **104**(4), 561–572 (2006)
103. Zhang, L., Zhou, W., Li, D.H.: Some descent three-term conjugate gradient methods and their global convergence. Optim. Methods Softw. **22**(4), 697–711 (2007)
104. Zhou, W., Zhang, L.: A nonlinear conjugate gradient method based on the MBFGS secant condition. Optim. Methods Softw. **21**(5), 707–714 (2006)
105. Zoutendijk, G.: Nonlinear programming computational methods. In: Abadie, J. (ed.) Integer and Nonlinear Programming, pp. 37–86. North-Holland, Amsterdam (1970)

Author Biography

Saman Babaie-Kafaki is an Associate Professor of Mathematics in Semnan University, Iran. He received his B.Sc. degree in Applied Mathematics from Mazandaran University, Iran, in 2003, and his M.Sc. and Ph.D. degrees in Applied Mathematics from Sharif University of Technology, Iran, in 2005 and 2010, respectively, under the supervision of Professor Nezam Mahdavi-Amiri. His research interests include numerical optimization, matrix computations and heuristic algorithms.

Numerical Methods for Large-Scale Nonsmooth Optimization

Napsu Karmitsa

Abstract Nonsmooth optimization refers to the general problem of minimizing (or maximizing) functions that are typically not differentiable at their minimizers (maximizers). NSO problems are encountered in many application areas: for instance, in economics, mechanics, engineering, control theory, optimal shape design, machine learning, and data mining including cluster analysis and classification. Most of these problems are large-scale. In addition, constantly increasing database sizes, for example in clustering and classification problems, add even more challenge in solving these problems. NSO problems are in general difficult to solve even when the size of the problem is small and problem is convex. In this chapter we recall two numerical methods for solving large-scale nonconvex NSO problems. Namely, the limited memory bundle algorithm (LMBM) and the diagonal bundle method (D-BUNDLE). We also recall the convergence properties of these algorithms. The numerical experiments have been made using problems with up to million variables, which indicates the usability of the methods also in real world applications with big data-sets.

Keywords Nondifferentiable optimization · Large-scale optimization · Bundle methods · Limited memory methods · Diagonal variable metric updates

1 Introduction

Nonsmooth optimization (NSO) refers to the general problem of minimizing (or maximizing) functions that are typically not differentiable at their minimizers (maximizers) (see e.g., [6]). NSO problems are encountered in many application areas: for instance, in economics [37], mechanics [35], engineering [34], control theory [13], optimal shape design [19], machine learning [22], and data mining [3, 9] including cluster analysis [14] and classification [1, 2, 7, 11]. Most of these problems are large-scale. In addition, constantly increasing database sizes, for example in clustering problems, add even more challenge in solving these problems.

N. Karmitsa (✉)
Department of Mathematics and Statistics, University of Turku, 20014 Turku, Finland
e-mail: napsu@karmitsa.fi

A. Emrouznejad (ed.), *Big Data Optimization: Recent Developments and Challenges*, Studies in Big Data 18, DOI 10.1007/978-3-319-30265-2_18

In this chapter, we are considering of solving the problem of the form

$$\begin{cases} \text{minimize} & f(x) \\ \text{subject to} & x \in \mathbb{R}^n, \end{cases} \tag{P}$$

where the objective function $f : \mathbb{R}^n \to \mathbb{R}$ is supposed to be locally Lipschitz continuous (LLC) and the number of variables n is supposed to be large. Note that no differentiability or convexity assumptions for problem (P) are made.

NSO problems are in general difficult to solve even when the size of the problem is small. In addition, besides problematics of nonsmoothness and size of the problem, nonconvexity adds another challenge; NSO is traditionally based on convex analysis and most solution methods rely strongly on the convexity of the problem.

In this chapter we recall two numerical methods for solving large-scale nonconvex NSO problems. Namely, the limited memory bundle algorithm (LMBM) [16–18, 25] and the diagonal bundle method (D-BUNDLE) [23]. The LMBM is a hybrid of the variable metric bundle methods [29, 38] and the limited memory variable metric methods (see e.g. [10]), where the first ones have been developed for small- and medium-scale NSO and the latter ones for smooth large-scale optimization. In its turn, the basic idea of the D-BUNDLE is to combine the LMBM with sparse matrix updating. The aim of doing so is to obtain a method for solving problem (P) with large numbers of variables, where the Hessian of the problem—if it exists—is sparse. The numerical experiments have been made using problems with up to million variables.

This chapter is organized as follows. In Sect. 2 we introduce our notation and recall some basic definitions and results from nonsmooth analysis. In Sects. 3 and 4, we discuss briefly the basic ideas of the LMBM and the D-BUNDLE, respectively, and remind their convergence properties. The results of the numerical experiments are presented and discussed in Sect. 5 and, finally, Sect. 6 concludes the chapter.

2 Notations and Background

We denote by $\|\cdot\|$ the Euclidean norm in $\mathbb{R}^n$ and by $a^T b$ the inner product of vectors a and b. In addition, we denote by $\|A\|_F$ the Frobenius norm of matrix $A \in \mathbb{R}^{n \times n}$. That is, we define

$$\|A\|_F = \sqrt{\sum_{i=1}^{n} \sum_{j=1}^{n} A_{i,j}^2}.$$

The *subdifferential* $\partial f(x)$ [12] of a LLC function $f : \mathbb{R}^n \to \mathbb{R}$ at any point $x \in \mathbb{R}^n$ is given by

$$\partial f(x) = \operatorname{conv}\{\lim_{i \to \infty} \nabla f(x_i) \mid x_i \to x \text{ and } \nabla f(x_i) \text{ exists}\},$$

where "conv" denotes the convex hull of a set. A vector $\xi \in \partial f(x)$ is called a *subgradient*.

The point $x^* \in \mathbb{R}^n$ is called *stationary* if $0 \in \partial f(x^*)$. Stationarity is a necessary condition for local optimality and, in the convex case, it is also sufficient for global optimality. An optimization method is said to be *globally convergent* if starting from any arbitrary point x_1 it generates a sequence $\{x_k\}$ that converges to a stationary point x^*, that is, $\{x_k\} \to x^*$ whenever $k \to \infty$.

3 Limited Memory Bundle Method

In this section, we describe the limited memory bundle algorithm (LMBM) by Karmitsa (née Haarala) et al. [16–18, 25] for solving general, possibly nonconvex, large-scale NSO problems. We assume that at every point x we can evaluate the value of the objective function $f(x)$ and one arbitrary subgradient ξ from the subdifferential $\partial f(x)$.

3.1 Method

As already said in the introduction, the LMBM is a hybrid of the variable metric bundle methods (VMBM) [29, 38] and the limited memory variable metric methods (see e.g. [10]), where the first ones have been developed for small- and medium-scale NSO and the latter ones for smooth large-scale optimization. The LMBM combines the ideas of the VMBMwith the search direction calculation of limited memory approach. Therefore, the time-consuming quadratic direction finding problem appearing in the standard bundle methods (see e.g. [21, 26, 31]) does not need to be solved, nor the number of stored subgradients needs to grow with the dimension of the problem. Furthermore, the method uses only a few vectors to represent the variable metric approximation of the Hessian matrix and, thus, it avoids storing and manipulating large matrices as is the case in the VMBM. These improvements make the LMBM suitable for large-scale optimization. Namely, the number of operations needed for the calculation of the search direction and the aggregate values is only linearly dependent on the number of variables while, for example, with the original VMBMthis dependence is quadratic.

Search Direction.
So, the LMBM exploits the ideas of the variable metric bundle methods, namely the utilization of null steps, simple aggregation of subgradients, and the subgradient locality measures, but the search direction d_k is calculated using the limited memory approach. That is,

$$d_k = -D^k \tilde{\xi}_k,$$

where $\tilde{\xi}_k$ is (an aggregate) subgradient and D^k is the limited memory variable metric update that, in the smooth case, represents the approximation of the inverse of the Hessian matrix. The role of matrix D^k is to accumulate information about previous subgradients. Note, however, that the matrix D^k is not formed explicitly but the search direction d_k is calculated using the limited memory approach (to be described later).

Line Search.
In NSO the search direction is not necessarily a descent one. In order to determine a new step into the search direction d_k, the LMBM uses the so-called *line search procedure* (see [18, 38]): a new iteration point x_{k+1} and a new auxiliary point y_{k+1} are produced such that

$$
\begin{aligned}
x_{k+1} &= x_k + t_L^k d_k \qquad \text{and} \\
y_{k+1} &= x_k + t_R^k d_k, \qquad \text{for } k \geq 1
\end{aligned}
\tag{1}
$$

with $y_1 = x_1$, where $t_R^k \in (0, t_{max}]$ and $t_L^k \in [0, t_R^k]$ are step sizes, and $t_{max} > 1$ is the upper bound for the step size. A necessary condition for a *serious step* to be taken is to have

$$
t_R^k = t_L^k > 0 \qquad \text{and} \qquad f(y_{k+1}) \leq f(x_k) - \varepsilon_L^k t_R^k w_k, \tag{2}
$$

where $\varepsilon_L^k \in (0, 1/2)$ is a line search parameter and $w_k > 0$ represents the desirable amount of descent of f at x_k. If the condition (2) is satisfied, we set $x_{k+1} = y_{k+1}$ and a serious step is taken.

On the other hand, a *null step* is taken if

$$
t_R^k > t_L^k = 0 \qquad \text{and} \qquad -\beta_{k+1} + d_k^T \xi_{k+1} \geq -\varepsilon_R^k w_k, \tag{3}
$$

where $\varepsilon_R^k \in (\varepsilon_L^k, 1/2)$ is a line search parameter, $\xi_{k+1} \in \partial f(y_{k+1})$, and β_{k+1} is the subgradient locality measure [27, 33] similar to standard bundle methods. That is,

$$
\beta_{k+1} = \max\{|f(x_k) - f(y_{k+1}) + (y_{k+1} - x_k)^T \xi_{k+1})|, \gamma \|y_{k+1} - x_k\|^2\}. \tag{4}
$$

Here $\gamma \geq 0$ is a distance measure parameter supplied by the user and it can be set to zero when f is convex. Using null steps gives more information about the nonsmooth objective function in the case the current search direction is "not good enough", that is, the descent condition (2) is not satisfied. In the case of a null step, we set $x_{k+1} = x_k$ but information about the objective function is increased because we store the auxiliary point y_{k+1} and the corresponding auxiliary subgradient $\xi_{k+1} \in \partial f(y_{k+1})$.

Under some semismoothness assumptions the line search procedure used with the LMBM is guaranteed to find the step sizes t_L^k and t_R^k such that exactly one of the two possibilities—a serious step or a null step—occurs [38].

Aggregation.
The LMBM uses the original subgradient ξ_k after the serious step and the aggregate subgradient $\tilde{\xi}_k$ after the null step for direction finding (i.e. we set $\tilde{\xi}_k = \xi_k$ if the previous step was a serious step). The *aggregation procedure* used in the LMBM utilizes only three subgradients and two locality measures. The procedure is carried out by determining multipliers λ_i^k satisfying $\lambda_i^k \geq 0$ for all $i \in \{1,2,3\}$, and $\sum_{i=1}^3 \lambda_i^k = 1$ that minimize the function

$$\begin{aligned}\varphi(\lambda_1, \lambda_2, \lambda_3) = [\lambda_1 \xi_m + \lambda_2 \xi_{k+1} + \lambda_3 \tilde{\xi}_k]^T D^k [\lambda_1 \xi_m + \lambda_2 \xi_{k+1} + \lambda_3 \tilde{\xi}_k] \\ + 2(\lambda_2 \beta_{k+1} + \lambda_3 \tilde{\beta}_k).\end{aligned} \tag{5}$$

Here $\xi_m \in \partial f(x_k)$ is the current subgradient (m denotes the index of the iteration after the latest serious step, i.e. $x_k = x_m$), $\xi_{k+1} \in \partial f(y_{k+1})$ is the auxiliary subgradient, and $\tilde{\xi}_k$ is the current aggregate subgradient from the previous iteration ($\tilde{\xi}_1 = \xi_1$). In addition, β_{k+1} is the current subgradient locality measure and $\tilde{\beta}_k$ is the current aggregate subgradient locality measure ($\tilde{\beta}_1 = 0$). The optimal values $\lambda_i^k, i \in \{1,2,3\}$ can be calculated by using simple formulae (see [38]).

The resulting aggregate subgradient $\tilde{\xi}_{k+1}$ and aggregate subgradient locality measure $\tilde{\beta}_{k+1}$ are computed by the formulae

$$\tilde{\xi}_{k+1} = \lambda_1^k \xi_m + \lambda_2^k \xi_{k+1} + \lambda_3^k \tilde{\xi}_k \quad \text{and} \quad \tilde{\beta}_{k+1} = \lambda_2^k \beta_{k+1} + \lambda_3^k \tilde{\beta}_k. \tag{6}$$

Due to this simple aggregation procedure only one trial point y_{k+1} and the corresponding subgradient $\xi_{k+1} \in \partial f(y_{k+1})$ need to be stored.

The aggregation procedure gives us a possibility to retain the global convergence without solving the quite complicated quadratic direction finding problem (see e.g. [6]) appearing in standard bundle methods. Note that the aggregate values are computed only if the last step was a null step. Otherwise, we set $\tilde{\xi}_{k+1} = \xi_{k+1}$ and $\tilde{\beta}_{k+1} = 0$.

Matrix Updating.
In the LMBM both the limited memory BFGS (L-BFGS) and the limited memory SR1 (L-SR1) update formulae [10] are used in calculations of the search direction and the aggregate values. The idea of limited memory matrix updating is that instead of storing large $n \times n$ matrices D^k, one stores a certain (usually small) number of vectors $s_k = y_{k+1} - x_k$ and $u_k = \xi_{k+1} - \xi_m$ obtained at the previous iterations of the algorithm, and uses these vectors to implicitly define the variable metric matrices. Note that, due to fact that the gradient does not need to exist for nonsmooth objective, these correction vectors are computed using subgradients. Moreover, due to usage of null steps we may have $x_{k+1} = x_k$ and thus, we use here the auxiliary point y_{k+1} instead of x_{k+1}.

Let us denote by $\hat{m}_c$ the user-specified maximum number of stored correction vectors ($3 \leq \hat{m}_c$) and by $\hat{m}_k = \min\{k-1, \hat{m}_c\}$ the current number of stored correction vectors. Then the $n \times \hat{m}_k$ dimensional correction matrices S_k and U_k are defined by

$$S_k = \left[s_{k-\hat{m}_k} \; \dots \; s_{k-1}\right] \qquad \text{and} \qquad (7)$$
$$U_k = \left[u_{k-\hat{m}_k} \; \dots \; u_{k-1}\right].$$

The inverse L-BFGS update is defined by the formula

$$D^k = \vartheta_k I + \left[S_k \; \vartheta_k U_k\right] \begin{bmatrix} (R_k^{-1})^T(C_k + \vartheta_k U_k^T U_k)R_k^{-1} & -(R_k^{-1})^T \\ -R_k^{-1} & 0 \end{bmatrix} \begin{bmatrix} S_k^T \\ \vartheta_k U_k^T \end{bmatrix},$$

where R_k is an upper triangular matrix of order $\hat{m}_k$ given by the form

$$(R_k)_{ij} = \begin{cases} (s_{k-\hat{m}_k-1+i})^T(u_{k-\hat{m}_k-1+j}), & \text{if } i \le j \\ 0, & \text{otherwise,} \end{cases}$$

C_k is a diagonal matrix of order $\hat{m}_k$ such that

$$C_k = \operatorname{diag}[s_{k-\hat{m}_k}^T u_{k-\hat{m}_k}, \dots, s_{k-1}^T u_{k-1}],$$

and ϑ_k is a positive scaling parameter.

In addition, the inverse L-SR1 update is defined by

$$D^k = \vartheta_k I - (\vartheta_k U_k - S_k)(\vartheta_k U_k^T U_k - R_k - R_k^T + C_k)^{-1}(\vartheta_k U_k - S_k)^T.$$

In the case of a null step, the LMBM uses the L-SR1 update formula, since this formula allows to preserve the boundedness and some other properties of generated matrices which guarantee the global convergence of the method. Otherwise, since these properties are not required after a serious step, the more efficient L-BFGS update is employed. In the LMBM, the individual updates that would violate positive definiteness are skipped (for more details, see [16–18]).

Stopping Criterion.
For smooth functions, a necessary condition for a local minimum is that the gradient has to be zero and by continuity it becomes small when we are close to an optimal point. This is no longer true when we replace the gradient by an arbitrary subgradient. Although, the aggregate subgradient $\tilde{\xi}_k$ is quite a useful approximation to the gradient, the direct test $\|\tilde{\xi}_k\| < \varepsilon_S$, for some $\varepsilon_S > 0$, is too uncertain as a stopping criterion, if the current piecewise linear approximation of the objective function is too rough. Therefore, we use the term $\tilde{\xi}_k^T D^k \tilde{\xi}_k = -\tilde{\xi}_k^T d_k$ and the aggregate subgradient locality measure $\tilde{\beta}_k$ to improve the accuracy of $\|\tilde{\xi}_k\|$ (see, e.g., [31]). Hence, the stopping parameter w_k at iteration k is defined by

$$w_k = -\tilde{\xi}_k^T d_k + 2\tilde{\beta}_k \qquad (8)$$

and the algorithm stops if $w_k \leq \varepsilon_S$ for some user specified $\varepsilon_S > 0$. The parameter w_k is also used during the line search procedure to represent the desirable amount of descent.

Algorithm.

The pseudo-code of the LMBM is the following:

```
PROGRAM LMBM
  INITIALIZE x_1 ∈ R^n, ξ_1 ∈ ∂f(x_1), m̂_c ≥ 3, and ε_S > 0;
  Set k = 1, m = 1, d_1 = −ξ_1, ξ̃_1 = ξ_1, and β̃_1 = 0;
  WHILE the termination condition w_k ≤ ε_S is not met
    Find step sizes t_L^k and t_R^k, and the subgradient locality
      measure β_{k+1};
    Set x_{k+1} = x_k + t_L^k d_k and y_{k+1} = x_k + t_R^k d_k;
    Evaluate f(x_{k+1}) and ξ_{k+1} ∈ ∂f(y_{k+1});
    Store the new correction vectors s_k = y_{k+1} − x_k and
      u_k = ξ_{k+1} − ξ_m;
    Set m̂_k = min{k, m̂_c};
    IF t_L^k > 0 THEN
      SERIOUS STEP
        Compute the search direction d_{k+1} using ξ_{k+1} and L-BFGS
          update with m̂_k most recent correction pairs;
        Set m = k + 1 and β̃_{k+1} = 0;
      END SERIOUS STEP
    ELSE
      NULL STEP
        Compute the aggregate values
          ξ̃_{k+1} = λ_1^k ξ_m + λ_2^k ξ_{k+1} + λ_3^k ξ̃_k      and
          β̃_{k+1} = λ_2^k β_{k+1} + λ_3^k β̃_k;
        Compute the search direction d_{k+1} using ξ̃_{k+1} and L-SR1
          update with m̂_k most recent correction pairs;
      END NULL STEP
    END IF
    Set k = k + 1;
  END WHILE
  RETURN final solution x_k;
END PROGRAM LMBM
```

3.2 Global Convergence

We now recall the convergence properties of the LMBM algorithm. But first, we give the assumptions needed.

Assumption 1 The objective function $f : \mathbb{R}^n \to \mathbb{R}$ is LLC,

Assumption 2 The objective function $f : \mathbb{R}^n \to \mathbb{R}$ is upper semismooth (see e.g. [8]),

Assumption 3 The level set $\{ x \in \mathbb{R}^n \mid f(x) \leq f(x_1) \}$ is bounded for every starting point $x_1 \in \mathbb{R}^n$.

Lemma 1 *Each execution of the line search procedure is finite.*

Proof See the proof of Lemma 2.2 in [38]. □

Theorem 1 *If the* LMBM *algorithm terminates after a finite number of iterations, say at iteration k, then the point x_k is a stationary point of f.*

Proof See the proof of Theorem 4 in [18]. □

Theorem 2 *Every accumulation point $\bar{x}$ generated by the* LMBM *algorithm is a stationary point of problem (P).*

Proof See the proof of Theorem 9 in [18]. □

Remark 1 If we choose $\varepsilon_S > 0$, the LMBM algorithm terminates in a finite number of steps.

4 Diagonal Bundle Method

The classical variable metric techniques for nonlinear optimization (see, e.g. [15]) construct a dense $n \times n$-matrix that approximates the Hessian of the function. These techniques require to store and manipulate this dense matrix, which in large-scale setting becomes unmanageable. In the limited memory variable metric methods (see, e.g. [10, 36]) the storage of this large matrix can be avoided, but still the formed approximation of the Hessian is dense. This is also true for the LMBM described in the previous section. Nevertheless, in many large-scale problems the real Hessian (if it exists) is sparse. In this section, we describe the diagonal bundle method (D-BUNDLE) by Karmitsa [23] for sparse nonconvex NSO.

4.1 Method

The idea of the D-BUNDLE method is to combine the LMBM with sparse matrix updating. As with the LMBM we assume that the objective function $f : \mathbb{R}^n \to \mathbb{R}$ is LLC and we can compute $f(x)$ and $\xi \in \partial f(x)$ at every $x \in \mathbb{R}^n$.

Matrix Updating and Direction Finding.
Similarly to the LMBM, the D-BUNDLE uses at most $\hat{m}_c$ most recent correction vectors to compute updates for matrices. These vectors and the corrections matrices are defined as in (7). The obvious difference between the D-BUNDLE and the LMBM is that with the D-BUNDLE we use the diagonal update formula to compute

the *diagonal variable metric update*. Although it would be possible to use real sparsity pattern of the Hessian, the diagonal update formula introduced in [20] is used, since for this formula it is easy to check the positive definiteness of generated matrices. Moreover, using diagonal update matrix requires minimum amount of storage space and computations. Particularly, the approximation of the Hessian B^{k+1} is chosen to be a diagonal matrix and the check of positive definiteness is included as a constraint to problem. Thus, the update matrix B^{k+1} is defined by

$$\begin{cases} \text{minimize} & \|B^{k+1}S_k - U_k\|_F^2 \\ \text{subject to} & B_{i,j}^{k+1} = 0 \text{ for } i \neq j \\ & B_{i,i}^{k+1} \geq \varepsilon \text{ for } i = 1, 2, \ldots, n \text{ and } \varepsilon > 0. \end{cases} \tag{9}$$

This minimization problem has a solution

$$B_{i,i}^{k+1} = \begin{cases} b_i/Q_{i,i}, & \text{if } b_i/Q_{i,i} > \varepsilon \\ \varepsilon, & \text{otherwise,} \end{cases}$$

where $b = 2\sum_{i=k-\hat{m}_k}^{k-1} \text{diag}(s_i)u_i$ and $Q_{i,i} = 2\sum_{i=k-\hat{m}_k}^{k-1}[\text{diag}(s_i)]^2$, and $\text{diag}(v)$, for $v \in \mathbb{R}^n$, is a diagonal matrix such that $\text{diag}(v)_{i,i} = v_i$.

Now, the search direction is computed by the formula

$$d_k = -D^k\tilde{\xi}_k, \tag{10}$$

where $\tilde{\xi}_k$ is (an aggregate) subgradient and D^k represents the diagonal update matrix such that $D^k = (B^k)^{-1}$ in (9). To ensure the global convergence of the D-BUNDLE, we have to assume that all the matrices D^k are bounded. Due to the diagonal update formula this assumption is trivially satisfied. In addition, the condition

$$\tilde{\xi}_k^T D^k \tilde{\xi}_k \leq \tilde{\xi}_k^T D^{k-1} \tilde{\xi}_k \tag{11}$$

has to be satisfied each time there occurs more than one consecutive null step. In the D-BUNDLE this is guaranteed simply by *skipping the updates*. That is, after a null step we set $D^{k+1} = D^k$, but the new aggregate values are computed.

Aggregation, Line Search and Stopping Criterion

The D-BUNDLE uses the aggregation procedure similar to the LMBM to guarantee the convergence of the method and to avoid unbounded storage, that is, the convex combination of at most three subgradients is used to form a new aggregate subgradient $\tilde{\xi}_{k+1}$ and a new aggregate subgradient locality measure $\tilde{\beta}_{k+1}$ (cf. (5) and (6)). Of course, the diagonal update matrix D^k is used in Eq. (5) instead of limited memory update.

In addition, the D-BUNDLE uses the line search procedure similar to the LMBM (cf. (1)–(3)) to determine a new iteration and auxiliary points x_{k+1} and y_{k+1}. That is, the step sizes $t_R^k \in (0, t_{max}]$ and $t_L^k \in [0, t_R^k]$ with $t_{max} > 1$ are computed such that either condition (2) for serious steps or condition (3) for null steps is satisfied.

Finally, also the stopping criterion of the D-BUNDLE algorithm is similar to that of the LMBM (cf. (8)). Similarly to the LMBM, the parameter w_k is used also during the line search procedure to represent the desirable amount of descent (cf. (2) and (3)).

Algorithm.

The pseudo-code of D-BUNDLE is the following:

```
PROGRAM D-BUNDLE
  INITIALIZE x_1 ∈ R^n, ξ_1 ∈ ∂f(x_1), D^1 = I, m̂_c ≥ 1, and ε_S > 0;
  Set k = 1, m = 1, m̂_k = 0, d_1 = −ξ_1, ξ̃_1 = ξ_1, and β̃_1 = 0;
  WHILE the termination condition w_k ≤ ε_S is not met
    Find step sizes t_L^k and t_R^k, and the subgradient
      locality measure β_{k+1};
    Set x_{k+1} = x_k + t_L^k d_k and y_{k+1} = x_k + t_R^k d_k;
    Evaluate f(x_{k+1}) and ξ_{k+1} ∈ ∂f(y_{k+1});
    IF t_L^k > 0 THEN
      SERIOUS STEP
        Store the new correction vectors s_k = y_{k+1} − x_k
          and u_k = ξ_{k+1} − ξ_m;
        Set m̂_k = min{k, m̂_c};
        Compute the new diagonal matrix D^{k+1} = (B^{k+1})^{-1}
          using m̂_k most recent correction vectors;
        Compute the search direction d_{k+1} = −D^{k+1} ξ_{k+1};
        Set m = k + 1 and β̃_{k+1} = 0;
      END SERIOUS STEP
    ELSE
      NULL STEP
        Compute the aggregate values
          ξ̃_{k+1} = λ_1^k ξ_m + λ_2^k ξ_{k+1} + λ_3^k ξ̃_k    and
          β̃_{k+1} = λ_2^k β_{k+1} + λ_3^k β̃_k;
        Set D^{k+1} = D^k;
        Compute the search direction d_{k+1} = −D^{k+1} ξ̃_{k+1};
      END NULL STEP
    END IF
    Set k = k + 1;
  END WHILE
  RETURN final solution x_k;
END PROGRAM D-BUNDLE
```

4.2 Global Convergence

We now recall the convergence properties of D-BUNDLE-algorithm. The assumptions needed are the same as with LMBM, that is, Assumptions 1–3. Under these assumptions, Lemma 1 is also valid.

Theorem 3 *If the* D-BUNDLE *algorithm terminates after a finite number of iterations, say at iteration k, then the point* x_k *is a stationary point of* f*. On the other hand, any accumulation point of an infinite sequence of solutions generated by* D-BUNDLE *is a stationary point of* f*.*

Proof See the proof of Theorem 3.1 in [23]. □

Thus, similarly to the LMBM, the D-BUNDLE algorithm either terminates at a stationary point of the objective function f or generates an infinite sequence (x_k) for which accumulation points are stationary for f. Moreover, if we choose $\varepsilon_S > 0$, the D-BUNDLE method terminates in a finite number of steps.

5 Numerical Experiments

In this section we compare LMBM, D-BUNDLE and some other existing methods for NSO. The test set used in our experiments consists of extensions of classical academic nonsmooth minimization problems [17]. We have tested these problems up to million variables.

5.1 Solvers

The tested optimization codes with references to more detailed descriptions of the methods and their implementations are presented in Table 1.

A brief description of each software and the references from where the code can be downloaded are in order.

Table 1 Tested pieces of software

Software	Author(s)	Method	References
`PBNCGC`	Mäkelä	Proximal bundle	[30, 31]
`QSM`	Bagirov and Ganjehlou	Quasi-secant	[4, 5]
`LMBM`	Karmitsa	Limited memory bundle	[17, 18]
`D-Bundle`	Karmitsa	Diagonal bundle	[23]

PBNCGC is an implementation of the most frequently used bundle method in NSO; that is, the *proximal bundle method*. The code includes constraint handling (bound constraints, linear constraints, and nonlinear/nonsmooth constraints) and a possibility to optimize multiobjective problems. The quadratic direction finding problem characteristic for bundle methods is solved by the PLQDF1 subroutine implementing the dual projected gradient method proposed in [28].

PBNCGC can be used (free for academic purposes) via WWW-NIMBUS-system (http://nimbus.mit.jyu.fi/) [32]. Furthermore, the Fortran 77 source code is available for downloading from http://napsu.karmitsa.fi/proxbundle/.

QSM is a *quasi-secant solver* for nonsmooth possibly nonconvex minimization. Although originally developed for small- and medium-scale problems, QSM has shown to be very efficient also in large-scale setting [6, 24].

The user can employ either analytically calculated or approximated subgradients in his experiments (this can be done automatically by selecting one parameter). We have used analytically calculated subgradients here.

The Fortran 77 source code is available for downloading from http://napsu.karmitsa.fi/qsm/.

LMBM is an implementation of the LMBM. In our experiments, we used the adaptive version of the code with the initial number of stored correction pairs used to form the variable metric update equal to 7 and the maximum number of stored correction pairs equal to 15.

The Fortran 77 source code and the mex-driver (for MatLab users) are available for downloading from http://napsu.karmitsa.fi/lmbm/.

D-Bundle is an implementation of the D-BUNDLE. The Fortran 95 source code of D-Bundle is available for downloading from http://napsu.karmitsa.fi/dbundle/.

All of the algorithms except for D-Bundle were implemented in Fortran77 using double-precision arithmetic. The experiments were performed on an Intel® Core™ 2 CPU 1.80 GHz. To compile the codes, we used gfortran, the GNU Fortran compiler.

5.2 Test Problems and Parameters

As already said the test set used in our experiments consists of extensions of classical academic nonsmooth minimization problems from the literature. That is problems 1–10 first introduced in [17]. Problems 1–5 are convex while problems 6–10 are nonconvex. These problems can be formulated with any number of variables. We have used here 1000, 10 000, 100 000 and 1 000 000 variables.

To obtain comparable results the stopping parameters of the codes were tuned by the procedure similar to [6]. In addition to the usual stopping criteria of the solvers, we terminated the experiments if the elapsed CPU time exceeded half an hour for problems with 1000 variables, an hour with 10 000 variables, and 2 h with 100 000 and million variables.

We say that a solver finds the solution with respect to a tolerance $\varepsilon > 0$ if

$$\frac{f_{best} - f_{opt}}{1 + |f_{opt}|} \leq \varepsilon,$$

where f_{best} is a solution obtained with the solver and f_{opt} is the best known (or optimal) solution. We have *accepted the results* with respect to the tolerance $\varepsilon = 10^{-3}$. In addition, we say that the result is *inaccurate*, if a solver finds the solution with respect to a tolerance $\varepsilon = 10^{-2}$. Otherwise, we say that a solver *fails*.

For LMBM, PBNCGC and QSM the maximum size of the bundle was set to 100. With D-Bundle the natural choice for the bundle size is *two*. For all other parameters we have used the default settings of the codes. In addition, for D-Bundle the number of stored correction pairs was set to *three*.

5.3 *Results*

The results are summarized in Tables 2, 3, 4 and 5. We have compared the efficiency of the solvers both in terms of the computational time (*cpu*) and the number of function and subgradient evaluations (*nfg*, evaluations for short). We have used bold-face text to emphasize the best results.

The results for problems with 1000 and 10 000 variables reveal similar trends (see Tables 2 and 3): In both cases PBNCGC was clearly the most robust solver. In addition, PBNCGC usually used either the least or the most evaluations, making it very difficult to say if it is an efficient method or not. The robustnesses of D-Bundle LMBM and QSM were similar.

Table 2 Summary of the results with 1000 variables

P	PBNCGC	QSM	LMBM	D-bundle
	nfg/*cpu*	*nfg*/*cpu*	*nfg*/*cpu*	*nfg*/*cpu*
1	19 738/789.59	18 263/7.19	fail	**6 136/0.60**
2	**46/0.57**	4 242/87.67	fail	inacc/6.47
3	24 424/1 800.00	2 326/4.60	6 540/0.32	**242/0.01**
4	16 388/1 800.13	2 036/0.88	**558/0.37**	6 843/1.12
5	**56**/0.07	667/0.10	228/**0.04**	3 643/0.66
6	272/0.05	**254/0.03**	1 062/0.94	1 126/0.20
7	**76/0.22**	inacc/0.93	352/0.37	6 119/3.24
8	28 318/1 800.09	2 836/6.83	**1 230**/0.67	7 974/**0.44**
9	**98**/0.10	inacc/0.02	200/0.06	569/**0.05**
10	**398/6.73**	inacc/11.74	inacc/0.31	fail

Table 3 Summary of the results with 10 000 variables

P	PBNCGC	QSM	LMBM	D-bundle
	nfg/*cpu*	*nfg*/*cpu*	*nfg*/*cpu*	*nfg*/*cpu*
1	fail	**160 910**/1 000.62	fail	179 502/**214.53**
2	**54/56.06**	fail	fail	fail
3	13 026/3 600.21	1 868/5.20	**796**/7.45	2 161/**0.99**
4	5 960/3 600.83	2086/**7.32**	**882**/8.81	fail
5	**54/0.53**	903/1.41	784/3.00	3 370/5.47
6	**74/0.26**	inacc/0.35	10 106/143.32	10 102/9.09
7	**244/23.36**	inacc/15.39	inacc/9.68	fail
8	8 826/3 600.48	2 383/8.53	**1 194**/6.11	5 311/**1.98**
9	**284**/2.20	fail	396/3.20	575/**0.34**
10	**2472/811.77**	inacc/164.68	fail	inacc/2.45

Table 4 Summary of the results with 100 000 variables

P	QSM	LMBM	D-bundle
	nfg/*cpu*	*nfg*/*cpu*	*nfg*/*cpu*
1	fail	fail	fail
2	fail	fail	fail
3	1 687/46.39	1 718/184.16	**145/0.99**
4	2 137/**77.20**	**1 258**/147.37	fail
5	1 252/111.28	**802**/63.27	827/**15.44**
6	fail	fail	fail
7	inacc/254.81	fail	**3 152/552.43**
8	**1 237/26.09**	2 300/245.10	3 810/38.67
9	inacc/83.64	**994**/99.41	1 159/**15.29**
10	inacc/93.09	inacc/157.20	inacc/23.22

D-Bundle usually used more evaluations than LMBM. However, in terms of cpu-time it was comparable or—especially with larger problem—even better than LMBM. Furthermore, D-Bundle succeed in solving P1 which has shown to be very difficult to LMBM due its sparse structure.

A problem with 100 000 variables can be considered as an extremely large NSO problem. With the problems of this size, the solver PBNCGC could not be compiled due to the arithmetic overflow. Moreover, the other solvers succeed in solving at most five problems out of ten (see Table 4). Thus, no very far-reaching conclusions can be made. As before, the robustnesses of D-Bundle, LMBM, and QSM were similar and they also used approximately same amount of evaluations. However, D-Bundle usually used clearly less cpu-time than LMBM and QSM.

Table 5 Summary of the results with million variables

P	LMBM	D-bundle
	nfg/*cpu*	*nfg*/*cpu*
1	fail	fail
2	fail	fail
3	**1 698**/168.18	2 431/**155.81**
4	14 074/2 212.79	**7 120/1 397.11**
5	1 692/414.81	**371/70.72**
6	fail	fail
7	fail	**6 258/2 896.71**
8	**4 970**/924.28	6 872/**742.20**
9	3 702/869.06	**3 319/540.67**
10	fail	inacc/7201.68

With million variable QSM could not be compiled either. Furthermore, the solver LMBM with the bundle size equal to 100 could be compiled but not run: the procedure was killed by the host for using too many resources. Thus, for million variables we changed the size of the bundle to *two* also for LMBM.

Now, D-Bundle was the most robust solver. It succeed in solving six problems out of ten with the desired accuracy while LMBM succeed in solving five problems. In addition, with D-Bundle some of these failures were inaccurate result: with the relaxed tolerance $\varepsilon = 10^{-2}$ it succeed in solving seven problems while with LMBM the number of failures is still five even if the relaxed tolerance was used. With this very limited set of test problems we can say that D-Bundle was little bit better that LMBM with these extremely large problems.

The numerical experiments reported confirm that proximal bundle method is the most robust method tested. However, it cannot be used with very large problems and the efficiency of the algorithm is highly volatile. In addition, the quasi-secant method used was unable to solve extremely large-scale problems. Although not developed for large-scale problems, it solved the problems quite efficiently.

The LMBM and the D-BUNDLE are efficient for both convex (problems 1–5) and nonconvex (problems 6–10) NSO problems. We can conclude that the LMBM and the D-BUNDLE are a good alternative to existing nonsmooth optimization algorithms and for extremely large-scale problems they might well be the best choices available.

6 Conclusions

In this chapter we have recalled the basic ideas of two numerical methods for solving large-scale nonconvex NSO problems. Namely, the limited memory bundle algorithm (LMBM) and the diagonal bundle method (D-BUNDLE). The numerical experiments have been made using problems with up to million variables, which indicates the usability of the methods also in real world applications with big data-sets.

Indeed, the main challenges in clustering and classification are big sizes of data-sets, possible outliers and noise, and missing data. Using NSO approach has been noticed to prevent the latter ones. In addition, when using NSO the L_1—penalization techniques in possible feature selection are readily usable with classification problems. One topic of the future research will be the usage of the LMBM and the D-BUNDLE in solving real world clustering problems with large data-sets, that is, for example in clustering of biomedical data, with a specific focus on subtype discovery in Type 1 diabetes and cancer.

References

1. Astorino, A., Fuduli, A.: Nonsmooth optimization techniques for semi-supervised classification. IEEE Trans. Pattern Anal. Mach. Intell. **29**(12), 2135–2142 (2007)
2. Astorino, A., Fuduli, A., Gorgone, E.: Nonsmoothness in classification problems. Optim. Methods Softw. **23**(5), 675–688 (2008)
3. Äyrämö, S.: Knowledge Mining Using Robust Clustering. Ph.D. thesis, University of Jyväskylä, Department of Mathematical Information Technology (2006)
4. Bagirov, A.M., Ganjehlou, A.N.: A secant method for nonsmooth optimization. Submitted (2009)
5. Bagirov, A.M., Ganjehlou, A.N.: A quasisecant method for minimizing nonsmooth functions. Optim. Methods Softw. **25**(1), 3–18 (2010)
6. Bagirov, A.M., Karmitsa, N., Mäkelä, M.M.: Introduction to Nonsmooth Optimization: Theory, Practice and Software. Springer (2014)
7. Bergeron, C., Moore, G., Zaretzki, J., Breneman, C., Bennett, K.: Fast bundle algorithm for multiple instance learning. IEEE Trans. Pattern Anal. Mach. Intell. **34**(6), 1068–1079 (2012)
8. Bihain, A.: Optimization of upper semidifferentiable functions. J. Optim. Theory Appl. **4**, 545–568 (1984)
9. Bradley, P.S., Fayyad, U.M., Mangasarian, O.L.: Mathematical programming for data mining: formulations and challenges. INFORMS J. Comput. **11**, 217–238 (1999)
10. Byrd, R.H., Nocedal, J., Schnabel, R.B.: Representations of quasi-Newton matrices and their use in limited memory methods. Math. Program. **63**, 129–156 (1994)
11. Carrizosa, E., Romero Morales, D.: Supervised classification and mathematical optimization. Comput. Oper. Res. **40**(1), 150–165 (2013)
12. Clarke, F.H.: Optimization and Nonsmooth Analysis. Wiley-Interscience, New York (1983)
13. Clarke, F.H., Ledyaev, Y.S., Stern, R.J., Wolenski, P.R.: Nonsmooth Analysis and Control Theory. Springer, New York (1998)
14. Demyanov, V.F., Bagirov, A.M., Rubinov, A.M.: A method of truncated codifferential with application to some problems of cluster analysis. J. Global Optim. **23**(1), 63–80 (2002)
15. Fletcher, R.: Practical Methods of Optimization, 2nd edn. Wiley, Chichester (1987)
16. Haarala, M.: Large-scale nonsmooth optimization: variable metric bundle method with limited memory. Ph.D. thesis, University of Jyväskylä, Department of Mathematical Information Technology (2004)
17. Haarala, M., Miettinen, K., Mäkelä, M.M.: New limited memory bundle method for large-scale nonsmooth optimization. Optim. Methods Softw. **19**(6), 673–692 (2004)
18. Haarala, N., Miettinen, K., Mäkelä, M.M.: Globally convergent limited memory bundle method for large-scale nonsmooth optimization. Math. Program. **109**(1), 181–205 (2007)

19. Haslinger, J., Neittaanmäki, P.: Finite Element Approximation for Optimal Shape, Material and Topology Design, 2nd edn. Wiley, Chichester (1996)
20. Herskovits, J., Goulart, E.: Sparse quasi-Newton matrices for large scale nonlinear optimization. In: Proceedings of the 6th Word Congress on Structural and Multidisciplinary Optimization (2005)
21. Hiriart-Urruty, J.-B., Lemaréchal, C.: Convex Analysis and Minimization Algorithms II. Springer, Berlin (1993)
22. Kärkkäinen, T., Heikkola, E.: Robust formulations for training multilayer perceptrons. Neural Comput. **16**, 837–862 (2004)
23. Karmitsa, N.: Diagonal bundle method for nonsmooth sparse optimization. J. Optim. Theory Appl. **166**(3), 889–905 (2015). doi:10.1007/s10957-014-0666-8
24. Karmitsa, N., Bagirov, A., Mäkelä, M.M.: Comparing different nonsmooth optimization methods and software. Optim. Methods Softw. **27**(1), 131–153 (2012)
25. Karmitsa, N., Mäkelä, M.M., Ali, M.M.: Limited memory interior point bundle method for large inequality constrained nonsmooth minimization. Appl. Math.Comput. **198**(1), 382–400 (2008)
26. Kiwiel, K.C.: Methods of Descent for Nondifferentiable Optimization. Lecture Notes in Mathematics, vol. 1133. Springer, Berlin (1985)
27. Lemaréchal, C., Strodiot, J.-J., Bihain, A.: On a bundle algorithm for nonsmooth optimization. In: Mangasarian, O.L., Mayer, R.R., Robinson, S.M. (eds.) Nonlinear Programming, pp. 245–281. Academic Press, New York (1981)
28. Lukšan, L.: Dual method for solving a special problem of quadratic programming as a subproblem at linearly constrained nonlinear minmax approximation. Kybernetika **20**, 445–457 (1984)
29. Lukšan, L., Vlček, J.: Globally convergent variable metric method for convex nonsmooth unconstrained minimization. J. Optim. Theory Appl. **102**(3), 593–613 (1999)
30. Mäkelä, M.M.: Multiobjective proximal bundle method for nonconvex nonsmooth optimization: Fortran subroutine MPBNGC 2.0. Reports of the Department of Mathematical Information Technology, Series B. Scientific Computing, B. 13/2003 University of Jyväskylä, Jyväskylä (2003)
31. Mäkelä, M.M., Neittaanmäki, P.: Nonsmooth Optimization: Analysis and Algorithms with Applications to Optimal Control. World Scientific Publishing Co., Singapore (1992)
32. Miettinen, K., Mäkelä, M.M.: Synchronous approach in interactive multiobjective optimization. Eur. J. Oper. Res. **170**(3), 909–922 (2006)
33. Mifflin, R.: A modification and an extension of Lemaréchal's algorithm for nonsmooth minimization. Math. Program. Study **17**, 77–90 (1982)
34. Mistakidis, E.S., Stavroulakis, G.E.: Nonconvex Optimization in Mechanics. Heuristics and Engineering Applications by the F.E.M. Kluwert Academic Publishers, Dordrecht, Smooth and Nonsmooth Algorithms (1998)
35. Moreau, J.J., Panagiotopoulos, P.D., Strang, G. (eds.): Topics in Nonsmooth Mechanics. Birkhäuser Verlag, Basel (1988)
36. Nocedal, J.: Updating quasi-Newton matrices with limited storage. Math. Comput. **35**(151), 773–782 (1980)
37. Outrata, J., Kočvara, M., Zowe, J.: Nonsmooth Approach to Optimization Problems With Equilibrium Constraints: Theory, Applications and Numerical Results. Kluwert Academic Publisher, Dordrecht (1998)
38. Vlček, J., Lukšan, L.: Globally convergent variable metric method for nonconvex nondifferentiable unconstrained minimization. J. Optim. Theory Appl. **111**(2), 407–430 (2001)

Author Biography

Dr. Napsu Karmitsa has been an adjunct professor in applied mathematics at the Department of Mathematics and Statistics at the University of Turku, Finland, since 2011. She obtained her M.Sc. degree in organic chemistry (1998) and Ph.D. degree in scientific computing (2004) both from the University of Jyvaskyla, Finland. At the moment, she holds a position of an academy research fellow granted by the Academy of Finland. Karmitsa's research is focused on nonsmooth optimization (NSO) and analysis. Special emphasis is given to nonconvex, global and large-scale cases. She is also developing numerical methods for solving possible nonconvex and large-scale NSO problems. Karmitsa is one of the authors of the book "Introduction to Nonsmooth Optimization: Theory, Practice and Software" (Springer 2014) together with Professors Bagirov and Mäkelä. The book is the first easy-to-read book in NSO. In addition, Karmitsa's webpage http://napsu.karmitsa.fi/ is likely to be one of the leading source of NSO tools available online. There one can find, e.g., a short introduction to NSO, source codes for some NSO solvers as well as links to some others, and Solver-o-matic—an online decision tree for choosing a NSO solver.

Metaheuristics for Continuous Optimization of High-Dimensional Problems: State of the Art and Perspectives

Giuseppe A. Trunfio

Abstract The age of big data brings new opportunities in many relevant fields, as well as new research challenges. Among the latter, there is the need for more effective and efficient optimization techniques, able to address problems with hundreds, thousands, and even millions of continuous variables. Over the last decade, researchers have developed various improvements of existing metaheuristics for tacking high-dimensional optimization problems, such as hybridizations, local search and parameter adaptation. Another effective strategy is the cooperative coevolutionary approach, which performs a decomposition of the search space in order to obtain sub-problems of smaller size. Moreover, in some cases such powerful search algorithms have been used with high performance computing to address, within reasonable run times, very high-dimensional optimization problems. Nevertheless, despite the significant amount of research already carried out, there are still many open research issues and room for significant improvements. In order to provide a picture of the state of the art in the field of high-dimensional continuous optimization, this chapter describes the most successful algorithms presented in the recent literature, also outlining relevant trends and identifying possible future research directions.

Keywords Large scale global optimization · Evolutionary optimization · Differential evolution · Memetic algorithms · Cooperative coevolution

1 Introduction

Nowadays, Evolutionary Algorithms (EAs) for global continuous optimization are used with great success in many relevant applications characterized by lack of detailed information on the objective function, complex fitness landscapes and constraints on computing time.

However, among the new challenges that the research in the field has to face, there are those raised by real-world problems requiring the optimization of a high number

G.A. Trunfio (✉)
DADU, University of Sassari, Alghero, Italy
e-mail: trunfio@uniss.it

A. Emrouznejad (ed.), *Big Data Optimization: Recent Developments and Challenges*, Studies in Big Data 18, DOI 10.1007/978-3-319-30265-2_19

of variables. Also fostered by the recent advances in the field of big data, relevant examples of such optimization problems are becoming increasingly common, such as in clustering [18, 33], data analytics [12], optimization of networks [16], engineering optimization [10], gene regulatory networks [65] and simulation of complex systems [3, 4].

Unfortunately, standard EAs suffer from the so-called *curse of dimensionality*, that is, their performance deteriorates rapidly as the number of variables to be optimized increases. For the above reasons, the study of suitable approaches for dealing with the continuous optimization of high-dimensional problems has been recently recognized as a relevant field of research, which is often referred to as 'large scale global optimization' (LSGO).

Over the years, various approaches have been proposed to devise algorithms suitable to address LSGO problems. On the one hand, many researchers have proposed various improvements of existing effective metaheuristics. Among the most relevant enhancements is worth mentioning the hybridization of different evolutionary operators, the use of local search (LS) and several approaches for parameter adaptation. On the other hand, there is the cooperative coevolutionary (CC) approach, based on the divide-and-conquer principle, which performs a decomposition of the search space in order to obtain sub-problems of smaller size. Another important line of research consists of using the effective approaches mentioned above together with High Performance Computing (HPC). In this case, the objective is to address very high-dimensional optimization problems in reasonable computing times.

Many of the developed strategies have proven really effective in the optimization of problems with a large number of variables. However, despite the relevant amount of research already done in the field, there are still many important research issues and room for significant improvements.

The main aim of this chapter is to describe the most successful algorithms presented in the LSGO literature. Moreover, it outlines some relevant trends and identifies possible future research directions in the field.

First, the following section describes some relevant optimization algorithms based on Differential Evolution (DE) [56], which is one of the most common component of LSGO metaheuristics. Then, Sect. 3 describes the use of a successful strategy for dealing with LSGO problems, namely LS. Subsequently, Sect. 4 describes the CC approach, including the most relevant related research problems. The chapter ends with Sect. 5, where the most promising approaches and research lines are summarized.

2 Differential Evolution and Applications to LSGO

According to the recent literature, several algorithms that successfully addressed LSGO problems are based on DE [56]. In brief, DE is an EA for optimizing an objective function $f : \mathbb{R}^d \rightarrow \mathbb{R}$ in which a population of n_p real vectors $\mathbf{x} \in \mathbb{R}^d$ are evolved using mutation, recombination and selection of best individuals.

In DE, at each generation and for each individual $\mathbf{x}_i$, a mutant vector $\mathbf{y}_i$ is first produced using a mutation operator. The most typical strategies are as follows [56]:

$$\begin{aligned}
\text{DE/rand/1}: &\quad \mathbf{y}_i = \mathbf{x}_{r_1} + F \cdot (\mathbf{x}_{r_2} - \mathbf{x}_{r_3}) \\
\text{DE/best/1}: &\quad \mathbf{y}_i = \mathbf{x}_{(best)} + F \cdot (\mathbf{x}_{r_1} - \mathbf{x}_{r_2}) \\
\text{DE/current to best/1}: &\quad \mathbf{y}_i = \mathbf{x}_i + F \cdot (\mathbf{x}_{best} - \mathbf{x}_i) + F \cdot (\mathbf{x}_{r_1} - \mathbf{x}_{r_2}) \\
\text{DE/best/2}: &\quad \mathbf{y}_i = \mathbf{x}_{best} + F \cdot (\mathbf{x}_{r_1} - \mathbf{x}_{r_2}) + F \cdot (\mathbf{x}_{r_3} - \mathbf{x}_{r_4}) \\
\text{DE/rand/2}: &\quad \mathbf{y}_i = \mathbf{x}_{r_1} + F \cdot (\mathbf{x}_{r_2} - \mathbf{x}_{r_3}) + F \cdot (\mathbf{x}_{r_4} - \mathbf{x}_{r_5})
\end{aligned}$$

where: the indexes r_k, $k = 1, \dots, 5$, are random and mutually different integers in the interval $[1, n_p]$, with $r_k \neq i$; the parameter $F \in [0, 2]$ is a *mutation scale factor* and $\mathbf{x}_{best}$ is the current best solution found by the algorithm. In the above notation '*DE/a/b*', *a* refers to the vector used to generate the mutant vectors and *b* is the number of difference vectors used in the mutation process.

After mutation, for each mutant vector $\mathbf{y}_i$, $i = 1, \dots, n_p$, a *trial vector* $\mathbf{z}_i$ is obtained through a crossover operator. The typical *binary crossover* (bin) is defined as:

$$z_{ij} = \begin{cases} y_{ij} & \text{if rand}(0,\ 1) \leq CR \quad \text{or} \quad j = j_{rand} \\ x_{ij} & \text{otherwise} \end{cases} \tag{1}$$

where rand(0, 1) is a random number in [0, 1] and $CR \in [0, 1)$ is a control parameter defining the probability of deriving variables from the mutant vector $\mathbf{y}_i$. The random index $j_{rand} \in [1, d]$ ensures that at least one variable of $\mathbf{y}_i$ is used (even if $CR = 0$). An alternative is represented by the *exponential crossover* (exp). In this case, starting from a random position chosen from $[1, \dots, d]$, the variables are copied from the mutant to the trial vector in a circular manner. After the copy of each variable, a random number *rand*(0, 1) is drawn. The copy process stops either when all variables are copied or the first time that $rand(0,\ 1) > CR$.

As soon as a trial vector $\mathbf{z}_i$ is created, the typical greedy selection of DE consists of replacing $\mathbf{x}_i$ with $\mathbf{z}_i$ only if the latter has a better fitness.

Including also the crossover scheme, the above 'DE/a/b' notation can be extended to identify the different base variants of DE. For example, 'DE/rand/1/bin' denotes a DE scheme with a mutation based on a random vector and one difference vector followed by a binary crossover.

Starting from the relatively simple mechanism outlined above, many variations have been developed to enhance its search efficiency. A major improvement, typically introduced in DE-based algorithms, consists of some form of adaptive or self-adaptive mechanisms to dynamically update the control parameters *F* and *CR* to the characteristics of optimization problems [1, 5, 7, 9, 23, 31, 51, 64, 73, 77–79].

As mentioned above, several DE algorithms proved very effective to address the scalability issue that characterizes LSGO. For example, in the 2011 Special Issue of *Soft Computing* journal, on 'Scalability of Evolutionary Algorithms and other Meta-heuristics for Large Scale Continuous Optimization Problems', six articles out of thirteen were based on DE. Moreover, in the recent literature on LSGO, DE has been often used with other techniques, such as Memetic Algorithms (MA) [27] and CC [41, 45, 75].

In the rest of this section we outline some DE-based approaches that have proved particularly effective in dealing with the scalability issue.

2.1 *The jDElscop Algorithm*

The jDElscop algorithm [8] is a self-adaptive DE that was the runner-up on the set of optimization problems proposed for the above mentioned 2011 special issue of *Soft Computing* journal.

Besides being an adaptive DE, jDElscop uses a population size reduction strategy, a sign-change control mechanism for the F parameter and three different DE strategies, namely 'DE/rand/1/bin', 'DE/rand/1/exp' and 'DE/best/1/bin'.

During the search, the latter are selected for each individual as follows. In the first half of the process (i.e., when the number of fitness evaluations is below the half of available budget), the 'DE/best/1/bin' is used with probability 0.1. Otherwise, 'DE/rand/1/bin' and 'DE/rand/1/exp' are chosen, the former for the first half of n_p individuals and the latter for the remaining individuals.

The sign changing mechanism affects the F parameter at each mutation with a certain probability (which is equivalent to swapping indexes r_2 and r_3).

As for the adaptation of parameters, instead of using constant F and CR for the whole population and optimization process, jDElscop adopts, for each DE strategy, a couple of dynamic parameters F_i and CR_i associated to each individual (i.e., six additional parameters are encoded into each individual). At each generation, F_i and CR_i are randomly changed, within predefined intervals, according to an heuristic depending on some fixed parameters. The idea, common to many other adaptive DE algorithms proposed in literature, is that better values of parameters will generate better individuals, which will more likely survive and propagate their parameters through the generations.

A particular feature of jDElscop is a mechanism of population size reduction, which is performed for a small number of times during the optimization process (e.g., four times). The new population size is equal to half of the previous one and the reduction is based on a sequence of pair-wise comparisons between individuals of the first and second halves of the population. The individual with a better fitness is placed in the first half of the current population, which represents the population for the next generation.

In [8], jDElscop was tested on 19 benchmark functions with dimension up to 1000. As mentioned above, the results were quite satisfactory as the algorithm was outperformed only by the MOS-based approach [27] described later in this chapter.

2.2 *The GaDE Algorithm*

Another DE algorithm that proved particularly effective on LSGO is the 'generalized adaptive differential evolution for large-scale continuous optimization' (GaDE), introduced in [78]. In GaDE, instead of using an heuristic, which introduces new parameters that may be difficult to set, the adaptation is based on a probability distribution. Such an approach was also previously used by other adaptive DEs, such as SaDE [50], SaNSDE [77] and JADE [79]. In particular, as done in JADE, also in GaDE the scale factors F_i are generated for each individual using a Cauchy distribution $F_i = C(F_m, 0.2)$, where the parameter F_m is computed using a new learning mechanism depending on the fitness improvements achieved during the process. The crossover rate CR adaptation is based on generating a values for each individual from a Gaussian distribution $CR_i = N(CR_m, 0.1)$, where the parameter CR_m is updated with the same mechanism of F_m.

Binary crossover and two different mutation schemes are used in GaDE, namely 'DE/rand/1' and 'DE/current to best/1'. The choice of the mutation operator is based on the same learning mechanism implemented in SaDE [50]. In practice, each type of mutation has an associated probability of being used. During a fixed learning period (e.g., the first 50 generations), the successes of each type of mutation are recorded and the probability of choosing each operator is updated accordingly.

In terms of effectiveness when dealing with LSGO problems, GaDE obtained the third place among all the algorithms included in the 2011 special issue of *Soft Computing* journal mentioned above.

2.3 *The jDEsps Algorithm*

A further DE algorithm that proved very effective on LSGO is the 'Self-adaptive differential evolution algorithm with a small and varying population size', called jDEsps [6] and representing an improved version of the jDElscop algorithm outlined in the above Sect. 2.1.

The jDEsps is still a self-adaptive DE algorithm with multiple strategies and a mechanism for varying the population size during the optimization. However, compared to jDElscop, in this case the algorithm begins with a small population size, which is increased later and then reduced. Other relevant differences are: (i) a strategy (called jDEw) that moves the best individual found so far using a large step movement; (ii) a simple probabilistic LS procedure (called jDELS) applied to the best individual. The jDELS is obtained as a 'DE/best/1' mutation with a very small and probabilistic parameter F. As discussed in the next section, LS has been widely adopted for developing optimization algorithms able to affectively address LSGO problems.

The jDEsps algorithm was the runner-up at the special session on LSGO within the 2012 IEEE Congress on Evolutionary Computation (CEC).

3 Memetic Algorithms for LSGO

According to the literature, an effective way to improve the efficiency of EAs consists of hybridizing the optimization process with LS techniques. The latter are typically applied to some selected and promising individuals at each generation. In such a search process, commonly referred to as Memetic Algorithm (MA) [38, 39], the EA mechanisms help to explore new search zones, while the LS exploits the current best solutions.

Clearly, since LS can be computationally expensive, a good balance between exploration (EA) and exploitation (LS) is a key factor for the success of a MA implementation. This is particularly true in case of problems with high dimensionality. In fact, in such cases LS concerns much challenging neighbourhoods and a suitable value of its *intensity* (i.e., the number of fitness function evaluations assigned to LS) should be carefully determined in order to achieve a satisfactory search efficiency. With this in mind, several hybrid EAs, endowed with various types of LS, have been developed in literature specifically to address LSGO problems. Below, two among the most effective approaches are briefly outlined.

3.1 *The MA with LS Chains Approach*

A successful MA, which proved very efficient in LSGO, is the *MA with LS chains* (MA-LS-Chains), proposed in [35, 36] and later developed in [37].

In brief, at each EA generation the LS chain method resumes, on some individuals, the LS exactly from its last state (i.e., that resulting from the LS at the previous generation). Such an approach allows to effectively adapt LS parameters, including its intensity, during the search process. A first version of MA-LS-Chains, namely MA-CMA-Chains [35], was based on the Covariance Matrix Adaptation Evolution Strategy (CMA-ES) [21] as LS method. However, although very effective, CMA-ES is also very computationally expensive, especially in case of high dimensional problems (i.e., it is based on many operations with complexity $O(d^3)$, being d the problem dimension). As a results, MA-CMA-Chains was not able to effectively tackle LSGO problems. Later, a new MA-LS-Chains algorithm was developed in [36] using a more scalable LS method, namely the Solis and Wets' algorithm (SW) [55]. Remarkably, the resulting MA-SW-Chains was the winner of the LSGO special session at CEC 2010.

The MA-SW-Chains algorithm is a steady-state GA, that is, only few new individuals replace parents in the population at each generation (e.g., only the worst parent is replaced by a better newly generated individual). It is worth noting that the resulting long permanence of individuals into the population, through generations, allows the MA-LS-Chains approach to resume LS on some individuals. The main characteristics of MA-SW-Chains are outlined below.

To achieve high population diversity, the search algorithm uses the Blend Alpha Crossover (BLX-α) as recombination operator, with $\alpha = 0.5$ [17]. The BLX-α crossover creates new offspring by sampling values for each direction i in the range $[min_i - \alpha I, max_i + \alpha I)$, where min_i and max_i are the smaller and larger values of the two parents along the ith direction, and $I = max_i - min_i$ [59].

Moreover, in MA-SW-Chains the authors adopted the *negative assortative mating strategy* [19, 20], where mating between individuals having similar phenotype is less frequent than what would be expected by chance. According to [19, 20], a way to implement such a mating scheme is as follows: (i) a first parent is chosen by the roulette wheel method and further n_{add} candidate parents are also selected ($n_{add} = 3$ in MA-SW-Chains); (ii) then, the similarity between each of the n_{add} individuals and the first parent is computed using a suitable metric; (iii) among the n_{add} individuals, the less similar is chosen as the second parent.

Another characteristic of MA-SW-Chains is the use of BGA mutation [40]. In the latter, when a variable x_i is selected for mutation, its value is changed into $x_i \pm r_i \delta$, where r_i is the *mutation range*, the sign + or − is randomly selected with the same probability and δ is a small random number in [0, 1].

As mentioned above, the LS in MA-SW-Chains is based on the SW algorithm [55], which is an adaptive randomized hill-climbing heuristic. The SW process starts with a solution $\mathbf{x}^{(0)}$ and explores its neighbourhood, through a step-by-step process, to find a sequence of improved solutions $\mathbf{x}^{(1)}$, $\mathbf{x}^{(2)}$ ... $\mathbf{x}^{(q)}$. More in details, in SW two neighbours are generated at each step by adding and subtracting a random deviate $\boldsymbol{\Delta}$, which is sampled from a normal distribution with mean $\mathbf{m}$ and standard deviation ρ. If either $\mathbf{x}^{(i)} + \boldsymbol{\Delta}$ or $\mathbf{x}^{(i)} - \boldsymbol{\Delta}$ is better than $\mathbf{x}^{(i)}$, the latter is updated and a *success* is registered. Otherwise, the value of $\boldsymbol{\Delta}$ is considered as a *failure*. On the basis of the number of successes, and depending on some fixed parameters, the values of ρ and $\mathbf{m}$ are updated during the process, in order to both increase the convergence speed and bias the search towards better areas of the neighbourhood. The process continues up to a certain number of fitness function evaluations (i.e., the LS intensity).

The main aspect of MA-SW-Chain consists of resuming LS through subsequent generations. In particular, thanks to the fact that the evolutionary process is steady state, the LS has very often the opportunity to continue LS on the same individuals and using the same state achieved when the LS was halted (e.g., in terms of ρ, $\mathbf{m}$ and number of successes and failures).

Another important characteristic of MA-SW-Chain is that, to avoid super exploitation, the total number of fitness function evaluations dedicated to LS is a fixed fraction of the total available budget. In addition, at each generation only one individual is chosen to be improved by LS using a strategy that allows to activate new promising chains and to exploit existing ones. However, when LS has been already applied to all available individuals without achieving adequate fitness improvements, the population is randomly reinitialized retaining only the best individual.

As mentioned above, MA-SW-Chain was very effective on the 20 test problems proposed for the LSGO special session in the CEC 2010. Such tests, however, were limited to a dimensionality of 1000.

Recently, a significant development of MA-SW-Chain was presented in [24], where a parallel implementation on Graphics Processing Units was applied to optimization problems up to the remarkable dimension of 3,000,000 variables. In this case, the study aimed at showing the speedup that can be obtained with the GPU-based MA-SW-Chains optimization technique and how it increases as the dimensionality grows. Interestingly, for dimensionality 3,000,000 the run time is reduced from 18 days to less than 2.5 h on the used hardware, which was composed of a recent NVIDIA GPU card and a workstation equipped with a standard CPU.

Considering the results of [24], at the current state of the technology it seems that the only actual possibility to address optimization problems with millions of dimensions is to couple highly scalable search algorithms with HPC.

3.2 The MOS-Based Algorithms

Other hybrid algorithms with LS that were very successful in dealing with LSGO problems are based on the Multiple Offspring Sampling (MOS) approach [25], in which different mechanism for creating new individuals are used in the optimization. During the process, the goodness of each involved mechanism is evaluated according to some suitable metrics. Then, the latter are used to dynamically adapt the participation, in terms of computational effort, of each technique.

A MOS algorithm that showed the best performance on the test set proposed for the 2011 SI of *Soft Computing* journal on LSGO, was based on DE and a LS strategy [27]. The approach, which is briefly described below, belongs to the so called high-level relay hybrid (HRH) category [60], where different metaheuristics are executed in sequence, each of which reusing the output population of the previous one.

The algorithm is composed of a fixed number of steps. At each step, a specific amount FEs of fitness function evaluations is distributed between the involved techniques $T^{(i)}$ according to some *Participation Ratios* Π_i.

Initially, all Π_i have the same value. At each step of the search process, a *Quality Function* (QF) attributes a quality value $Q^{(i)}$ to each technique $T^{(i)}$ on the basis of the individuals produced in the previous step. In [27], the QF is defined in a way to account both for the achieved average fitness increment and the number of such improvements. Then, the participation ratios Π_i of the involved techniques $T^{(i)}$ are updated accounting for the $Q^{(i)}$ values. In particular, the Π_i are computed using the relative difference between the quality of the best technique and the remaining ones. However, regardless of its $Q^{(i)}$ value, a minimum threshold is established, in order to prevent a technique to exclude the others in case of its much greater efficiency at the early steps of the search process. As soon as the Π_i are computed, they are used to determine the number of allowed fitness function evaluations in the next step for all the involved techniques.

An additional strategy included in the algorithm is the population reset, preserving the best solution, in case of convergence of the whole population to the same solution.

As for the adopted LS, referred to as MTS-LS1, it was based on one of the methods included in the Multiple Trajectory Search (MTS) algorithm [68]. In brief, it searches separately along each direction using a deterministic search range (*SR*) initialized to a suitable value SR_0. The value of *SR* is reduced to one-half if the previous LS does not lead to improvements. When *SR* falls below a small threshold, it is reinitialized to SR_0. Along each search direction, the solution's coordinate is first subtracted by *SR* to look for fitness improvements. In case of improvement, the search proceeds to consider the next dimension. Otherwise, the variable corresponding to the current direction is restored and then is added by *SR*/2, again to check if the fitness improves. If it is, the search proceeds to consider the next dimension. If it is not, the variable is restored and the search proceeds to consider the next direction. The detailed pseudo-code can be found in [68].

The MOS algorithm outlined above, was tested on the 19 test functions up to 1000 dimensions, proposed for the above mentioned SI of *Soft Computing* journal. The scalability of the algorithm was very satisfactory, as 14 out of the 19 functions of the benchmark were solved with the maximum precision.

Another MOS-based hybrid algorithm was later presented at the CEC 2012 LSGO session, where it outperformed all the competitors [28]. In this case, the approach combined two LS techniques without a population-based algorithm, namely the MTS-LS1 [68] and the SW algorithm [55] already outlined above. In addition, the QF was only based on the fitness increment achieved at each generation by each technique. Such an approach, besides achieving the best result on the CEC 2012 special session on LSGO, in a comparison combining the results of CEC 2010 and CEC 2012 sessions, also outperformed many other algorithms [62].

Later, in [29] a new MOS-based hybrid algorithm was presented, which combines a GA with two strategies of LS: the SW algorithm [55] and a variation of MTS-LS1 [68], called MTSLS1-Reduced. The latter operates as the MTS-LS1 outlined above. However, it spends more computational effort on the most promising directions. The GA used a BLX-α crossover, with $\alpha = 0.5$, and a Gaussian mutation [22]. Such a MOS-based approach was the best performing algorithm in the test set proposed for LSGO session at the CEC 2013 (i.e., a set of 15 large-scale benchmark problems, with dimension up to 1000, devised as an extension to the previous CEC 2010 benchmark suite).

4 Cooperative Coevolution

Cooperative Coevolution (CC), introduced in [48], is another effective approach for addressing LSGO problems. CC can be classified as a divide-and-conquer technique, in which the original high-dimensional problem is decomposed into lower-dimensional *subcomponents*, which are easier to solve.

Typically, each subcomponent is solved using an ordinary optimization metaheuristic. During the process, the only cooperation takes place at the evaluation of the fitness through an exchange of information between the search processes operating on the different subcomponents.

CC was first applied to a GA by Potter and De Jong in [48]. Subsequently, the approach has been successfully tested with many different search algorithms such as Ant Colony Optimization [14], Particle Swarm Optimization (PSO) [15, 30, 47, 69], Simulated Annealing [54], DE [75], Firefly Algorithm [66] and many others.

More in details, a CC optimization is based on partitioning the d-dimensional set of search directions $G = \{1, 2, \dots, d\}$ into k sets $G_1 \dots G_k$. Each group G_i of directions defines a subcomponent whose dimension can be significantly lower than d. By construction, a candidate solution found by a subcomponent contains only some elements of the d-dimensional vector required for computing the corresponding fitness function f. Thus, to evaluate the latter, a common d-dimensional *context vector* $\mathbf{b}$ is built using a representative individual (e.g., the best individual) provided by each subcomponent. Then, before its evaluation, each candidate solution is complemented through the appropriate elements of the context vector. In this framework, the cooperation between sub-populations emerges because the common vector is used for the fitness evaluation of all individuals.

In their original paper, Potter and De Jong [48] proposed to decompose a d-dimensional problem into d sub-populations (i.e., $G_i = \{i\}$). The fitness of each individual was computed by evaluating the d-dimensional vector formed by the individual itself and a selected member (e.g., the current best) from each of the other sub-populations.

Subsequently, the idea was applied to PSO by Van den Bergh and Engelbrecht in [69], where the authors introduced the decomposition of the original d-dimensional search space into k subcomponents of the same dimension $d_k = d/k$. In other words, in such an approach the groups of dimensions associated to the subcomponents are defined as:

$$G_i = \{(i-1) \times d_k + 1, \dots, i \times d_k\}$$

and the context vector is:

$$\mathbf{b} = (\underbrace{b_1^{(1)}, \dots, b_{d_k}^{(1)}}_{\mathbf{b}^{(1)}}, \underbrace{b_1^{(2)}, \dots, b_{d_k}^{(2)}}_{\mathbf{b}^{(2)}}, \dots, \underbrace{b_1^{(k)}, \dots, b_{d_k}^{(k)}}_{\mathbf{b}^{(k)}})^T$$

where $\mathbf{b}^{(i)}$ is the d_k-dimensional vector representing the contribution of the ith subcomponent (e.g., its current best position):

$$\mathbf{b}^{(i)} = (b_1^{(i)}, b_2^{(i)}, \dots, b_{d_k}^{(i)})^T$$

Given the jth individual $\mathbf{x}^{(i,j)} \in S^{(i)}$ of the ith subcomponent:

$$\mathbf{x}^{(i,j)} = (x_1^{(i,j)}, x_2^{(i,j)}, \dots, x_{d_k}^{(i,j)})^T$$

its fitness value is given by $f(\mathbf{b}^{(i,j)})$, where $\mathbf{b}^{(i,j)}$ is defined as:

$$\mathbf{b}^{(i,j)} = (\underbrace{b_1^{(1)}, \ldots, b_{d_k}^{(1)}}_{\mathbf{b}^{(1)}}, \ldots, \underbrace{x_1^{(i,j)}, \ldots, x_{d_k}^{(i,j)}}_{\mathbf{x}^{(i,j)}}, \ldots, \underbrace{b_1^{(k)}, \ldots, b_{d_k}^{(k)}}_{\mathbf{b}^{(k)}})^T$$

In other words, the fitness of $\mathbf{x}^{(i,j)}$ is evaluated on the vector obtained from **b** by substituting the components provided by the *i*th sub-population with the corresponding components of $\mathbf{x}^{(i,j)}$.

Algorithm 1: CC(f, n)

```
1  𝒢 = {G_1, ..., G_k} ← grouping(n);
2  pop ← initPopulation();
3  contextVector ← initContextVector(pop);
4  fitnessEvaluations ← 0;
5  while fitnessEvaluations < maxFE do
6      foreach G_i ∈ 𝒢 do
7          pop_i ← extractPopulation(pop, G_i);
8          best_i ← optimizer(f, pop_i, contextVector, G_i, maxFESC);
9          pop ← storePopulation(pop_i, G_i);
10         fitnessEvaluations ← fitnessEvaluations + maxFESC;
11         contextVector ← updateContextVector(best_i, G_i);
12 return contextVector and f(contextVector);
```

Except for the evaluation of individuals, the optimization is carried out using the standard optimizer in each subspace. Algorithm 1 outlines a possible basic CC optimization process for population-based metaheuristic. First, a decomposition function creates the k groups of directions. Then the population and the context vector are randomly initialized. The optimization is organized in *cycles*. During each cycle, the optimizer is activated in a round-robin fashion for the different subcomponents and the context vector is updated using the current best individual of each subcomponent. A budget of *maxFESC* fitness evaluations is allocated to each cycle and subcomponent. The CC cycles terminate when the number of fitness evaluations reaches the value *maxFE*. Note that several variants to this scheme can be possible. For example, the context vector could be updated in a synchronous way at the end of each cycle.

The CC framework has attracted a significant amount of research with the aim of addressing relevant design aspects that typically affect its optimizing performance. In the following, some of the main results reported in the literature are outlined.

4.1 Random Grouping

A major issue with the CC approach, early recognized in [48, 49], is that when interdependent variables are assigned to different subcomponents the search efficiency can decline significantly.

The interdependency between decision variables is usually referred to as *non-separability* [53] or *epistasis*, that is gene interaction. Basically, separability means that the influence of a variable on the fitness value is independent of any other variables. More formally, following [2] a function $f : \mathbb{R}^d \mapsto \mathbb{R}$ is separable iff:

$$\arg \min_{x_1, \ldots, x_d} f(x_1, \ldots, x_d) = \left(\arg \min_{x_1} f(x_1, \ldots), \ldots, \arg \min_{x_d} f(\ldots, x_d) \right) \quad (2)$$

otherwise the function $f(\mathbf{x})$ is non-separable.

The level of separability has long been regarded as one of the measures of difficulty of an evolutionary optimization algorithm. For example, in [53], Salomon showed that the performance of a simple GA can decrease significantly in case of non-separable problems.

In the CC case, it is clear that the simple decomposition method outlined above, when applied to non-separable problems, can lead to slow convergence. In fact, interdependent variables are likely to be located in different subcomponents during the whole optimization process.

To cope with this problem, in the Random Grouping (RG) approach, proposed by Yang et al. in [75, 76], the directions of the original search space are periodically grouped in a random way to determine the CC subcomponents. Such an approach was successfully applied to DE, on high dimensional non-separable problems with up to 1000 dimensions. Subsequently, the RG idea was integrated into several CC optimizers. For example, in [30] the authors applied RG to PSO to solve optimization problems up to 2000 variables. Such a cooperative PSO outperformed some state of the art evolutionary algorithms on complex multi-modal problems. Also, in [47] the author used the same cooperative approach with RG in a micro-PSO, showing that even using a small number of individuals per subcomponent the algorithm is very efficient on high-dimensional problems. Moreover, in [66] the RG approach was successfully applied to a CC version of the Firefly Algorithm (FA) [74].

Compared with the simple linear decomposition described above, it has been proved that RG increases the probability of having two interacting variables in the same sub-population at least for some iteration of the search algorithm [43, 75]. More in details, in the linear decomposition proposed in [32, 48, 69] the ith sub-population operates on the group of directions G_i defined as the interval:

$$G_i = [(i-1) \times d_k + 1, \ldots, i \times d_k]$$

In addition, the decomposition $\mathcal{G} = \{G_1, \ldots, G_k\}$ of Algorithm 1 is *static*, in the sense that it is defined before the beginning of optimization cycles. Instead, a RG approach assigns to the ith group $d_k = d/k$ directions q_j, with j randomly selected without replacement from the set $\{1, 2, \ldots, d\}$.

To motivate their approach, in [75] the authors showed that RG leads to a relatively high probability to optimize in the same subcomponent two interacting variables for at least some cycles.

Later, in [43] was shown that the probability of having all the interacting variables grouped into the same subcomponent for sufficient number of cycles can be very low. To mitigate this problem, the authors suggested to increase the frequency of RG. A way to maximize the RG frequency consists of executing only one generation of the evolutionary optimizer per CC cycle. In [43] a higher frequency of RG provided significant benefits on some non-separable high dimensional problems.

However, according to the analytical result showed in [43], to group at least once many interacting variables together the RG approach would require an infeasible number of cycles. Nevertheless, even when only some of such variables are grouped together in turns, according to results in the literature, the RG approach can be beneficial.

4.2 Adaptive Decomposition

When using the CC approach on some optimization problems, there exists an optimal value of the chosen size $d_k = d/k$ that maximizes the performance of a given optimizer [46]. Unfortunately, according to [46, 76] it is not easy to predict such an optimal size before the optimization, since it strongly depends on both the problem and the characteristics of the adopted optimizer. Small group sizes can be suitable for separable problems, making easier the optimization of each subcomponent. On the other hand, large group sizes may increase the probability of grouping together interacting variables in non-separable problems.

Among the techniques that have been proposed for the automatic adaptation of the subcomponent sizes, there is the *Multilevel Cooperative Coevolution* (MLCC) framework [76]. The MLCC idea is to operate with a predefined set of decomposers, that is, with a set of allowed group sizes $\mathscr{V} = \{d_{k_1}, d_{k_2}, \ldots, d_{k_m}\}$. At the beginning of each cycle, MLCC selects a decomposer d_{k_i} from $\mathscr{V}$ on the basis of its performance during the past cycles. To such purpose, the algorithm attributes a performance index r_i to each decomposer as follows:

1. initially, all the $r_i \in \mathbb{R}$ are set to 1;
2. then, the r_i are updated the basis of the gain of fitness associated to their use trough the equation $r_i = (f^{(prev)} - f^{(cur)})/|f^{(prev)}|$, where $f^{(prev)}$ is the best fitness at the end of the previous cycle and $f^{(cur)}$ is the best fitness achieved at the end of the current cycle in which the decomposer d_{k_i} has been used.

At the beginning of each cycle, the performance indexes are converted into probabilities using a Boltzmann *soft max* distribution [58]:

$$p_i = \frac{e^{r_i/c}}{\sum_{j=1}^{t} e^{r_j/c}} \tag{3}$$

where c is a suitable constant. The latter should be set in such a way to associate a high probability of being selected to the best decomposers (exploitation), still giving some chances to all the available decomposers (exploration). The above mechanism allows to self-adapt the problem decomposition to the particular objective problem and also to the evolution stages. In [76], the MLCC adaptation method was tested, using a RG strategy, on a suite of benchmark functions. The authors found that in some cases the self-adaptive strategy outperformed the corresponding methods based on the static selection of d_k and on the random selection of the group sizes at each cycle.

An improvement of the MLCC approach, named MLSoft, was later introduced in [46]. The authors noted that MLCC can be seen in a perspective of a reinforcement learning (RL) approach [58], where the improvement of fitness is the reinforcement signal and the actions consist in the choice of the decomposer. However, instead of selecting actions on the basis of its long-term utility, as typically done in RL, in the MLCC their immediate reward is used. Instead, MLSoft replaced r_i with a suitable *value function* V_i, which was intended as an estimate of the long term utility associated to the use of a decomposer. In [46], V_i was simply defined as the arithmetic mean of all rewards r_i received by the decomposer d_{k_i} during the optimization process. The MLSoft algorithm was tested on eight fully-separable functions using a rich set of decomposers and different values of the parameter c in Eq. (3). According to the results, although MLSoft outperformed MLCC, it was not able to outperform the corresponding CC framework with a fixed and optimal subcomponent size. This suggests that there is still room for improvement in the way in which the size of decomposers are determined in the CC approach.

A major issue with the adaptive approach outlined above can be easily recognized. In both the MLCC and MLSoft approaches, a decomposer is randomly drawn at each cycle according to its current probability, which is computed on the basis of its value function. The latter reflects the rewards obtained by the decomposer at the end of the cycles in which it has been used. Unfortunately, in such a learning scheme the rewards obtained by the different decomposers may be strongly affected by the state of the environment in which they have operated. This is because of the expected evolution of the population on the fitness landscape, which can be significantly complex. In other words, an hypothetical agent that has to choose a decomposer operates on a non-stationary and history-dependent environment, for which RL schemes conceived for Markovian environments are not guaranteed to converge to the optimal policy (although they can still be used with acceptable results in some cases [58]). Therefore, an effective automatic decomposition approach should be able to learn on a dynamic environment.

Such a problem has been recently addressed in [67] using a new approach in which, during short learning phases, the decomposers of a predefined set are concurrently applied starting from the same state of the search, including the same context vector. In other words, they are concurrently executed on the same initial environment in order to achieve an unbiased estimate of their value functions. The experimental results on a set of large-scale optimization problems showed that the method can lead to a reliable estimate of the suitability of each subcomponent size. Moreover, in some cases it outperformed the best static decomposition.

4.3 Automatic Grouping

Another important line of research, which significantly contributed to the enhancement of the CC approach, concerns the possibility of identifying and grouping together interacting variables. In contrast to the RG approach, the objective is to discover the underlying structure of the problem in order to devise and adapt a suitable decomposition.

The first technique to identify interacting variables in a CC framework was proposed in [72]. The authors observed that if a candidate solution where two directions have been changed achieves a better fitness than the same solution where only one of the directions was changed, then this may indicate the presence of an interdependency. The creations of groups was carried out during the optimization process (i.e., *online*) exploiting some additional fitness evaluations for each individuals. The technique proved effective, although the approach was tested only on few functions with dimensionality up to 30.

Following the idea proposed in [72], which basically consists of observing the changes of the objective function due to a perturbation of variables, more effective methods have been later developed. In most cases, the decomposition stage is performed *offline*, that is the groups are created before the optimization starts. Other approaches presented in the literature for automatically grouping variables in CC are based on learning statistical models of interdependencies [57] or on the correlation between variables [52]. However, as noted in [44], correlation coefficients are not a proper measure for separability in the CC optimization context.

An important step in the development of an automatic grouping strategy for CC optimizations has been the Delta Grouping (DeG) approach proposed in [44]. The DeG algorithm is based on the concept of *improvement interval* of a variable, that is the interval in which the fitness value could be improved while all the other variables are kept constant [44, 53]. It has been observed that in non-separable functions, when a variable interacts with other variables, its improvement interval tends to be smaller. Therefore, in the DeG approach the identification of interacting variables was based on measuring the amount of change (i.e., the *delta value*) in each of the decision variables during the optimization process. In particular, the DeG algorithm sorts the directions according to the magnitude of their *delta values* in order to group the variables with smaller delta values in the same subcomponent. Clearly, as pointed out in [44], not always a small improvement interval implies a variable interdependency. However, when tested on both the CEC 2008 [63] and CEC 2010 LSGO [61] benchmark functions, the DeG method performed better than other relevant CC methods.

A drawback of DeG is its low performance when there is more than one non-separable subcomponent in the objective function [44]. On the other hand, being an online adaptation technique, the DeG approach has the ability to adapt itself to the fitness landscape. Such a property can be valuable when the degree of non-separability changes depending on the current region of the search space explored by the individuals in the population.

A different grouping technique, proposed in [11], is the Cooperative Coevolution with Variable Interaction Learning (CCVIL), which can be viewed as a development of the method presented in [72]. In CCVIL, the optimization is carried out trough two stages, namely *learning* and *optimization*, in the first of which the grouping structure is discovered. According to [11], an interaction between any two variables x_i and x_j is taken under consideration if the following condition holds:

$$\begin{aligned} &\exists\ \mathbf{x}, x'_i, x'_j : \\ &f(x_1, \dots x_i, \dots, x_j, \dots, x_d) < f(x_1, \dots x'_i, \dots, x_j, \dots, x_d) \wedge \\ &f(x_1, \dots x_i, \dots, x'_j, \dots, x_d) > f(x_1, \dots x'_i, \dots, x'_j, \dots, x_d) \end{aligned} \tag{4}$$

The learning stage of CCVIL starts by placing each direction in a separate subcomponent, that is by separately optimizing the variables in sequence. During this process, CCVIL tests if the currently and the previously optimized dimensions interact by using Eq. 4. The latter can be applied because only two dimensions changed. Before each learning cycle, the order of optimization of variables is randomly permutated, so that each two dimensions have the same chance to be processed in sequence. After the convergence of the learning stage in terms of grouping, CCVIL starts the optimization stage.

In [11], the authors tested the CCVIL approach using the CEC 2010 benchmark functions on LSGO [61]. According to the results, CCVIL improved the underlying CC algorithm in most of the benchmark functions. However, a significant issue to be solved concerns the distribution of computational effort between learning and optimization stages of CCVIL.

Another recent approach for adaptive grouping, named *Differential Grouping* (DG) algorithm, has been proposed in [41] for *additively separable* (AS) functions $f : \mathbb{R}^d \mapsto \mathbb{R}$, which can be expressed as the sum of k independent nonseparable functions. In this case, there exists an ideal problem decomposition $\mathcal{G}_{id}$ composed of k groups of variables G_i such that if $q \in G_i$ and $r \in G_j$, with $i \neq j$, then q and r are independent. However, it is worth noting that $\mathcal{G}_{id}$ is not necessarily the best decomposition for a CC optimization algorithm, as can be inferred from the results presented in [46]. In fact, depending on both the problem and the optimizer, could be effective to split some groups G_i into sub-groups with lower dimension.

The DGA approach was founded on the formal proof that for AS functions, if the forward differences along x_p:

$$\Delta f_{x_p}(\mathbf{x}, \delta)|_{x_p=a, x_q=b} \quad \text{and} \quad \Delta f_{x_p}(\mathbf{x}, \delta)|_{x_p=a, x_q=c}$$

are not equal, with $b \neq c$ and $\delta \neq 0$, then x_p and x_q are non-separable. The forward difference with interval δ, in a point $\mathbf{x}$ and along the direction x_p is defined as:

$$\Delta f_{x_p}(\mathbf{x}, \delta) = f(\dots, x_p + \delta, \dots) - f(\dots, x_p, \dots)$$

and requires two function evaluations to be estimated. The DG presented in [41], exploits the above property to create groups of interacting variables. The algorithm operates by checking the interactions trough pairwise comparisons among variables. However, DG does not necessarily require all the comparisons. In fact, when an interaction is detected between two variables, one of the two is placed on a group and excluded by further comparisons. For example, for $d = 1000$, with $m = 50$ only 21,000 function evaluations are required, while with $m = 1$ (i.e., fully separable problem) DG requires 1,001,000 additional fitness evaluations [41]. It is important to note that such an additional computational effort reduces the budget available for the subsequent optimization phase.

The DG approach was tested in a CC optimizer using CEC 2010 benchmark functions [61] showing a good grouping capability. Also, DG outperformed the CCVIL approach in most functions, both in terms of grouping accuracy and computational cost. However, on some test cases, such as those derived by the Rosenbrock function, the decomposition accuracy was low. This suggests that the DG approach is not very effective in case of indirect interactions between variables and that further research is needed to develop a better method of automatic grouping. Likely, a more accurate algorithm would require even more function evaluations than DG, which would be subtracted by the available computational budget.

4.4 Dealing with Unbalanced Subcomponents

Given the ability of automatic decomposition described above, in [45] the authors noted that there is often an imbalance between the contribution to the fitness of the different subcomponents. In particular, in CC there are situations in which the improvements in some of the subcomponents are not apparent simply because they are negligible in comparison to the fitness variation caused by other subcomponents. Thus, according to [45], in most cases devoting the same amount of computational resources to all subcomponents (i.e., the value *maxFESC* in Algorithm 1) in a round-robin fashion, can result in a waste of fitness evaluations. In order to mitigate this issue, in [45] the Contribution Based Cooperative Co-evolution (CBCC) algorithm was proposed, where:

1. the contribution ΔF_i of each subcomponent is estimated by measuring the changes in global fitness when it undergoes optimization. Such contributions are accumulated from the first cycle during the optimization.
2. each cycle is composed of a round-robin *testing phase*, where the contributions ΔF_i are updated, and a subsequent stage in which the subcomponent with the greatest ΔF_i is iteratively selected for further optimization;
3. when there is no improvement in the last phase, the algorithm starts a new cycle with a new testing phase.

Clearly, the CBCC algorithm must be integrated with an effective grouping strategy, which should be able to decompose the problem into independent groups as much as possible.

The CBCC has proved to be promising when tested on the LSGO benchmark functions which have been proposed for the CEC 2010 [61]. However, the experiments showed that CBCC is too much influenced by historical information in the early stages of evolution. For example, it may happen that the subcomponent that is initially recognized as the major fitness contributor, reaches convergence very soon. In this case, the CBCC approach presented in [45] does not switch immediately to the subcomponent with the largest contribution, due to the influence of the initial assessment of contributions. From this point of view, there is still room for developing an adaptive procedure that can cope effectively with the problem of imbalance between the contribution to the fitness of the different subcomponents.

4.5 Successful Algorithms Based on CC

As already noted, the CC approach was adopted in conjunction with a number of optimization algorithms. Below, we mention the most successful ones at the CEC special sessions on LSGO.

- **DECC-G**. Originally presented in [75], DECC-G is a CC approach based on the 'Self-Adaptive with Neighborhood Search Differential Evolution' (SaNSDE) algorithm [77]. The latter is a self-adaptive DE in which the mutation operator is replaced by a random neighborhood search. The DECC-G algorithm was the runner-up at the 2013 CEC special session on LSGO.
- **CC-CMA-ES**. The 'Scaling up Covariance Matrix Adaptation Evolution Strategy using cooperative coevolution' (CC-CMA-ES) is a CC approach based on CMA-ES [21]. This algorithm obtained the third position at the 2013 CEC special session on LSGO.
- **2S-Ensemble**. The 'Two-stage based ensemble optimization for Large-Scale Global Optimization' (2S-Ensemble) was proposed in [70, 71]. It divides the search procedure into two different stages: (i) in the first stage, a search technique with high convergence speed is used to shrink the search region on a promising area; (ii) in the second stage, a CC based search technique is used to exploit such a promising area to get the best possible solution. The CC uses randomly three different optimizers, based on the above mentioned SaDE [50], on a GA and a DE. Moreover, the size of the decomposition is adaptive. The 2S-Ensemble algorithm was the runner-up at the 2010 CEC special session on LSGO.

5 Conclusions

Among the relevant research challenges that characterize the era of big data, there is the need for optimization techniques able to effectively tackle problems with hundreds, thousands, and even millions of variables. Ideal candidates for dealing with such complex optimization tasks are EAs, which, however, can suffer form significant scalability problems as the dimensionality of the problem increases. For these reasons, the field of LSGO attracted a great amount of research over the last decade.

Rather than presenting a comprehensive review (which can be found in [26, 34]), this chapter focused on outlining some relevant trends and important research issues concerning LSGO, also describing the most successful algorithms presented in literature. By examining the latter, we can recognize the following recurring components and promising approaches:

- **Hybridization**. According to the literature, hybridization was often a key factor for developing highly scalable optimization algorithms [27, 29, 70].
- **Adaptation**. Different forms of adaptation were successfully adopted in all the most scalable optimization algorithms. For example, adaptation concerned DE parameters in [8, 75, 78], problem decomposition in [46] and metaheuristic selection in [27, 29].
- **Differential Evolution**. Several adaptive versions of DE, developed in recent years, proved very effective and scalable. Many hybrid algorithms with significant performance in addressing LSGO problems were based on (or included) DE variants [6, 8, 78].
- **Local search**. The most relevant algorithms for tackling LSGO problems included LS strategies [24, 27, 29, 36, 37]. Often, simpler LS algorithms, such as the SW [55], were also more effective [29, 37].
- **Diversity promotion**. Avoiding premature convergence is another aspect that was explicitly addressed by some successful algorithms. Typically, this is obtained trough a diversity control mechanism triggered by suitable indicators (e.g., the reduction of variance in the population) [13]. For example, the MOS algorithm used in [27] includes a population reset mechanism, preserving the best solution, to be activated in case of convergence of the whole population to the same point. Also in the MA-SW-Chain algorithm, the population is randomly reinitialized, retaining only the best individual, when LS has been already applied to all available individuals without achieving fitness improvements [37].
- **Cooperative Coevolution**. The CC approach proved very promising as a framework for optimizing high-dimensional problems. Important research problems in this case concern the automatic and adaptive decomposition [41, 46] and the issue of unbalanced subcomponents [45].
- **High Performance Computing**. To address very high-dimensional optimization problems (e.g., millions of variables) it is mandatory the adoption of HPC, besides efficient algorithms. However, according to the literature, it seems that the parallelization of algorithms specifically designed for LSGO is still in its infancy [24, 47].

As a final remark, it should be noted that accurate benchmarking on problems with a number of variables exceeding 1000 were rarely tackled in the literature. In addition, the typical approach used so far for evaluating new metaheuristics consisted of using only benchmarks test functions rather than real-world optimization problems. This suggests that the design of effective benchmarks is another critical aspect of the research in the field [42].

In conclusion, the LSGO area is rich in research directions which could potentially result in significant advances of the current state of the art.

References

1. Abbass, H.: The self-adaptive pareto differential evolution algorithm. In: Proceedings of the 2002 Congress on Evolutionary Computation, 2002. CEC'02, vol. 1, pp. 831–836 (2002)
2. Auger, A., Hansen, N., Mauny, N., Ros, R., Schoenauer, M.: Bio-inspired continuous optimization: the coming of age. Piscataway, NJ, USA, invited talk at CEC2007 (2007)
3. Blecic, I., Cecchini, A., Trunfio, G.A.: Fast and accurate optimization of a GPU-accelerated CA urban model through cooperative coevolutionary particle swarms. Proc. Comput. Sci. **29**, 1631–1643 (2014)
4. Blecic, I., Cecchini, A., Trunfio, G.A.: How much past to see the future: a computational study in calibrating urban cellular automata. Int. J. Geogr. Inf. Sci. **29**(3), 349–374 (2015)
5. Brest, J., Boskovic, B., Greiner, S., Zumer, V., Maucec, M.S.: Performance comparison of self-adaptive and adaptive differential evolution algorithms. Soft Comput. **11**(7), 617–629 (2007)
6. Brest, J., Boskovic, B., Zamuda, A., Fister, I., Maucec, M.: Self-adaptive differential evolution algorithm with a small and varying population size. In: 2012 IEEE Congress on Evolutionary Computation (CEC), pp. 1–8 (2012)
7. Brest, J., Greiner, S., Boskovic, B., Mernik, M., Zumer, V.: Self-adapting control parameters in differential evolution: a comparative study on numerical benchmark problems. IEEE Trans. Evol. Comput. **10**(6), 646–657 (2006)
8. Brest, J., Maucec, M.S.: Self-adaptive differential evolution algorithm using population size reduction and three strategies. Soft Comput. **15**(11), 2157–2174 (2011)
9. Brest, J., Zumer, V., Maucec, M.: Self-adaptive differential evolution algorithm in constrained real-parameter optimization. In: IEEE Congress on Evolutionary Computation, 2006. CEC 2006, pp. 215–222 (2006)
10. Chai, T., Jin, Y., Sendhoff, B.: Evolutionary complex engineering optimization: opportunities and challenges. IEEE Comput. Intell. Mag. **8**(3), 12–15 (2013)
11. Chen, W., Weise, T., Yang, Z., Tang, K.: Large-scale global optimization using cooperative coevolution with variable interaction learning. In: Parallel Problem Solving from Nature. PPSN XI, Lecture Notes in Computer Science, vol. 6239, pp. 300–309. Springer, Berlin, Heidelberg (2010)
12. Cheng, S., Shi, Y., Qin, Q., Bai, R.: Swarm intelligence in big data analytics. In: Yin, H., Tang, K., Gao, Y., Klawonn, F., Lee, M., Weise, T., Li, B., Yao, X. (eds.) Intelligent Data Engineering and Automated Learning—IDEAL 2013. Lecture Notes in Computer Science, vol. 8206, pp. 417–426. Springer, Berlin, Heidelberg (2013)
13. Cheng, S., Ting, T., Yang, X.S.: Large-scale global optimization via swarm intelligence. In: Koziel, S., Leifsson, L., Yang, X.S. (eds.) Solving Computationally Expensive Engineering Problems, Springer Proceedings in Mathematics & Statistics, vol. 97, pp. 241–253. Springer International Publishing (2014)
14. Doerner, K., Hartl, R.F., Reimann, M.: Cooperative ant colonies for optimizing resource allocation in transportation. In: Proceedings of the EvoWorkshops on Applications of Evolutionary Computing, pp. 70–79. Springer-Verlag (2001)

15. El-Abd, M., Kamel, M.S.: A taxonomy of cooperative particle swarm optimizers. Int. J. Comput. Intell. Res. **4** (2008)
16. Ergun, H., Van Hertem, D., Belmans, R.: Transmission system topology optimization for large-scale offshore wind integration. IEEE Trans. Sustain. Energy **3**(4), 908–917 (2012)
17. Eshelman, L.J., Schaffer, J.D.: Real-coded genetic algorithm and interval schemata. In: Foundation of Genetic Algorithms, pp. 187–202 (1993)
18. Esmin, A.A., Coelho, R., Matwin, S.: A review on particle swarm optimization algorithm and its variants to clustering high-dimensional data. Artif. Intell. Rev. 1–23 (2013)
19. Fernandes, C., Rosa, A.: A study on non-random mating and varying population size in genetic algorithms using a royal road function. In: Proceedings of the 2001 Congress on Evolutionary Computation, 2001, vol. 1, pp. 60–66 (2001)
20. Fernandes, C., Rosa, A.: Self-adjusting the intensity of assortative mating in genetic algorithms. Soft Comput. **12**(10), 955–979 (2008)
21. Hansen, N., Ostermeier, A.: Completely derandomized self-adaptation in evolution strategies. Evol. Comput. **9**(2), 159–195 (2001)
22. Hinterding, R.: Gaussian mutation and self-adaption for numeric genetic algorithms. In: IEEE International Conference on Evolutionary Computation, 1995, vol. 1, pp. 384–389 (1995)
23. Huang, V., Qin, A., Suganthan, P.: Self-adaptive differential evolution algorithm for constrained real-parameter optimization. In: IEEE Congress on Evolutionary Computation, 2006. CEC 2006, pp. 17–24 (2006)
24. Lastra, M., Molina, D., Bentez, J.M.: A high performance memetic algorithm for extremely high-dimensional problems. Inf. Sci. **293**, 35–58 (2015)
25. LaTorre, A.: A framework for hybrid dynamic evolutionary algorithms: multiple offspring sampling (MOS). Ph.D. thesis, Universidad Politecnica de Madrid (2009)
26. LaTorre, A., Muelas, S., Pea, J.M.: A comprehensive comparison of large scale global optimizers. Inf. Sci. (in press) (2015)
27. LaTorre, A., Muelas, S., Peña, J.M.: A mos-based dynamic memetic differential evolution algorithm for continuous optimization: a scalability test. Soft Comput. **15**(11), 2187–2199 (2011)
28. LaTorre, A., Muelas, S., Pena, J.M.: Multiple offspring sampling in large scale global optimization. In: 2012 IEEE Congress on Evolutionary Computation (CEC), pp. 1–8 (2012)
29. LaTorre, A., Muelas, S., Pena, J.M.: Large scale global optimization: experimental results with MOS-based hybrid algorithms. In: 2013 IEEE Congress on Evolutionary Computation (CEC), pp. 2742–2749 (2013)
30. Li, X., Yao, X.: Cooperatively coevolving particle swarms for large scale optimization. IEEE Trans. Evol. Comput. **16**(2), 210–224 (2012)
31. Liu, J., Lampinen, J.: A fuzzy adaptive differential evolution algorithm. Soft Comput. **9**(6), 448–462 (2005)
32. Liu, Y., Yao, X., Zhao, Q.: Scaling up fast evolutionary programming with cooperative coevolution. In: Proceedings of the 2001 Congress on Evolutionary Computation, Seoul, Korea, pp. 1101–1108 (2001)
33. Lu, Y., Wang, S., Li, S., Zhou, C.: Particle swarm optimizer for variable weighting in clustering high-dimensional data. Mach. Learn. **82**(1), 43–70 (2011)
34. Mahdavi, S., Shiri, M.E., Rahnamayan, S.: Metaheuristics in large-scale global continues optimization: a survey. Inf. Sci. **295**, 407–428 (2015)
35. Molina, D., Lozano, M., García-Martínez, C., Herrera, F.: Memetic algorithms for continuous optimisation based on local search chains. Evol. Comput. **18**(1), 27–63 (2010)
36. Molina, D., Lozano, M., Herrera, F.: Ma-sw-chains: Memetic algorithm based on local search chains for large scale continuous global optimization. In: 2010 IEEE Congress on Evolutionary Computation (CEC), pp. 1–8 (2010)
37. Molina, D., Lozano, M., Sánchez, A.M., Herrera, F.: Memetic algorithms based on local search chains for large scale continuous optimisation problems: Ma-ssw-chains. Soft Comput. **15**(11), 2201–2220 (2011)
38. Moscato, P.: On evolution, search, optimization, genetic algorithms and martial arts: towards memetic algorithms. Technical Report, Caltech Concurrent Computation Program Report 826, Caltech, Pasadena, California (1989)

39. Moscato, P.: New ideas in optimization. In: Memetic Algorithms: A Short Introduction, pp. 219–234. McGraw-Hill Ltd., UK, Maidenhead, UK, England (1999)
40. Mühlenbein, H., Schlierkamp-Voosen, D.: Predictive models for the breeder genetic algorithm I. continuous parameter optimization. Evol. Comput. **1**(1), 25–49 (1993)
41. Omidvar, M.N., Li, X., Mei, Y., Yao, X.: Cooperative co-evolution with differential grouping for large scale optimization. IEEE Trans. Evol. Comput. **18**(3), 378–393 (2014)
42. Omidvar, M.N., Li, X., Tang, K.: Designing benchmark problems for large-scale continuous optimization. Inf. Sci. (in press) (2015)
43. Omidvar, M.N., Li, X., Yang, Z., Yao, X.: Cooperative co-evolution for large scale optimization through more frequent random grouping. In: Proceedings of the IEEE Congress on Evolutionary Computation, pp. 1–8. IEEE (2010)
44. Omidvar, M.N., Li, X., Yao, X.: Cooperative co-evolution with delta grouping for large scale non-separable function optimization. In: IEEE Congress on Evolutionary Computation, pp. 1–8 (2010)
45. Omidvar, M.N., Li, X., Yao, X.: Smart use of computational resources based on contribution for cooperative co-evolutionary algorithms. Proceedings of the 13th Annual Conference on Genetic and Evolutionary Computation. GECCO'11, pp. 1115–1122. ACM, New York, NY, USA (2011)
46. Omidvar, M.N., Mei, Y., Li, X.: Effective decomposition of large-scale separable continuous functions for cooperative co-evolutionary algorithms. In: Proceedings of the IEEE Congress on Evolutionary Computation. IEEE (2014)
47. Parsopoulos, K.E.: Parallel cooperative micro-particle swarm optimization: a master-slave model. Appl. Soft Comput. **12**(11), 3552–3579 (2012)
48. Potter, M.A., De Jong, K.A.: A cooperative coevolutionary approach to function optimization. In: Proceedings of the International Conference on Evolutionary Computation. The Third Conference on Parallel Problem Solving from Nature: Parallel Problem Solving from Nature, PPSN III, pp. 249–257. Springer-Verlag (1994)
49. Potter, M.A., De Jong, K.A.: Cooperative coevolution: an architecture for evolving coadapted subcomponents. Evol. Comput. **8**(1), 1–29 (2000)
50. Qin, A., Huang, V., Suganthan, P.: Differential evolution algorithm with strategy adaptation for global numerical optimization. IEEE Trans. Evol. Comput. **13**(2), 398–417 (2009)
51. Qin, A.K., Suganthan, P.N.: Self-adaptive differential evolution algorithm for numerical optimization. In: Proceedings of the IEEE Congress on Evolutionary Computation, CEC 2005, 2–4 Sept 2005, Edinburgh, UK, pp. 1785–1791. IEEE (2005)
52. Ray, T., Yao, X.: A cooperative coevolutionary algorithm with correlation based adaptive variable partitioning. In: Proceedings of the IEEE Congress on Evolutionary Computation, pp. 983–989. IEEE (2009)
53. Salomon, R.: Reevaluating genetic algorithm performance under coordinate rotation of benchmark functions—a survey of some theoretical and practical aspects of genetic algorithms. BioSystems **39**, 263–278 (1995)
54. Snchez-Ante, G., Ramos, F., Frausto, J.: Cooperative simulated annealing for path planning in multi-robot systems. MICAI 2000: Advances in Artificial Intelligence. LNCS, vol. 1793, pp. 148–157. Springer, Berlin, Heidelberg (2000)
55. Solis, F.J., Wets, R.J.B.: Minimization by Random Search Techniques. Math. Oper. Res. **6**(1), 19–30 (1981)
56. Storn, R., Price, K.: Differential evolution a simple and efficient heuristic for global optimization over continuous spaces. J. Glob. Optim. **11**(4), 341–359 (1997)
57. Sun, L., Yoshida, S., Cheng, X., Liang, Y.: A cooperative particle swarm optimizer with statistical variable interdependence learning. Inf. Sci. **186**(1), 20–39 (2012)
58. Sutton, R.S., Barto, A.G.: Reinforcement Learning: An Introduction. MIT Press (1998)
59. Takahashi, M., Kita, H.: A crossover operator using independent component analysis for real-coded genetic algorithms. In: Proceedings of the 2001 Congress on Evolutionary Computation, 2001, vol. 1, pp. 643–649 (2001)
60. Talbi, E.G.: A taxonomy of hybrid metaheuristics. J. Heuristics **8**(5), 541–564 (2002)

61. Tang, K., Li, X., Suganthan, P.N., Yang, Z., Weise, T.: Benchmark functions for the CEC'2010 special session and competition on large-scale global optimization. http://nical.ustc.edu.cn/cec10ss.php
62. Tang, K., Yang, Z., Weise, T.: Special session on evolutionary computation for large scale global optimization at 2012 IEEE World Congress on Computational Intelligence (cec@wcci-2012). Technical report, Hefei, Anhui, China: University of Science and Technology of China (USTC), School of Computer Science and Technology, Nature Inspired Computation and Applications Laboratory (NICAL) (2012)
63. Tang, K., Yao, X., Suganthan, P., MacNish, C., Chen, Y., Chen, C., Yang, Z.: Benchmark functions for the CEC'2008 special session and competition on large scale global optimization
64. Teo, J.: Exploring dynamic self-adaptive populations in differential evolution. Soft Comput. **10**(8), 673–686 (2006)
65. Thomas, S., Jin, Y.: Reconstructing biological gene regulatory networks: where optimization meets big data. Evol. Intell. **7**(1), 29–47 (2014)
66. Trunfio, G.A.: Enhancing the firefly algorithm through a cooperative coevolutionary approach: an empirical study on benchmark optimisation problems. IJBIC **6**(2), 108–125 (2014)
67. Trunfio, G.A.: A cooperative coevolutionary differential evolution algorithm with adaptive subcomponents. Proc. Comput. Sci. **51**, 834–844 (2015)
68. Tseng, L.Y., Chen, C.: Multiple trajectory search for large scale global optimization. In: IEEE Congress on Evolutionary Computation, 2008. CEC 2008 (IEEE World Congress on Computational Intelligence), pp. 3052–3059 (2008)
69. van den Bergh, F., Engelbrecht, A.P.: A cooperative approach to particle swarm optimization. IEEE Trans. Evol. Comput. **8**(3), 225–239 (2004)
70. Wang, Y., Huang, J., Dong, W.S., Yan, J.C., Tian, C.H., Li, M., Mo, W.T.: Two-stage based ensemble optimization framework for large-scale global optimization. Eur. J. Oper. Res. **228**(2), 308–320 (2013)
71. Wang, Y., Li, B.: Two-stage based ensemble optimization for large-scale global optimization. In: 2010 IEEE Congress on Evolutionary Computation (CEC), pp. 1–8 (2010)
72. Weicker, K., Weicker, N.: On the improvement of coevolutionary optimizers by learning variable interdependencies. In: 1999 Congress on Evolutionary Computation, pp. 1627–1632. IEEE Service Center, Piscataway, NJ (1999)
73. Xue, F., Sanderson, A., Bonissone, P., Graves, R.: Fuzzy logic controlled multi-objective differential evolution. In: The 14th IEEE International Conference on Fuzzy Systems, 2005. FUZZ'05, pp. 720–725 (2005)
74. Yang, X.S.: Firefly algorithms for multimodal optimization. In: Stochastic Algorithms: Foundations and Applications, 5th International Symposium, SAGA 2009, Sapporo, Japan, 26–28 Oct 2009. Proceedings, LNCS, vol. 5792, pp. 169–178. Springer (2009)
75. Yang, Z., Tang, K., Yao, X.: Large scale evolutionary optimization using cooperative coevolution. Inf. Sci. **178**(15), 2985–2999 (2008)
76. Yang, Z., Tang, K., Yao, X.: Multilevel cooperative coevolution for large scale optimization. In: IEEE Congress on Evolutionary Computation, pp. 1663–1670. IEEE (2008)
77. Yang, Z., Tang, K., Yao, X.: Self-adaptive differential evolution with neighborhood search. In: IEEE Congress on Evolutionary Computation, 2008. CEC 2008. (IEEE World Congress on Computational Intelligence), pp. 1110–1116 (2008)
78. Yang, Z., Tang, K., Yao, X.: Scalability of generalized adaptive differential evolution for large-scale continuous optimization. Soft Comput. **15**(11), 2141–2155 (2011)
79. Zhang, J., Sanderson, A.: Jade: adaptive differential evolution with optional external archive. IEEE Trans. Evol. Comput. **13**(5), 945–958 (2009)

Author Biography

Giuseppe A. Trunfio works as a researcher in computer science at the University of Sassari (Italy). He holds a Ph.D. degree in computational mechanics from the University of Calabria (Italy) and has served as a research associate at the Italian National Research Council. His current research interests include modelling and simulation, optimization metaheuristics and high performance computing. In particular, he has been involved in several interdisciplinary research projects, where his contributions mainly concerned the development of simulation models, the design and implementation of decision support systems, and advanced applications of computational methods. Moreover, he is the author of many research studies published in international journals, conference proceedings and book chapters.

Convergent Parallel Algorithms for Big Data Optimization Problems

Simone Sagratella

Abstract When dealing with big data problems it is crucial to design methods able to decompose the original problem into smaller and more manageable pieces. Parallel methods lead to a solution by concurrently working on different pieces that are distributed among available agents, so that exploiting the computational power of multi-core processors and therefore efficiently solving the problem. Beyond gradient-type methods, that can of course be easily parallelized but suffer from practical drawbacks, recently a convergent decomposition framework for the parallel optimization of (possibly non-convex) big data problems was proposed. Such framework is very flexible and includes both fully parallel and fully sequential schemes, as well as virtually all possibilities in between. We illustrate the versatility of this parallel decomposition framework by specializing it to different well-studied big data problems like LASSO, logistic regression and support vector machines training. We give implementation guidelines and numerical results showing that proposed parallel algorithms work very well in practice.

Keywords Decomposition algorithm • Parallel optimization • LASSO • Logistic regression • Support vector machine

1 Introduction

Finding a solution of a big data problem could be a very prohibitive task if we use standard optimization methods since time and memory consumption could be unsustainable. Therefore many decomposition methods were proposed in the last decades. Decomposition methods are able to solve a big data problem by solving a sequence of smaller sub-problems that, due to their medium sized data, can be

S. Sagratella (✉)
Department of Computer, Control and Management Engineering,
Sapienza University of Rome, Via Ariosto 25, 00185 Rome, Italy
e-mail: sagratella@dis.uniroma1.it

A. Emrouznejad (ed.), *Big Data Optimization: Recent Developments and Challenges*, Studies in Big Data 18, DOI 10.1007/978-3-319-30265-2_20

processed by using standard optimization methods. Decomposition methods can be divided into sequential (Gauss-Seidel or Gauss-Southwell) ones, in which sub-problems are solved sequentially since the input of one is the output of another, and into parallel (Jacobi) ones, in which some sub-problems can be solved simultaneously since they do not need as input the output of the others. Of course parallel decomposition methods are more efficient since they can exploit the computational power of multi-core processors. Gradient-type methods can be easily seen as parallel decomposition methods [1], but, in spite of their good theoretical convergence bounds, they suffer from practical drawbacks since they do not exploit second order information.

In this chapter we present a parallel decomposition framework for the solution of many big data problems that is provably convergent, can use second order information and works well in practice. Main features of the proposed algorithmic framework are the following:

- it is parallel, with a degree of parallelism that can be chosen by the user and that, given a block decomposition of the variables, can go from a complete simultaneous update of all blocks at each iteration to the case in which only one block is updated at each iteration (virtually all possibilities in between are covered);
- it is provably convergent to a solution of smooth and non-smooth convex problems and it is provably convergent to a stationary point of non-convex problems;
- it can exploit the original structure of the problem as well as second order information;
- it allows for inexact solution of the sub-problems, a feature that can be very useful in practice since the cost of computing the exact solution of all sub-problems could be high in some applications;
- it appears to be numerically efficient on many big data applications.

Many real big data applications can be tackled by using the parallel methods described in this chapter. In the following sections, we will discuss on specific optimization problems arising from big data machine learning applications. In particular we define parallel algorithms for LASSO, logistic regression and support vector machine training, that can be effectively used with classification and regression purposes in many fields such as compressed sensing and sensor networks.

In Sect. 2 we present the parallel algorithmic framework for smooth problems with separable feasible sets. In Sect. 3 we specialize it for non-smooth problems by giving some specific implementations for big data applications. In Sect. 4 we extend the parallel algorithmic framework to solve big data problems with non-separable feasible sets. Technical details for topics treated in Sects. 2 and 3 can be found in [2], while those treated in Sect. 4 can be found in [3].

2 Smooth Problems with Separable Feasible Sets

In this section we consider optimization problems in which the feasible set is convex and separable in blocks of variables. Exploiting the block-separability property, we can define different agents each controlling one or more of such blocks of variables. In our parallel framework all agents can act concurrently optimizing over their own blocks of variables, in a way that will be clarified below, in order to achieve a solution of the original problem. The key factors yielding effectiveness of the parallel framework are:

1. the structure of the sub-problems solved by each agent;
2. a phase of fitting in which variables from all agents are gathered and the current point is updated.

Before introducing the parallel algorithmic framework for different types of optimization problems stemmed from big data applications, we need to formalize all principal aspects.

Let $X := X_1 \times \ldots \times X_N \subseteq \Re^n$ be a feasible set defined as a Cartesian product of lower dimensional non-empty, closed and convex sets $X_i \subseteq \Re^{n_i}$, and let $\mathbf{x} \in \Re^n$ be partitioned accordingly: $\mathbf{x} := (\mathbf{x}_1, \ldots, \mathbf{x}_N)$. Therefore we obtain N different blocks of variables and we say that the ith block contains variables $\mathbf{x}_i$. All problems described in this section encompass these features.

Now let us consider a possibly non-convex smooth function $F: \Re^n \to \Re$, and suppose to minimize it over X:

$$\min_{\mathbf{x}} F(\mathbf{x}), \quad \text{s.t. } \mathbf{x} \in X. \tag{1}$$

Clearly if F were separable as X, that is $F(\mathbf{x}) := \sum_{i=1}^{N} f_i(\mathbf{x}_i)$, then solution of Problem (1) would be equivalent to solving N smaller optimization problems: $\min_{\mathbf{x}_i} f_i(\mathbf{x}_i)$, s.t. $\mathbf{x}_i \in X_i$, $i = 1, \ldots, N$. Such favourable structure is hardly encountered in real big data problems. Thus here we consider F non-separable, and we assume that ∇F is Lipschitz continuous on X.

In order to define the sub-problems solved by agents, we have to introduce an approximating function $P_i(\mathbf{z}; \mathbf{w}): X_i \times X \to \Re$ for each block of variables having the following properties (we denote by ∇P_i the partial gradient of P_i with respect to the first argument $\mathbf{z}$):

- $P_i(\bullet; \mathbf{w})$ is convex and continuously differentiable on X_i for all $\mathbf{w} \in X$;
- $\nabla P_i(\mathbf{x}_i; \mathbf{x}) = \nabla_{\mathbf{x}_i} F(\mathbf{x})$ for all $\mathbf{x} \in X$;
- $\nabla P_i(\mathbf{z}; \bullet)$ is Lipschitz continuous on X for all $\mathbf{z} \in X_i$.

Such function P_i should be regarded as a simple convex approximation of F at the point $\mathbf{x}$ with respect to the block of variables $\mathbf{x}_i$ that preserves the first order properties of F with respect to $\mathbf{x}_i$. Minimizing the approximating function P_i instead of the original function F may be preferable when optimizing F is too costly or difficult (e.g. if F is not convex). Therefore using P_i in some situations can facilitate the task performed by agents.

We are ready to reveal the first of the two key factors stated above, that is the structure of the sub-problems solved by each agent. At any point $\mathbf{x}^k \in X$ the sub-problem related to the ith block of variables is the following:

$$\min_{\mathbf{x}_i} P_i(\mathbf{x}_i; \mathbf{x}^k) + \frac{\tau_i^k}{2} \|\mathbf{x}_i - \mathbf{x}_i^k\|^2, \quad \text{s.t. } \mathbf{x}_i \in X_i, \tag{2}$$

where τ_i^k is a non-negative parameter. In each Sub-problem (2) F is replaced by the corresponding approximating function P_i and a proximal term is added to make the overall approximation strongly convex. Note that if $P_i(\bullet; \mathbf{x}^k)$ is already uniformly strongly convex, one can avoid the proximal term and set $\tau_i^k = 0$. There are two reasons for which Sub-problems (2) must be strongly convex: on the one hand it is useful since then Sub-problems (2) has a unique solution and then the tasks of all agents are well defined, and on the other hand it is necessary in order to produce at each iteration a direction that is "sufficiently" descent. For the sake of notational simplicity, we denote by $\mathbf{y}_i(\mathbf{x}^k, \tau_i^k)$ the unique solution of Sub-problem (2) at $\mathbf{x}^k$ when using a proximal parameter τ_i^k.

We are now ready to formally introduce the parallel algorithmic framework.

Algorithm 1 (Parallel Algorithmic Framework for separable sets)

Step 0: (Initialization)
Set $\mathbf{x}^0 \in X$, $\tau_i^0 \geq 0$ for all i and $k = 0$.

Step 1: (Termination test)
If $\mathbf{x}^k$ is a stationary point of Problem (1) then STOP.

Step 2: (Blocks selection)
Choose a subset of blocks $J^k \subseteq \{1, \ldots, N\}$.

Step 3: (Search direction)
For all blocks $i \in J^k$ compute in parallel $\mathbf{y}_i(\mathbf{x}^k, \tau_i^k)$ (solution of (2)),
set $\mathbf{d}_i^k = \mathbf{y}_i(\mathbf{x}^k, \tau_i^k) - \mathbf{x}_i^k$ for all blocks $i \in J^k$ and
set $\mathbf{d}_i^k = \mathbf{0}$ for all blocks $i \notin J^k$.

Step 4: (Stepsize)
Choose $\alpha^k \geq 0$.

Step 5: (Update)
Set $\mathbf{x}^{k+1} = \mathbf{x}^k + \alpha^k \mathbf{d}^k$, set $\tau_i^{k+1} \geq 0$ for all i, set $k = k + 1$, and go to Step 1.

At Step 0 Algorithm 1 starts from a feasible point $\mathbf{x}^0$, and if necessary set the initial proximal parameters.

At Step 1 a check on the stopping criterion is performed. This is typically based on merit evaluations using first order information.

At Step 2 a subset J^k of the blocks of variables is selected. Note that only blocks in J^k are used at Step 3 in order to compute the search direction and then at Step 5 to update the corresponding blocks. This possibility of updating only some blocks has been observed to be very effective in practice. As stated below, the algorithmic framework is guaranteed to converge simply if J^k contains at least one block j whose computed direction satisfies the following condition:

$$\left\|\mathbf{d}_j^k\right\| \geq \rho \max_{i \in \{1,\dots,N\}} \left\|\mathbf{d}_i^k\right\|, \tag{3}$$

for any $\rho > 0$. It is worth noting that, depending on specific applications, condition (3) can be guaranteed by using simple heuristics.

At Step 3 one search direction for each block is computed by solving Sub-problem (2) for those blocks selected at Step 2. This is typically the main computational burden and, in this algorithmic framework, it can be parallelized by distributing blocks among the agents.

At Step 4 a stepsize α^k is computed. The choice of a suitable stepsize α^k is crucial for the convergence of the method. As explained below, α^k can be computed by using a line-search procedure or a diminishing rule. This is the core of the fitting phase mentioned above as a key factor for the effectiveness of the method.

At Step 5 current point and proximal parameters are updated.

Algorithm 1 is very flexible since we can always choose $J^k \equiv \{1, \dots, N\}$ resulting in the simultaneous update of all the blocks and obtaining a full Jacobi scheme, or we can always update one single block per iteration and obtaining a Gauss-Southwell kind of method. All possibilities in between and classical Gauss-Seidel methods can also be derived.

Theorem 1 *Let* $\{\mathbf{x}^k\}$ *be the sequence generated by Algorithm 1 and let* $\rho > 0$. *Assume that Problem (1) has a solution and that the following conditions are satisfied for all* k:

(a) J^k *contains at least one element satisfying (3);*
(b) *for all* $i \in J^k$: τ_i^k *are such that Sub-problem (2) is strongly convex;*
(c) *stepsize* α^k *is such that*

$$\textit{either } F(\mathbf{x}^k + \alpha^k \mathbf{d}^k) \leq F(\mathbf{x}^k) + \theta \alpha^k \nabla F(\mathbf{x}^k)^T \mathbf{d}^k, \quad \theta \in (0,1),$$
$$\textit{or } \alpha^k \in (0,1],\ \lim_{k\to\infty} \alpha^k = 0 \textit{ and } \textstyle\sum_{k=0}^{\infty} \alpha^k = +\infty.$$

Then either Algorithm 1 converges in a finite number of iterations to a stationary point of Problem (1) or every limit point of $\{\mathbf{x}^k\}$ *(at least one such points exists) is a stationary point of Problem (1).*

Theorem 1 gives simple conditions for the convergence of Algorithm 1 to a stationary point of Problem 1. Note that in case F is convex all stationary points are solutions and therefore Algorithm 1 converges to a solution of Problem (1).

While we discussed on conditions (a) and (b) of Theorem 1 above, we must spend some words on condition (c). We have two possibilities to compute a "good" stepsize α^k: with an Armijo line-search or with a diminishing rule. If we implement the first strategy we obtain an effective method with a monotone decreasing of the objective function. This is very often the best choice if the numerical burden needed to evaluate the objective value is not too heavy. Moreover if F is rather "easy" (e.g. it is quadratic) then performing an exact line-search is for sure a mandatory task since the numerical effort done to compute the best stepsize is counterbalanced by large decreases of the objective value. However in some cases it could be convenient to use the diminishing stepsize rule, since it does not need any objective function evaluation.

In many big data problems it can be useful to further reduce the computational effort of the agents by solving Sub-problems (2) with an approximate method that yields inexact solutions. Without burdening the discussion with further technical details, it is important to say that convergence of Algorithm 1 can be proved also if Sub-problems (2) are solved inexactly and this just implies only few additional technical conditions.

Another important feature of Algorithm 1 is that it can incorporate hybrid parallel-sequential (Jacobi–Gauss-Seidel) schemes wherein blocks of variables are updated simultaneously by agents, but blocks assigned to an agent are updated sequentially in a Gauss-Seidel fashion for its point of view. This procedure seems particularly well suited when the number of blocks N is greater than the number of available agents. We denote this hybrid method as Gauss-Jacobi method, whose convergence can be proved under conditions similar to those of Theorem 1.

3 Non-smooth Problems with Separable Feasible Sets

A type of non-smooth non-convex optimization problem that arises in many big data applications has the following form:

$$\min_{\mathbf{x}} \ F(\mathbf{x}) + G(\mathbf{x}), \quad \text{s.t. } \mathbf{x} \in X, \tag{4}$$

where $X := X_1 \times \ldots \times X_N \subseteq \mathfrak{R}^n$ is a feasible set defined, as in Sect. 2, as a Cartesian product of lower dimensional non-empty, closed and convex sets $X_i \subseteq \mathfrak{R}^{n_i}$, $F: \mathfrak{R}^n \to \mathfrak{R}$ is a (possibly non-convex) smooth function, and $G: \mathfrak{R}^n \to \mathfrak{R}$ is a non-smooth continuous and convex function which is block separable according to X. Therefore letting $\mathbf{x} \in \mathfrak{R}^n$ be partitioned accordingly, that is $\mathbf{x} := (\mathbf{x}_1, \ldots, \mathbf{x}_N)$, we can write: $G(\mathbf{x}) := \sum_{i=1}^{N} g_i(\mathbf{x}_i)$, where all functions g_i are non-smooth, continuous and convex.

Problem (4) arises in many fields of engineering, so diverse as compressed sensing, neuro-electromagnetic imaging, sensor networks, basis pursuit denoising, machine learning, data mining, genomics, meteorology, tensor factorization and completion, geophysics, and radio astronomy. Usually the non-smooth term G is used to promote sparsity of the solution, therefore obtaining a parsimonious representation of some phenomenon at hand. Just to mention some examples, we list typical instances of Problem (4):

- $F(\mathbf{x}) = (1/2)\,\|\mathbf{A}\mathbf{x} - \mathbf{b}\|^2$, $G(\mathbf{x}) = c\,\|\mathbf{x}\|_1$ and $X = \Re^n$, with $\mathbf{A} \in \Re^{m \times n}$, $\mathbf{b} \in \Re^m$, $c > 0$ and $m > 0$ given constants; this is the renowned and much studied LASSO problem [4] for which we report below an implementation of parallel algorithm;
- $F(\mathbf{x}) = (1/2)\,\|\mathbf{A}\mathbf{x} - \mathbf{b}\|^2$, $G(\mathbf{x}) = c\sum_{i=1}^{N} \|\mathbf{x}_i\|$ and $X = \Re^n$, with $\mathbf{A} \in \Re^{m \times n}$, $\mathbf{b} \in \Re^m$, $c > 0$ and $m > 0$ given constants; this is the group LASSO problem [5];
- $F(\mathbf{x}) = \sum_{j=1}^{m} \log\left(1 + \exp\left(-a_j\,\mathbf{y}_j^T\,\mathbf{x}\right)\right)$, $G(\mathbf{x}) = c\,\|\mathbf{x}\|_1$ (or $G(\mathbf{x}) = c\sum_{i=1}^{N} \|\mathbf{x}_i\|$) and $X = \Re^n$, with $\mathbf{y}_i \in \Re^n$, $a_i \in \Re$, $c > 0$ and $m > 0$ given constants; this is the sparse logistic regression problem [6, 7] for which we give below implementation details of a parallel algorithm;
- $F(\mathbf{x}) = \sum_{j=1}^{m} \max\left\{0,\, 1 - a_j\,\mathbf{y}_j^T\,\mathbf{x}\right\}^2$, $G(\mathbf{x}) = c\,\|\mathbf{x}\|_1$ and $X = \Re^n$, with $\mathbf{y}_i \in \Re^n$, $a_i \in \{-1, 1\}$, $c > 0$ and $m > 0$ given constants; this is the L1-regularized L2-loss Support Vector Machine problem [8];
- other problems that can be cast in the form (4) include the Dictionary Learning problem for sparse representation [9], the Nuclear Norm Minimization problem, the Robust Principal Component Analysis problem, the Sparse Inverse Covariance Selection problem and the Nonnegative Matrix Factorization problem [10].

The algorithmic framework for the solution of Problem (4) is essentially the same as Algorithm 1. The only differences are at Step 1 in which iterations must be stopped if $\mathbf{x}^k$ is a stationary point of Problem (4) (instead of Problem (1)), and at Step 3 in which agents have to compute (in parallel) $\mathbf{y}_i\left(\mathbf{x}^k, \tau_i^k\right)$. In particular, in order to solve Problem (4), $\mathbf{y}_i\left(\mathbf{x}^k, \tau_i^k\right)$ must be the solution of a sub-problem slightly different from (2):

$$\min_{\mathbf{x}_i}\; P_i\left(\mathbf{x}_i; \mathbf{x}^k\right) + g_i(\mathbf{x}_i) + \frac{\tau_i^k}{2}\left\|\mathbf{x}_i - \mathbf{x}_i^k\right\|^2, \quad \text{s.t. } \mathbf{x}_i \in X_i, \tag{5}$$

note that the only difference from (2) is the presence of function g_i, since all assumptions and considerations made on P_i in Sect. 2 are still valid.

Convergence conditions are the same as those in Theorem 1, the only one difference is about condition (c) in case one uses the line-search. In particular standard line-search methods proposed for smooth functions cannot be applied to objective function of Problem (4) due to the non-smooth part G, and then we need to rely on slightly different and well known line-search procedures for non-smooth functions like this one: $F\left(\mathbf{x}^k + \alpha^k\mathbf{d}^k\right) + G\left(\mathbf{x}^k + \alpha^k\mathbf{d}^k\right) \le F\left(\mathbf{x}^k\right) + G\left(\mathbf{x}^k\right) - \theta\,\alpha^k\left\|\mathbf{d}^k\right\|^2$, with $\theta \in (0, 1)$.

We are now in a position to give specific algorithms for two well-known L1-regularized big data problems that have been mentioned above: LASSO and sparse logistic regression. First of all we introduce a valid optimality measure function $\|Z(\mathbf{x})\|_\infty$ that is used to obtain an approximate termination test for L1-regularized problems:

$$Z(\mathbf{x}) := \nabla F(\mathbf{x}) - \prod_{[-c,c]^n} (\nabla F(\mathbf{x}) - \mathbf{x}), \tag{6}$$

where $\prod_{[-c,c]^n}(\mathbf{z})$ is the projection of $\mathbf{z}$ over the set $[-c,c]^n$ (we recall that c is the parameter before the L1-norm). Note that such projection can be efficiently computed since it acts component-wise on $\mathbf{z}$, i.e. $[-c,c]^n := [-c,c] \times \ldots \times [-c,c]$, and note that $Z(\mathbf{x}) = \mathbf{0}$ is equivalent to the standard necessary optimality conditions for the unconstrained L1-regularized problems.

Algorithm 2 (Parallel Algorithm for LASSO)

Step 0: (Initialization)

Set $\mathbf{x}^0 = \mathbf{0}$, $\tau_i^0 = \frac{\mathrm{tr}(\mathbf{A}^T\mathbf{A})}{2n}$ for all i, $\alpha^0 = 0.9$, $\varepsilon > 0$, $\rho \in (0,1]$, and set $k = 0$.

Step 1: (Termination test)

If $Z(\mathbf{x}^k) \leq \varepsilon$ then STOP.

Step 2-3: (Blocks selection and Search direction)

Send $\mathbf{x}^k$ and all τ_i^k to available agents;

get back from the agents all $\mathbf{y}_i(\mathbf{x}^k, \tau_i^k)$ (solutions of (5));

set $M(\mathbf{x}^k) := \max_{i \in \{1,\ldots,N\}} \|\mathbf{y}_i(\mathbf{x}^k, \tau_i^k) - \mathbf{x}_i^k\|$;

set $J^k := \{i \in \{1,\ldots,N\} : \|\mathbf{y}_i(\mathbf{x}^k, \tau_i^k) - \mathbf{x}_i^k\| \geq \rho M(\mathbf{x}^k)\}$;

set $\mathbf{d}_i^k = \mathbf{y}_i(\mathbf{x}^k, \tau_i^k) - \mathbf{x}_i^k$ for all blocks $i \in J^k$ and

set $\mathbf{d}_i^k = \mathbf{0}$ for all blocks $i \notin J^k$.

Step 4-5: (Stepsize and Update)

Set $\mathbf{z}^k = \mathbf{x}^k + \alpha^k \mathbf{d}^k$;

if $F(\mathbf{z}^k) + G(\mathbf{z}^k) < F(\mathbf{x}^k) + G(\mathbf{x}^k)$ then

set $\mathbf{x}^{k+1} = \mathbf{z}^k$ and $\tau_i^{k+1} = 0.9\tau_i^k$ for all i

else

set $\mathbf{x}^{k+1} = \mathbf{x}^k$ and $\tau_i^{k+1} = 2\tau_i^k$ for all i;

set $\alpha^{k+1} = \alpha^k(1 - 10^{-7}\alpha^k)$;

set $k = k + 1$, and go to Step 1.

At Step 0 variables and parameters are initialized to some values that proved to work well in practice: the starting point is set to zero, the proximal parameters are all initialized to the half of the mean of the eigenvalues of $\nabla^2 F$, and the initial stepsize is set to 0.9 (the diminishing rule is used).

At Step 1 the stopping criterion uses the merit value (6) and the tolerance parameter ε. Typical values for ε are 10^{-2} or 10^{-3}.

Step 2–3 produces the search direction. First of all, current point and proximal parameters are sent to agents in order to let them properly construct their sub-problems. Then agents compute in parallel solutions of Sub-problems (5). In particular in (5) directly function F is used since it is quadratic and an approximating function P is unnecessary. It is worth mentioning that if blocks contain only one single variable, then solution of (5) can be computed by using a simple analytical soft thresholding formula [1]. After all agents send back all solutions $\mathbf{y}_i(\mathbf{x}^k, \tau_i^k)$, subset J^k is defined according to rule (3). Finally the search direction is computed.

At Step 4–5 current point, proximal parameters and stepsize are updated. In particular if current iteration produces a decreasing in the objective function then current point is updated with the new guess and all proximal parameters are decreased (good iteration), else current point is not updated and proximal parameters are increased (bad iteration). Finally stepsize is updated by using a diminishing rule. These updating procedures are crucial for the good behaviour of the algorithm (see below for practical guidelines).

Algorithm 2 is essentially the same presented in [11, 2]. Numerical results, developed on a big cluster with many parallel processes (i.e. many agents), reported in these references, showed that Algorithm 2 outperforms other existing methods for LASSO.

Practical guidelines:

- Initializing all variables to zero can be very useful for L1-regularized big data problems, like LASSO, since a big amount of variables at solution are equal to zero, and therefore the starting point is quite close to the solution.
- Setting properly the stepsize and the proximal parameters is crucial in order to let the algorithm work well. In particular these two types of parameters collaborate to obtain a decrease of the objective value. If the stepsize is close to 1 then we obtain long steps but we could also obtain disastrous increments of the objective value, while if the stepsize is close to 0 then we obtain tiny steps but objective value always decreases. Conversely high values of the proximal parameters produce small steps with objective value decreasing, while small values do the contrary. Therefore we have to balance well all these parameters. The rule is: the bigger N is, the smaller must the stepsize be (or the bigger must the proximal parameters be). Values proposed here proved to be very good in practice for many different applications, see [2].
- Different values of ρ can produce different performances of the algorithm. Good values of ρ proved to be 0.5 or 0.75, since only variables that are sufficiently far from solution are updated, while other variables that probably are already near their optimum (typically equal to zero) are not updated.

- Unconstrained L1-regularized problems, like LASSO, have G and X separable element-wise. A good practical choice is to set all blocks containing one single variable in order to solve each sub-problem analytically. However different settings work as much well.
- Data transmissions with agents can be implemented by using a Message Passing Interface (MPI) paradigm. Only Send/Receive and Reduce operations are needed.

A specific parallel algorithm for solving sparse logistic regression problems is not too different from Algorithm 2. The main difference is the use of an approximating function P in the objective function of the sub-problems solved by the agents. For example, if we use the second order approximation of F as approximating function P, then we obtain sub-problems having the same structure of those of LASSO. This is a very useful issue since all considerations made above on LASSO problems solvability can be exploited also for logistic regression problems, see [2] for details.

4 Problems with Non-separable Feasible Sets

In this section we consider Problem (1) with a non-separable convex feasible set X. If we want to solve this problem we have to enrich Algorithm 1 with some additional features. In fact if we decomposed the problem in some fixed blocks of variables (in the same manner as done in Sects. 2 and 3) and then we used Algorithm 1 as it is, then we would obtain a generalized Nash equilibrium of a particular potential game, but not a solution of Problem (1) since, here, we are considering the case in which X is non-separable, see [3]. Details on this issue go beyond the scope of this book, we refer the interested reader to [12–14] for background material on generalized Nash equilibria.

The additional key feature, allowing the parallel algorithmic framework to lead to a solution of the problem (and not to a "weaker" Nash equilibrium), is very simple: blocks of variables must be not fixed, and they must vary from an iteration to another. The choice of which variables to insert in each block during iterations is crucial for the convergence of the algorithm and must be ruled by some optimality measure.

In order to better understand this additional feature, it is useful to introduce the support vector machine (SVM) training problem, whose dual is a big data problem with non-separable feasible set [15, 16], and to describe the parallel algorithm for finding one of its solutions.

The dual SVM training problem is defined as $F(\mathbf{x}) = (1/2)\,\mathbf{x}^T\mathbf{Q}\mathbf{x} - \mathbf{e}^T\mathbf{x}$ and $X = \{\mathbf{x} \in \Re^n\colon\ \mathbf{0} \leq \mathbf{x} \leq C\mathbf{e},\ \mathbf{y}^T\mathbf{x} = 0\}$, with $\mathbf{Q} \in \Re^{n \times n}$ positive semi-definite, $C > 0$ and $\mathbf{y} \in \{-1, 1\}^n$ given constants and $\mathbf{e} \in \Re^n$ vector of all ones. It is easy to see that constraint $\mathbf{y}^T\mathbf{x} = 0$ makes feasible set X non-separable. Many real SVMs applications are characterized by so big data that matrix $\mathbf{Q}$ cannot be entirely stored in memory and it is convenient to reconstruct blocks of it when dealing with the corresponding blocks of variables. For this reason, classical optimization methods that use first and second order information cannot be used and decomposition methods (preferably parallel ones) are the one way.

Given $\mathbf{x}^k$ feasible and given one block of variables $\mathbf{x}_i$, let us define the sub-problem of Problem (1), with non-separable set X, corresponding to block $\mathbf{x}_i$:

$$\min_{\mathbf{x}_i} F(\mathbf{x}_i, \mathbf{x}^k_{-i}) + \frac{\tau_i^k}{2} \|\mathbf{x}_i - \mathbf{x}_i^k\|^2, \quad \text{s.t. } (\mathbf{x}_i, \mathbf{x}^k_{-i}) \in X, \tag{7}$$

where $\mathbf{x}^k_{-i}$ are all elements of $\mathbf{x}^k$ that are not in block $\mathbf{x}_i$, for simplicity we use directly function F (but we could use the approximating function P as well), and the feasible set is defined as the projection of $\mathbf{x}^k_{-i}$, using the set X, on the space of variables $\mathbf{x}_i$. In the SVM case the feasible set of the sub-problem is simply defined as $\left\{ \mathbf{x}_i \in \Re^{n_i}:\ \mathbf{0} \le \mathbf{x}_i \le C\mathbf{e}_i,\ (\mathbf{y}_i^k)^T \mathbf{x}_i = (\mathbf{y}_i^k)^T \mathbf{x}_i^k \right\}$.

Algorithm 3 (Parallel Algorithmic Framework for non-separable sets)

Step 0: (Initialization)
Set $\mathbf{x}^0 \in X$, $\tau_i^0 \ge 0$ for all i and $k = 0$.
Step 1: (Termination test)
If $\mathbf{x}^k$ is a stationary point of Problem (1) then STOP.
Step 2: (Blocks definition and selection)
Define N^k blocks of variables (in order to make a partition),
choose a subset of blocks $J^k \subseteq \{1, \ldots, N^k\}$.
Step 3: (Search direction)
For all blocks $i \in J^k$ compute in parallel $\mathbf{y}_i(\mathbf{x}^k, \tau_i^k)$ (solution of (7)),
set $\mathbf{d}_i^k = \mathbf{y}_i(\mathbf{x}^k, \tau_i^k) - \mathbf{x}_i^k$ for all blocks $i \in J^k$ and
set $\mathbf{d}_i^k = \mathbf{0}$ for all blocks $i \notin J^k$.
Step 4: (Stepsize)
Choose $\alpha^k \ge 0$.
Step 5: (Update)
Set $\mathbf{x}^{k+1} = \mathbf{x}^k + \alpha^k \mathbf{d}^k$, set $\tau_i^{k+1} \ge 0$ for all i, set $k = k + 1$, and
go to Step 1.

It is easy to see that, the only one difference from Algorithm 1 lays in Step 2. In fact, as said above, when dealing with non-separable feasible sets, decomposition of variables must be not fixed during iterations. Then, at Step 2, Algorithm 3 redefines at each iteration N^k blocks of variables. Note that such blocks decomposition must be a partition and therefore each variable is in one and only one block.

Now we give a convergence result for Algorithm 3 in the SVM training case, however similar results could be given for other different problems with non-separable feasible sets. First of all we have to introduce an optimality measure, that is the most violating pair step: $M(\mathbf{x}^k) := \|\mathbf{z}^k - \mathbf{x}^k\|$, where $\mathbf{z}^k$ is the new point after the problem has been optimized over the most violating pair (which is the pair of variables that violates the most first order conditions at $\mathbf{x}^k$). Details on this issue can be found in [3], however it is worth noting that such optimality measure M can be computed by using some simple analytical formula.

Theorem 2 (Convergence for SVM training)
Let $\{\mathbf{x}^k\}$ be the sequence generated by Algorithm 3 and let $\rho > 0$. Assume that the following conditions are satisfied for all k:

(a) J^k *contains at least one element i satisfying* $\|\mathbf{d}_i^k\| \geq \rho M(\mathbf{x}^k)$;
(b) *for all* $i \in J^k$: τ_i^k *are such that Sub-problem (7) is strongly convex;*
(c) *stepsize* α^k *is such that*

$$\text{either } F(\mathbf{x}^k + \alpha^k \mathbf{d}^k) \leq F(\mathbf{x}^k) + \theta \alpha^k \nabla F(\mathbf{x}^k)^T \mathbf{d}^k, \quad \theta \in (0,1),$$
$$\text{or } \alpha^k \in (0,1], \lim_{k \to \infty} \alpha^k = 0 \text{ and } \sum_{k=0}^{\infty} \alpha^k = +\infty.$$

Then either Algorithm 3 converges in a finite number of iterations to a solution of the SVM training problem or every limit point of $\{\mathbf{x}^k\}$ (at least one such points exists) leads to a solution of the SVM training problem.

Only some words on condition (a) are in order, since other assumptions are very close to those of Theorem 1. It is important to say that condition (a) can be guaranteed by simply including in one block the most violating pair or any other pair of variables that sufficiently violates first order conditions. Therefore we can say that solving problems with non-separable feasible sets is just a little bit harder than solving those with separable ones. In fact we could easily give convergence results for other problems like the ν-SVM training [15] that is characterized by two linear equality constraints (instead of only one like for SVM training). The only one thing to do is to find a global optimality measure and to guarantee condition (a) of Theorem 2.

Finally, in order to take a picture of the good behaviour of Algorithm 3 for SVM training, in Fig. 1 we report numerical results for two problems taken from the LIBSVM data set (www.csie.ntu.edu.tw/~cjlin/libsvm): cod-rna (59535 training data and 8 features) and a9a (32561 training data and 123 features). In particular we implemented Algorithm 3 with $\mathbf{x}^0 = \mathbf{0}$ and $\tau_i^k = 10^{-8}$ for all i and all k, at Step 2 we constructed the blocks by grouping the most violating pairs (that is the first block contains the most violating pairs, the second block contains the remaining most violating pairs, and so on), and at Step 4 we used an exact line-search. In Fig. 1 we report the relative error (which is equal to (F − F*)/F*, where F* is the optimal value) versus iterations, and we consider three different cases: the case in which we put in J^k only the first block (1 block), the case in which we put in J^k only the first and the second blocks (2 blocks), and the case in which we put in J^k the first, the

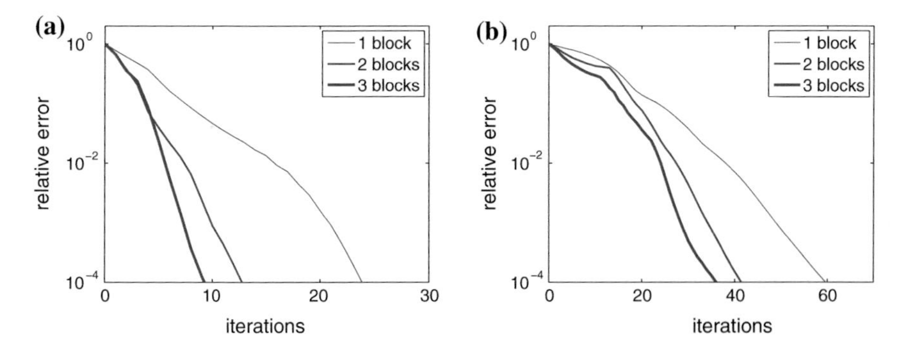

Fig. 1 Relative error VS iterations for 1, 2 and 3 simultaneous blocks updating, **a** cod-rna with block size of 10,000 variables, **b** a9a with block size of 1000 variables

second and the third blocks (3 blocks). Both two problems used a gaussian kernel with gamma equal to 1 and C = 1, block size for cod-rna is of 10,000 variables, while for a9a is of 1000 variables.

It is easy to see that using more blocks is better since (ideally) CPU-time per iteration is the same in the three cases, but we obtain faster decreases in the 3 blocks case.

5 Conclusions and Direction for Future Research

We have defined easy-to-implement decomposition methods for several classes of big data optimization problems. We have shown that these methods can be efficiently parallelized and we have specialized them for important big data machine learning applications.

Interesting lines for future research, on the one hand, would aim at following the general guidelines described here to develop new parallel algorithms for solving other real-world applications in engineering, natural sciences and economics. On the other hand, they would aim at defining new parallel decomposition schemes for constrained optimization problems with non-smooth or implicit (non-separable) feasible sets, such as bilevel and hierarchical optimization problems.

Acknowledgments The work of the author was partially supported by the grant: Avvio alla Ricerca 397, Sapienza University of Rome.

References

1. Beck, A., Teboulle, M.: A fast iterative shrinkage-thresholding algorithm for linear inverse problems. SIAM J. Imaging Sci. **2**, 183–202 (2009)
2. Facchinei, F., Scutari, G., Sagratella, S.: Parallel selective algorithms for nonconvex big data optimization. IEEE Trans. Sig. Process. **63**, 1874–1889 (2015)

3. Manno, A., Palagi, L., Sagratella, S.: A class of parallel decomposition algorithms for SVMs training (2015). arXiv:1509.05289
4. Tibshirani, R.: Regression shrinkage and selection via the lasso. J. Roy. Stat. Soc. Ser. B **58**, 267–288 (1996)
5. Yuan, M., Lin, Y.: Model selection and estimation in regression with grouped variables. J. Roy. Stat. Soc. Ser. B **68**, 49–67 (2006)
6. Meier, L., Van Der Geer, S., Buhlmann, P.: The group lasso for logistic regression. J. Roy. Stat. Soc. Ser. B **70**, 53–71 (2008)
7. Shevade, S.K., Keerthi, S.S.: A simple and efficient algorithm for gene selection using sparse logistic regression. Bioinformatics **19**, 2246–2253 (2003)
8. Yuan, G.X., Chang, K.W., Hsieh, C.J., Lin, C.J.: A comparison of optimization methods and software for large-scale l1-regularized linear classification. J. Mach. Learn. Res. **11**, 3183–3234 (2010)
9. Mairal, J., Bach, F., Ponce, J., Sapiro, G.: Online dictionary learning for sparse coding. In: Proceedings of 26th International Conference on Machine Learning, Montreal (2009)
10. Goldfarb, D., Ma, S., Scheinberg, K.: Fast alternating linearization methods for minimizing the sum of two convex functions. Math. Prog. **141**, 349–382 (2013)
11. Facchinei, F., Sagratella, S., Scutari, G.: Flexible parallel algorithms for big data optimization. In: 2014 IEEE Int Conf on Acoust Speech Signal Process (ICASSP), pp. 7208–7212 (2014)
12. Dreves, A., Facchinei, F., Kanzow, C., Sagratella, S.: On the solution of the KKT conditions of generalized Nash equilibrium problems. SIAM J. Optim. **21**, 1082–1108 (2011)
13. Facchinei, F., Piccialli, V., Sciandrone, M.: Decomposition algorithms for generalized potential games. Comput. Optim. Appl. **50**, 237–262 (2011)
14. Facchinei, F., Sagratella, S.: On the computation of all solutions of jointly convex generalized Nash equilibrium problems. Optim. Lett. **5**, 531–547 (2011)
15. Chang, C.C., Lin, C.J.: LIBSVM: a library for support vector machines. ACM Trans. Intell. Syst. Technol. **2**, 27 (2011)
16. Cortes, C., Vapnik, V.: Support-vector networks. Mach. Learn. **20**, 273–297 (1995)

Author Biography

Simone Sagratella received his B.Sc. and his M.Sc. in management engineering respectively in 2005 and in 2008, all from Sapienza University of Rome, Italy.

From 2009 to 2012 he followed a Ph.D. program in operations research at Department of Computer, Control, and Management Engineering Antonio Ruberti at Sapienza University of Rome, and received his Ph.D. degree in 2013 with a dissertation on quasi-variational inequalities.

He is currently a research fellow at the same department.

His research interests include optimization, Nash games, variational inequalities, bilevel programming and big data optimization.

Index

Note: Page numbers followed by *f* and *t* refer to figures and tables, respectively

A. Emrouznejad (ed.), *Big Data Optimization: Recent Developments and Challenges*, Studies in Big Data 18, DOI 10.1007/978-3-319-30265-2

C